W9-CJQ-419

Biological Assessment and Criteria

Tools for Water Resource Planning and Decision Making

Edited by
Wayne S. Davis
Thomas P. Simon

LEWIS PUBLISHERS
Boca Raton   London   Tokyo

Library of Congress Cataloging-in-Publication Data

Biological assessment and criteria : tools for water resource planning
  and decision making / edited by Wayne S. Davis, Thomas P. Simon.
     p. cm.
  Includes bibliographical references and index.
  ISBN 0-87371-894-1
  1. Water quality biological assessment. 2. Water quality management.
  3. Environmental monitoring. 4. Indicators (Biology). I. Davis, Wayne S. II. Simon, Thomas P.
  QH96.8.B5B54 1994
  363.73′94—dc20   94-27369
                                                      CIP

This book contains information obtained from authentic and highly regarded sources. Reprinted material is quoted with permission, and sources are indicated. A wide variety of references are listed. Reasonable efforts have been made to publish reliable data and information, but the author and the publisher cannot assume responsibility for the validity of all materials or for the consequences of their use.

Neither this book nor any part may be reproduced or transmitted in any form or by any means, electronic or mechanical, including photocopying, microfilming, and recording, or by any information storage or retrieval system, without prior permission in writing from the publisher.

All rights reserved. Authorization to photocopy items for internal or personal use, or the personal or internal use of specific clients, may be granted by CRC Press, Inc., provided that $.50 per page photocopied is paid directly to Copyright Clearance Center, 27 Congress Street, Salem, MA 01970 USA. The fee code for users of the Transactional Reporting Service is ISBN 0-87371-894-1/95/$0.00+$.50. The fee is subject to change without notice. For organizations that have been granted a photocopy license by the CCC, a separate system of payment has been arranged.

CRC Press, Inc.'s consent does not extend to copying for general distribution, for promotion, for creating new works, or for resale. Specific permission must be obtained in writing from CRC Press for such copying.

Direct all inquiries to CRC Press, Inc., 2000 Corporate Blvd., N.W., Boca Raton, Florida 33431.

© 1995 by CRC Press, Inc.
Lewis Publishers is an imprint of CRC Press

No claim to original U.S. Government works
International Standard Book Number 0-87371-894-1
Library of Congress Card Number 94-27369
Printed in the United States of America 1 2 3 4 5 6 7 8 9 0
Printed on acid-free paper

Acknowledgments

This book was made possible by many contributors to this discipline, past and present. We gratefully acknowledge the chapter authors for their hard work and dedication to this effort. We also thank the many reviewers listed that enabled this book to meet scientific peer review standards. Chapters affiliated with the USEPA Environmental Research Laboratory in Corvallis (where much of this discipline was refined) were part of the official USEPA peer review process and those reviewers are also acknowledged. Special thanks to those involved in organizing the first national biocriteria workshop in 1987 and fostering the development and implementation of biological criteria long before that workshop: James L. Plafkin (to whom this book is dedicated), Robert Hughes, John Maxted, James Luey, and Thom Whittier. This book was edited by Wayne Davis and Thomas Simon in their private capacity. No official support or endorsement by the Environmental Protection Agency or any other agency of the Federal government is intended or should be inferred.

Reviewers

Robert W. Adler
University of Utah
College of Law
Salt Lake City, UT 84112

Paul Angermeier
VA Coop. Fish & Wildlife
Research Unit
Virginia Polytechnic Institute and State
 University
Blacksburg, VA 24061

Peter B. Bayley
Illinois State Natural History Survey
607 E. Peabody Drive
Champaign, Illinois 61820

Arthur C. Benke
Department of Biology
University of Alabama
Tuscaloosa, AL 35487

Kevin Bestgen
Department of Fish. & Wildlife Biology
Colorado State University
Ft. Collins, CO 80523

Dan Beyers
Larval Fish Laboratory
Colorado State University
Ft. Collins, CO 80523

Robert W. Bode
New York Department of Environmental
 Conservation
50 Wolf Road
Albany, NY 12233-3503

Dennis L. Borton
National Council of the Paper Industry for Air
 and Stream Improvement, Inc.
P.O. Box 12868
New Bern, NC 28561–2868

G. Allen Burton, Jr.
Wright State University
Biological Sciences Dept.
Dayton, OH 45435

Donald Charles
Patrick Center for Environmental Research
Academy of Natural Sciences
1099 Benjamin Franklin Parkway
Philadelphia, PA 19103

Deborah Coffey
ManTech Environmental Technology, Inc
200 Southwest 35th Street
Corvallis, OR 97333

David L. Courtemanch
Maine Department of Environmental Protection
State House #17
Augusta, ME 04333

Kenneth W. Cummins
South Florida Water
Management District
P.O. Box 24680
West Palm Beach, FL
33416-4680

Don Duff
U.S. Department of Agriculture — Forest
 Service
Wasatch-Cache National Forest
Federal Building, 125 S. State
Salt Lake City, Utah 84138

Joseph J. Dulka
DuPont — Ecological Safety and Toxicology
Experiment Station
Wilmington, DE 19880–0402

Eleanor Ely
Editor, The Volunteer Monitor
1318 Masonic Avenue
San Francisco, CA 94117

Eldon Franz
Program in Env. Sciences
& Regional Planning
Washington State University
Pullman, WA 99164–4430

James E. Gammon
Department of Biological Sciences
DePauw University
Greencastle, IN 46135

George R. Gibson
U.S. EPA
Health and Ecological Criteria Division
401 M Street, SW
Washington, D. C. 20460

Martin E. Gurtz
USGS-Water Resource Division
P.O. Box 2857
300 Fayetteville Street Mall
Raleigh, NC 27602

John Harrington
Department of Geography
Indiana State University
Terre Haute, IN 47809

James E. Harrison
U.S. EPA Region 4
345 Courtland St.
Atlanta, GA 30365

David D. Hart
Academy of Natural Sciences
19th and the Parkway
Philadelphia, PA 19103

Gretchen Hayslip
U.S. EPA Region 10
Environmental Services Division
1200 Sixth Avenue
Seattle, WA 98101

Judy Helgen
Minnesota Pollution Control Agency
520 Lafayette Road
St. Paul, MN 55155

Brian Hill
U.S. EPA
3411 Church Road
Newtown, OH 45244

William L. Hilsenhoff
University of Wisconsin
Department of Entomology
Madison, WI 53706

Robert Hite
Illinois EPA
2209 West Main
Marion, IL 62959

C. Evan Hornig
U.S. EPA Region 6
1445 Ross Avenue
Dallas, Texas 75202–2733

Hoke Howard
U.S. EPA Region 4
College Station Rd.
Athens, GA 30613

Robert M. Hughes
ManTech Environmental Technology, Inc.
200 SW 25th Street
Corvallis, OR 97333

Philip Johnson
U.S. EPA Region 8
999 18th St., Suite 500
Denver, CO 80202–2405

James R. Karr
Director, Institute for Environmental Studies
University of Washington
Engineering Annex — FM 12
Seattle, WA 98195

John S. Kopec
Ohio Department of Natural Resources
Division of Natural Areas & Preserves
1889 Fountain Square Ct.
Columbus, OH 43224

Richard Langdon
VT Dept. Env. Conservation
Waterbury, VT 05676

David R. Lenat
North Carolina DEHNR
Env. Mgmt., Water Quality
Archdale Bldg., 27687
Raleigh, NC 27611

Edwin Liu
U.S. EPA Region 9
75 Hawthorne Street
San Francisco, CA 94105

John Lyons
Wisconsin DNR
1350 Femrite Drive
Monona, WI 53716

F. Douglas Martin
4168 Quinn Drive
Evans, GA 30809–9166

William Matthews
Univ. of Oklahoma
Biological Station
Star Route
Kingston, OK 73439

John Maxted
Delaware Dept Nat. Res. & Env. Control
P.O. Box 1401
Dover, DE 19903

Frank McCormick
U.S. EPA
103 South Main Street
Cincinnati, OH 45268

Richard W. Merritt
Department of Entomology
Michigan State University
East Lansing, MI 48824

Reuven Ortal
ESE Department
The Hebrew University of Jerusalem
Givat Ram
Jerusalem 91904
ISRAEL

Lawrence M. Page
Illinois Natural History Survey
607 East Peabody Drive
Champaign, IL 61820

David Penrose
North Carolina DEHNR
Env. Mgmt., Water Quality
Archdale Bldg.
Raleigh, NC 27611

Frank Pezold
Northeast Louisiana Univ.
Department of Biology
Monroe, LA 71209–0504

Mark Pifher
Anderson, Johnson and Gianunzio
104 South Cascade Avenue, Suite 204
Colorado Springs, CO 80903

Edward T. Rankin
Ohio EPA
Ecological Assessment Section
1685 Westbelt Drive
Columbus, OH 43228

Vincent H. Resh
Department of Environmental Science
Policy and Management
University of California
Berkeley, CA 94720

Barry H. Rosen
South Florida Water Management District
3301 Gun Club Road
West Palm Beach, FL 33416-4680

Steven T. Ross
Department of Biology
Univ. Southern Mississippi
Box 5018
Hattiesburg, MS 39401

Larry Shepard
U.S. EPA Region 7
726 Minnesota Avenue
Kansas City, KS 66101

Timothy Simonson
Wisconsin DNR
1350 Femrite Drive
Monona, WI 53716

Bernhard Statzner
Laboratoire de Biologie animale et Ecologie
Université Claude Bernhard
43 Boulevard du 11 Novembre 1918
69622 Villeurbanne Cedex
FRANCE

Robert J. Steedman
Ontario Ministry of Natural Resources
Center for Northern Forest Ecosystem Research
Lakehead University Campus
955 Oliver Road
Thunder Bay, Ontario
Canada P7B 5E1

Donald Stevens
ManTech Environmental Technology, Inc
200 Southwest 35th Street
Corvallis, OR 97333

R. Jan Stevenson
Department of Biology
University of Louisville
139 Life Sciences Building
Louisville, KY 40292

James B. Stribling
Tetra Tech, Inc.
10045 Red Run Blvd.
 Suite 110
Owings Mills, MD 21117

Stanley W. Szczytko
University of Wisconsin
College of Natural Resources
Steven Point, WI 54481

Kazumi Tanida
Department of Life Sciences
University of Osaka Prefecture
College of Integrated Arts and Sciences
Mozu-ume-machi
Sakai, Osaka 591

Carl Weber
University of Maryland — Baltimore County
Department of Biology
Catonsville, MD 21228

Denis White
Oregon State University
c/o U.S. EPA
Environmental Research Lab.
200 SW 35th Street
Corvallis, OR 97333

Torgny Wiederholm
Statens Naturvardsverk
EIA Dept. Box 7050
S-750 07 Uppsala
SWEDEN

Contributors

Robert W. Adler
University of Utah
College of Law
Salt Lake City, UT 84112

Michael T. Barbour
Tetra Tech, Inc.
10045 Red Run Boulevard, Suite 110
Owings Mills, MD 21117

Robert W. Bode
New York State Department of Environmental
   Conservation
50 Wolf Road
Albany, New York 12233-3503

David L. Courtemanch
Maine Department of Environmental Protection
State House #17
Augusta, Maine 04333

Wayne S. Davis
U.S. Environmental Protection Agency
401 M. Street, S.W. (2162)
Washington, D.C. 20460

Jeffrey E. DeShon
Ohio EPA
Ecological Assessment Section
1685 Westbelt Drive
Columbus, OH 43228-3809

Kenneth L. Dickson
Director, Institute of Applied Sciences
University of North Texas
P.O. Box 13078
Denton, Texas 76203-3078

C. Evan Hornig
U.S. EPA Region 6
Environmental Services Division
1445 Ross Avenue
Dallas, Texas 75202-2733

Robert M. Hughes
ManTech Environmental Technology, Inc.
200 SW 35th Street
Corvallis, OR 97333

James R. Karr
Director, Institute for Environmental Studies
University of Washington
Engineering Annex — FM 12
Seattle, WA 98195

David P. Larsen
U.S. EPA Environmental Research Laboratory
200 SW 35th Street
Corvallis, OR 97333

Joyce E. Lathrop
University of Maryland-Baltimore County
Department of Biological Sciences
Catonsville, Maryland 21228

John Lyons
Wisconsin Department of Natural Resources
Bureau of Research
1350 Femrite Drive
Monona, Wisconsin 53716

Abby Markowitz
Maryland Save Our Streams
258 Scotts Manor Drive
Glen Burnie, MD 21061

Margaret A. Novak
New York State Department of Environmental
   Conservation
50 Wolf Road
Albany, New York 12233-3503

James M. Omernik
U.S. EPA Environmental Research Laboratory
200 Southwest 35th Street
Corvallis, Oregon 97333

Edward T. Rankin
Ohio EPA
Ecological Assessment Section
1685 Westbelt Drive
Columbus, OH 43228

Robin J. Reash
American Electric Power Service Corporation
Environmental Engineering Division
1 Riverside Plaza
Columbus, OH 43215

Vincent H. Resh
Department of Environmental Science
Policy and Management
University of California
Berkeley, California 94720

Barry H. Rosen
South Florida Water Management District
3301 Gun Club Road
West Palm Beach, FL 33416-4680

Thomas P. Simon
U.S. Environmental Protection Agency
Water Division
Water Quality Branch
77 W. Jackson Blvd.
Chicago, Illinois 60604

Mark T. Southerland
Versar Inc.
9200 Rumsey Road
Columbia, Maryland 21045-1934

James B. Stribling
Tetra Tech, Inc.
10045 Red Run Blvd.
 Suite 110
Owings Mills, MD 21117

Chris O. Yoder
Ohio EPA
Ecological Assessment Section
1685 Westbelt Drive
Columbus, OH 43228

Foreword

This book is dedicated to James L. Plafkin, a friend and colleague, whose commitment to the environment and creative thinking stimulated much of the recent developments in biological assessment of water quality. Jim developed this life long interest at his beloved University of Michigan's Biological Station on Douglas Lake and during graduate training in environmental biology at Virginia Tech. He had many career options upon receiving his Ph.D., but decided to pursue the U.S. Environmental Protection Agency because he felt that it was where he could make the greatest contribution. He was right. Jim was instrumental in the conception, development, and implementation of programs in biological assessment, biocriteria, and ecological risk assessment within EPA. His enthusiasm for life, sense of humor, and ability to make things happen are dearly missed by all who knew him.

The next decade promises to be an era where biological approaches for water quality management and protection are truly integrated with more traditional chemical and physical based approaches. A variety of factors have contributed to the emergence of biological considerations. Emphasis has shifted from a technology based approach for addressing water pollution to a water quality based approach. One of the major goals of a water quality based approach is ensuring that designated uses are achieved. Aquatic life uses are valued by the public and embodied in the Clean Water Act. Perhaps of equal importance has been the advances made in methods and techniques for biological assessment. Rapid bioassessment methods for streams and rivers have been developed and tested and now make it possible to economically assess the status of aquatic life resources. Thus it is now practical to directly assess, rather than indirectly estimate by chemical and physical surrogate measures, the end points of concern. Bioassessments allow measurement of the condition of resources at risk from point and nonpoint source stressors. Aquatic organisms serve as continuous monitors and provide an integrated assessment. Because of this attribute they may detect problems that other methods miss or underestimate. By establishing biocriteria for our nation's rivers and streams and monitoring whether or not these criteria are met, it is possible to directly assess whether or not progress is being made in reaching the fishable and swimmable goals of the Clean Water Act.

This book presents a state-of-the-art overview of applying biological assessments and biocriteria for water quality management in freshwaters. Case studies are presented which illustrate how different states have used bioassessment to identify and diagnose water quality problems. Examples are also provided of the use of narrative and quantitative biocriteria as regulatory tools to complement water quality criteria and standards. Thus the book provides useful and timely information for water quality managers.

From my perspective, the use of bioassessments and biocriteria is long overdue. Logic dictates that setting goals and regulations based on attainment of biological criteria makes sense. However, it is not the end all. As we begin considering water quality management for the 21st century, let us not forget the basic assumption that environmental quality control practices are apt to be ineffective in the absence of informational feedback about the responses of the system being protected. While use of bioassessments and biocriteria are a step in the right direction, they still are not real time. The challenge is to evolve physical, chemical and biological assessment technologies and methods so that we can accomplish real time assessment and feedback.

Kenneth L. Dickson
Director
Institute of Applied Sciences
 and Regents Professor
University of North Texas
Denton, Texas

Contents

SECTION I:
CONCEPTUAL FRAMEWORK
FOR BIOCRITERIA DEVELOPMENT

## CHAPTER 1

Introduction

Wayne S. Davis and Thomas P. Simon

1.0 WHAT ARE BIOLOGICAL CRITERIA?

The ecological goal for the restoration and protection of freshwater ecosystems in the United States is based upon the Clean Water Act and Great Lakes Water Quality Agreement objectives of "biological integrity." Biological integrity is defined as "...the ability of an aquatic ecosystem, to support and maintain a balanced, integrated, adaptive community of organisms having a species composition, diversity, and functional organization comparable to that of the natural habitats of a region" (Karr and Dudley 1981). Although unimpaired waterbodies may no longer exist, an estimate of expected biological integrity in surface waters based upon "least-impacted" conditions can be used to guide restoration and protection programs and answer the ecological equivalent of "how clean is clean."

Least-impacted regional reference conditions form the basis for developing regional biological goals, or biological criteria. Biological criteria are "numeric values or narrative expressions that describes the reference [least-impacted] biological integrity of aquatic communities inhabiting waters of a given designated aquatic life use" (USEPA 1990a). Although "pristine" waters will never be an attainable goal, it is reasonable to base water resource goals upon regional reference conditions that describe the "best" attainable biological integrity. Principles for successful development and implementation of numeric biological criteria are based upon developing a **reference condition** (Hughes, Chapter 4) from a **regional framework** (e.g., Omernik, Chapter 5), where appropriate (e.g., Yoder and Rankin, Chapter 9; Hornig et al., Chapter 10), a **multiple metric** characterization of the aquatic community (Barbour et al., Chapter 6), and a **habitat evaluation** (e.g., Rankin, Chapter 13). Successful programs which do not include a regional framework are presented by Bode and Novak (Chapter 8) and Courtemanch (Chapter 20).

2.0 BIOCRITERIA AS ENVIRONMENTAL INDICATORS

Karr (Chapter 2) and Adler et al. (1993) emphasize the problems with the traditional water resource management approaches of the past 20 years. One prominent issue is the continued dependence of regulatory and natural resource agencies upon using administrative measures (e.g., number of permits issued and other "bean counts") to determine environmental progress rather than using direct measurements of the living resources (Beardsley 1992). Recently, there have been some attempts to manage for results based upon environmental, or ecological, indicators (SAB 1990; McKenzie et al. 1992; Gurtz 1994; Paulsen and Linthurst 1994).

Cairns et al. (1993) suggest that there are two ways to document conditions of the environment at the community and ecosystem level. One way is a "bottom-up" approach in which laboratory data are used to model effects expected in natural systems. The second way is a "top-down" approach based upon direct measurements of ecosystem health followed by diagnostic testing for sources and causes of any problems.

Biological criteria are an important component of this "top-down" approach. USGAO (1988) recognized the importance of regionalized biocriteria as environmental indicators and concluded that "the biological indicators and ecoregion findings appear to be useful tools that could help the U.S. Environmental Protection Agency (USEPA) manage for environmental results." They explained that "[b]ecause these measures target the effects of pollution and contaminants on living organisms, they can provide the direct measures of environmental health that have been heretofore lacking. In addition, they may allow EPA to better target monitoring activities, especially in attempting to relate program activities to environmental changes." USEPA agreed with the conclusions and recommendations of this USGAO report and proceeded to identify environmental indicators for surface waters (USEPA 1990b; Abe et al. 1992). As a result, USEPA has decided that the best environmental indicators for surface waters are those that relate directly to measuring biological integrity (USEPA 1991d, 1990b; Paulsen and Linthurst 1994). Cairns et al. (1993) provide a well-thought framework for developing and applying indicators of ecological health.

3.0 INTEGRATING BIOCRITERIA WITH OTHER APPROACHES

Freshwater biological monitoring and assessment has generally been focused on identifying structural changes in the species present and their abundances resulting from the introduction of a stressor or alteration of the physical habitat (Norris and Georges 1993). Such "classical" or "traditional" studies have employed both inferential and descriptive statistics to document observed changes. In some instances, a cause-and-effect relationship can be determined among single stressors and biotic response. However, one issue that has continually arisen despite quantitative and statistically sound results is the ecological significance of those results (McBride et al. 1993). Biological criteria can address two major components to this issue: (1) the selection of an appropriate reference, and (2) statistical and ecological (practical) significance.

Two other often independent approaches to the protection and restoration of freshwater ecosystems in which biological assessment and criteria can play an important role include ecological risk assessment and conserving biological diversity.

3.1 Selecting a Reference

Pollution impact assessments for aquatic ecosystems are based on estimates of change from natural conditions using a "control" or "reference." The reference represents the desired state of the living resource based on undisturbed habitat representative of the region. It is often assumed that the upstream site, or "before," condition is the desired state. It is in the selection of the reference that many professionals have ultimately failed to successfully characterize the site conditions (e.g., good or poor, meeting or failing goals or standards, nominal or subnominal, healthy or impaired, etc.). Biological criteria provide this quantitative estimate of a desired state from an appropriate reference benchmark which can then be used to measure departure of a reference site and test site from a regional reference condition (Hughes, Chapter 4).

3.2 Statistical and Ecological Significance

Pollution impact studies often use inferential statistics to determine significant differences (i.e., hypothesis testing) of aquatic community sample results upstream and downstream, or before and after, of a suspected perturbation (Resh and McElravy 1993). However, there is sometimes a question whether statistically significant differences are actually ecologically significant (McBride et al. 1993). This question is due to "inherent" natural variations of populations and sometimes improper sample replication (Voshell et al. 1989). Confounding effects from natural variability of populations are reduced by using an index period for sampling (Hughes et al. 1992; Barbour et al., Chapter 6), application of consistent methodology, and relying upon community measures reflecting ecologically important attributes (e.g., trophic structure, taxa richness, etc.; Karr 1991).

Hurlbert (1984) identified improper replication in field studies using statistical hypothesis testing to be a potentially serious source of error. Whereas laboratory studies can replicate results by using multiple treatments and a control, field studies rarely have this potential. The downstream and upstream conditions are viewed as a single treatment (downstream) and control (upstream), and what many refer to as "replicate" samples are really subsamples from within the treatment and control, or pseudoreplicates (Hurlbert 1984). Stewart-Oaten et al. (1986), Voshell et al. (1989), and McBride et al. (1993) have thoroughly discussed pseudoreplication and Stewart-Oaten et al. (1986) proposed combining a "before/after" and "control/impact" approach (BACI) as a solution.

Biological criteria allow the results of pollution impact studies to be put in the context of natural conditions and goals (Bode and Novak, Chapter 8). The ecological significance of a statistically signifi-cant change in the biota from paired sites, or the same site through time, can be interpreted by the biocriterion (whether the "impact" resulted in impairment of the capability to support a healthy aquatic community). In this manner, numeric biological criteria can be used for planning and managing the restoration and protection of freshwater ecosystems. Studies can then be undertaken to find the source(s) and stressor(s) causing any changes, and the level of restoration that may be required.

3.3 Ecological Risk Assessment

Ecological risk assessment (ERA) is a process that evaluates the likelihood that adverse ecological effects may occur, or are occurring, as a result of exposure to one or more stressors (USEPA 1992b). Simply, ERA is "based upon the probability of harm to ecological integrity" (Cairns and McCormick 1992). Inherent in the ERA framework is the stressor-response analyses that quantifies the relationship between the stressor and the environmental value to be protected. Biological assessment and criteria fit well in this conceptual framework by providing a measurable representation of ecosystem integrity, quantifying environmental values to be protected, and responsiveness to the effects of nonchemical stressors. Similarly, the ERA framework is useful in diagnosing the causes of observed changes in the multiple biological metrics.

3.4 Biocriteria and Conserving Biodiversity

The link between biological assessment and criteria (biointegrity) and biological diversity is crucial (Hughes and Noss 1992; Karr, Chapter 2). In fact, the biological integrity objective of the Clean Water Act can also be a viewed as a mandate to protect biological diversity (Blockstein 1992). Conserving biological diversity (e.g., endangered, threatened, and special protection species) has been a sustaining principle for many natural resource agencies but has received little practical attention from environmental regulatory agencies. Biological diversity has been traditionally viewed as "the variety of organisms at all levels" (Wilson 1992), but the failure of past efforts to protect biological diversity is resulting in an expansion of this definition to also include the natural processes affecting biological diversity (Cairns and Lackey 1992; Hughes and Noss 1992).

Definitions of biological integrity (Karr and Dudley 1981) and biological diversity (Noss 1990) both contain measures of function, composition, and different levels of biological organization (Hughes and Noss 1992). The need to link biological criteria and biological diversity efforts is essential, whether biocriteria is viewed as a measure of biodiversity (Hughes and Noss 1992) or biodiversity as a component of ecological or biological integrity (Karr 1993a). Without integrating program efforts, we may not achieve the goals of either, especially if these programs are viewed competitively.

4.0 LIMITATIONS OF BIOLOGICAL CRITERIA

The USEPA Science Advisory Board (SAB) recently reviewed the draft "Technical Guidance on Biological Criteria for Streams and Small Rivers," which puts forth the primary components of USEPA's biocriteria development efforts (USEPA 1990a; SAB 1993; Gibson 1994). The SAB agreed that biocriteria, based upon a multiple metric and regional reference condition approach, can be used to: (1) evaluate and demonstrate the accuracy of current regulatory and management activities in protecting aquatic ecosystems, (2) serve as a site-specific assessment of ecological degradation and ecosystem response to remediation or mitigation activities, and (3) assess biological resource trends in well-characterized watersheds (SAB 1993).

The SAB also clearly defined the limitations of the current approach which prevents its use for establishing point source criteria or permit limitations. Those limitations include: (1) the degree to which subtle impacts can be detected, (2) the current lack of diagnostic tools to determine the causes of observed impacts, and (3) the state of the science in defining ecoregions and reference areas (SAB 1993). Approaches to lessen some of these limitations are presented in this book such as use of multiple assemblages to detect more subtle impacts (Yoder and Rankin, Chapter 9), subregionalization of ecoregions (Omernik, Chapter 5), selecting the appropriate reference (Hughes, Chapter 4; Reash, Chapter 11) and biological response signatures (Yoder and Rankin, Chapter 17). However, with the rapid growth and use of biological assessment and criteria approaches in the United States (Southerland and Stribling, Chapter 7) and through-out the world (Davis, Chapter 3), more research is needed in these areas, and others, to continue developing scientifically valid biological assessment methods and criteria for all freshwater resources.

5.0 STRUCTURE AND LAYOUT OF THE BOOK

The purpose of this book is to present and discuss the conceptual framework for determining biological expectations (biocriteria) for freshwater systems and document the programmatic use of biological criteria. Although freshwater rivers and streams are the focus of this book, the same framework is being applied for other resource types. The intended audience includes water resource managers, planners, engineers and scientists, citizens and environmental groups, and regulatory and natural resource agency scientists and managers who are interested in furthering the tools needed for ecologically and economically sound water resource planning and decision-making. We attempt to provide the reader with examples of framework issues and a review of technical, programmatic, and legal aspects of the use of biological criteria. We illustrate how biological assessments and criteria development have been success-ful in addressing many regulatory *and* nonregulatory water resource quality concerns, but also the compilation of concepts into acceptable tools for a wide spectrum of application.

This book is comprised of four main sections: (1) fundamental concepts supporting biological numerical criteria, (2) program application, experiences, and recommendations, (3) technical needs and advances to support planning and decision-making, and (4) perspectives on implementing biological criteria. The first section begins with a retrospective look at the water pollution control policies and rationale followed by a historical perspective on the scientific development influencing the development of numerical biological assessments. Fundamental concepts defining reference condition, using ecologi-cal regions for spatial stratification, and the multiple metric/multiple assemblage approach. The second section begins with a detailed update on state agency use of biological assessment and criteria, including several case studies followed by government and industry experiences and recommendations for their use. The third section features technical advances in characterizing habitat, periphyton, macroinvertebrates, and fish for biological criteria development. The final section closes with discussion of important issues including the role of volunteer citizen monitoring, state policy implications, and legal perspectives.

Many of the authors in this book are associated with federal and state regulatory agencies because these agencies have advanced the development and implementation of numerical biological criteria and have legal responsibility for restoration and protection of freshwater ecosystems. The actions taken by these agencies affect programs and activities in academia, industry, nonregulatory government agencies, citizens, and environmental groups, and sometimes programs in other countries. The scientific rationale for biological criteria are made possible from previous studies conducted throughout the world. It is our attempt to document the rationale for numerical biological criteria development and implementation for freshwater ecosystems in the United States.

6.0 ACKNOWLEDGMENTS

We wish to express our appreciation to all of the authors associated with this effort. They have enriched our appreciation for applying biological measures in water resource programs based on conversations of issues, careful review of their manuscripts, and knowing them professionally and personally. Our career paths were enlightened by our friend and colleague to whom this book is dedicated, Jim Plafkin. His efforts at EPA headquarters were pioneering and collaborative in difficult political times. His desire to embrace rapid bioassessment into the regulatory culture at EPA was innovative and fostered the use of consistent and quantitative biological information. Our gratitude and respect for Jim inspired this book.

## CHAPTER 2

Protecting Aquatic Ecosystems:
Clean Water Is Not Enough

James R. Karr

1.0 INTRODUCTION

Two climate variables, ambient temperature and availability of water, are the primary limiting factors to biological systems. Technological innovation has allowed humans, to some extent, to break free of the limits that these factors impose on other species, but not without an extraordinary output of human energy. Simply to keep warm, for example, we must burn fuels and insulate ourselves with clothing and shelters. Thus have we been able to spread throughout the globe, as long as water was readily available.

For millennia, people could satisfy their need for water by locating permanent settlements next to rivers, lakes, or oceans. But those settlements, and the agriculture they depend on, have depleted and contaminated critical water supplies. Moreover, demand for water is increasing; current usage exceeds long-term availability in many geographic areas, even as the distances over which we move water are stretched. In the Middle East for example, conflict over water began well before contemporary oil crises and will likely continue after global oil supplies are exhausted (Myers 1993). Global water consumption more than doubled between 1940 and 1980 and is likely to double again in the next 20 years. As many as 80 countries already suffer serious water deficits (Myers 1993). Even in the United States, some regions are not self-sufficient in water supply, and regions thought to be self-sufficient have experienced severe water crises in recent years; the "rainy city" of Seattle, Washington, imposed water rationing for several months in 1992. Water is no longer free to the first who claims it; late arrivals cannot go elsewhere to find water. As a result, the prior appropriation doctrine that dominates western water law, disenfranchises future generations.

Thoughtful societal action is needed at local, regional, national, and global levels to ensure future supplies of water and associated aquatic resources. As Norman Myers (1993, p. 38) notes, "Our future will be deeply compromised unless we learn to manage water as a critical ingredient of our lives." The decline in the distribution, abundance, and quality of water and aquatic ecosystems thus represents a threat to the sustainability of all living systems and the quality and long-term viability of human society. The complex reasons for this decline center on the hubris of a modern society that operates as if it had unlimited latitude to alter water resource systems without degrading the ability of those resources to meet human needs, or as if it could replace those lost services through application of increasingly complex technology. Degradation results from the acceleration of environmental change by humans and the sensitivity of water resources to change.

Two major dimensions of the water crisis must be dealt with immediately. First, the supply of water and water-associated resources is not adequate for current, let alone future, demand. Second, allocation of existing supplies is based on a narrow conceptual understanding of the value of water to human society, especially a tendency to chronically undervalue aquatic ecosystems. The time to undertake a critical evaluation of the status and trends in water and associated aquatic resources in the United States is now.

The condition of rivers in the United States illustrates the message to be derived from that evaluation. Rivers are, in many ways, the lifeblood of human society; they provide water for domestic and industrial use, serve as transportation corridors, and offer food, recreation, and scenic beauty. But in addition to this direct and obvious importance to society, the status of rivers sends important signals about the health of the surrounding landscape, much as blood samples tell a great deal about the health of an individual human.

2.0 CHANGING PERSPECTIVES ON WATER QUALITY

Twenty-one years ago, I was asked to join a research team to evaluate the influence of agriculture on water quality in Allen County, Indiana. Several events in the early months of that project led me to believe that existing programs would never succeed in protecting the integrity of water resources. At one point, for example, I asked for evidence that "best management practices" selected to control soil erosion actually improved water quality. To my surprise, an EPA official responded, "We are not interested in the effects of our actions, only that the actions are taken." In short, actions, and not measurable end points, were the goal. That philosophy persists today in counts of permits issued and fines levied (Adler et al. 1993).

On another occasion I suggested that a critical measure of project success would be our ability to reduce sediment input to the study stream. I was told that sediment was not a pollutant, a position defended with the observation that EPA had never defined for sediment a criterion level that signaled violation of water quality standards. Finally, I was regularly told that biological conditions were not within the mandate of the Clean Water Act, and thus of EPA. One writer during that period maintained that "the various forms of life in a river are purely incidental, compared with the main task of a river, which is to conduct water runoff from an area toward the oceans" (Einstein 1972).

A geomorphologist thinking in terms of millions of years, or an engineer attempting to control a river, might be tempted to ignore biology. But these views do not protect societal interests. Society ignores biology at its own detriment because human society is itself a living system, and other living systems are critical to the success of human society. The biosphere, including crops, livestock, forests, and the plants and animals of rivers and urban and suburban environments, are intimately tied to, and in turn determine, the quality of water resource systems; these components can hardly be considered "incidental." In fact, the 1972 Clean Water Act is explicit in its call for action "to restore and maintain the physical, chemical, and biological integrity of the nation's waters."

3.0 WHY BE CONCERNED ABOUT BIOLOGY?

The Clean Water Act and other environmental legislation — indeed all concerns about environmental degradation — derive from recognition that current programs fail to support healthy biological systems. Biology is of concern whether one cherishes the presence of wild egret populations in the southeastern United States or white feathers in ladies hats; presence of spawning salmon in rivers of the Pacific Northwest or old-growth forest as a source of timber profit. Similarly, we prohibit exploitation of children in industrial sweatshops or seek to prevent unethical individuals, corporations, or governments from subjecting minorities to disproportionate risk from handling or disposing of toxic chemicals out of concern for human biology.

Despite the importance of biological and ecological systems to society, we have managed water resources as if the biological systems associated were indeed incidental to society. Fortunately, many water resource professionals and private citizens are recognizing the need for change in the science, policy, and politics of water resources. Five realities point to the need for such changes: (1) the biological components of water resources are in steep decline; (2) the cause of the decline is broader than simple chemical contamination, the primary focus of conventional water quality programs; (3) the legal and regulatory framework in place today cannot respond in a timely manner to these two realities; (4) long-term success in protecting water resources requires careful thought about assessment endpoints, including development of biological criteria; and 5) the quantitative expectations for attributes that constitute biological health or the integrity of a water resource system vary geographically.

1. Water resources, especially their biological components, are in steep decline. Despite strong mandates and massive expenditures to protect "the physical, chemical, and biological integrity," of the nation's waters, signs of continuing degradation are pervasive whether one examines conditions at the scale of individual rivers (Karr et al. 1985b), states (Moyle and Williams 1990; Jenkins and Burkhead 1994), North America (Williams et al. 1989; Frissell 1993), or the globe (Hughes and Noss 1992; Moyle and Leidy 1992; Williams and Neves 1992; Allan and Flecker 1993). Aquatic systems have been and continue to be degraded on scales that are unprecedented for most terrestrial environments.

Devastation is obvious, even to the untrained eye, in the destruction of river channels by dams or channelization (Table 1), but degradation goes beyond destruction of stream channels (Table 2). As a result, society risks losing forever many of the goods and services provided by those systems. For example, biological diversity in aquatic habitats is threatened, and commercial fish harvests in major US rivers have fallen by at least 80%, to a fraction of their levels early in this century. Other signs of degradation include declining genetic diversity, fish consumption advisories, and homogenization of aquatic biota. Habitat loss and fragmentation, invasion of exotic species, excessive water withdrawals, effects of chemical contamination, and overharvest by sport and commercial fishers all contribute to this degradation. The aquatic biota of North America has been decimated.

The pervasive nature of this biological degradation demonstrates that dependence on a technology-based approach to protecting the quality of water resources has failed. Under Section 305(b) of the Clean Water Act, for example, states report on the status of water resources within their boundaries, usually based on chemical analyses that chronically underestimate the extent of degradation. In Ohio, the proportion of the state's waters assessed as degraded doubled as a result of the more comprehensive, sensitive, and objective assessment provided by the use of biological rather than chemical criteria (Yoder 1991a). More conventional chemical criteria failed to detect 50% of the impairment of surface waters.

Current programs are not protecting rivers or their biological resources because the Clean Water Act is being implemented as if crystal clear distilled water running down concrete conduits were the goal. How would we respond as a society if our agricultural productivity declined by more than 80%? How can we continue to ignore declines of that magnitude in river resources that are essential to the economic and ecological health of human society?

2. Degradation is caused by a broader array of factors than chemical contamination, the primary focus of conventional water-quality programs. We have assumed that water protected from chemical contaminants assures chemical, physical, and biology integrity, but the cumulative impacts of human actions go well beyond chemical contamination. We waste money and degrade resources because decisions based on chemical criteria do not adequately protect water quality; priority lists of chemicals do not accurately reflect ecological risks; point-source approaches do not effectively control the influence of nonpoint sources or the cumulative effects of numerous contaminants; and, finally, the chemical-contaminant approach fails to diagnose water resource problems caused by other human influences.

Degradation begins in upland areas of a watershed or catchment as a result of human actions that alter the plant cover of the land surface. These changes, combined with alteration of stream corridors, alters the quality of water delivered to the stream channel as well as the structure and dynamics of those channels and their adjacent riparian environments. The cumulative effects of these degrade water resources by influencing one or more of five primary classes of variables (Table 3), with potentially devastating and often undetected effects on water quality (Karr 1991a). Many of these effects can be seen in the decline of salmonid populations in the Pacific Northwest today (Table 4). Efforts to protect the quality of water resources are doomed unless they explicitly incorporate this range of factors into a comprehensive planning and assessment process.

In short, the technology-based approaches of the past 20 years concentrated on a narrow range of human actions while equally serious threats were ignored. Recognizing this weakness does not diminish our success in regulating contamination from some point sources, but we should not allow those accomplishments to permit us to overlook continuing degradation in resource condition.

3. The legal and regulatory framework in place today does not respond in a timely manner to continued degradation. Degradation of aquatic systems continues because government agencies have been weak, inappropriately focused, and therefore largely ineffective at reversing resource declines. Underfunding — the chronic complaint from all bureaucracies and scientists — is not, however, the most important problem. The most important problem is that we do not see water and associated aquatic resources as integrated and complex natural resource systems. The failure to adopt an integrative

Table 1.   Degradation of United States Rivers and Their Floodplains

- Of 3.2 million miles of rivers in the continental United States, only 2% are healthy enough to be considered high quality and worthy of protection (Benke 1990).
- Only one large river (longer than 600 miles), the Yellowstone, has not been severely altered.
- Of medium-sized rivers (between 120 and 600 miles long), only 42 studied in the National Rivers Inventory (Benke 1990) have not been dammed.
- Sixty to eighty percent of riparian corridors have been destroyed throughout the United States (Swift 1984).
- More than 70% of the original floodplain forests in the United States have been converted to urban and agricultural use (Hunt 1992).

Table 2.   Degradation of Aquatic Biota: Examples from Rivers of the United States

- A larger proportion of aquatic organisms (34% of fish, 75% of unionid mussels, and 65% of crayfish) than terrestrial organisms (from 11 to 14% of birds, mammals, and reptiles; Master 1990) is classed as rare to extinct.
- Twenty percent of native fishes of the western United States are extinct or endangered (Miller et al. 1989).
- Thirty-two percent of Colorado River native fish are extinct, endangered, or threatened (Carlson and Muth 1989).
- In the Pacific Northwest, 214 native, naturally spawning Pacific salmon and steelhead stocks face "a high or moderate risk of extinction, or are of special concern" (Nehlsen et al. 1991).
- Since 1910, naturally spawning salmon runs in the Columbia River have declined by more than 95% (Ebel et al. 1989).
- During the twentieth century, the commercial fish harvests of major U.S. rivers have declined by more than 80% (Missouri and Delaware Rivers), more than 95% (Columbia River), and 100% (Illinois River) (see Karr et al. 1985b; Ebel et al. 1989; Hesse et al. 1989; Patrick 1992).
- In 1910, more than 2600 commercial mussel fisherman operated on the Illinois River; virtually none remain today.
- Since 1850, many fish species have declined or disappeared from rivers in the United States (Maumee River, Ohio: 45%; Illinois River, Illinois: 67% [Karr et al. 1985b]; California rivers: 67% [Moyle and Williams 1990]). This decline, combined with the introduction of alien species, has homogenized the aquatic biota of many regions (an average of 28% of the fish species in major drainages of Virginia are introduced; Jenkins and Burkhead 1994).
- Since 1933, 20% of molluscs in the Tennessee River system have been lost (Williams et al. 1989); 46% of the remaining molluscs are endangered or seriously depleted throughout their range.
- Thirty-eight states reported fish consumption closures, restrictions, or advisories in 1985; 47 states in 1992. Contaminated fish pose health threats to wildlife and people (Colborn et al. 1990; Cunningham et al. 1990), including intergenerational consequences such as impaired cognitive functioning in infants born to women who consume contaminated fish (Jacobson et al. 1990).
- Riparian corridors have been decimated (Swift 1984).

Table 3.   Five Primary Classes of Water Resource Variables Altered by Cumulative Effects of Human Actions

- *Water quality*: temperature, turbidity, dissolved oxygen, acidity, alkalinity, organic and inorganic chemicals, heavy metals, toxic substances
- *Habitat structure*: substrate type, water depth and current velocity, spatial and temporal complexity of physical habitat
- *Flow regime*: water volume, temporal distribution of flows
- *Energy source*: type, amount, and particle size of organic material entering stream, seasonal pattern of energy availability
- *Biotic interactions*: competition, predation, disease, parasitism, mutualism

perspective, and especially the failure to deal with this issue in the ongoing Clean Water Act reauthorization, is unacceptable on legal, scientific, economic, and ethical grounds.

Implementation of the Clean Water Act has concentrated on (1) the effectiveness of wastewater treatment technology to control point sources of pollution and (2) human cancer risk. The dominance of these two issues has prevented program managers, political leaders, and the public at large from tracking the actual condition of water resources. We need to shift the vision and mandate of the Clean Water Act — and societal focus as well — to a much broader concept: the biological integrity or ecological health of all water resources.

Table 4. Resource Degradation Typical of Problems in Northwest Watersheds for Each of the Five Classes of Variables Influenced by Human Action

Characteristic	Degradation
Food (energy) source	Altered supply of organic material from riparian corridor
	Reduced or unavailable nutrients from the carcasses of adult salmon after spawning
Water quality	Increased temperatures
	Oxygen depletion
	Chemical contaminants
Habitat structure	Sedimentation and loss of spawning gravel
	Obstructions that interfere with movement of adult or juvenile salmonids
	Lack of coarse woody debris
	Destruction of riparian vegetation and overhanging banks
	Lack of deep pools
	Altered abundance and distribution of constrained and unconstrained channel reaches
Flow regime	Altered flows that limit survival rates during any phase of the salmon life cycle
Biotic interactions	Increased predation on young by native or exotic species
	Overharvest by sport or commercial fishers

Another shift will have to come in our use of the word *pollution*. In conventional usage, and in the technical jargon that permeates the Clean Water Act and its implementing regulations, *pollution* means chemical contamination. A more appropriate definition, present in the 1987 Clean Water Act but little used, states that pollution is any "[hu]manmade or [hu]man-induced alteration of the physical, chemical, biological, or radiological integrity of water." Under this definition, humans may degrade or pollute by withdrawing water for irrigation, overharvesting fish populations, or introducing exotic species or chemical contaminants. For this reason, we must redefine *risk* to the integrity of water resources, and reformulate how we assess risk.

No agency currently focuses on the integrity of rivers. For historical reasons, the agencies are ill equipped to conduct the kind of assessment that is required. None adopts a holistic perspective that encompasses the physical, chemical, and biological elements and processes critical to the quality and quantity of water resource systems. Narrowly focused status assessments have been initiated under the Clean Water Act (e.g., Section 305[b]) or by agencies such as the US Geological Survey. Other legislative initiatives, such as the Farm Bill, touch on the status of river resources.

Under the current narrow regulatory scheme, massive expenditures to protect water quality have failed to halt degradation. At least $473 billion has been spent to build, operate, and administer water-pollution control facilities since 1970 (W Q 2000, 1991), yet the decline of many rivers continues. Have those funds been spent efficiently? In one recent study, for example, the construction of expensive tertiary denitrification facilities for wastewater treatment had no beneficial impact on biological integrity in the rivers receiving the wastewater (Karr et al. 1985a). The same study showed that a major factor degrading the river was the chlorine coming from those same wastewater treatment plants. In other words, too often we build expensive treatment facilities with inadequate analysis of the factors actually responsible for local degradation. More careful review of plans to protect river resources could be both less expensive and more effective.

Only integrative approaches are likely to protect societal interests on the spatial and temporal scales demanded by present circumstances. Over the past half-century, laws, government agencies, and citizen groups have proliferated, each with a narrow, idiosyncratic perspective; each driven by special-interest or commodity-group goals. Only an integrative approach within each level of government (USEPA, U.S. Forest Service, U.S. Fish and Wildlife Service, U.S. Geological Survey), across levels of government (local, state, regional, national, and international), and among special-interest citizen groups can treat water resources at the appropriate level of sophistication. Policy makers and regulatory agencies tend to retreat to the safety of existing law, but all groups should keep in mind that much existing water law is outdated because it evolved when knowledge was less, societal needs and values were different, and ample water supplies were available for human use.

Finally, we remain blissfully ignorant of the actual benefits and costs of regulatory actions ostensibly designed to protect water quality. For decades, we have operated water quality programs as if we fully understood the relationships between societal action and resource condition. In fact, however, our policies are untested hypotheses and in hindsight, clearly continue to permit resource degradation. We can — indeed should — actively use each management programs as experiments to test hypotheses about system responses to human actions (Lee 1993). Only then can we modify these programs to benefit society, both economically and environmentally.

4. Long-term success in protecting water resources requires careful thought about assessment end points, including development of biological criteria. Using chemical criteria, USEPA acknowledges that water resources throughout the United States are significantly degraded (USEPA 1992e). In 1990, the states reported that 998 waterbodies had fish advisories in effect, and 50 waterbodies had fishing bans imposed. More than one third of river miles assessed did not fully support designated uses as defined under the Clean Water Act. More than half of assessed lakes, 98% of the assessed Great Lakes shore-miles, and 44% of assessed estuary area did not fully support designated uses. Yet these assessments still underestimate the magnitude of real degradation.

Water resources are not simply water; their quality and value as resources depend on more than water quality and quantity alone. They also depend on biological components (species) and the underlying biological processes that sustain those species. Assessments of species richness, species composition, population size, and trophic composition of resident organisms (biota) are the most direct possible measures in support of the Clean Water Act's goal of maintaining biological integrity (Karr 1993b). EPA defines biological integrity as "the condition of the aquatic community inhabiting unimpaired waterbodies of a specified habitat as measured by community structure and function" (USEPA 1990a). An aquatic community is "an association of interacting populations of aquatic organisms in a given waterbody or habitat." The emphasis is on the interaction and association of species and the structure and function of resident aquatic communities.

We must begin to track the health of rivers as we track the status of local and national economies. We must not limit our assessment and protection goals to large "fishable and swimmable" rivers. Riverine ecosystems encompass an interactive mosaic of environments extending from headwater streams and wet meadows at the upper limit of drainage basins to the ocean. This mosaic includes small streams and large rivers, shoreline (riparian) and streambed (hyporheic) zones, terrestrial watersheds, and groundwater. Water, plants, animals, nutrients, debris, and by-products of human society move among these environments without reference to political boundaries.

Because rivers reflect the status and quality of their landscapes, river assessment programs must make the connections between water quality and quantity, ground- and surface water, and the interdependence of the aquatic biota on water and the landscape. Because biology is the ultimate integrator of these complex interactions, biology provides the most direct and most effective assessment of the status of rivers. Integrating biological assessment into existing assessment programs — if done comprehensively and in a scientifically sophisticated fashion — will provide the information critical to charting a course of federal and state programs to protect the economic and ecological interests of society.

Biological criteria (biocriteria) provide sensitive tracking of resource condition, particularly because impairment of waters is predominantly caused by nontoxic and nonchemical factors. Additional strengths of biological monitoring include the ability to assess and characterize resource status; diagnose and identify chemical, physical, and biological impacts as well as their cumulative effects; serve a broad range of environmental and regulatory programs when integrated with chemical and toxicity assessments; and provide a cost-effective approach to resource protection. Ambient biological monitoring is less likely than the current chemical approach to underprotect water resources. Furthermore, ample evidence exists to show that chemical approaches can waste economic and environmental resources.

In addition, the impediments to effective biological monitoring have been largely overcome. Recent studies by state and federal agencies and by university-based scientists show that the classic arguments against biological monitoring and biological criteria carry little weight relative to the resource protection benefits that result from use of biological monitoring and biological criteria (Karr 1991; Yoder 1991a). Moreover, unlike technology-based approaches, biological monitoring increases the likelihood that unanticipated effects on living systems of water use or management programs will be detected earlier rather than later.

Biological criteria, goals, and end points critical to the health of water resources include protection of indigenous species, protection of whole biological assemblages in waterbodies, and protection of the structure and function of those biological assemblages. Exotic and introduced species, such as the much-publicized zebra mussel or purple loosestrife, threaten valued native populations of fish, shellfish, and other aquatic organisms. The introduction of such exotics often has demonstrably more ominous and long-lasting influences than chemical contaminants on water resources of importance to human society.

Virtually all leading scientific methods proposed to evaluate biological integrity explicitly recognize exotic species as indicators of water resource degradation.

The issue goes beyond single species, however. Certain taxa may have obvious importance to humans because of their value as commodities, but these species do not exist in isolation. We cannot predict which other organisms are critical to the persistence of commercial or otherwise cherished species. Exclusion of insects, zooplankton, phytoplankton, higher plants, bacteria, and fungi from protection ignores the important contribution of these taxa to the structure and function of an ecologically healthy biotic community, a community that is essential to maintaining valuable aquatic resources. No matter how important a particular species is to human society, it cannot persist outside the biological context that sustains it. Because we cannot presume to know today which species will be important tomorrow, the ecological integrity of our waterways rests on the well-being of all the biological components.

In short, programs to protect the biological integrity of aquatic resources should be broadly conceived to include measures of higher-order attributes that provide a more meaningful characterization of the condition of an aquatic resource than conventional measures of water quality or quantity.

5. The quantitative expectations for attributes that constitute biological health or the integrity of a water resource system vary geographically. Criteria developed for chemical contaminants have been applied uniformly for diverse waterbodies; conversely, early advocates of biological criteria were criticized because biological conditions vary geographically and were therefore considered inappropriate monitoring tools. Yet more detailed work has shown that water chemistry also shows considerable geographic variation, and different quantitative criteria are often required to accommodate that variation. In fact, far from being a weakness, regional variation in chemical and biological conditions (Omernik 1987) reflects actual conditions in water resources. Upon reflection, the idea that the same chemical or biological criteria should apply to all waters seems ludicrous.

4.0 TOWARD REAL PROTECTION

Because past water management programs have not been grounded in the preceding five realities, the quality of river resources continues to decline. In many respects, we have been lulled into believing that our rivers have been protected by the Clean Water Act, "wild and scenic" designation, or local laws and regulations when, in fact, their health and integrity have been progressively compromised by human actions. Failure to treat water quality and quantity, ground and surface water, navigable and nonnavigable waters, fish-bearing and non-fish-bearing streams as inextricably connected segments of broad aquatic systems is responsible for this continued degradation.

The degraded state of American rivers reflects the failure of agencies to adopt the conceptual framework necessary to identify the problem and failure of resource policy to focus on the correct problem. Altering that situation requires an approach that replaces conventional assessments with an understanding of rivers as integrated chemical, physical, and biological systems. Adoption of an integrated approach to the study and protection of river resources is essential to prevent additional damage. Further, the degraded condition of most rivers requires an active program of restoration to preserve the long-term economic and ecological interests of society (Doppelt et al. 1993).

Perhaps the most important step in a transition to real protection of aquatic resources is to stop talking about water quality and start formulating policy focused on protecting the ecological health of aquatic resource systems. If we wish to reverse the trends outlined here, we must adopt a broader concept of "water resource"; redefine societal goals based on that concept; forge partnerships among scientists, policy makers, resource managers, and citizens to develop approaches for attaining those goals; revise the legal framework guiding water resource policy; and redouble our efforts to protect existing resources and restore those that are degraded.

ACKNOWLEDGMENTS

Many individuals and agencies have provided financial and intellectual support for my research in water resources over the past 20 years. I express my appreciation to all of them. Finally, thanks to Dr. Ellen Chu for providing insightful editorial comments on several early drafts of this manuscript.

## CHAPTER 3

Biological Assessment and Criteria: Building on the Past

Wayne S. Davis

1.0 INTRODUCTION

Use of ambient biological communities, assemblages, and populations to protect, manage, and even exploit water resources have been developing for the past 150 years. Although precise analytical tools available to the water resource scientist have become more sophisticated, direct measurements of plants, invertebrates, fish, and microbial life are still used as indicators for sanitation, potable water supplies, protection of fisheries, and recreation (McKenzie et al. 1993). With increasing natural resource demands and ecological protection needs anticipated for the twenty-first century, water resource scientists, engineers, managers, and planners must use ambient biological assessments and criteria to protect water resources in an economically and environmentally sound manner (USGAO 1991; ITFM 1992; Adler et al. 1993; NRC 1993).

Since the 1960s, many natural resource, land management, and regulatory agencies have recognized the importance of ambient biological assessments for managing and protecting water resource quality (Burgess 1980; Weber 1980). Unfortunately, without a widely acceptable technical framework for using biological assemblage data, the majority of water resource management decisions relied solely upon surrogate measures of the aquatic community. Such measures include toxicity testing, tissue chemistry and comparisons with chemical criteria through direct measurements and modeled predictions. Despite recent demonstrations of the successful uses of ambient biological criteria (USEPA 1991a; Southerland and Stribling, Chapter 7), there is still a great deal of misunderstanding and concern regarding the application of biological assessments at many levels of management (Suter 1993; Karr 1993a,b). Much of this skepticism comes from personal and professional experience with using, or trying to use, biological assemblage data to assess water resource quality and the great difficulty biologists had (which many still do) with expressing the results in a meaningful, objective, and consistent manner. In other cases little attempt was made to use biological assessments and the potential value of the data were not recognized.

This chapter is intended to provide a better understanding and appreciation of how far numeric biological assessment and criteria development has progressed since the early days of stream pollution biology. This chapter also introduces four tools developed in the past decade that now allow us to transform biological assemblage data into numeric criteria and standards: (1) a functional definition of biological integrity to serve as an understandable water resource goal; (2) minimizing the problems with interpreting the natural geographic and temporal variability of data by aggregating within regions of ecological similarity (Omernik, Chapter 5); (3) using multiple reference sites within ecological or faunal regions to obtain assemblage expectations, or reference condition, for specific geographic areas (Hughes, Chapter 4); and (4) combining several assemblage attributes (or metrics) to produce a single numeric

Table 1.   Milestones in Biocriteria Development

Year	Event
1894	First biological stream monitoring station established to study pollution effects (Illinois River)
1901	Concept of biological classification systems proposed
1908	Indicator organisms for water pollution established
1913	Report linking river pollution with effects on aquatic life
1949	Histogram approach for numerically displaying aquatic community response to pollution
1954–5	Numeric biotic indices for benthic macroinvertebrate assemblages (Beck's biotic index and saprobic index of Pantle and Buck)
1966	Shannon–Wiener diversity index applied to pollution biology
1976	Index of Well-Being for fish assemblages (combined the Shannon-Wiener diversity index with the numbers and weight of the fish)
1981	Index of Biotic Integrity for fish assemblages (multiple metric approach using attributes of the fish community)
1984	Maximum species richness lines to set metric criteria for IBI
1987	Publication of ecoregions (although used as early as 1982)
	Ten metric Invertebrate Community Index developed for Ohio
	Regional reference conditions for biocriteria established in Ohio
1989	Rapid Bioassessment Protocols for benthos and fish
1990	Numeric biological criteria adopted into Ohio water quality standards
1993	EPA Science Advisory Board Review of Draft technical guidance document for biocriteria development in wadable rivers and streams

measure of biological integrity (Barbour et al., Chapter 6). Table 1 presents some of the key conceptual or technical advances that have occurred in this century and shows how recent many of our most important advances have been.

2.0 EARLY INDICATORS OF WATER POLLUTION

Water pollution is not a phenomenon of the twentieth century. Detailed observations of the effects of pollution upon aquatic life and human health have been made for over 150 years. This section highlights some early works of scientists who tried to make others aware of the effects of civilization's rapid progress upon not only natural systems, but also on human health.

2.1 Recognizing Pollution

The classic studies of Chadwick (1842; Flinn 1965), Hassall (1850) and Cohn (1853) have often been credited as the first to use aquatic organisms as indicators of environmental pollution. These researchers documented the relationship among human illness and poor sanitary and drinking water conditions in the mid-1800s. Their efforts led to the development of the first set of national legislation addressing water pollution in Great Britain (Alexander 1876). These early efforts were the basis for the bacteriological tests still used today for protecting against human illness due to contaminated potable and recreational waters (Wilson and Miles 1946; APHA et al. 1993).

Severe reduction of river and stream fisheries were also recorded in England. Fisheries in the rivers Mersey and Irwell declined until the early 1800s when all aquatic life virtually disappeared (Klein 1957). Previously abundant salmon were sufficiently depleted in the River Thames by 1833 due to river pollution causing all commercial fishing to cease by 1850 (Fitter 1945). Recognition of cause and effect between sources of pollution and impacts upon aquatic life led to early attempts to remedy the problems. One of the first documented cases of stream recovery based on ambient biological communities occurred in the river Soar at Leicester in England. The River Soar was reported to be a "common sewer for the drainage of this town" in the late 1700s and in the 1830s "the Soar became so corrupt that fish could not live in them and consequently disappeared entirely" (Chesbrough 1858). After altering the drainage patterns of the town by installing a new sewer system, which discharged wastes further downstream, the river started its recovery and "such a purification of the river as to restore the water to its original clearness, and cause the reappearance, in the summer of 1856, of fish which had not been seen in it for twenty-years before" (Chesbrough 1858).

2.2 Contributions of Early Naturalists

Along with the exploration and westward expansion of the United States during the late 1700s through the middle 1800s, detailed studies and faunistic surveys provided the groundwork for some of the first intentional studies using aquatic life to gauge existing and encroaching pollution. Without documenting the distribution and abundance of the living resources it would have been difficult to determine the effect of natural or man-made influences. Naturalists tediously recorded the presence and distribution of existing and new species of terrestrial, aquatic, and semiaquatic life, most notably through the exploration of the Ohio River valley by the original "boatload of knowledge" (Frost and Mitsch 1989; Pitzer 1989). One of the first ichthyologists to visit the Ohio River basin was C.A. LeSueur in 1817 and 1818 (Trautman 1981). LeSueur (1827) later published the classic *American Ichthyology*. Rafinesque (1820) described over 100 "new" species, although many of them were later shown to be the same as other known species. Kirtland (1838) began his studies on the fishes of Ohio. Jordan surveyed the eastern states and published several reviews of midwestern fish between 1876 and 1891 (Trautman 1981) and Forbes began surveying the fishes of Illinois in the 1870s (Forbes and Richardson 1908).

While adverse effects upon the fish, wildlife, vegetation, and other aquatic life were documented, their use as environmental indicators was purely qualitative. For example, Hildreth (1848) noted the destruction of riparian vegetation in the midwestern United States due to habitat alteration from farming and logging as well as the influence of agricultural water use on Ohio rivers and streams. Kirtland (1850) documented the westward advancement of "progress" in northeastern Ohio between 1797 and 1850 stating that "the whole face of nature has been changed." He described the reduction in the great sturgeon and muskellunge populations already occurring by 1850 and commented that "many smaller species have increased in all our waters…the slaughter houses about the river, afford them large supplies of food and contribute to their increase." The relationship between pollution and undesirable effects upon fisheries was firmly understood by American naturalists by the mid-1800s.

Charles Darwin's (1859) account of his travels on H.M.S. *Beagle* facilitated the movement of natural biologists away from simply enumerating the distribution of populations in faunistic surveys to a fascination with natural selection and how populations respond to natural and anthropogenic changes. It was Stephen Forbes* who initiated some of the most important and fascinating work in aquatic biology and pollution effects. Forbes (1887) built upon Darwin's concept of adaptation and natural selection to develop ecological principles based upon the interrelationships of aquatic populations in his classic presentation entitled "The lake as a Microcosm."

2.3 Illinois River Biological Station

Forbes' insight and application of the principle of natural selection led to the establishment of a biological station on the shores of the Illinois River in 1894 (Bennett 1958). This was the first laboratory designed to assess the effects of pollution on aquatic life. "The general objects of our Station are to provide additional facilities and resources for the natural history of the state…especially with reference to the improvement of fish culture and to the prevention of a progressive pollution of our streams and lakes" (Forbes 1895). In retrospect, after nearly 25 years of operating the Havana Biological Station, Forbes (1928) specifically defined the stations two main objectives: "…one the effect on the plant and animal life of a region produced by the periodic overflow and gradual recession of the waters of great rivers…and the other the collection of materials for a comparison of chemical and biological conditions of the water of the Illinois River at the time then present and after the opening of the sewage canal of the Sanitary District of Chicago which occurred five years later" (in 1900). The biological station would later provide the data and information to document the damage to the health of the Illinois River due to the opening of the Chicago Drainage Canal (Kofoid 1903, 1908; Purdy 1930) despite "authoritative" claims to the contrary (Leighton 1907; Randolph 1921).

---

* Director of the Illinois State Laboratory of Natural History from 1872 to 1917, later renamed the Illinois Natural History Survey, which he directed from 1917 until his death in 1930.

3.0 BIOLOGICAL STREAM CLASSIFICATION

While the Illinois State Laboratory of Natural History began its systematic collection of aquatic life in 1894, the use of indicator organisms to help classify trophic status of rivers and streams was also developing in Europe. Development and application of the first stream classification system was based upon the responses of aquatic life to organic enrichment forming distinct longitudinal "zones" in streams. This "saprobien system" eventually resulted in the development of numerical indices for describing community and assemblage structure (and sometimes function) based upon tolerance or sensitivity to types of pollution.

3.1 Saprobien System

Robert Lauterborn (1901) is credited with originating "the conception of the sapropelic* world of life" by defining the saprobic zone in running waters as the "zone of processes of decomposition." The following year, Kolkwitz and Marsson (1902) presented a historical account of a "new" saprobien system based on indicator organisms (in this case plankton) to classify streams. These German biologists identified three specific stages, which they referred to as saprobic zones, of progressive decomposition to classify slow-moving waters affected by sewage: polysaprobien, mesosaprobien, and oligosaprobien. A fourth zone, the katharobien, was defined as unaffected by the decomposition and essentially "pristine." Kolkwitz and Marsson (1908, 1909) published the first extensive lists of indicator organisms associated with the individual zones of contamination. They first reported the associated saprobien zones (i.e., tolerances) of about 300 species, predominantly benthic and planktonic plants. They later added to the list over 500 planktonic and benthic animals (mostly zooplankton and bacteria) to the list.

Many studies have confirmed the value of this type of a classification system. Strong supporters of the saprobien system in the United States included Stephen Forbes and Robert Richardson (1928) who began publishing reports on the conditions of the Illinois River. They defined the river's degradation via pollutional zones (septic, polluted, contaminate, and clean water) similar to those of Kolkwitz and Marsson. Forbes and Richardson's zones were based upon "integrated" studies of water chemistry, plankton, benthic macroinvertebrate, and fish populations. Their pollutional surveys conducted prior to 1911 documented that 107 miles of river were polluted below the mouth of the Chicago Drainage Canal (Forbes and Richardson 1913).

When the United States Public Health Service Act of 1912 required the Public Health Service to conduct studies on the sanitary condition of interstate waters, Cumming (1916) initiated this effort in 1913. He used three "stages of impurity" and "clean water" based on the saprobien system to describe the aquatic health of the Potomac River watershed. Demonstration of the severity and sources of pollution in the river led to wastewater treatment for the Washington, D.C. area. The Public Health Service studied the Ohio River in 1914 to 1915 and used a classification system based on plankton abundance resulting in four possible ratings of pollution for each sample site: pollution abundant, moderate pollution, slight pollution, and pollution absent (Purdy 1922).

The Public Health Service later studied the Illinois River in 1921 and 1922 but relied upon Richardson's studies to fully document the effects of pollution on the aquatic life. Richardson (1921a, b) defined pollution zones and tolerances for the biota, focusing primarily on the benthic macroinvertebrate community. He documented 146 miles of polluted river between 1913 and 1915, and 226 miles of polluted conditions in 1920, including 146 miles of near anoxic conditions. The last report on the pollution biology of the Illinois River conducted by Richardson was published in 1928, which also marked the change of responsibility for the river surveys within Illinois to the State Water Survey (Bennett 1958). In this classic study, Richardson (1928) detected shifts in water quality based on observations of the benthos alone, although he also used chemical data to better define the pollutional zones. He further refined the pollutional, or saprobic, zones (called septic, pollutional, subpollutional, and clean water) based on "index values" of the benthos using specific taxa as indicators.

Many early authors used indices similar to the saprobien system to classify running waters according to biological effects from pollution (Ingram et al. 1966; Mackenthun and Ingram 1967). Some authors

---

* From the Greek *sapros* meaning rotten or dead. *Webster's New World Dictionary* defines saprobic as "of or pertaining to organisms living in highly polluted water."

recommended using communities of microorganisms (Liebmann 1951; Fjerdingstad 1965) or adopting more levels to reflect additional pollution ratings such as toxic and radioactive conditions in addition to organic enrichment and decomposition (Fjerdingstad 1960; Slàdeček 1965). For example, Kolkwitz (1950) published a revised saprobien system resulting in a total of seven saprobic zones. [For a critical review of the saprobien system, see Bick (1963), Fjerdingstad (1964), Friedrich (1990), Friedrich et al. (1992), and Ghetti and Ravera (1994).]

3.2 Critical Issues with Indicator Organisms

Although widely used, the saprobien system and the general concept of indicator organisms met with a great deal of criticism. Doudoroff and Warren (1957) stated their doubts about the saprobien system and the indicator organisms proposed because: (1) they reflected only pollution from sewage wastes, (2) the relationship among organisms and pollution tolerances were not well studied, and most importantly (3) the indicator organisms used were generally not reflective of economic value. The authors were concerned about the lack of attention biologists paid to economically valuable species: "[t]hey [some biologists] seem to have curiously attached at least as much importance to the elimination of any species of diatom, protozoan, rotifer, or insect as to the disappearance of the most valuable food or game fish" (Doudoroff and Warren 1957). They strongly advocated the study of economically valuable fish species and the use of toxicity testing as better indicators of water quality.

There were also doubts about assigning a single saprobic zone, or indicator value, to a species that could also be found outside of that zone. Brinkhurst (1969) stated that "I can see no way in which different saprobity values could be given to each species to account for its reactions to the many different forms of polluting materials...I find the systems...less efficient than the opinion of a qualified biologist expressed in plain language." Fjerdingstad (1964), Hynes (1965), and Cairns (1974) were also critical of using indicator species and they all advocated a community-based approach instead. Hynes favored using benthos while Fjerdingstad was convinced that attached algae (diatoms) were the superior group since they were not subject to stream drift.

Hawkes (1957) summarized concerns regarding the saprobien system and the manner in which specific biological indicators were perceived to be used as follows:

> [i]n using this system it must be borne in mind that factors other than pollution affect the nature of stream communities. The absence of organisms may be a more important indication than the presence of other species. The community of organisms should be taken into consideration rather than the presence or absence of one or few 'indicator organisms'. In some cases specific identification is essential, in others the genera or the whole family may be indicative. Knowledge of stream ecology is continually advancing and no doubt the list will be modified and extended in light of this knowledge.

Even as we have progressed in our understanding of the limitations of using indicator organisms to measure water resource quality, indicator species continue to be used as "integrators" (Ryder and Edwards 1985) and as components of modern ecological health indices (Cairns 1993; Karr 1993a).

3.3 Importance of the Saprobien System

Although Bartsch and Ingram (1966) felt that the saprobien system was purely a European tool that was not used "to determine the existence and magnitude of pollution, but to obtain colorful biological data for an anti-pollution campaign," the saprobien system was very important in focusing biologists on measuring water resource quality by the presence or absence of a wide range of indicator biota. This approach became more sophisticated by eventually relying upon communities or specific assemblages (e.g., benthos) and attention to their relative abundance and distributions. In fact, the saprobien system is widely used outside the United States and is no longer dependent (if it ever was) upon only a few indicator species (Friedrich et al. 1992; Ghetti and Ravera 1994). Much of the disagreement and criticism of using indicator species, and even communities, was largely based on the presentation and interpretation of the results and what was defined as pollution. The saprobien system and the other similar stream classification systems led to the investigation of the "pollutional status" and eventually to the definition of pollution as applied to the aquatic resources. Perhaps the most remarkable scientific contribution of the saprobien system was the direct development of numeric biotic (or saprobic) indices, which are discussed in the following section.

4.0 NUMERIC BIOLOGICAL INDICES

One of the biologist's challenges is to present information that is understandable, meaningful, and helpful to associated disciplines, to administrators, and to the general public who are the financial supporters as well as the benefactors of a pollution abatement program. (Ingram et al. 1966)

This challenge resulted in a search for numerical expressions in a form simpler to understand than long species lists and well-thought but lengthy technical explanations of the data.

4.1 Early Indices of Pollution

The work of Wright and Tidd (1933) was considered by some to be the "original numeric index" (Myslinksi and Ginsburg 1977). Abundance of oligochaetes was used to assess the degree of pollution as follows: values of less than 1000 m^{-2} indicated negligible pollution, between 1000 to 5000 m^{-2} indicated mild pollution, and over 5000 m^{-2} indicated severe pollution. However, Richardson (1928) found that numerical abundances of each index group was not as significant as their relative abundances and overall occurrences. He reported the number of pollution tolerant Tubificid worms in the Illinois River to range from under 1000 to over 350,000 per square yard in pollutional zones, and chironomid midge larvae to range from zero to over 1000 per square yard. Richardson concluded that seasonal and habitat changes were responsible for much of the variability of species abundance at a given site, supporting the use of relative abundance as the better index measure.

Several attempts were made in the next two decades to numerically characterize the biological data in a meaningful and understandable manner. A variety of schemes were used including indices based on trophic function, structural ratios of taxa, feeding requirements, and other inventions (Washington 1984). Not satisfied with how aquatic biological field data was presented, Ruth Patrick (1950) developed a "histogram" approach based upon seven taxonomic groups. She used (I) blue-green algae, (II) oligochaetes, leeches, and pulmonate snails, (III) protozoa, (IV) diatoms, red algae, and most green algae, (V) other rotifers, clams, prosobranchia snails and tricladid worms, (VI) all insects and crustacea, and (VII) fish. Patrick's work portrayed the results in a graphical and numeric format, but also recognized the importance of community composition rather than a population-based analyses. Stream classes of healthy, semihealthy, polluted, very polluted, and atypical were assigned based upon comparing the predominance of cleaner water groups IV, VI, and VII compared with the other four groups. Cairns (1974) felt that

The particular importance of this paper was that it showed that biological data used to assess pollution could be presented numerically and that one need not depend upon the usual unwieldy (and for nonbiologists incomprehensible) list of species to make an estimate of the degree of pollution…Thus, the method had both scientific merit and, perhaps more importantly, results could be easily communicated to people to whom species lists were meaningless.

Aquatic community assessments using the seven taxonomic groups was an ecological strength of Patrick's method but it was feared that state agencies would have difficulty with routinely collecting this wide array of data (Beck 1954). In addition, Cairns (1974) acknowledged one drawback of Patrick's method, similar to the saprobien system, in which "it required a highly skilled professional to make the species determinations necessary to properly categorize the system and its response to pollution stress." Although ecologically significant, Patrick's approach was too burdensome on most field biologists who also preferred the comfort of dealing with the one assemblage that they knew best. Thus the need for a cost-effective "rapid" biological assessment method was established. Biologists and other water resource scientists continued to struggle to derive a suitable index, or numerical expression, of the aquatic indicator organisms response to assess pollution effects. One method that many thought would solve this dilemma measured structural diversity of the community or assemblage rather than the functional or indicator role of the populations.

4.2 Diversity Indices

Species diversity, or the evenness of the distribution of individuals in a community assemblage, has been widely used since the 1960s as a measure of stream community response to pollution (Norris and Georges 1993). Cairns (1977) saw great potential for diversity indices and felt that it was "probably the best single means of assessing biological integrity in freshwater streams and rivers." Diversity indices gained favor with sanitary engineers and biologists as easy numeric indices that, as Hawkes feared, would be "entered into neat columns alongside analytical results" of chemical and physical parameters. One of the most popular diversity indices used for water resource quality assessment, H' was published by C.E. Shannon (Shannon and Weaver 1949). It is correctly termed the Shannon–Wiener index because Wiener (1948) independently published a similar measure at approximately the same time (Washington 1984). Possibly the first use of the Shannon–Wiener diversity index for assessing water quality was by Wilhm and Dorris (1966) who used diversity to describe longitudinal variation in the benthic community structure of an Oklahoma creek affected by municipal and industrial wastes. They found a severe decrease in the diversity index immediately downstream from a pollution source and an increase in the diversity index as recovery occurred. Wilhm and Dorris (1970) described the ranges of H' (calculated as d) associated with clean, moderately polluted, and substantially polluted streams.

Although the Shannon–Wiener diversity index has been used most often with benthic macroinvertebrates, it has also been used for many other assemblages. An example of this is the Index of Well-Being (Iwb) for fish. James Gammon (1976) developed the Iwb based upon both abundance and biomass measures as follows:

$$\text{Iwb} = 0.5 \ln N + 0.5 \ln B + H'_N + H'_B$$

where: N = number of individuals caught per kilometer
  B = biomass of individuals caught per kilometer
  H' = Shannon-Wiener diversity index calculated based on individuals per kilometer (H'_N) and biomass per kilometer (H'_B)

This combination of diversity and biomass has been highly successful for assessing fish assemblages in large rivers (Hughes and Gammon 1987; Plafkin et al. 1989; Yoder and Rankin, Chapter 9).

Despite the popularity and apparent success of the Shannon–Wiener index, it has also been severely criticized in the United States (Bilyard and Brooks-McAuliffe 1987; Fausch et al. 1990) and Europe (Metcalfe 1989; Friedrich et al. 1992; Ghetti and Ravera 1994) resulting in diminished use. The greatest criticisms of diversity indices include: (1) their inability to reflect ecological significance, (2) total reliance upon structural (abundance) measures that vary greatly depending upon the time of year sampled, the collecting gear used, and the level of taxonomic resolution, and (3) the loss of community composition information by using a single index value (Washington 1984; Metcalfe 1989; Fausch et al. 1990). Cairns et al. (1993) explained that since "the identity of the species is ignored in the calculation of diversity indices, these measures are not sensitive to compensatory changes in the community...which alter the taxonomic composition of the community but have little effect on community diversity." Hilsenhoff (1977) found little ecological significance when using diversity indices and concluded that "the diversity index does not accurately assess the water quality of streams, ranking some of the cleanest undisturbed wilderness streams with moderately enriched or polluted streams."

Despite these limitations, the popularity of the Shannon–Wiener diversity index, among others, is quite high (Norris and Georges 1993). The diversity index is currently used to characterize a variety of aquatic assemblages in different aquatic resource types throughout the world (Friedrich et al. 1992; Ghetti and Ravera 1994).

4.3 Beck's Biotic Index

Beck (1954) developed a biotic index that produced a numeric end point that could be easily interpreted by sanitary engineers and other water resource managers. Although he conceded the popularity of Patrick's method, Beck criticized Patrick's histograms because it was "hardly within economic reach of the average state regulatory agency or industry, and the information obtained is not available to the general research worker." Beck's index was originally based upon three classes of benthos — Class I (intolerant), Class II (facultative), and Class III (pollution tolerant) — but he decided not to use the tolerant organisms since they were sometimes found in cleaner waters although at a much lower abundance (Beck 1955). Beck's index, which ranged from 0 to 40, did not rely upon organism abundance but assigned numeric values (weights) of 2 and 1 for the different Class I and Class II taxa, respectively. The final index values were calculated by the formula with S representing the number of taxa within each group:

$$\text{Biotic index} = 2(S \times \text{Class I}) + (S \times \text{Class II})$$

Although this index did not achieve prominence and widespread use as an assessment tool among state biologists, it was considered a successful advance in the field of aquatic biology in the United States and was credited with popularizing the term "biotic index." This index is currently used by the Soil Conservation Service (Terrell and Perfetti 1989) as one of several water quality indicators. Perhaps the most widespread use of this index is by citizen volunteer monitoring organizations, which uses three classes of indicator organisms as originally proposed (Kopec 1989; Lathrop and Markowitz, Chapter 19).

4.4 Saprobic Index

There was another "biotic" that which was developed in central Europe at the same time Beck's biotic index was presented in the United States. Pantle and Buck's (1955) biotic index was based directly upon the saprobien system. Its simplicity and numeric relationship to the original four zones of stream pollution lead to the development of a widely used biotic index in the United States (Hilsenhoff 1982a) and could be considered the true predecessor of today's biotic indices (Friedrich et al. 1992; Ghetti and Ravera 1994). The authors directly used the saprobien system and assigned each zone a number from 1 to 4: 1 was oligosaprobic, 2 was beta-mesosaprobic, 3 was alpha-mesosaprobic, and 4 was polysaprobic. Each organism associated with the various zones based upon Liebmann's (1962) revised list of indicator organisms were assigned the respective indicator value (s) multiplied by a relative abundance weight (h) of either 1 (species only found by chance), 3 (species occurring frequently), or 5 (species occurring in abundance). These weighted values were then averaged to derive the saprobic rating, S, as follows:

$$S = \frac{\Sigma s \times h}{\Sigma h}$$

Saprobic ratings of 1 to 1.5 indicated very slight impurity, 1.5 to 2.5 was moderate impurity, 2.5 to 3.5 revealed heavy impurity, and 3.5 to 4.0 showed very heavy impurity.

Tümpling (1962) established regression lines for S with biochemical oxygen demand (BOD), oxygen deficit in percent saturation, and concentration of ammonium ion. There was a great concern that any valid index should be correlated with BOD loadings and instream dissolved oxygen. Tümpling (1969) also showed the index (S) could be used to determine saprobity with a 95% confidence interval to a level of 0.2 to 0.3 units of S. He cited Liebmann's (1962) support of this index and his own 1962 work as verification (Tümpling 1962). Guhl (1986) also found that surface waters could be defined as biologically different if the saprobic index varied by more than 0.2 units if sampled by the same investigator and 0.5 units if sampled by different investigators.

The saprobic index has been modified in many ways since it was first proposed by Pantle and Buck. However, the conceptual modifications made by Zelinka and Marvan (1961) were the most substantial and forever changed the use of the saprobic index. They addressed many of the criticisms of the original index, as well as the general use of indicator organisms by adding a saprobic valency and indicator weight to the original index. The saprobic valency expressed the relative frequency of the species in different degrees of saprobity on a scale that totaled 10. By establishing the saprobic valency, Zelinka and Marvan satisfied a major criticism of the saprobic index — it's dependance upon arbitrarily assigning a single saprobic zone to a species which is likely to be in more than one zone. They also felt that some species were more useful as indicator organisms and assigned an indicative weight from 1 (poor indicator) to 5 (very good indicator). The best indicators were those species that had been assigned a saprobic valency of 8, 9, or 10 within any given zone (Slàdeček 1991). This showed that the best indicators were representative of a single saprobic zone. The lowest indicator weights were given to species found throughout many or most of the saprobic zone. These conceptual modifications of the saprobic index have been quite popular and ensured widespread use not only throughout central Europe, but also other parts of the world. Bick (1963), Fjerdingstad (1964), Friedrich (1990), Slàdeček (1965, 1985, 1988, 1991), Friedrich et al. (1992), and Ghetti and Ravera (1994) provide a great deal of insight into the specific changes and uses of the Saprobic Index.

4.5 Selected Macroinvertebrate Indices

Woodiwiss (1964) presented the biological system of stream classification that was used by the Trent River Board in England. The Trent biotic index varied from 0 to 10, with 10 representing clean water, based upon the relative abundance of representative benthos groups. This index greatly influenced the development of the Chandler biotic index, Belgian biotic index, Extended biotic index, and Indice Biologique (Metcalfe 1989).

Metcalfe (1989) also categorized European indices into saprobic, diversity, and biotic but did not view modern biotic indices as an extension or modification of the original saprobic indices, contrary to Friedrich et al. (1992) and Ghetti and Ravera (1994). This was because saprobic indices had an early focus on plankton and periphyton whereas biotic indices were based on benthic macroinvertebrates. Friedrich et al. (1992) differentiated the indices by labeling biotic indices as those that utilize only some taxa from an assemblage and the saprobic indices as using as many species of the community as possible.

In the United States, Hilsenhoff (1977, 1982a) also recognized the direct relationship among the early saprobic indices and biotic indices. He credits Pantle and Buck (1955) and Chutter's (1972) index as predecessors of the Hilsenhoff biotic index. Chutter (1972) developed a biotic index for South African streams and assigned specific tolerance values for various taxa ranging from 0 to 10. His index accounted for both the number of individuals and the number of taxa, but contained only limited quality (i.e., tolerance) values.

Hilsenhoff's (1977, 1982a) biotic index (see below), originally scaled from 0 (clean) to 5 (polluted) was based on a wide array of aquatic insect taxa from Wisconsin identified to genus or species.

$$\text{Biotic index} = \frac{\Sigma n_i \times a_i}{N}$$

where n_i = number of individuals of each taxon
a_i = tolerance value assigned to that taxon
N = total number of individuals in the sample

He compared the biotic index with Shannon's diversity index based upon several physical and chemical parameters using rank correlation analysis. Hilsenhoff (1977) found that the biotic index correlated much better than the diversity index in distinguishing among pollution gradient in streams. Hilsenhoff revised the index in 1982 and in 1987 to reflect new index values, expanded the biotic index scale to 0 to 10, and included several new taxa. He then developed a popular family-level biotic index that has also been widely used for screening water resource quality (Hilsenhoff 1988). A Hilsenhoff biotic index, with tolerance values modified for specific geographic regions, is used for water quality assessment in many states (Lenat 1993; Bode and Novak, Chapter 8; Southerland and Stribling, Chapter 7) and has become incorporated into the new generation of numeric multimetric indices (Plafkin et al. 1989; Barbour et al., Chapter 6; Resh, Chapter 12). Resh and Jackson (1993) provide a comprehensive review of rapid bioassessment techniques using benthic macroinvertebrates.

4.6 Multiple Metric Indices

During this time, debate continued regarding the use of numerical biological indices based upon indicator organisms. Brinkhurst (1969) stated that "the value of biological methods of pollution detection is now widely accepted, but there is still considerable debate about the means of providing inexperienced biologists with simple standard procedures and of reporting biological data to non-biologists." The advantage of both diversity and biotic indices is that they reduced complex interactions and pollution responses of an aquatic community into a single number for water quality management purposes. However, neither of these indices were successful in describing the overall "health" or condition of the aquatic ecosystem under a variety of conditions. It was clear that a better tool was need to more consistently and accurately characterize the aquatic communities.

Karr (1981) published the Index of Biotic Integrity (IBI) to provide a more accurate and consistent approach towards measuring the societal goal of "biological integrity" (see Section 5.2 in this chapter). The IBI includes discrete measurements of 12 fish assemblage attributes, or metrics, based on species composition, trophic composition, abundance, and condition. Each metric was assigned a score (5, 3, or 1) based upon specific ecological expectations. The 12 metric scores were summed to provide a cumulative site assessment. The scores result in "integrity classes" for streams of excellent, good, fair, poor, very poor, or no fish (Karr et al. 1986). The IBI is called a composite or multiple metric index because it combines several community attributes into a single index value without losing the information from the original measurements. A number of natural resource and regulatory agencies have demonstrated this to be a very successful tool for water resource quality evaluations (Simon and Lyons, Chapter 16; USEPA 1991a; Abe et al. 1992). [Please refer to Simon and Lyons (Chapter 16) and Yoder and Rankin (Chapter 9) for more information on the application and regional and local modification of the IBI metrics.]

It did not take long before multiple metric indices were developed for benthic macroinvertebrates and periphyton. The Ohio Environmental Protection Agency developed an Invertebrate Community Index (ICI) in 1986 (DeShon, Chapter 15) that is based on ten structural and functional metrics that quantify subjective judgements that had been used for a number of years. Shackelford (1988) developed a multiple metric benthic index for Arkansas that combined seven measures of community diversity, indicator organism, and functional groups. USEPA then published a set of composite indices called Rapid Bioassessment Protocols (RBPs) for benthic macroinvertebrate and fish communities (Plafkin et al. 1989). The RBP benthic community metrics are based on very general structural and trophic relationships that could be applied nationally. The primary fish assessment methods were Karr's IBI and Gammon's Iwb. Hayslip (1993) recently compared the benthic metrics used by states in the Pacific Northwest and found a great deal of metric modification of the original RBPs.

Periphyton assemblages were also described using multiple metrics. A periphyton biotic index (Kentucky DEP 1992) was developed for use in the State of Kentucky to complement fish and macroinvertebrate water resource assessments. The metrics used included taxa richness, relative abundance of sensitive and tolerant taxa, percent community similarity compared with reference sites, and biomass. Bahls (1993) developed a periphyton index for Montana streams using three metrics for soft-bodied taxa (dominant phylum, indicator taxa, and number of genera) and four metrics for diatoms (Shannon–Wiener diversity index, pollution index, siltation index, and a similarity index compared with a reference condition). Rosen (Chapter 14) further discusses the periphyton indices and metrics used for developing biocriteria.

Indices of biotic integrity have not been without criticism. Suter (1993) outlined the following exhaustive list of potential faults of what he called "indexes of heterogenous variables" such as the IBI: (1) ambiguity, (2) eclipsing, (3) arbitrary combining functions, (4) arbitrary variances, (5) unreality, (6) post-hoc justification, (7) unitary response scale, (8) no diagnostic results, (9) disconnected from testing and modeling, (10) nonsense results, and (11) improper analogy to other indices. Suter (1993) explained that indices like the IBI "are justified on the basis of field studies rather than any theory of ecosystem health or any societal or ecological value of the index or its components." I am not going to present a response to each of these items, but allow these perceived issues to be addressed by the many qualified chapter authors in this book. However, I do agree that more research and testing is needed before the concepts of these indices can be expanded to develop truly ecosystem "health" indices. Karr (1993a), Simon and Lyons (Chapter 16), and Yoder (Chapter 21) present detailed responses to Suter's criticisms.

5.0 FRAMEWORK FOR CRITERIA DEVELOPMENT

There have been several key areas in which the scientific thought and application of biological tools significantly advanced water resource assessment and criteria development. Those already mentioned include the use of indicator organisms originated in the saprobien system, numeric biological indices, and the aggregation of several numeric biological attributes into multiple metric indices for measuring biological integrity. The remaining substantial developments are a combination of technical achievement and conceptual implementation for describing societal and ecological goals from which progress towards meeting those goals could be measured. They include defining pollution through beneficial use assessments for aquatic life support based on measures of biological integrity and using multiple reference sites to define attainable (reference) conditions within a regional framework (i.e., ecoregions).

5.1 Debating Ecological and Societal Goals for Water Resources

The legal authority for water rights and ensuring the water resource is fit to serve private and public uses has long been recognized (Warren 1971). However, the uses or combination of uses, and their priorities were quite different depending upon the needs of the user. In the 1800s and early 1900s, the focus on water uses were primarily as conveyances of municipal and industrial wastes and for drinking water, a very distasteful combination! When the water resource was not required for a potable water supply, the common "standard" was one that avoided a public nuisance. An example of the concern for economic use of the water resource without regard for how the aquatic life or downstream communities were affected was found in the city of Chicago during the early 1900s. Consider the following excerpts from a report delivered to the Board of Trustees of the Sanitary District of Chicago by its Commissioner (Wisner 1911):

> The question has always been considered from the standpoint of nuisance, and not as to whether or not the water was so polluted as to destroy fish life.... Our investigations lead to the conclusion that a nuisance may not occur, even though all the fish be dead through the lack of sufficient dissolved oxygen necessary to fish life.... From an inspection of the available data on the condition in the Illinois River, and in the Des Plaines River, prior to the opening of the Drainage Canal in 1900, it is evident that a marked improvement took place. The foul conditions had been tolerated for years. Fish life has decreased in the main river. Since the opening of the canal the fish catch is said to have improved, although no definite data are available. Owing to the great extent of the fish industry in the Illinois River it is essential that the condition of the river, insofar as the Sanitary District is concerned be kept as good as possible.... It is necessary that immediate steps be taken to ascertain the conditions along the river...and that the continued examination be made year after year by the Sanitary District in order to have the data in hand to refute possible law suits for damages to fishing, and the possible reopening of the St. Louis case. This is not only a matter 'of sanitation, but a question of self-defense, protecting the root and purpose of the Sanitary District.

Additional information about the early history of Chicago's water quality experiences can be found in Cain (1978) and Davis (1990).

Potential beneficial uses were clarified in the 1948 Federal Water Pollution Control Act (Public Law 845) which stated, "[i]n the development of such comprehensive program due regard shall be given to the improvements which are necessary to conserve such waters for public water supply, propagation of fish and aquatic life, recreational purposes, and agriculture, industrial, and other legitimate uses" (Mackenthun and Ingram 1967). Defining an independent use solely for the propagation of fish and aquatic life was a very important advance. However, it was difficult to determine whether this use was being attained, or was even attainable. The problem with using biological assessments and indices for water resource quality assessment was not just with the numeric interpretation of the data, but also with understanding what the measurements meant with respect to the desired condition of the resource. This problem was reflected by Doudoroff and Warren's (1957) comment that "[a]lthough most authors evidently have recognized the economic significance of pollution, it appears that when devising their biological indices and measures of water pollution and its severity some biologists have completely disregarded all economic considerations."

Many water resource quality specialists disagreed as to whether to emphasize economically important populations such as gamefish, coastal invertebrates, and freshwater mussels or all aquatic life equally. For example, the Ohio River Valley Sanitation Commission (ORSANCO) Compact of 1948 called for waters that are "capable of maintaining fish and other aquatic life" (Cleary 1955). However, in 1954 the Aquatic Life Advisory Committee to ORSANCO concluded that their mission was concerned with only the "production of fish crops" measured by bioassays (Cleary 1955). Although there was a great deal of information available on benthic macroinvertebrates, phytoplankton, attached algae, and fish assemblages, ORSANCO's focus turned only to using toxicity test results for criteria development.

Confusion and disagreement with defining societal goals for clean water and the control of pollution was reduced (or at least redirected) with the 1972 Amendments to the Federal Water Pollution Control Act, commonly referred to as the Clean Water Act (USGPO 1989). The general objective of the Act is to "restore and maintain the chemical, physical, and biological integrity of the Nation's waters" [Section 101(a)]. The second national goal of the Act [Section 101(a)(2)] was "wherever attainable, an interim goal of water quality which provides for the protection and propagation of fish, shellfish, and wildlife and provides for recreation in and on the water." (Section 101(a)(2) has commonly become known as the "fishable and swimmable goal".) USEPA's written opposition to the controversial objective of the act firmly established the relationship and relevance of water quality to support the beneficial uses:

> The pursuit of natural integrity of water for its own sake without regard to the various beneficial uses of water is unnecessary, uneconomical, and undesirable from a social, economic, or environmental point of view. We believe the purpose of water pollution control is the achievement and protection of water quality for beneficial uses. (USGPO 1972)

5.2 Defining Biological Integrity

Legislation to protect aquatic life first appeared in 1876 and has continued to develop in its scope and intent (Table 2). Most of the early legislation was geared toward the protection of waters for human use (beneficial uses). The 1972 Clean Water Act represented a change in that its objective and new interim goal went far beyond the application of mere beneficial uses. It was viewed as having an "ecological" beneficial use for the sake of the environment, independent of any readily available economic benefits. This language was not trivial and caused a great deal of concern regarding how to define "integrity" (especially biological integrity) and the measurements to be applied. The 1972 House Committee on Public Works (USGPO 1972a) defined integrity as a "concept that refers to a condition in which the natural structure and function of ecosystems is maintained." Continuing, they stated "[o]n that basis we could describe that ecosystem whose structure and function is 'natural' as one whose systems are capable of preserving themselves at levels believed to have existed before irreversible perturbations caused by man's activities. Such systems can be identified with substantial confidence by scientists." It is rewarding that Congress had such a high a opinion of our discipline. The 1972 Senate Public Works Committee (USGPO 1972b) stated that "The 'natural...integrity' of the waters may be determined partially by consultation of historical records or comparable habitats; partially from ecological studies of the area or comparable habitats; partially from modelling studies which make estimations of the balanced natural ecosystems on the information available".

The National Commission on Water Quality (USGPO 1976), which was appointed to make a full and complete investigation and study of all aspects of the Clean Water Act requirements (Section 315), had difficulty in setting the course of its study because of the ambiguity of the term "biological integrity" (USGPO 1975). Their difficulty was obvious when they concluded that "[t]he most quantifiable indicators of biological health in an aquatic system are the physical and chemical parameters assessed at each of the study sites. Individually and together, they provide a broad picture of existing water quality and projected progress toward a quality that will support the purposes listed in the interim goal" (USGPO 1976). Based on this more comfortable and traditional position, they declared that the objective of the act really meant to focus on a combination of all three integrity measures as a single concept of ecosystem integrity (USGPO 1976).

Still not satisfied with the answers provided by Congress or the National Commission on Water Quality, EPA hosted a national forum on the Integrity of Water in 1975. Both qualitative and quantitative concepts of chemical, physical, and biological integrity were reviewed. Two definitions of biological integrity were informally proposed at the forum. The first was by Cairns (1977) who felt that "biological integrity may be defined as the maintenance of community structure and function characteristic of a particular locale or deemed satisfactory to society." The second definition was proposed by Frey (1977) who defined the integrity of water as "the capability of supporting and maintaining a balanced, integrated, adaptive community of organisms having a composition and diversity comparable to that of the natural

Table 2. Important Legislation and Agreements Facilitating Biological Criteria

Legislation	Year	Key Elements for Biocriteria
River Pollution Prevention Act (England)	1876	First legislation intended to provide protection to fisheries as well as prevention of nuisance
Public Health Service Act	1912	First national investigations of pollution and aquatic life in major U.S. river systems
Federal Water Pollution Control Act (PL 80–845)	1948	Established federal authority for interstate water pollution control. Recognized propagation of fish and aquatic life as a legitimate beneficial use of waters
Ohio River Valley Sanitation Commission Compact	1948	Established mechanism for developing water quality criteria to protect aquatic life via toxicity testing
FWPCA Amendments (PL 84–660)	1956	Fish and aquatic life protection formalized as a beneficial use. Began national water quality monitoring network requiring systematic collection of aquatic life
FWPCA Amendments (PL 89–234)	1965	Federal authority to review state water quality standards
National Environmental Policy Act (PL 91–190)	1969	Submission of Environmental Assessment and Impact Statements on proposed federal actions
Clean Water Act (PL 92–500)	1972	Objective of the Act focused on maintaining and restoring biological integrity of surface waters. Started large movement towards defining and measuring biological integrity
Endangered Species Act (PL 93–205)	1973	Provides protection for special status species
Water Quality Act (PL 100–4)	1987	Shift from technology-based to water quality-based approach
Great Lakes Water Quality Agreement	1987	Adopted a biological integrity objective moving towards ecological integrity and measuring indicators of ecosystem health
Water Pollution Prevention and Control Act (Draft S 1114)	1993	Supports "biological discharge criteria" based upon establishing the biological conditions of the waterbody
Nonpoint Source Water Pollution Act (Draft HR 2543)	1993	Expands ecosystem integrity protection approach and supports biological criteria
Biological Survey Act	1993	Requires a national biological survey to assess the status and trends of the biological resources in the U.S. and establishes an new agency to carry out the survey

habitats of the region." Apparently, neither of these definitions were widely accepted at that time and it was evident that defining biological integrity, and hence the methods to measure it, was a complex task that required a more focused effort.

USEPA's Water Office asked the Corvallis Environmental Research Laboratory to review the definition of biological integrity and to suggest ways it might be monitored (Hurley 1981). As a result of that request, the USEPA Corvallis Laboratory assembled a team of experts from within USEPA, academia, and the Fish and Wildlife Service to tackle the problem (Hughes et al. 1982). This team provided the breakthrough to assemble a functional definition and framework for describing biological integrity. At a national workshop in 1981, they presented: (1) a definition of biological integrity establishing base (reference) conditions within faunal regions (ecoregions), (2) methods comparing base-line conditions with impacted conditions to determine relative well-being, (3) the use of multiple sites to establish reference condition, and (4) the foundation of Karr's Index of Biotic Integrity (Hughes et al. 1982). They concluded that

> a definition of biological integrity has been adopted that established base biological conditions as those found in the least-disturbed typical reaches of large, relatively homogeneous faunal regions. Once these base biological conditions have been established, data gathered at other locations within faunal regions will be compared with the base to determine the relative well-being of each non-base location. They suggested the use of fish assemblages as the indicator of biointegrity. (Hughes et al. 1982)

Karr and Dudley (1981) further defined biological integrity with the ecosystem perspective of Frey (1977) by adding functional organization to the desirable characteristics of the aquatic community. This differed from the narrower "fishable and swimmable" Clean Water Act goal and also met the intent of the House Committee (i.e., natural structure and function of an ecosystem) as well as the Senate Committee (i.e., comparable habitat). Karr and Dudley's (1981) definition has become widely accepted within the regulatory and scientific community (Schneider 1992). Karr (1981) recommended that biological integrity be used to "assess the degree to which waters provide for beneficial uses," especially aquatic life support. It did not take long before these concepts were tested for state programs (Southerland and Stribling, Chapter 7). Currently, the term biological integrity has been used synonymously with attaining the beneficial use for aquatic life protection (Yoder and Rankin, Chapter 9). [Please see Karr (1991; Chapter 2) and Adler (Chapter 22) for additional perspectives of the legislative and technical activities relating to ultimate focus on biological integrity.]

5.3 Reference Condition and Regionalization

When Hughes et al. (1982) recommended a definition of biological integrity and ways to measure it, they had in mind developing sets of least impacted reference (or attainable) conditions within faunal regions (i.e., ecoregions) to compare with each test location. The rationale for using ecological regions was to establish reference conditions based upon patterns in community attributes that had previously been found to vary naturally among geographic regions (Hughes and Larsen 1988). Without accounting for natural geographic variability it would be difficult to establish numerical indices that were comparable from one part of a State to another, much less nationally. Therefore, using ecoregional reference conditions allow an unbiased estimate of the surface water's attainable (least impacted) conditions (Hughes et al. 1986). These concepts were the turning point in finally developing defensible biological (and some chemical) criteria for state water quality standards and other programs. Hughes (Chapter 4) and Omernik (Chapter 5) discuss reference condition and ecoregions in detail.

The first complete application of the framework for biocriteria development occurred as a result of the Ohio Stream Regionalization Project (SRP). This cooperative effort among Ohio EPA, USEPA's Environmental Research Laboratory in Corvallis, Oregon, and USEPA's Region 5 office in Chicago was conducted in 1983 and 1984 (Whittier et al. 1987). The SRP identified and delineated five ecoregions in Ohio and then focused on selecting least-impacted reference watersheds and sites to determine the best attainable condition in those waters. Field sampling was conducted for over a year and included physical habitat, fish and macroinvertebrate assemblages, and chemical water quality in 109 streams. The fish assemblage was measured by several means including the Index of Biotic Integrity and the Index of Well-Being. The results were displayed in box plots and the attainable conditions were based upon the 50th percentiles of each of the attributes. Ohio EPA later refined the attainable conditions for aquatic life (warmwater biocriteria) based on a 25th percentile of the ecoregional reference site conditions of each measurable attribute related to drainage area. The success of this demonstration project led the State of Ohio to adopt numeric biological criteria in 1990 based on results of over 236 reference sites throughout the state (Yoder and Rankin, Chapter 9).

6.0 FUTURE PROSPECTS

Many natural resource, land management and regulatory agencies are beginning to implement biological assessments and even criteria development as essential tools to protect water resource quality and biodiversity (ITFM 1992, 1994; CEQ 1993; NRC 1993; USEPA 1993b, c,d,e,f) and to reduce the uncertainty in applying the traditional chemical criteria to protect those resources (USEPA 1990a, 1991c). Although these efforts have been sustained by state agencies for the past decade, federal agencies are also beginning to actively participate in biological assessments which will likely lead to criteria development (NRC 1993, Table 3). USEPA has issued guidance on developing biological criteria programs (USEPA 1990, 1993c), developing narrative biocriteria (USEPA 1992c), and is finalizing a technical guidance document for streams (Gibson 1994). EPA has hosted several biocriteria workshops and national meetings in cooperation with state agencies (e.g., USEPA 1987b; Davis 1990a; Hayslip 1993; see Southerland and Stribling, Chapter 7). However, there is no guarantee that these efforts will be successful. We must continually educate ourselves and our colleagues regarding the benefits of biological assessments and criteria in environmental restoration and protection. We must improve our existing methods and applications and maintain the necessary research on new and promising techniques. We also must not forget the dedication and philosophy of those scientists who brought us to the present.

Table 3.   U.S. Federal Agencies Involved in National Biological Assessments

Agency	Program
Department of Interior	
Fish and Wildlife Service	National Contaminant Biomonitoring Program
	Biomonitoring of Environmental Status and Trends
	Waterfowl Breeding Population and Habitat Survey
	National Survey of Fish, Hunting and Wildlife
Geological Survey	National Water Quality Assessment
National Park Service	Watershed Protection Program: Park-Based Water Quality Data Management Program
	National Wild and Scenic Rivers System
Bureau of Land Management	Federal Land Policy and Management Act Assessments
	BLM Initiatives
Bureau of Reclamation	National Irrigation Water Quality Program
National Biological Survey	National Biological Survey
Department of Commerce	
National Oceanic and Atmospheric	National Status and Trends Program
Administration	Fisheries Statistics Program
	Living Marine Resources Program
	Classified Shellfishing Waters
Environmental Protection Agency	Great Lakes Fish Monitoring Program
	National Water Quality Monitoring Program
	Environmental Monitoring and Assessment Program
Tennessee Valley Authority	Water Resources and Ecological Monitoring Program
Department of Agriculture	
Forest Servce	Resource Planning Act Assessments
	Forest Service Water Quality Program
	Watershed improvement program
	Nonpoint source pollution management program
Soil Conservation Service	President's Water Quality Initiative

Source: USEPA 1993d.

ACKNOWLEDGMENTS

I deeply thank Joyce Lathrop for the detailed editing of this chapter, as well as Thomas Simon and three anonymous reviewers. Special thanks to my family (Joyce, Michael, Rachael, and Nathan) who put up with my long hours at the computer, and my dad, who helped me collect benthos from the Chicago River that started my interest in aquatic ecology. This effort is dedicated to Jim Plafkin, Michael Glorioso, J. Pat Abrams, and Karl Simpson who all provided guidance and encouragement with my professional endeavors.

## CHAPTER 4

Defining Acceptable Biological Status by Comparing with Reference Conditions

Robert M. Hughes

1.0 INTRODUCTION

An essential component of biological assessments and criteria is how to determine whether a water body is healthy or unhealthy. What benchmarks or information do we use for comparison and what are acceptable and unacceptable deviations from those benchmarks? This paper describes several ways to define reference conditions and proposes acceptable and unacceptable departures from them. An expanded description of the process for selecting regional reference sites, and some cautionary notes concerning them, are provided because of continued interest in their application.

If we are to assess the state of water body health or to implement rational biological criteria adequately, then a reference condition needs to be determined for each type of water body or population of water bodies. Conditions approximating presettlement physical, chemical, and biological conditions are presumed to have ecological integrity, and data on these systems can serve as a foundation for developing biological criteria. In the past, scientists characterized acceptable condition by means of scattered information in the scientific literature, by professional judgment or individual ideas of what constituted a healthy or natural water body, by field experience, or by all three conceptual methods. Often no reference condition was established, and only raw physical, chemical, or biological data were presented; interpretation of water body health was left to persons far less qualified than the biologists who collected and reported results.

Today, inadequate data interpretation remains the case with many ambient physical and chemical habitat data, and it remains a key reason why biological monitoring often is held in low esteem (Ward 1989) and why decision makers can make little sense of environmental monitoring results (Messer et al. 1991). We can no longer afford to base our assessments of aquatic ecosystem integrity solely on professional judgement or to report uninterpreted data. Two facts necessitate a change: the status of aquatic resources continues to deteriorate (Hughes and Noss 1992) and federal regulation has required states to develop and apply biological criteria (U.S. EPA 1990a). Taking a similar approach as biological criteria, the EPA's Environmental Monitoring and Assessment Program for Surface Waters (EMAP-SW) requires that a water body's biological condition be designated as acceptable or unacceptable. A desirable outcome of having EPA's biocriteria and EMAP-SW programs embrace similar standards and processes for determining reference conditions is that the United States is likely to develop a more environmentally protective, defensible, and consistent monitoring and regulatory network.

The goals of this chapter are to (1) explain the purpose and objectives of determining a reference condition for populations of aquatic ecosystems, (2) examine six approaches for determining a reference condition, (3) describe the process for selecting regional reference sites, and (4) propose a way to determine acceptable and unacceptable quantitative scores from reference condition data.

1.2 Basic Characteristics of Suitable Reference Conditions

If we accept the need for quantitative measures of aquatic ecosystem health, we must decide what sort of a reference condition is desirable? There are at least five interrelated aspects must be considered.

1. The reference condition must be politically palatable and reasonable. In other words, it must be acceptable and understandable by persons most concerned with nature for its own sake and those unconcerned with nature or only concerned with what it can provide humans. If the process for determining the reference condition is acceptable and understandable by only one of these groups, it will not be broadly implemented by the majority of persons who fall between these extremes.
2. The reference condition for a general type of lake, stream, or wetland should represent large numbers of defined populations of lakes, streams, and wetlands. That is, we should not require a separate reference condition for each lake, stream reach, or wetland, despite recognizing that each ecosystem and various habitats within each ecosystem are somewhat different. On the other hand, we should not expect that a single reference condition will be satisfactory for all the streams, lakes, or wetlands in a major river basin or state, let alone the entire nation. Some sort of regional or waterbody classification or both is needed.
3. A reference condition, or process for determining it, must represent important aspects of "natural" or pre-Columbian conditions. Of course, no area on earth can be considered pristine because of global air pollution and climate change. There is also considerable evidence that the pre-Columbian humans of North America extensively modified the vegetation and wildlife of the continent (Martin 1967; Denevan 1992). Our forest lands little resemble pre-Columbian conditions because of 100 or more years of extensive logging, fire supression, and introduced diseases; the lands where urban areas, croplands, and rangelands have been even more extensively and intensively modified since the arrival of Europeans. These altered landscape conditions are reflected in altered surface water extent and condition. Widespread channel and flow modifications, point source pollution, and introduced species directly modify substantial numbers of lake, stream, and wetland ecosystems. Despite considerable difficulty in defining and quantifying "natural," our charge from the Federal Water Pollution Control Act is to restore and maintain the natural structure and function of aquatic ecosystems. How can reference conditions best represent such states?
4. Given the difficulty of empirically determining what conditions would be like in the absence of humans, and the fact that humans have resided here for thousands of years, we are forced to use minimal disturbance as a reference condition. Six approaches to determine reference condition are discussed in Sections 2 and 3. Generally, these approaches must be politically acceptable and scientifically defensible. This means that they cannot be based solely on current conditions where those conditions reflect fundamentally altered landscapes.
5. Wherever possible, states should share reference condition information when they share interstate or boundary waterbodies, or when they have similar types of waterbodies occurring in similar landscapes or ecological regions. Such coordination has both political and scientific value. Sharing information and approaches increases the amount of information and expertise available to set biological criteria. It thereby increases the probability that the reference conditions will be widely representative and that they will represent minimally disturbed conditions. Interstate cooperation on defining reference conditions also increases the political acceptability of reference conditions or biological criteria and reduces opportunities for dischargers and land owners to avoid regulatory controls.

1.3 Site Specific vs. Population-Based Reference Condition

This chapter focuses on determining reference conditions for a sizable population of waterbodies, vs. a single site of concern. Site-specific biological criteria and reference conditions are desirable, but state biologists are responsible for all waters, not just selected sites. The prevalence of diffuse pollution, extensive channel and flow modifications, multiple point sources, changes in physical habitat structure, and species introductions require more general approaches. Also, it seems that upstream/downstream, nearfield/farfield, or paired site reference conditions are intuitive concepts with little need for general guidance.

There are three additional problems with site-specific reference conditions: (1) because they typically lack any broad study design, site-specific reference conditions possess limited capacity for extrapolation — they have only site-specific value; (2) usually site-specific reference conditions allow limited variance estimates; there are too few sites for robust variance evaluations because each site of concern is typically represented by one to three reference sites; and (3) they involve a substantial assessment effort when considered on a statewide basis. Assuming a state must assess 50 to 100 sites annually and that each requires a minimum of three site-specific reference sites, a total of 150 to 300 reference sites must be monitored and assessed each field season. This is a major commitment of field and laboratory time, and the reference sites are inapplicable to the majority of aquatic ecosystems because they were not selected to be so. Certainly, site-specific references are needed where there are concerns with specific point sources. However, in a statistical survey of 1303 reaches in the conterminous United States, Judy et al. (1984) found that point sources were a major concern in only 5% of the nation's waters, whereas diffuse sources (38%), water diversions (13%), siltation (28%), erosion (18%), and channelization (12%) were more often listed as adversely affecting fish assemblages. Similarly, Miller et al. (1989), in an evaluation of North American fish extinctions, listed physical habitat alteration (73%) and introduced species (68%) as causal factors more often than chemical habitat alteration (38%). Williams et al. (1993) considered physical habitat destruction and species introductions as the chief causes of decline in freshwater mussels. Despite marked reductions in point source pollution over the past 20 years, aquatic ecosystems continue to deteriorate (Williams et al. 1989; Hughes and Noss 1992). Thus, a focus on point source pollution and site-specific reference conditions seem unwise at this time given the severity and diffuse nature of aquatic ecosystem problems; the limited fiscal and staff resources of states; the immediate requirement for biological criteria; and the thousands of lakes, stream miles, and wetlands per state that need some sort of monitoring and assessment (see Larsen, Chapter 18, for an approach for conducting such a monitoring program).

1.4 The Rationale for, and a Brief History of, Water Body Classification

Given an interest in large populations of waterbodies (lakes, streams, and wetlands), how can we hope to make sense out of their apparent diversity? In other words, what waterbodies are represented by a given set amounts of reference condition information? Classification forms the foundation of science and management because neither can deal with all objects and events as individuals; some system for generalization and prediction is needed. The opposing concepts of diversity and similarity are the bases of most classifications. All things are somewhat different, yet some things are more similar than others, offering a possible solution to the apparent chaos.

Modern taxonomic classification was initiated by von Linne and rapidly became widely implemented in the 18th century, giving biologists a consistent, widely applicable framework for study and communication. Soon after, in the early 19th century, German vegetation ecologists began classifying the world's vegetation assemblages (from Mueller-Dombois and Ellenberg 1974), while Clements (1916) was the earliest strong proponent for the usefulness of macro-level vegetation classification in North America. The maps and descriptions that these scientists produced greatly facilitated our ability to understand and manage our terrestrial environment, although they were intensely criticized by others calling for more accurate classification (Gleason 1926; Whittaker 1962).

Soon after European vegetation ecologists began classifying vegetation patterns, European aquatic ecologists began classifying aquatic ecosystems. Classical limnologists categorized lakes by geomorphology or origin, nutrient status, mixing patterns, or their epilimnion/hypolimnion ratio (see Cole 1975 or Wetzel 1975 for descriptions). None of these single-variable classifications adequately explains varying lake character. One of the better attempts at developing a conceptual model for hierarchical lake classification is that of Rawson (1939), who included the above variables plus size, temperature, and geographic location. However, little further work has occurred in synthesizing such information into a lake classification system from which biological criteria can be developed.

Stream classification (see reviews in Hawkes 1975; Hynes 1979; Naiman et al. 1992) has mostly focused on stream zonation, based first on fish assemblages and later on benthos, habitat, or stream order (e.g., Strahler 1957; Vannote et al. 1980). However, classifications based on biology and habitat require considerable sampling effort, and stream order incorporates large differences in stream discharge (Hughes and Omernik 1983; Minshall et al. 1985).

Clearly, a more robust stream classification was needed. Likens and Bormann (1974) and Hynes (1975) argued that streams are closely linked with their catchments, and Warren (1979) suggested that grouping similar landscapes should group similar streams. He also explained that it was necessary to base stream classification on a range of potential states rather than on a single current and often temporary condition. Minshall et al. (1983) demonstrated the need to link the River Continuum Concept of Vannote et al. (1980) with geographic regions. Platts (1980) and Frissell et al. (1986) presented models for hierarchical classifications ranging from regions to habitat types, but creation of such detailed maps are decades away because of the labor-intensive nature of stream reach classification. Brussock et al. (1985) classified streams of the United States into seven regions, but these are not considered sufficiently precise for application to biological criteria. Ichthyologists continue to use drainage basins to explain ichthyogeographic patterns, but basins fail to account for changes resulting from substantial intrabasin landscape changes (Omernik and Griffith 1991; Rahel and Hubert 1991; Hughes et al. 1994).

Wetland classification, like our scientific and regulatory interest in wetlands, has a much more recent history. As with lakes and streams there are typological (Cowardin et al. 1979) and regional (National Wetlands Working Group 1986) classifications. Both examples are also hierarchical.

In an attempt to classify lakes, streams, and wetlands, Omernik (1987) and Omernik and Gallant (1990) delineated ecoregions of the conterminous United States based on mapped landscape characteristics. Although these maps have fairly low resolution (76 and 14 regions, respectively) the methodology is transferable to higher resolution maps. For example, Clarke et al. (1991) resolved Omernik's eight Oregon ecoregions into 23 subregions and Omernik and his associates are developing several state-level maps that transform one ecoregion into 3 to 13 higher resolution ecoregions. For some purposes these hierarchical levels may still contain insufficient detail or too much variability. [For further information on this process see Omernik (Chapter 5).]

2.0 DETERMINING THE REFERENCE CONDITION

Given a suitable resource classification, there are at least six approaches for determining the reference condition for biological criteria (Hughes et al. 1993). They are not mutually exclusive; in fact, some are closely related or dependent on others. Each one has its own particular strengths and weaknesses and makes use of ecosystem classification to some degree.

2.1 Regional Reference Sites

Regional reference sites were proposed by Hughes et al. (1981, 1986, 1993) for streams and for New England lakes. They also form the logical basis for assessing range and forest health (U.S. Soil Conservation Service 1981; U.S. Forest Service 1984). Ohio EPA (1987a) and Arkansas Department of Pollution Control and Ecology (1988) have both found regional reference sites useful for setting stream biological criteria. Biggs et al. (1990) described why the New Zealand Department of Scientific and Industrial Research plans to use ecoeregional reference sites for establishing biocriteria nationally, based on a study of hydrology, water quality, periphyton, benthos, and fish in 144 catchments. Warrey and Hanau (1993) found regional reference catchments appropriate for establishing regional expectations for Ontario stream chemistry. Ecoregions were shown to be a useful stratification tool for streams in Arkansas (Rohm et al. 1987), Nebraska (Bazata 1991), Ohio (Larsen et al. 1986), Oregon (Hughes et al. 1987; Whittier et al. 1988), Washington (Plotnikoff 1992), and Wisconsin (Lyons 1989). Lyons, however, noted that finer resolution classification by gradient, temperature, and substrate would improve classification accuracy for fish assemblages.

Regional reference sites have also been found useful for developing biological criteria for lakes. Heiskary and Wilson (1989) described how ecoregional reference lakes were useful in developing trophic state goals for the four major lake regions of Minnesota. Hartig and Zarull (1992) proposed using reference sites for portions of the Great Lakes, and Malley and Mills (1992) documented the role of reference lakes in defining and understanding aquatic ecosystem health. Ecoregions were found effective for classifying fish assemblages in Tennessee River reservoirs (Hughes et al. 1994) and for developing reference conditions (along with temperature and area) for multiple assemblages in New England lakes (Hughes et al. 1993).

Although regional reference sites have proven useful for developing biological criteria for lakes and streams in various parts of the United States, they still have several limitations. Ecoregions have not been widely evaluated for their ability to classify reference conditions for wetlands and further regionalization is needed for agencies desiring local-scale reference conditions. Aquatic ecoregions are developed to be applicable to the entire aquatic community; if an agency is concerned only with a particular assemblage or particular species, assemblage- or species-specific regions may be more appropriate. Ecoregions at the 16 and 76 region level of resolution incorporate considerable variability because of scale; in some regions this level of variability may be unsatisfactory. Ecoregional reference sites still require some level of habitat classification at the site scale. The degree to which a set of reference sites represents a regional population of surface waters must be carefully examined through ordination and indicator analyses. Special care must be taken to exclude anomalous sites. Finally, one of the major concerns with regional reference sites is their acceptable level of disturbance, because sites that are moderately disturbed can skew biocriteria, thereby producing mediocre expectations.

2.2 Historical Data

The reference condition can be based on historical data and such data are useful for describing trends. For example, Trautman (1981) described how fish data through time indicated deterioration in Ohio landscapes and waters. Karr et al. (1985) determined from historical fish collections that the Maumee and Illinois River drainages originally supported 98 and 140 fish species, respectively. [See Hughes and Noss (1992) for a review of additional examples and Baker et al. (1993) for an example of fish losses in Adirondack lakes.] Such data are useful for particular sites that are resampled periodically, or for making general statements about the condition of waters in a state or nation; however, historical data have some serious limitations for application to biocriteria. Typically they are limited to only a single assemblage (most often fish) and they must incorporate a large number of samples to be very useful. Abundance data are usually lacking, so they are sensitive to species extirpation or range retraction or expansion only. This means that the situation must deteriorate seriously before problems are detected. Frequently the data are intractable, either because they are difficult to obtain or difficult to work with. Historical data in many cases simply are not available in sufficient quantity or quality. Often historical data were collected with different methods and for different purposes (neither of which may be described). Fish collections in particular suffer from a focus on game species, if collected by a fishery agency (Figures 1 and 2), or on unusual species, if collected by museums. Such differences make comparisons with current samples questionable. Some collections may have been taken from historically disturbed sites, but often such habitat information is missing or incomplete. Historical physical habitat data are particularly poor, but Sedell and Frogatt (1984) provide an example of the quality of data that can be obtained from sufficient searching. When historical data can be obtained at minimal cost, they should be examined, but carefully.

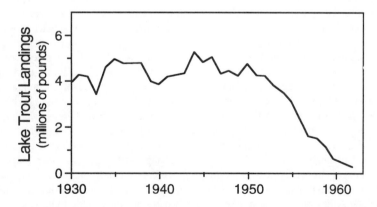

Figure 1. Commercial catches of lake trout from Lake Superior. A possible reference condition based on these data is 4 million pounds per year. (Adapted from Smith, S. H. 1972. *Journal of the Fisheries Research Board of Canada* 29:951–957. With permission.)

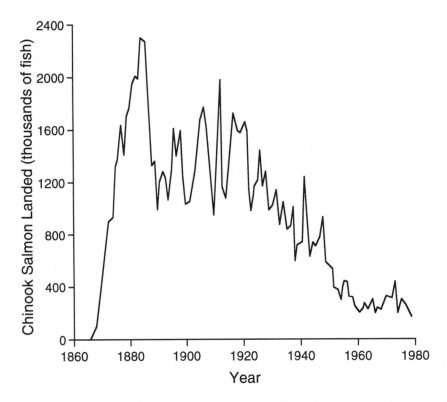

Figure 2. Commercial catches of Columbia River chinook salmon. These data suggest a reference condition of
1 million salmon per year. (Adapted from Ebel, W. J., C. D. Becker, J. W. Mullan, and H. L. Raymond.
1989. Pages 205–219 in D. P. Dodge (editor). *Proceedings of the Large River Symposium.* Canadian
Special Publication Fisheries and Aquatic Sciences 106. With permission.)

2.3 Paleoecological Data

A special case of historical data is the information stored in lake sediments. Data drawn from the
bottoms of lake sediment cores, particularly diatom assemblage data, may best represent the conditions
described as natural in Section 1.2. Such information has proven very useful for assessing changes in
climate, salinity, pH, and nutrients (Fritz 1990; Dixit et al. 1992; Smol and Glew 1992). Unfortunately,
diatoms are most affected by water quality changes, whereas some of our most serious aquatic ecosystem
problems result from changes in habitat structure and species introductions. The paleoecology of benthos
and zooplankton is less advanced than for diatoms, meaning that this method for determining reference
conditions is limited to diatom assemblages for all practical purposes. In addition, paleoecology is limited
to lakes and poorly suited to streams and most wetlands and reservoirs. Finally, it is essential to obtain
a sufficiently long core and date it to ensure that the reference condition represents the human presettlement
period. Dixit et al. (1992) describe the processes for sampling and data analysis and Hughes et al. (1993)
suggest how such data might be converted into biological criteria. Figure 3 shows several lakes whose
diatom assemblages have changed markedly, and others that have changed relatively little. Of course,
such changes require further interpretation if one does not assume that the lake has deteriorated with time.
A different method of analysis relates the amount of change in the diatom assemblage to particular
stressors or indicators of stress (Figure 4). In this case the reference condition might be equated with those
lakes experiencing minimal nutrient enrichment. Because of their retrospective potential, it seems
advisable to include sedimentary diatom assemblages when developing lake biocriteria.

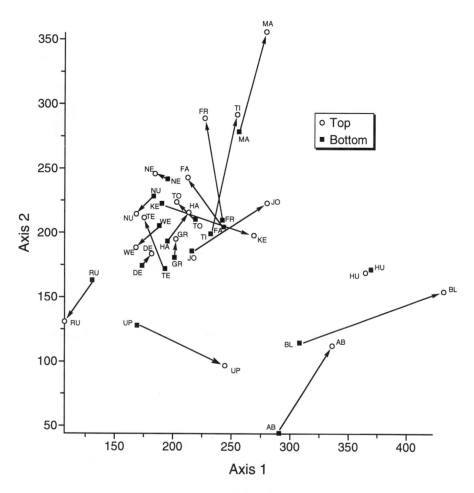

Figure 3. An ordination of diatom assemblages from 19 New England lakes. Line lengths represent the amount of change between surface and bottom sections of a 40-cm long sediment core. Lakes with twice the change of others in the same lake class could be considered unacceptable. (From Hughes, R. M., C. Burch Johnson, S. S. Dixit, et al. 1993. Pages 7–90 in D. P. Larsen and S. J. Christie (editors). *EMAP-Surface Waters 1991 Pilot Report.* EPA/620/R-93/003. USEPA, Corvallis, Oregon.)

2.4 Experimental Laboratory Data

Laboratory data, although not typically viewed as useful in determining reference conditions, can be valuable. Particularly, well-known relationships between test species and specific toxins, temperatures, nutrients, and dissolved oxygen concentrations indicate stressor levels that are probably harmful to aquatic communities. If field measurements suggest potential problems based on experimental data or previous field experience, such sites should not be considered as indicating reference conditions. Classic examples of such situations are hot springs in national parks, or acidified lakes and streams in areas lacking other human stressors. In general, though, experimental data do not offer sufficiently robust information from which to develop reference conditions for entire assemblages or communities in a population of waterbodies. Water quality data are especially unsuited for establishing reference conditions for systems disturbed by other stressors, such as structural, hydrological, and biological alterations. Instead, such data are most useful for screening out sites selected by other means or for improving model predictions.

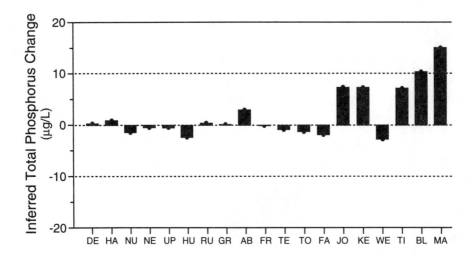

Figure 4. Diatom-inferred nutrient enrichment of 19 New England lakes. Lakes experiencing >10 µg/l increase in total phosphorus could be considered unacceptable. (From Hughes, R. M., C. Burch Johnson, S. S. Dixit, et al. 1993. Pages 7–90 in D. P. Larsen and S. J. Christie (editors). *EMAP-Surface Waters 1991 Pilot Report.* EPA-620-R-93-003. USEPA, Corvallis, Oregon.)

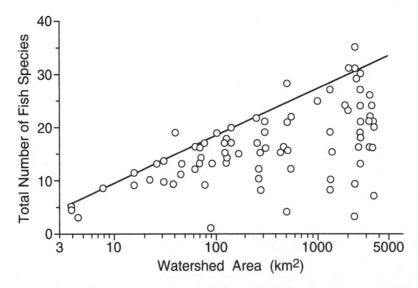

Figure 5. A maximum fish species richness line. This line, or a 25 to 33% deviation from it, may be considered the reference condition. (Adapted from Fausch, K. D., J. R. Karr, and P. R. Yant. 1984. *Transactions of the American Fisheries Society* 113: 39–55. With permission.)

2.5 Quantitative Models

Ultimately, we must seek to develop quantitative models from field and historical data. Such models become increasingly useful and accurate as the database size and complexity increase. By plotting metric and index values against well-distributed disturbance values or natural variables, one can estimate reference conditions through curve fitting. Fausch et al. (1984) used such an approach to develop maximum species richness lines for stream fish assemblages by plotting against stream size (Figure 5). Note that in their original paper, historical samples from 1930 produced fewer species than those from

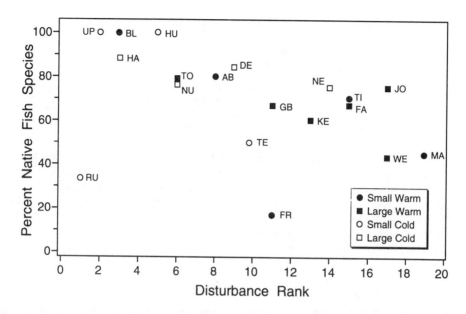

Figure 6. Percent native fish species as a function of canopy complexity and extent, shoreline disturbance intensity, and fish cover complexity. A possible reference condition is 100% natives. (From Hughes, R. M., C. Burch Johnson, S. S. Dixit, et al. 1993. Pages 7–90 in D. P. Larsen and S. J. Christie (editors). *EMAP-Surface Waters 1991 Pilot Report.* EPA/620/R-93/003. USEPA, Corvallis, Oregon.)

1963. Hughes et al. (1993) plotted percent native fish species against physical habitat quality to determine expected conditions at minimal disturbance (Figure 6).

This general approach can also be expanded to multivariate linear regression models, as long as the relationships are linear or transformable and the data are of sufficient quality. Nonlinear models are also feasible. Outliers, uneven distribution of data, and the absence of data from minimally disturbed sites can greatly distort such models. It is also useful to examine residuals and patterns in the data that may reveal other predictor variables. Models developed in such a manner should not be extended much beyond the data, the waterbody and habitat type, or the region from which the data were collected.

2.6 Best Professional Judgement

Best professional judgement is critical to all the above options for determining reference conditions, whether it involves selection of regional reference sites, the evaluation of historical data and experimental results, or the treatment of outliers and data patterns in models. A special case of professional judgement is the convening of expert panels to determine reference condition, similar to that of Thomas et al. (1990) for Pacific Northwest forests. Another example is peer review of the data and evaluations. As with quantitative models, the quality of the judgement is a function of the expertise of the scientists and the quality of the data supplied them. It is especially useful to select panelists and reviewers with opposing biases and different professional backgrounds to ensure different points of view, and to increase the credibility of the product in the eyes of the public.

2.7 Recommendation

The approach with the greatest potential for success combines regional reference sites and historical data (where they are available and of sufficient quality), interpreted through use of linear models and expert judgement. Paleolimnological data should be included for natural lakes and some wetlands. These approaches are applicable to a number of aquatic and riparian assemblages and were also recommended by Cairns et al. (1993).

Table 1.   Major Steps in Selecting Regional Reference Sites

1. Define areas of interest on maps
2. Define waterbody types, sizes, and classes of interest
3. Delineate candidate reference catchments
4. Conduct aerial or photo evaluation
5. Conduct field reconnaissance
6. Subjectively evaluate quality of candidate reference sites
7. Determine number of reference sites desired
8. Evaluate biological health of candidate reference sites

The reference sites should be classified and ranked through use of landscape and habitat character- istics, but evaluated biologically. There is no single scale at which ecological phenomena (including reference sites and biocriteria) should be studied. Ideally, states will implement a hierarchical classifica- tion system that incorporates major regional patterns as well as more immediate physical and chemical habitat information when these are appropriate. Such a hierarchical approach is crucial because the mechanisms governing the natural character of aquatic ecosystems, as well as their stressors, often operate at different scales than those at which the patterns are observed (Tonn 1990; Levin 1992).

3.0 REFERENCE SITE SELECTION

Because it is central to determining reference condition and often incorrectly applied, a general method for selecting reference sites is described in this section for lakes, streams, and wetlands. There are eight steps for selecting regional reference sites (Table 1). Although they are listed in order of recommended occurrence, the process is not linear, but iterative. One must often return to earlier steps to surmount obstacles or to cull choices.

3.1 Define Areas of Interest on Maps

Typically, the area of interest is an entire state, or a particular river basin or region of a state or multistate region. Depending on the size of the area, delineate ecoregions and subregions on 1:250,000 (or larger) scale maps. Use larger scale (higher resolution) maps if the area of interest is considerably smaller than a state or if many high resolution regions are desired. Whenever possible, base the area on natural, vs. political, boundaries to maximize applicability and the probability of selecting the least disturbed sites possible. (See Omernik, Chapter 5 for an explicit description of this process.) Next, delineate river basins or other landscape features known to support markedly different assemblages. This is most important only in areas where the assemblages have long been separated, producing species depauperate and speciose assemblages on different sides of mountain ranges, for example. It is also important if researchers are very concerned with particular species, subspecies, or stocks. Combine the maps of ecoregions and known faunal regions to produce potential natural regions; avoid "slivers" but delineate transitional areas and fuzzy boundaries between regions (see Clarke et al. 1991 for an example of fuzzy boundaries). If the importance of faunal regions is unknown, sites may best be allocated as described in Section 3.2 to ensure that they are represented.

3.2 Define Water Bodies of Interest

This step can become confusing if not done with considerable care. As with regions, it is best done hierarchically, considering the most important and most common waterbody types (lakes, streams, and wetlands), classes (within a type), and sizes of interest first. These waterbodies may be subsumed by the ecoregions or faunal regions; if not, a concerted effort is made to ensure that they are represented among the reference sites. Factors to consider include temperature (coldwater vs. warmwater lakes and streams), size (area, depth, volume, discharge, and exchange rate), river basins (if not included in the previous step), extremes in stream gradient or velocity, bed materials (coarse, fine, and bedrock), connectivity (including constrained and unconstrained stream channels, proximity of the system to mainstem rivers, barriers to migration such as dams, and the shoreline development ratio of lakes and wetlands), lake mixis class, and chemistry (especially the salinity or alkalinity of systems in arid and semiarid regions, the acid-neutralizing capacity of lakes and streams in areas sensitive to acidic deposition, turbidity, nutrient status, and color or concentration of dissolved organic carbon).

3.3 Delineate Candidate Reference Watersheds

Delineate candidate reference watersheds through use of maps, available data, remote sensing, and calls to or visits with local experts/managers. Generally, it is best to precede personal contacts with considerable data analyses. The focus at this step is twofold: to locate and reject disturbed areas (unless they will be used for indicator development purposes), and to seek and retain minimally disturbed areas. Areas disturbed by intense land use, such as logged or burned forests, intensively grazed or cropped lands, high road densities, urban/industrial/commercial/residential concentrations, and transportation and utility corridors, are identified from land-use data obtained from land management agencies or remote sensing. These areas are then designated on maps. Public lands are distinguished from private, as are the predominant management practices of each. Information on chemical use, pollutant discharges, hazardous waste sites, landfills, mines, oil fields, feedlots, poultry farms, and fish hatcheries is examined and such sites are designated as disturbed. Maps and water resource data are also useful for locating channel and flow manipulations (channelization, dams, and water withdrawals). Where atmospheric deposition is a concern, areas of high or frequent loadings are mapped. As the number of candidate catchments is narrowed, information on road proximity and livestock and population densities is used to cull areas. Finally, fish stocking and harvesting records and data on introduced species are used to winnow sites.

At the same time as some areas are being rejected, others are sought. The latter are minimally disturbed, typical areas or potential natural landscapes that are likely to endure. Possible places are agricultural or range oases and exclosures, old-growth forests and woodlots, roadless areas, preserves, refuges, areas close to terrestrial refuges, and sites distant from roads and the other types of disturbances described above. The goal is to locate candidate catchments or sites that are minimally disturbed and likely to remain so.

3.4 Conduct Field Reconnaissance

It is necessary to conduct field reconnaissance to screen potential sites. This step is essential because map and remote sensing data are often out-of-date and inaccurate. Also, information from local experts is inconsistent. During this phase, it is advisable to converse with local resource users when such opportunities arise; long-time residents can be especially illuminating. Scientists at our laboratory have found aerial observation, at about 1500 m above ground level, from a small high-wing airplane very cost effective for covering large areas in a short time; this perspective is especially valuable for assessing riparian and catchment conditions. If flying is out of the question, existing air photographs taken 1500 to 3000 m above ground are informative. Initially, candidate sites should be examined at three to four places on the ground to obtain a fairly comprehensive visual assessment. Actual site inspections should cover a length of at least 40 channel widths for each candidate stream (Angermeier and Karr 1986; Lyons 1992; Kaufmann 1993); lakes and wetlands should be viewed from a sufficient number of points to assess most of the shoreline and surface.

Information that is useful to collect is outlined in Table 2; most are common to all aquatic ecosystems. It is recommended that field assessments be conducted initially through use of a qualitative habitat evaluation form (see Plafkin et al. 1989 and Gibson 1994).

The first six variables are macro-level concerns. Riparian growth is considered extensive when it occurs all along the shoreline and is capable of shading the waterbody and buffering it from nearby human influences. It is old when it is depositing large woody debris or overhanging the waterbody. The extent of accumulated organic matter is often a good measure of the age and productivity of a wetland. Complex riparian vegetation is characterized by the presence of a canopy, understory, and ground cover. Shoreline complexity is a function of riparian and littoral complexity and integrity; it is reduced by vegetation clearing and bank control structures. Channel complexity is most relevant to streams and includes two scales: (1) channel sinuosity, and (2) macrohabitat structural complexity (large turbulent riffles, glides, undercut banks, large deep pools, runs, backwaters, sand/gravel bars, islands, side channels, backwater

Table 2. Reconnaissance Data for Evaluating Reference Sites

Variable	Streams	Lakes	Wetlands
Roads Distant	X	X	X
Riparian vegetation extensive and old	X	X	X
Riparian structure complex	X	X	X
Shoreline complex	X	X	X
Channel complex	X		
Basin complex		X	X
Water level recession minimal		X	X
Shoreline modification minimal	X	X	X
Habitat structure complex	X	X	X
Chemical stressors minimal	X	X	X
Channel/flow manipulation minimal	X	X	X
Sedimentation and turbidity minimal	X	X	X
Odors, films, scums, and slicks minimal	X	X	X
Pipes, drains, ditches and tile absent	X	X	X
Wildlife and benthos evident	X	X	X
Human and livestock activity minimal	X	X	X

pools/lakes, and wetlands in or near rivers). Basin complexity is most relevant to wetlands and lakes and includes the same two scales as channels, and many of the same variables. It is increased by basin heterogeneity that produces deep and shallow areas, shoals, islands, bays, points, shelves, and beaches. Wetland complexity is especially sensitive to subtle changes in water level, but all three types of waterbodies suffer if water levels fluctuate such that riparian or littoral vegetation fail to become established and if bank erosion is thereby accelerated.

The last nine categories are somewhat more localized in scale. Shoreline modification refers to the development of docks, human settlements, riprap, and bulwarks; the removal of natural objects such as saplings, brush, aquatic macrophytes, and snags; and the introduction of exotic plants. Habitat structural complexity includes substrate heterogeneity (variable substrate especially gravel, cobble, and boulders), large woody debris (snags, root wads, log jams, brush piles, and overhanging vegetation), undercut banks, and macrophytes. In wetlands, subtle differences in substrate size have great influence on vegetation. Chemical stressors are evaluated at the site by searching for sources such as unmapped pipes, trash dumps, landfills, lawns, and croplands. The presence of dams, irrigation canals, and field drains indicate flow or water level modifications. Channelization is most pertinent to streams and includes straightening, levees, riprap, snagging, and other channel control structures. Such streams and rivers are typically very homogeneous with an oversized channel. Although some systems are naturally turbid, increased amounts of fine particles (silt, sand, and clay) indicate catchment or channel disturbance. Aquatic ecosystems that look or smell unpleasant or unusual suggest natural or anthropogenic overloads. If semiaquatic and aquatic wildlife appear abundant there is a greater probability that the system is healthy than if none are detected at all. Finally, the waterbody may be a solid candidate for a reference site if there is no, or little, evidence of livestock grazing or human artifacts, such as lawns, public beaches, campgrounds, resorts, marinas, or motorcraft.

3.5 Subjectively Evaluate Quality of Candidate Reference Sites

Qualitative evaluation of the health of individual candidate reference sites has been one of the most problematic aspects of this approach. Several states have chosen fundamentally altered ecosystems to serve as reference sites. However, if reference sites are the benchmark sites upon which biocriteria are based, it is essential that they be disturbed as little as possible. Generally this means that they are not near roads and are somewhat difficult to access. For example, in western range and forest lands, they are watersheds in wilderness areas, unroaded old-growth forests, or canyonlands inaccessible to livestock. In agricultural regions with much greater road densities and devegetation, they can be found as sites in woodlots (Marsh and Luey 1982) or native prairie remnants; at least they should have intact old natural riparian vegetation or some type of buffer from the prevailing land use. In some extensively disturbed regions it takes considerable effort to locate minimally disturbed sites. The process of doing so, described above, takes more effort than site sampling and data analysis. It is not a trivial undertaking.

If the candidate reference sites in one of the 76 ecoregions of Omernik (1987) are unsuitably disturbed, reference sites in a similar region or in the lower resolution aggregate region (Omernik and Gallant 1990) should also be included to estimate the reference condition. Subregions also can be clustered into regions to increase the probability of locating minimally disturbed reference sites. In a similar manner, it may be necessary to select reference sites from another class of stream, lake, or wetland if all sites in a particular class are highly disturbed. Obviously, these choices must be made with considerable care and documented so that they do not include fundamentally different communities (such as warmwater and coldwater, or high gradient and low gradient). In many cases they can be avoided by very careful geographic analyses and reconnaissance. However, when data and professional judgement indicate that anthropogenic disturbances have a greater effect on candidate reference sites than differences in regions and waterbody classes, the biologically conservative approach is to include data from additional reference sites in similar regions and from similar waterbody classes.

A related problem arises with the selection of anomalous sites as reference sites. Occasionally, minimally disturbed sites are less disturbed because they are markedly different than the majority for natural reasons. For example, sites located on a ridgetop differ from those on the surrounding plains and an exceptionally deep, cold lake differs from the shallow, warm lakes characteristic of agricultural plains. Such anomalous sites should not serve as reference sites for waterbodies lacking the same natural potentials.

These two problems of anomalies and excessive disturbance indicate the importance of professional judgement tempered by extensive field experience. In most regions, there will be trade-offs between accepting or rejecting higher levels of disturbance than desired, deciding the ideal level and type of waterbody classification, rejecting or accepting atypical sites, and locating a sufficient number of reference sites. The reasons for making such decisions should be documented.

3.6 Determine the Number of Ecoregional Reference Sites Needed

As with all sampling designs, the ideal number of reference sites must be balanced against budget realities. In addition, managers are faced with questions about anomalies and excessive disturbance, which further reduce options. Clearly, it is best to begin the reference site selection process with as large a number of candidates as can be examined at each step in the procedure. These are then winnowed to an ever smaller number.

The actual number of regional reference sites needed is a function of regional variability and size, the desired level of detectable change, resources, and study objectives. Three is a commonly used minimum number for any particular waterbody class and size in a homogeneous region. When multiplied by the number of classes, sizes, and regions of interest, this can quickly produce a very large number of sites. Walters et al. (1988) recommended one reference site for four to five test sites. This seems excessive if a state must monitor hundreds of sites annually, and insufficient if an agency only monitors a few each year. Survey statisticians recommend a minimum of 20 randomly selected sites, but this number depends on desired detection levels. For example, using a model developed by Loftis et al. (1989), Hughes et al. (1992) showed that a change in four to six IBI units was detectable with a sample size of 20 sites if the sample standard deviation was 5.0.

Thus, one possibility would be to randomly select 20 reference sites from the candidate reference sites in each of the regions of interest. If there are particular waterbodies sizes and classes of interest, one could create categories to ensure their inclusion in the sample. This approach would provide an estimate of the reference condition for the population of reference sites and it would likely require less sampling than evaluating three of each size and class in each ecoregion.

3.7 Quantitatively Evaluate Biological Health of Reference Sites

The final step in selecting regional reference sites is to survey a minimum of two assemblages in them and evaluate the results. This is an important step because it is often the only way to determine if species (especially fish) were introduced. It is also a way to learn if some unknown stressor is acting on a system.

Rigorous evaluation of biological data from sites representing a gradient of anthropogenic disturbance is needed to develop quantitative indicators of disturbance or their scoring criteria. Although a number of metrics and indices have been proposed in recent years, their responsiveness to various types of stressors, their efficacy outside the region of their development, and their theoretical foundations often have not been thoroughly examined. For examples of this process, see Karr et al. 1986, Kerans and Karr 1994, Hughes et al. 1993, and Simon and Lyons (Chapter 16).

3.8 Refinements of the General Approach for Large Rivers, Lakes, and Wetlands

There are several general issues peculiar to the different waterbody types and sizes. One of the most obvious is the effect of size on the problem of defining reference condition. Horton (1945) demonstrated that there is an inverse logarithmic relationship between stream order and abundance, and a direct one with stream length. Similarly, there are many more small lakes and wetlands than the next larger size class, although most total wetland and lake areas are in the largest size classes. Because there are so many more, it is usually much easier to find small minimally disturbed waterbodies than large ones. This aids us in determining reference conditions for small ecosystems, but hinders locating reference sites for large waterbodies. Fortunately, though, there are few large waterbodies, often enabling us to treat them as individuals.

We must classify large ecosystems further (e.g., rivers into reaches, lakes into subasins or subunits, and wetlands into vegetation types). Minimally disturbed reference sites are selected after examining existing data, evaluating potential stressors, and reconnoitering the ecosystem(s). If there are no major distinguishing characteristics, sites can be located through use of a systematic random design. In large rivers, reference sites may be upstream of major sources of disturbance (Hughes and Gammon 1987) or as far as possible from upstream sources, cities, and dams (Randy Sanders, Ohio EPA, Columbus, personal communication). Frequently, large rivers cross ecoregion boundaries, but are little affected by the change for some distance downstream. In other words, there is a lag effect, especially when montane rivers enter plains carrying their cold water and coarse substrate considerable distances into the other ecoregion (Hughes and Gammon 1987; Hughes et al. 1987). Similar situations exist for small rivers that cross regional boundaries, but the changes occur much more quickly (Omernik and Griffith 1991).

Anomalous lakes can occur if the lake is very deep and obtains much of its water from deep groundwater. Such water may be quite different than that from surface runoff or shallow groundwater and expectations should be weighted accordingly; careful comparisons with other deep lakes in neighboring ecoregions may be preferable and provide more realistic expectations.

Proximity to other aquatic ecosystems can also confound accurate estimates of reference conditions. This is best shown with small streams that join large rivers and consequently include fish species from the large river, unlike headwater streams of the same size (Osborne et al. 1992). Small streams may influence larger rivers in a comparable manner at confluences. Riparian wetlands of lakes and streams support a somewhat different fauna than the same type of wetland vegetation that is isolated from an open waterbody.

At an even finer scale of resolution, small springs, seeps, riffles, rock outcroppings, and other atypical substrates can alter the microhabitat of macrobenthos and algae, which necessitates more thorough sampling designs than typically occur for those assemblages. For example, preliminary data indicate that at least nine systematically distributed samples, composited by macrohabitat type, are needed to characterize benthos and periphyton species richness in small wadable streams (Judy Li and Alan Herlihy, Oregon State University, Corvallis, personal communication).

Finally, it is important to state that natural gradients occur between waterbody types, classes, and sizes, ecoregions, and levels of disturbance. All our attempts at classification are human constructs that help us understand a spatially and temporally varying landscape. Consequently, we should not always expect distinct differences, even though we seek them.

4.0 DETERMINING ACCEPTABLE FROM UNACCEPTABLE CONDITION

Whatever methods are used to determine the reference condition, it is necessary to restate what it is that we are attempting to protect. Leopold (1949) stated that "A thing is right when it tends to preserve the integrity, stability, and beauty of the biotic community. It is wrong when it tends otherwise." Thus, actions that make surface waters appear less healthy, less stable, and more ugly from a biological perspective should not be considered acceptable. Biological integrity is "the ability to support and maintain a balanced, integrated, adaptive community with a biological diversity, composition, and functional organization comparable to those of natural aquatic ecosystems in the region" (Frey 1977; Karr and Dudley 1981). This means that the system should be self-regulating through time and that the parts fit together such that there are few wild gyrations in populations (though fluctuations are normal). An adaptive system can absorb many natural, and some of the anthropogenic, changes that occur. The waterbody's biological, taxonomic, ecological, and functional appearance should resemble the presettlement conditions of waterbodies in the same region. Such systems are not adversely impaired by humans. They do not necessarily support excellent fishing, crystal clear waters, or nonbiting insects.

It is important for biologists to determine whether the ecosystems they manage are in acceptable or unacceptable condition if they are to protect and restore them, or communicate with the public about them. Unfortunately, there are several problematic aspects of doing so.

1. Ecological, and therefore statistical, variability precludes clear-cut measures of ecosystem health. Considerable professional judgement is required.
2. Distinguishing healthy from unhealthy ecosystems is obscured by political, economic, and ethical issues. There are persons with opposing political reasons desiring that the actual situation be painted better or worse. The economic costs of doing nothing or doing something fall unequally on various interest groups. Ultimately, our assessments of ecosystem integrity are affected by our ethics and whether we view ecosystems and other forms of life as resources or fellow passengers on spaceship earth.
3. There is a concern with significance vs. power considerations. That is, with what certainty do we know the system is not healthy, vs. how confident are we that it is?
4. Are ecosystem processes or structures more appropriate indicators of integrity? Ford (1989) and Schindler (1987) demonstrate that structures are more sensitive, but Karr et al. (1986) show how elements of both can be applied.
5. Another old point of contention among aquatic biologists is the varying sensitivity of assemblages and their ease and cost of sampling and analysis. Competent biologists can probably use any group to assess health, but thorough comparisons of such selection criteria as societal value, sensitivity to a wide range of stressors, total implementation costs, knowledge of species guilds, and variance components (measurement, crew, spatial, and temporal) have not occurred.
6. Aquatic scientists and managers have yet to reach consensus on quantifying and defining naturalness. It is necessary to resolve this if biocriteria are to have any consistent meaning. Perhaps this book will help.

Despite the obstacles to setting numerical expectations for health of surface waters, there are encouraging precedents. Ohio (Yoder and Rankin, Chapter 9), Vermont, and Maine (Courtemanch, Chapter 20) have overcome the above six impediments, and others, to set objective biological criteria. The process is similar to that of toxicity tests. Initially, aquatic toxicity tests were viewed with considerable skepticism over their value as indicators of anything; now they are broadly accepted, regularly used, standardized, and often diagnostic, despite the lack of any ecological significance to an LC-50.

Given any reference condition data set, how might an agency designate the number signifying acceptable or unacceptable health? As with most reference condition issues, there is not a single best answer.

1. Certain low and high percentiles are clearly unacceptable or acceptable, but they leave a large number of ecosystems in the gray middle ground. Most persons would agree that 90% of the reference condition is still high quality and perhaps within the range of natural and measurement variability. Many would also agree that 25% or 50% of the reference condition is unacceptable; but is 51% of the reference condition, when determined from extensively disturbed reference sites, equivalent to 51% determined from nearly pristine sites? How do we evaluate a percent of an indicator that varies widely among reference sites as opposed to one that is relatively invariant?

2. Curve inflections are occasionally used to determine acceptable or unacceptable index values. However, these are highly variable among different assemblages and occasionally the inflections are very weak. They may also occur at much different places on the distribution curve or histogram.
3. Curve breaks are similar to curve inflections, but they occur frequently in small data sets. In such cases, which break should represent the acceptable condition?

Based on the earlier definition of biological integrity, it seems that we must set the acceptable condition as high as possible, say 75% of the reference condition. A second category would be considered clearly unacceptable, for example 50% of the reference condition. A third category from 50 to 75% could be considered marginal, admitting that we are uncertain about this midrange. If the reference condition represents a considerable amount of disturbance itself, the acceptable condition could be increased to 90% of the reference. If the indicator value at reference sites is highly variant, either the indicator is a poor one, or the reference sites are inadequately classified or receiving variable levels of disturbance. Highly variant indicators and highly disturbed reference sites should not be used, while appropriate ecosystem classification is at the root of the reference site concept. We must also acknowledge that the acceptable, marginal, and unacceptable categories are arbitrary, but that they represent our best professional judgement of what constitutes unacceptable and acceptable deviation from natural conditions.

5.0 CONCLUDING COMMENTS

The process of determining reference conditions should play a role in other management issues related to biocriteria. Best management programs for diffuse pollution can be modeled after reference sites and historical conditions. Formal aquatic preserves designed to protect species of concern, gene pools, or critical ecosystem types may be reference sites and aid in their selection (Henjum et al. 1994). The growing fields of restoration ecology and conservation biology may gain insights from, and guide, reference condition research. In a related vein, reference sites indicate an obvious need for a protection strategy; if the sites we expect to use to estimate reference condition are increasingly disturbed, our ability to make such estimates will be increasingly compromised (see Covich et al. 1992).

If sites are selected along disturbance gradients (minimally, moderately, and highly disturbed) of key stressors, they can be used to develop and evaluate indicator assemblages, metrics, and indices. As suggested above, much more research is needed in such basic areas as sampling effort, assemblage responsiveness, and indicator sensitivity and variance. Initial research on such a range of site conditions can provide valuable data for indicator development and comparisons (Hughes et al. 1993).

Again, I stress the importance of selecting regional reference sites that are disturbed as little as possible. This is a challenging process where there are few or nominimally disturbed sites in a region. The process requires conscientious map analysis and reconnaissance, and often some degree of compromise among ecoregions and number of reference sites. Experience to date suggests that too little effort is usually dedicated to this process; convenient reference sites along roads or at bridges are frequently more disturbed than is desirable. Reference sites have little value when selected from extensively disturbed ecoregions, such as the Huron Erie Lake Plain of Ohio (Ohio EPA 1990a) and the valley subregions of the Central Appalachian Ridges and Valleys (Gerritsen et al. 1994), where only highly disturbed "reference" sites were located. Although it would seem obvious, sites must at least appear healthy to serve as reference sites. Channelized agricultural ditches or ponds with livestock instead of natural riparian vegetation are just as inappropriate as waters draining urban or industrial areas. Similarly, selection of reference sites from randomly selected sites, especially when those sites are drawn from a population of disturbed sites as depicted in DNREC (1992) and Larsen (Chapter 18), will yield a set of disturbed reference sites and weak biological criteria. However, a set of 94 randomly selected lakes in the northeast U.S. yielded 22 reference candidates after analysis of maps, air photos, and lake management information (S. Thiele, ManTech, Corvallis, unpublished data). Whether sites are selected subjectively or randomly, the selection criteria described in Section 3 must be carefully followed. To paraphrase Leopold (1949), a reference site is right when it tends to preserve the integrity and beauty of the ecosystem. It is wrong when it doesn't.

I urge increased objectivity and thought in determining reference condition; however, I recognize that considerable professional judgement is involved in site selection, data analysis, and determining acceptable, marginal, and unacceptable indicator scores. This is as it should be, but these judgements should be those of experienced field biologists, documented, and widely peer reviewed. Biological judgements should not be left to administrators, policy boards, courts, or other politicians. Managers' jobs are to help citizens decide what they will do about scientists' best assessments of existing conditions and trends. Too often biologists have acquiesced and let policy matters influence their assessments. This muddies the biological assessments, obscures the difficult policy decisions needed, and typically results in continued degradation of biological integrity, without explicitly deciding to allow it. Therefore, I recommend setting biological criteria as high as the reference condition data allow. It is the manager's responsibility, not the biologist's, if a majority of sites are in unacceptable condition or experiencing unacceptable trends. We are not fulfilling our responsibility to the aquatic ecosystems or to the public if we state ecosystems have biological integrity when they do not. Two of our professional responsibilities are to alert managers and the public if our data indicate ecosystem deterioration and to advocate changes in management practices to correct those conditions. The fundamental question of environmental biologists today is not whether we can turn the clock back to an era when most systems had biological integrity, but whether we can slow or reverse the continued rapid degradation of the environment and the concomitant loss of biological integrity and diversity.

I support using multiple approaches to determine reference condition. We must apply the best information possible, whether it comes from ecoregional reference sites, historical data, paleolimnology, or modeling results. However, the one area that could provide the greatest returns for increased effort is hierarchical ecoregional and waterbody classification. Our historical data are limited and we cannot improve it (we can only make quality collections now for future historical analyses). Paleolimnology is largely limited to lakes and diatoms; paleolimnological data on vertebrates, many invertebrates, and other systems is inadequate for estimating reference conditions. Modeling can be applied to any quantitative information, and would benefit from more reference site data. A cohesive hierarchical classification framework, perhaps linked with that developed by the Canada Committee on Ecological Land Classification (Wiken 1986), is needed to improve communication and assessment within and among state agencies.

A final caution: the methodology outlined in this chapter is not an ultimate or permanent solution. It was not written to be a cookbook. Given the variety of agencies and environments in the United States, considerably more development, research, and teamwork are necessary for effective biological assessment to occur. The longer we delay determining appropriate reference conditions and selecting and preserving relevant reference sites, the more difficult it is to define acceptable condition and use biological assessments to prevent further deterioration of our aquatic ecosystems.

ACKNOWLEDGMENTS

I thank Wayne Davis and Tom Simon for inviting me to write this chapter, and for having the patience to wait for me to make the time to do so. The ideas expressed herein are also the product of many years of discussions with Phil Larsen and Jim Omernik; in fact many of the ideas may have originated with them. I must also thank Phil Kaufmann for examples of other methods for estimating reference condition and for further broadening my perspective. I acknowledge Sandy Thiele who helped outline this chapter, evaluated the process of finding reference lakes, and who should have been a coauthor but lacked the time. Finally, I am grateful to Glenn Griffith, Sandy Thiele, Thom Whittier and three anonymous reviewers who made the time to critically comment on an earlier version of this manuscript. This chapter was completed at the U.S. EPA laboratory in Corvallis, Oregon, through contract 68C80006 with ManTech Environmental Technology, Inc., subjected to peer and administrative review, and approved for publication.

## CHAPTER 5

Ecoregions: A Spatial Framework for Environmental Management

James M. Omernik

1.0 BACKGROUND

In recent years there has been an increasing awareness that effective research, inventory, and management of environmental resources must be undertaken with an ecosystem perspective. Resource managers and scientists have come to realize that the nature of these resources (their quality, how they are interrelated, and how we humans impact them) varies in an infinite number of ways, from one place to another and from one time to another. However, there *are* recognizable regions within which we observe particular patterns (Frey 1975). These regions generally exhibit similarities in the mosaic of environmental resources, ecosystems, and effects of humans and can therefore be termed ecological regions or ecoregions. Definition of these regions is critical for effectively structuring biological risk assessment, which must consider the regional tolerance, resilience, and attainable quality of ecosystems.

There is general agreement that these ecological regions exist, but there is considerable disagreement about how to define them (Gallant et al. 1989; Omernik and Gallant 1990). Some of this disagreement stems from differences in individual perceptions of ecosystems, the uses of ecoregions, and where humans fit into the picture. Most, however, agree with a general definition that ecoregions comprise regions of relative homogeneity with respect to ecological systems involving interrelationships among organisms and their environment. Rowe (1990, 1992) has argued that ecological regions subsume patterns in the quality and quantity of the space these organisms (including humans) occupy. He implied that the organisms as a group, or singly, are no more central to the system than the space they occupy. Each is a part of the whole, which is different in pattern in space as well as time. This more holistic definition appears to be gaining acceptance (Barnes 1993).

Canadian resource managers have been at the forefront of developing ecoregional frameworks and stressing the need for an ecoregional perspective (Government of Canada 1991). They have argued that the majority of environmental research is of the single-medium/single-purpose type (Figure 1), whereas much of the focus and concern of environmental management has recently been on the entire ecosystem, including biodiversity, effects of human activities on all ecosystem components, and the attainable conditions of ecosystems (Wiken, personal communication). Efforts to assess, research, and manage the ecosystems (multipurpose/multimedia, or lower right-hand portion of Figure 1) are normally carried out via extrapolation from data gathered from single-medium/single-purpose research (e.g., effects of logging road construction on salmonid production in streams) or in some cases through single-medium/multipurpose studies such as using indicator species. The problem is that little effort is being expended on studying ecosystems holistically and attempting to define differences in patterns of ecosystem mosaics. Wiken is not suggesting, nor am I, that this is an either/or situation or that the balance should be reversed. Certainly we must continue basic research on processes and the effects specific human activities (and human activities in aggregate) have on environmental resources. However, in order to maximize the meaningfulness

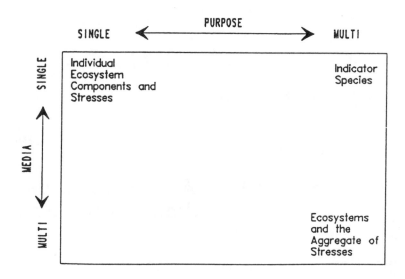

Figure 1. Types of environmental research.

of extrapolations from these studies and the use of data collected from national or international surveys, we must develop a clearer understanding of ecosystem regionalities.

A large barrier to developing a clear understanding of ecosystem regionalities is the common belief that to be scientifically correct, regions must be quantitatively developed and that they are objective realities (Hart 1982a). In a review of regionalization, Grigg (1967) appeared to conclude that to be effective regionalization must be quantitative or objective, apparently based on the assumption that certain processes determine spatial patterns of geographical phenomena at certain scales. His principles of regionalization clearly reflect this belief. Grigg noted that this line of thinking was especially true of the work of geographers in eastern Europe and Russia, although he foresaw a change in this point of view. The idea that subjective approaches might be appropriate was generally Western (particularly North American). To date, attempts to define ecological regions using only quantitative techniques have met with little success. However, efforts to understand and even map ecosystem regionalities in a "scientifically correct", purely quantitative way, appear to be growing. Levin (1992) has acknowledged that "the problem of pattern and scale is the central problem in ecology," but he stressed that to gain an understanding of the patterns of ecosystems in time and space and the causes and consequences of patterns, we must develop the appropriate measures and quantify these patterns. The "patterns" Levin referred to are doubtless the regionalities of ecosystems, or ecological regions, that occur at all scales. What we must also realize, however, is that valuable as this type of research is, it is likely to bear fruit only if sufficient, complementary qualitative geographic research on ecosystems and the aggregate of ecosystem components and human imposed stresses is also conducted.

The development of ecological regions has been, and will probably continue to be, challenging and controversial. Until quite recently, another criticism that has impeded geographers in the development of effective regions has been that to base an ecological approach on the assumption that different regions have different capacities for organisms (including humans) was believed to be subscribing to "environmental determinism". Although this belief is being defused with increased understanding of ecosystems, the need to combine art with science in regional geographic research, including the development of ecoregions, continues to meet resistance (Golledge et al., 1982; Hart 1982a, 1982b, 1983; Healey 1983). This resistance is not universal, however, particularly in applied areas such as military intelligence. Military geographers, when tasked to define regions within which broad-scale military operations or specific types of operations may be conducted, have long employed qualitative techniques to filter such aspects as the relative inaccuracies and differences in levels of generality in mapped information (Omernik and Gallant 1990). In this case, the focus is on defining areas within which there is likely to be similarity in general or particular combinations of conditions regarding such factors as physiography, climate, geology, soil type, vegetation, and land use. Knowledge of spatial relationships between geographic phenomena, the relative accuracy and level of generality of mapped information, and differences and appropriateness of classifications on maps of similar subjects, allow the geographer to screen each piece of intelligence (data source) and delineate the most meaningful regions. The test of these regions is in their ultimate usefulness, rather than in the scientific rigor of a particular qualitative mapping technique. Advances in remote sensing and geographic information systems have obviously greatly increased the efficiency of the regionalization efforts, but qualitative analyses continue to be invaluable in providing meaning to regional responses in remote sensing products and map interpretation.

2.0 ECOREGION DEFINITION

Ecoregions occur and can be recognized at various scales. If one is viewing the conterminous United States from a satellite, one can recognize broad ecoregions, including the semiarid to arid basin, range, and desert areas of the West and Southwest, and the rugged mountains of the West. The latter typically contain a mosaic of characteristics ranging from alpine glaciated areas at or above timberline to dense coniferous forests, to near xeric conditions at lower elevations and rain shadow areas. Other such broad ecological regions include the glaciated corn belt and associated nutrient-rich intensively cultivated areas in the central United States and Upper Midwest, and the contrasting nutrient-poor glaciated regions of forests and high-quality lakes and streams in the Northeast and northern Upper Midwest. At a larger scale (closer to the earth), one can recognize regions within these regions, and at successively larger scales, regions within those regions.

The recognition of these regions is nothing new. They have long been perceived by people from all walks of life — from the earliest explorers in whose logs we read descriptions of the different mosaics in flora, fauna, climate, and physiography in the different regions they traveled, to present-day ecologists and resource managers who are attempting to understand the effects human activities are having on ecosystems. The problem has been in defining the regions. Although most resource managers have a general understanding of the spatial complexities in ecosystems and how they can be perceived at various scales, they tend to use inappropriate frameworks to research, assess, manage, and monitor them. One reason for this is that until recently there have been no attempts to map ecosystem regions, so rather than make interpretations, managers have chosen surrogates. These surrogates have often comprised single-purpose frameworks of a particular characteristic believed to be important in causing ecosystem quality to vary from one place to the next. The most commonly used single-purpose frameworks have been potential natural vegetation (e.g., Küchler 1964, 1970), physiography (e.g., Fenneman 1946), hydrology (e.g., USGS 1982), climate (e.g., Trewartha 1943), and soils (U.S. Department of Agriculture 1981). Another reason for using single-purpose frameworks, as mentioned in the preceding section, stemmed from the belief that a scientifically rigorous method for defining ecological regions must address the processes that cause ecosystem components to differ from one place to another and from one scale to another.

Several classifications have been developed to address biotic regions, or biomes, but with the implication that these classifications define ecosystem regions as well. This is understandable, because the perception that ecosystems comprise more than differences in biota and their capacities and interrelationships, although not new, has gained wide acceptance only relatively recently. Most of these mapped classifications reflect patterns in vegetation and climate and have been regional in scale (e.g., Dice 1943; Holdridge 1959; Brown and Lowe 1982; Brown and Reichenbacher, in press). Very few have been global (e.g., International Union for Conservation of Nature and Natural Resources 1974; Udvardy 1975). Bailey's ecoregions (Bailey 1976, 1989, 1991; Bailey and Cushwa 1981), although based on a number of landscape characteristics, rely on the patterns of a single characteristic at each hierarchical level. These regions have been developed at regional and global scales. A more detailed explanation of Bailey's approach and its limitations with respect to attempts to use it to frame aquatic ecosystems are given later in this chapter.

The need for an ecoregional/reference site framework to facilitate the development of biological criteria was recognized in the late 1970s. This need was part of a larger concern for a framework to structure the management of aquatic resources in general (Warren 1979) and was coupled with an increasing awareness that there was more to water quality management than addressing water chemistry, which had been the primary focus. The biota must be considered as well, as must the physical habitat and the toxicity (Karr and Dudley 1981).

The earliest attempts within the Environmental Protection Agency (EPA) to classify streams and other aquatic resources adapted Bailey's (1976) ecoregion classification. It was felt that the character of streams reflects the aggregate of the characteristics in the watersheds they drain. Because Bailey's scheme incorporated a number of these characteristics and was intended to show differences in patterns of ecosystems, it appeared to provide a logical framework. However, an attempt to use the scheme for classifying aquatic ecosystems proved unsuccessful and resulted in the development of a different framework believed to be more effective (Hughes and Omernik 1981a; Omernik et al. 1982) Although Bailey's approach, which is based on the work of Crowley (1967), considers a number of characteristics, it depends largely on a single characteristic at each hierarchical level, and therein appears to be the problem. At the "section" level, for example (there are roughly 53 sections in the conterminous United States), the regions are based primarily on Küchler's (1964, 1970) potential natural vegetation. At Bailey's next more detailed "district" level, Hammond's (1970) land surface form regions are used. These characteristics are helpful in identifying ecoregions in some parts of the country, but not others. The Sand Hills of Nebraska, a relatively large, homogeneous ecological region, recognized on nearly every small-scale map of soils, physiography, geology, vegetation, and land use, was not identified at Bailey's section level because of the way in which Küchler's classification was applied. Although Hammond's land surface form is useful for defining ecoregions in some areas at the scale of Bailey's districts, it is very ineffective in others such as the Southern Rockies where elevational and vegetative differences are far more important. Here, Hammond land surface form map units, based on physiographic characteristics such as high mountains with greater than 3000 feet of local relief and less than 20% of the area gently sloping, often cover the gamut of ecosystem variations in the larger ecological region they occupy.

The first compilation of ecoregions of the conterminous United States by EPA was performed at a relatively cursory 1:3,168,000 scale and was published at a smaller 1:7,500,000 scale (Omernik 1987) (Figure 2). The approach recognized that the combination and relative importance of characteristics that explain ecosystem regionality vary from one place to another and from one hierarchical level to another (Gallant et al. 1989; Omernik and Gallant 1990). This is similar to the approach used by Environment Canada (Wiken 1986). In describing ecoregionalization in Canada, Wiken (1986) stated:

> Ecological land classification is a process of delineating and classifying ecologically distinctive areas of the earth's surface. Each area can be viewed as a discrete system which has resulted from the mesh and interplay of the geologic, landform, soil, vegetative, climatic, wildlife, water and human factors which may be present. The dominance of any one or a number of these factors varies with the given ecological land unit. This holistic approach to land classification can be applied incrementally on a scale-related basis from very site-specific ecosystems to very broad ecosystems.

Hence, the difference between this approach to defining ecoregions and most preceding methods is that it is based on the hypothesis that ecological regions gain their identity through spatial differences in a combination of landscape characteristics. The factors that are more or less important vary from one place to another at all scales. One of the strengths of the approach lies in the analysis of multiple geographic characteristics that are believed to cause or reflect differences in the mosaic of ecosystems, including their potential composition. All maps of particular characteristics (e.g., soils, physiography, climate, vegetation, geology, and land use) are merely representations of aspects of that characteristic. Each map varies in level of generality (regardless if at the same scale), relative accuracy, and classification used. Subjective determinations must be made in the compilation of all maps regarding the level of generality, the classification to be used, and what can be represented and what cannot, whether the map is hand drawn or computer generated. Everything about a particular subject cannot be shown once the map scale becomes smaller than 1:1. Hence, an ecoregion that exhibits differences in characteristics such as physiography or soils, may not be depicted by a map of one of those subjects because of the classification and level of generality chosen, as well as the accuracy of the author's source materials. On the other hand, because ecosystem regions reflect differences in a combination of characteristics, use of multiple sources of mapped information permit the detection of these regions. It is simply a matter of safety in numbers.

Although the approaches used by EPA and Environment Canada are remarkably similar, particularly regarding their use of qualitative, or subjective analyses, the initial compilation of ecoregions maps in both countries was completely independent. Authors of the maps in both countries were unaware of the other's ongoing work until after the maps had been compiled. This situation has subsequently changed

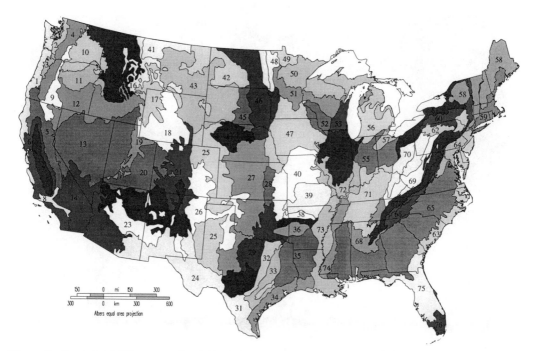

Figure 2. Ecoregions of the conterminous United States. 1 Coast Range, 2 Puget Lowland, 3 Willamette Valley, 4 Cascades, 5 Sierra Nevada, 6 Southern and Central California Plains and Hills, 7 Central California Valley, 8 Southern California Mountains, 9 Eastern Cascades Slopes and Foothills, 10 Columbia Plateau, 11 Blue Mountains, 12 Snake River Basin/High Desert, 13 Northern Basin and Range, 14 Southern Basin and Range, 15 Northern Rockies, 16 Montana Valley and Foothill Prairies, 17 Middle Rockies, 18 Wyoming Basin, 19 Wasatch and Uinta Mountains, 20 Colorado Plateaus, 21 Southern Rockies, 22 Arizona/New Mexico Plateau, 23 Arizona/New Mexico Mountains, 24 Southern Deserts, 25 Western High Plains, 26 Southwestern Tablelands, 27 Central Great Plains, 28 Flint Hills, 29 Central Oklahoma/Texas Plains, 30 Edwards Plateau, 31 Southern Texas Plains, 32 Texas Blackland Prairies, 33 East Central Texas Plains, 34 Western Gulf Coastal Plain, 35 South Central Plains, 36 Ouachita Mountains, 37 Arkansas Valley, 38 Boston Mountains, 39 Ozark Highlands, 40 Central Irregular Plains, 41 Northern Montana Glaciated Plains, 42 Northwestern Glaciated Plains, 43 Northwestern Great Plains, 44 Nebraska Sand Hills, 45 Northeastern Great Plains, 46 Northern Glaciated Plains, 47 Western Corn Belt Plains, 48 Red River Valley, 49 Northern Minnesota Wetlands, 50 Northern Lakes and Forests, 51 North Central Hardwood Forests, 52 Driftless Area, 53 Southeastern Wisconsin Till Plains, 54 Central Corn Belt Plains, 55 Eastern Corn Belt Plains, 56 Southern Michigan/Northern Indiana Till Plains, 57 Huron/Erie Lake Plain, 58 Northeastern Highlands, 59 Northeastern Coastal Zone, 60 Northern Appalachian Plateau and Uplands, 61 Erie/Ontario Lake Plain, 62 North Central Appalachians, 63 Middle Atlantic Coastal Plain, 64 Northern Piedmont, 65 Southeastern Plains, 66 Blue Ridge Mountains, 67 Central Appalachian Ridges and Valleys, 68 Southwestern Appalachians, 69 Central Appalachians, 70 Western Allegheny Plateau, 71 Interior Plateau, 72 Interior River Lowland, 73 Mississippi Alluvial Plain, 74 Mississippi Valley Loess Plains, 75 Southern Coastal Plain, 76 Southern Florida Coastal Plain. (Adapted from Omernik, J. M., *Ann. Assoc. Am. Geographers,* 77: insert, 1987.)

and those responsible for the design and development of both ecoregion frameworks are now collaborating in a multicountry, multiagency effort [including the U.S. Geological Survey/Earth Resources Observation Satellite (USGS/EROS)] to develop an ecoregional framework for the circumpolar arctic-subarctic region. At the time of this writing, a draft of ecoregions of Alaska, consistent with the ecoregions of Canada, has been completed. Publication of this map is planned for 1994. An additional goal of this group is to develop a consistent ecoregional framework for North America.

Needs for ecoregional frameworks exist at all scales. Global assessments require the coarsest levels and national assessments require more detailed levels such as are provided by EPA's Ecological Areas of the Conterminous United States (Figure 3) [a revision of Aggregations of Ecoregions of the Conterminous United States by Omernik and Gallant (1990)] or Environment Canada's Ecozones (Wiken 1986). The scale of state level needs is more appropriately addressed using EPA's Ecoregions (Omernik 1987) or subregions (Gallant et al., 1989; Clarke et al., 1991), and Environment Canada's Ecoprovinces or

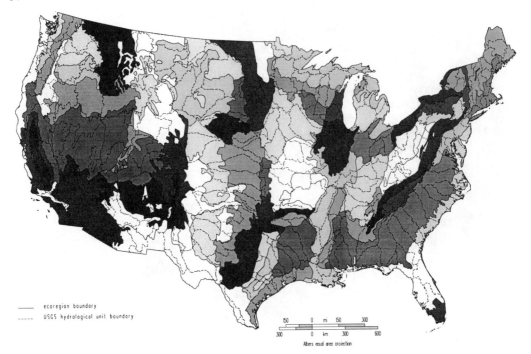

_____ ecoregion boundary
----- USGS hydrological unit boundary

Figure 3. Ecological areas of the conterminous United States. (Adapted from Omernik, J. M. and Gallant, Proc. Global Nat. Resource Symp: Preparing for the 21st Century, Vol. 2, p. 943, 1990.)

Ecoregions. Because of the confusion with other meanings of the terms province, zone, district, etc., EPA has not adapted that scheme of naming different hierarchical levels. Development of a less confusing classification scheme is currently being discussed for use with the planned North American ecoregion framework and will probably use different Roman numerals, with the lowest being the most general and the highest, the most detailed. Regions are simply regions regardless of their scale, but some means of identifying different hierarchical levels is no doubt needed. More detailed ecoregions that would be helpful at local levels, such as defined by Thiele (personal communication) for a part of the Grande Ronde Basin in Oregon, have not been developed for the United States. Obviously, the more detailed the hierarchical level (the larger the scale), the more time consuming the chore of completing ecoregions on a per unit area basis.

3.0 REFINEMENT OF ECOREGIONS AND DELINEATION OF SUBREGIONS

A number of states, notably Ohio, Arkansas, and Minnesota, have used the first approximation of ecoregions published in 1987 to develop biological criteria, and to set water quality standards and lake management goals. Most states, however, found the resolution of regions delineated on Omernik's (1987) 1:7,500,000-scale map to be of insufficient detail to meet their needs. This has led to several collaborative projects with states, EPA regional offices, and the EPA Environmental Research Laboratory in Corvallis, Oregon, to refine ecoregions, define subregions, and locate sets of reference sites within each region and subregion. This work is being conducted at a larger scale (1:250,000) and includes the determination of ecoregion and subregion boundary transition widths. These projects currently cover Iowa, Florida, and Massachusetts, and parts of Alabama, Mississippi, Virginia, West Virginia, Maryland, Pennsylvania, Oregon, and Washington. Results of much of this work is in varying stages of completion; some maps with accompanying texts have been submitted to journals for consideration of publication and others are being prepared for publication as state and EPA documents.

The process of refining ecoregions and defining subregions is similar to the initial ecoregion delineation. The main difference, besides doing the work at a larger scale, is in the collaborative nature of the projects, which include scientists and resource managers from the states and EPA regions covered (and in many cases other governmental agencies), as well as geographers at the EPA Environmental Research Laboratory in Corvallis, Oregon. This particular mix of expertise is necessary to maximize consistency from one part of the country to another and to insure that the final product is useful. The process merely documents the spatial patterns that effective resource managers already recognize. Therefore, interacting with scientists and resource managers who know local conditions is essential in the delineation of ecoregions, particularly at lower hierarchical (larger scale) levels.

Although some of these ecoregionalization projects have involved only one state, a number have focused on delineation of subregions within one or more ecoregions covering more than one state. One such project encompasses the portions of the Blue Ridge, Central Appalachian Ridges and Valleys, and Central Appalachian Ecoregions that cover Pennsylvania, West Virginia, Virginia, and Maryland (Figure 4). The advantage of this type of project involving more than one state covering similar ecological regions and subregions is that it encourages data sharing across state lines and calibration of sampling methods by ecoregion rather than political unit. It also provides a reality check regarding the quality of data collected by different states within the same region. Because natural ecological regions rarely correspond to spatial patterns of state boundaries, or any other political unit, there are numerous cases where a state covers only a small portion of an ecoregion or subregion that has its greatest extent in neighboring states. The distinctly different subregions of the Central Appalachian Ridges and Valleys ecoregion provide a case in point (Figures 4 and 5). Most of these discontinuous subregions are in Pennsylvania and Virginia and only small parts of each subregion occur in West Virginia and Maryland. Where a number of reference sites within each subregion are needed to determine within-region variability, realistically attainable quality (discussed later in this chapter and in greater detail in Hughes, Chapter 4), and between-region differences, the number of sites available within either West Virginia or Maryland is likely to be insufficient.

Typically a project to refine ecoregions, define subregions, and locate sets of reference sites begins with a data collection meeting. This meeting should include those people who have spatial perceptions of the environmental resources and ecosystem patterns in the particular regions covered, those who have knowledge of data sources, and those who will be eventual users of the framework. It is important to include representation from the various state and federal agencies that have mutual interests in resource quality (aquatic and terrestrial) in the particular ecoregions covered by the project. Data needs include medium-scale (generally 1:250,000 to 1:1,000,000) mapped information on causal and integrative factors such as bedrock and surficial geology, soils, land use, hydrology, physiography, and existing and potential vegetation, as well as available interpretations (written or mapped) of biomes or ecosystems. Some of the most important sources of information are the "mental maps" of ecosystem patterns held by scientists and resource managers who have studied the area.

Remote sensing data, particularly data from National Oceanic and Atmospheric Administration (NOAA) and Advanced Very High Resolution Radiometer (AVHRR), are also helpful in defining ecological regions. The wide swath width per scene (up to 27,000 km), pixel resolution of approximately 1 km^2, and suitability for combination into average, seasonal, or annual classes such as "vegetation greenness" (Loveland et al. 1991) make AVHRR data appropriate for broad-scale regional analysis of ecosystem patterns. These data appear to be especially helpful in attaining consistency across international borders and other areal units where mapped resource materials on landscape characteristics vary in quality, availability, and type.

Based on the information gathered, and using the methods outlined by Omernik (1987), Gallant et al. (1989), and the publications in preparation covering the recent state and multi-state subregionalization projects, a first approximation of refined ecoregions and subregions is then complied. However, the methods are continually being refined. Each new project reveals its own unique set of problems and challenges, but much of the knowledge gained from involvement in the variety of geographic areas covered and the variety of scales on which the work has been conducted can be applied to new projects and areas.

Critical to the process of interpreting and integrating the source material is the care that must be taken to avoid defining regionalities of particular ecoregion components such as fish or macroinvertebrate characteristics, or patterns in a single, or a set of, chemical parameters. At the onset of each project, and at the initial idea and data gathering meetings, the question of whether ecoregions or special purpose regions are desired is always asked. When the interest is on a particular subject such as eutrophication, sensitivity of surface waters to acidification, or nutrient concentrations in streams, special purpose maps rather than ecoregion maps are appropriate and can be developed (e.g., Omernik 1977; Omernik and Powers 1983; Omernik and Griffith 1986; Omernik et al. 1988). But response to the question in nearly every case has been that ecoregions are the desired regional framework. The primary interest of most state environmental resource management agencies has been in developing biological criteria, but there has also been concern for a mechanism to structure the assessment and management of nonpoint source

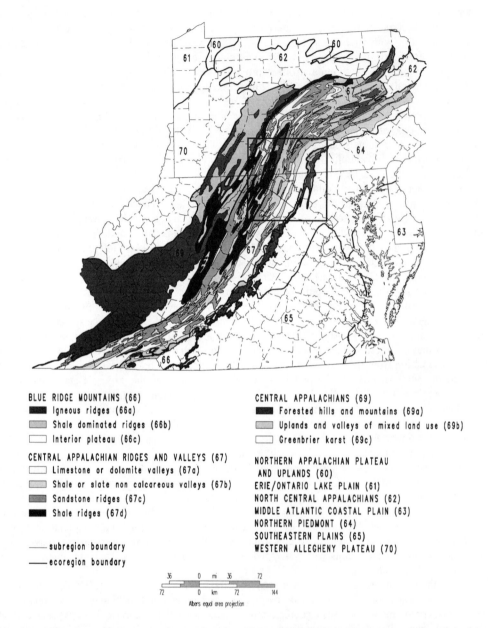

BLUE RIDGE MOUNTAINS (66)
▪▪▪ Igneous ridges (66a)
▨ Shale dominated ridges (66b)
☐ Interior plateau (66c)

CENTRAL APPALACHIAN RIDGES AND VALLEYS (67)
☐ Limestone or dolomite valleys (67a)
▨ Shale or slate non calcareous valleys (67b)
▨ Sandstone ridges (67c)
▪▪▪ Shale ridges (67d)

—— subregion boundary
—— ecoregion boundary

CENTRAL APPALACHIANS (69)
▪▪▪ Forested hills and mountains (69a)
▨ Uplands and valleys of mixed land use (69b)
☐ Greenbrier karst (69c)

NORTHERN APPALACHIAN PLATEAU
 AND UPLANDS (60)
ERIE/ONTARIO LAKE PLAIN (61)
NORTH CENTRAL APPALACHIANS (62)
MIDDLE ATLANTIC COASTAL PLAIN (63)
NORTHERN PIEDMONT (64)
SOUTHEASTERN PLAINS (65)
WESTERN ALLEGHENY PLATEAU (70)

36   0 mi 36   72
72   0 km 72   144
Albers equal area projection

Figure 4. Ecoregions and subregions of the Blue Ridge, Central Appalachian Ridges and Valleys, and Central Appalachians of EPA Region 3. (Boxed portion is enlarged in Figure 5)

pollution as well as a variety of environmental resource regulatory programs. The attractiveness of an ecoregional framework is that, although not fitting any one purpose perfectly, it has general applicability to many environmental resource management needs, and facilitates reporting and transfer of information between subject areas (e.g., wetlands, surface waters, forestry, soils, and agriculture).

4.0 REFERENCE SITES

Upon completion of the initial revision of ecoregions and delineation of subregions, sets of reference sites are identified for each subregion. As with the regionalization, this process is collaborative, but

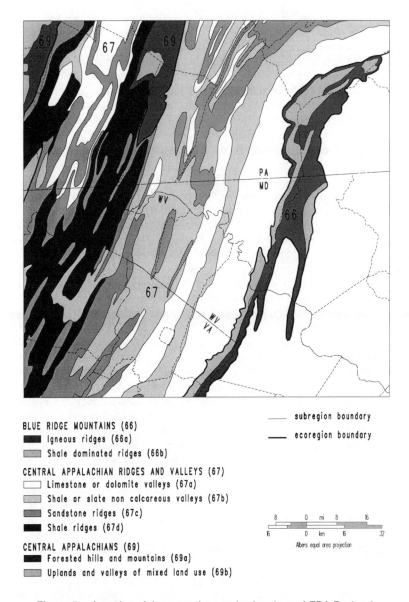

BLUE RIDGE MOUNTAINS (66)
▰ Igneous ridges (66a)
▰ Shale dominated ridges (66b)

CENTRAL APPALACHIAN RIDGES AND VALLEYS (67)
▱ Limestone or dolomite valleys (67a)
▰ Shale or slate non calcareous valleys (67b)
▰ Sandstone ridges (67c)
▰ Shale ridges (67d)

CENTRAL APPALACHIANS (69)
▰ Forested hills and mountains (69a)
▰ Uplands and valleys of mixed land use (69b)

―――― subregion boundary

―――― ecoregion boundary

Albers equal area projection

Figure 5. A portion of the ecoregions and subregions of EPA Region 3.

normally with heavier involvement of state biologists and ecologists. In general, sets of reference sites are selected for each region and subregion to get a sense of the regionally attainable conditions regarding aquatic ecosystems. Attainable quality refers to those conditions that are realistic, rather than "pristine", which implies the unrealistic turning back of the clock and the absence of humans in the ecosystem. Candidate stream sites must be "relatively undisturbed" yet representative of the ecological region they occupy (Hughes et al. 1986; Gallant et al. 1989; Hughes, Chapter 4).

An initial selection of reference sites is normally accomplished by interpreting 1:100,000- and 1:250,000-scale maps with guidance from state resource managers as to minimum stream sizes for each subregion and locations of known problem areas and point sources. The probable relative lack of disturbance can be interpreted from topographic maps, particularly the recent 1:100,000-scale series. General determinations of the extent of recent channelization, woodland or forest, urbanization, proximity of roads to streams, and mining and other human activities can be made using these maps. USGS flow records can be consulted to approximate the minimum watershed size necessary for each subregion, but state water resource managers and regional biologists generally have a better idea about when a stream becomes a stream of interest because of their intimate understanding of their own areas. Intermittent streams are often considered valued resources if the enduring pools are of sufficient size. State and regional experts should also be consulted regarding the minimum number of sites necessary for each region or subregion. The minimum number is generally a function of the size and complexity of the subregion. For some small or very homogeneous regions, the point of diminishing returns may be reached with a number of five or six, whereas in other complex regions and in areas where reference sites representing different stream sizes are a concern, a much larger number would be desirable.

4.1 Field Verification of Reference Sites, Ecoregions, and Subregions

Once sets of candidate reference sites have been identified for each region, they should be reviewed by state biologists and regional experts. Based on their personal knowledge of the region, these regional experts may choose to add or delete potential sites. Then field verification of the ecoregion and subregion delineations is conducted coupled with visits to representative sets of reference sites within each ecoregion and subregion. Ideally, this field work is conducted by the entire group collaborating on the particular regionalization/reference site project. Hence, it should include the geographers responsible for delineating the regions, subregions, and boundary transition widths, as well as compiling the initial list and map of candidate watersheds. Also included should be the regional biologists and water resource managers who provided information used to define the regions and locate sets of reference sites, and who will eventually use the framework. The best test of the regional framework and sets of reference sites is their ultimate usefulness. The regions must make sense to those who know and manage the resources in the area and are developing the biological criteria. Lastly, it is useful to include in the field verification exercises experts from other agencies and biologists from adjacent states who are considering use of the ecoregion/reference site approach in their assessment and regulatory programs.

Visits to a number of stream reference sites in each region allow a visual subjective analysis of within- and between-region similarities and differences regarding stream habitat conditions as well as landscape characteristics of the ecoregions and subregions the reference sites/watersheds occupy. Here it is important to maximize the number of sites visited and to spend only as much time at each site as is necessary to evaluate regionalities in site characteristics and the natural and anthropogenic factors that may cause within-region differences. Sampling at each site at this stage should not go beyond turning over a few rocks and/or roughly sorting a leaf pack. Not only is this a helpful cursory method of evaluating stream and habitat quality, it would be practically impossible to restrain most biologists from the activity when they are at the stream site. Final selection of sets of stream reference sites is made by state resource managers and biologists after they have visited and evaluated all candidate reference sites.

4.2 The Concept of Pristine and Least-Disturbed Conditions

It must be understood that reference sites do not represent "pristine" conditions, conditions that would exist if humans were removed from the scene, or pre-European settlement conditions. To select such sites is impossible. There are no pristine areas in the United States, or in the world for that matter, if the term is to imply an absence of human impact. Even sites in remote mountainous areas have been impacted by human-caused atmospheric pollutants. Reference sites representing least-disturbed ecosystem conditions are a moving target of which humans and natural processes are a part. The idea that conditions were pristine in North America prior to European settlement has been convincingly challenged in the past couple of decades (Denevan 1992). Humans have probably played a major role in shaping landscape pattern and molding ecosystem mosaics for thousands of years. It is unrealistic to attempt to map the ecosystem regions and reference site conditions that we believe would exist if humans were removed from the scene, unless of course we are all willing to move to another planet. It is also inadvisable (perhaps stupid or self-destructive would be better words) to fail to recognize the impact we humans are having on the overall system of which we are but one part, and what we must do to maintain the integrity of the system.

Like the mosaic of geographic conditions that shape ecosystem patterns, that which can be categorized as "least disturbed" is relative to the region in which a set of reference sites is being selected. In the Boston Mountains Ecoregion (in Arkansas and Oklahoma), minimally impacted reference sites comprise streams having watersheds without point sources, little grazing activity, and a relative lack of recent logging activity and road building. In this region, stream reference sites and their watersheds come close to mirroring the present perception that most people have of high-quality stream conditions. In the Huron/Erie Lake Plain Ecoregion (in Ohio, Indiana, and Michigan), on the other hand, there are no streams with watersheds that are not almost completely in cultivated agriculture. Many are also heavily impacted by urbanization and industries, and all streams relative to watersheds of 30 mi^2 or more have been channelized at one time or another. However, there are some streams that are relatively free of impact from point sources, industries, and major urbanization, that have not been channelized for many years so that the riparian zones have been allowed to grow back into woody vegetation with the channels becoming somewhat meandering. These types of streams and watersheds would comprise relatively undisturbed references for the region. Although the quality of the set of streams reflects the range of best attainable conditions given the current land use patterns in the regions, this does not imply that the quality cannot be improved. An analysis of the differences in the areal patterns of water quality from reference sites (the biota in particular) with patterns in natural landscape characteristics (such as soil and geology, and human stresses including agricultural practices), should provide a sense for the factors that are responsible for within-region differences in quality. A measure of how much the quality can be improved can then be derived through changing management practices in selected watersheds where associations were determined.

4.3 Selecting Reference Sites for Small and/or Disjunct Subregions

The approach for selecting sets of reference sites for subregions is the same as for the larger ecoregions. The maximum stream and watershed sizes of sites representative of subregions are normally smaller, of course, because the subregions are smaller and in many cases discontinuous, such as subregions of the Central Appalachian Ridges and Valleys (Figures 4 and 5). Where subregions represent bands of different mosaics of conditions, as is the case in some western mountainous ecoregions, it may be necessary to choose reference sites that comprise watersheds containing similar proportions of different subregions. Subregions of the Southern Rocky Mountains Ecoregion are, for example, characterized by different combinations of vegetation, elevation, land use, and climate characteristics (Gallant et al. 1989). Although factors such as geology and soils are also important, the other factors appear to be the most important in this ecoregion. One subregion consists of disjunct areas at or above timberline with heavy snowpack and most of the alpine glacial lakes in the ecoregion. Another comprises the areas generally at lower elevations with coniferous forest, steep gradient streams, and little to no grazing activity, because of limitations such as soil productivity. Still another subregion consists of the areas, generally at even lower elevations, where mixed forest and grazing are common. Yet another subregion is made up of the drier portions of the ecoregion, generally bordering adjacent predominantly xeric ecoregions. For the most part, only very small streams have watersheds completely within any one of these subregions. Larger streams more closely meeting size criteria for reference sites tend to drain areas in two or more subregions. Sets of reference sites for these types of subregions must therefore consist of watersheds that have similar proportions in different subregions. When selecting these sites one must account for the fact that minimally disturbed conditions often vary considerably from one subregion to another. Streams/watersheds within each set should be similar to one another regarding "relative disturbance" and should reflect higher water quality than streams with watersheds with similar proportions in each subregion but with greater human impact.

4.4 Anomalous Sites

In selecting reference sites, care must be taken to avoid including anomalous stream sites and watersheds. This can be particularly difficult when such streams are very attractive and represent the best conditions in a region. For example, an ecoregion or subregion typified by flat topography and deep soils, where minimally impacted streams with low gradients, no riffles, and sand or mud bottoms are normal, may also include a small area of rock outcrops and gravels in which streams have some riffles and gravel substrate. Obviously, the habitat in these streams is different than elsewhere and, therefore, the quality regarding biological diversity and assemblages cannot be expected in other parts of the region. Certainly streams such as this one should be protected and not be allowed to degrade to standards and expectations set for streams typical of most of the region, but neither should the typical streams be expected to attain the quality of an anomaly.

5.0 AERIAL RECONNAISSANCE OF ECOREGIONS

Visits to ecoregions are critical for verifing the regions and the approximations of boundary transition widths. Although often prohibitive because of cost, and time consuming because of visibility limitations in some regions, overflights are invaluable. By visualizing regions from different distances above the ground we can more easily distinguish the "forest from the trees." When we are too close to a subject, our attention is drawn to details. When we attempt to regionalize from this vantage point we have a tendency to define regions based on all of the details and relationships we have observed. When standing back away from the subject, we are able to observe the patterns in the sum of these details and interrelationships. Certainly many factors that are important in molding ecosystem regionalities miss the eye when we visualize the earth from a distance, but many if not most of the general characteristics that affect the quality, quantity, and distribution of ecosystem components can be perceived from a distance. These interrelated characteristics include land surface form, vegetation, and land use. Differences in patterns of these characteristics reflect differences in soils, geology (bedrock and surficial), climate, hydrology, and biological diversity.

As in the interpretation of any map, there are caveats to consider when verifying ecosystem patterns from the air. One must take into consideration the season, as well as precipitation and temperature deviations (both long and short term) and how they may have affected vegetation and land cover patterns. Patterns in human activities must be considered as well, and these often vary as a function of ownership or political unit, as well as ecoregion, which reflect differences in potential and capacity. When flying over the Southern Rocky Mountains along the Wyoming/Colorado state line in mid-1980, I noticed differences in timber management practices between states. North of the line in Wyoming, patterns of logging activity were apparent, whereas south of the line they were not. This may not have been a difference in state practices. It may have reflected differences in ownership, say, between federal and state or federal and private. Regardless, such within-ecoregion differences in land use and land cover must be distinguished from ecoregional characteristics.

Remote sensing data such as that from AVHRR data are often useful in sorting these patterns. Vegetation greenness classes derived from periodic AVHRR NDVI (normalized difference vegetation index) composites during the growing season are particularly useful in revealing differences in combinations of land cover characteristics (Loveland et al. 1991). These tools should be especially helpful in ultimately quantifying the landscape characteristics that make up ecoregions at various scales. Large-scale remote sensing data, such as high-altitude aerial photography and Landsat or SPOT satellite imagery, can be useful as well. However, the use of larger-scale (covering smaller areas) materials is also expensive and they must be carefully evaluated for representativeness. For example, whereas many thematic maps are the products of interpretations that include consideration of seasonal and year-to-year differences, Landsat imagery and high-altitude aerial photography are snapshots of conditions at a particular time. The real value of this larger-scale imagery may be in screening reference sites, where determining the relative extent of human disturbance is a major issue.

6.0 WATERSHEDS AND ECOREGIONS

One of the most common spatial frameworks used for water quality management and the assessment of ecological risk and nonpoint source pollution has been that of hydrologic units (or basins or watersheds). The problem with using this type of framework for geographic assessment and targeting is that it does not depict areas that correspond to regions of similar ecosystems or even regions of similarity in the quality and quantity of water resources (Omernik and Griffith 1991). Patterns in Major Basins and USGS Hydrologic Units (USGS 1982), which comprise groupings of major basins with adjacent smaller watersheds and interstices, have no similarity to patterns of ecoregions, which do reflect patterns in aquatic ecosystem characteristics (Figure 6). Many, if not most, major basins drain strikingly different ecological regions.

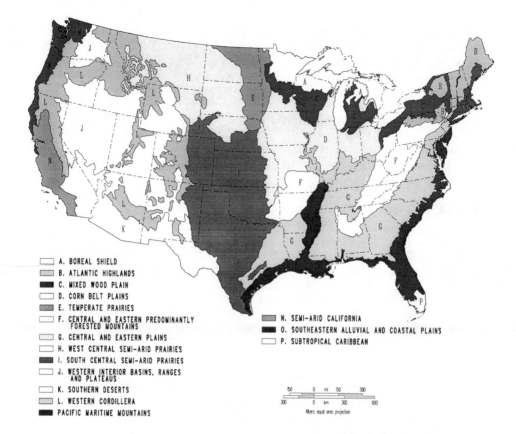

A. BOREAL SHIELD
B. ATLANTIC HIGHLANDS
C. MIXED WOOD PLAIN
D. CORN BELT PLAINS
E. TEMPERATE PRAIRIES
F. CENTRAL AND EASTERN PREDOMINANTLY FORESTED MOUNTAINS
G. CENTRAL AND EASTERN PLAINS
H. WEST CENTRAL SEMI-ARID PRAIRIES
I. SOUTH CENTRAL SEMI-ARID PRAIRIES
J. WESTERN INTERIOR BASINS, RANGES AND PLATEAUS
K. SOUTHERN DESERTS
L. WESTERN CORDILLERA
M. PACIFIC MARITIME MOUNTAINS

N. SEMI-ARID CALIFORNIA
O. SOUTHEASTERN ALLUVIAL AND COASTAL PLAINS
P. SUBTROPICAL CARIBBEAN

50   0   mi   50   300
300   0   km   300   600
Albers equal area projection

Figure 6. Omernik ecoregions and USGS hydrologic units.

The recent stress on "a watershed approach," although an excellent idea in that it changes the focus from dealing with predominantly point types of environmental problems to including those of a spatial nature, carries the implication of geographic targeting. The perception is that by looking at ecosystems and individual environmental resources within a watershed context, we are taking a giant step forward toward understanding ecological risk, ecosystem potential, and ultimately more effective ecosystem management. Although the rhetoric may be better, in reality what is being done may be little different than what has been done before. We now call case studies watershed studies. The real problem is that we may be fooling ourselves into believing that by adopting a "watershed approach" we are providing a spatial context within which to better understand and manage ecosystems. Use of watersheds is critical for ecosystem research, assessment, and management, but it should be done within a natural ecoregional framework that subsumes patterns in the combination of geographical characteristics (e.g., soils, geology, physiography, vegetation, and land use) associated with regional differences in ecosystems. We must develop an understanding of ecosystem regionalities at all scales, in order to make meaningful extrapolations from site-specific data collected from case studies or watershed studies, or whatever they are called. A recent national conference titled "Watershed '93," at which approximately 230 presentations were given over a week-long period, included no papers with titles addressing the applicability and limitations of a watershed or basin framework for ecological risk assessment and resource management (USEPA, 1994). Watershed studies are a necessity, but equally important is the development of an understanding of the spatial nature of ecosystems, their components, and the stresses we humans put upon them.

In most areas, the use of watersheds is an obvious necessity in defining and understanding spatial patterns of aquatic ecosystem quality and addressing ecological risk. It should be noted that in major portions of the country, topographic watersheds either cannot be defined or their approximation has little meaning (Hughes and Omernik 1981). Regions characterized by karst topography, extensive sandy soils, lack of relief, or excessive aridity are examples of areas where watersheds are less important. Reference streams draining watersheds that are completely within a particular region tend to be similar to one another when compared to reference streams in adjacent regions. Larger streams draining more than one ecoregion will reflect characteristics from each of the regions, with the relative influence of each region depending on its proportion of the total watershed, as well as differences in flow contributions from each region and, of course, point sources. Streams in arid regions that have large proportions of their watersheds in adjacent mesic or hydric regions will tend to have a different attainable quality than streams with similar watershed sizes that have smaller proportions of their basins in well-watered ecosystems. Water quality expectations for streams that have watersheds completely within the arid regions will generally be different than those for the other stream types. However, in many arid areas, spatial differences in subsurface watershed characteristics have a stronger influence on water quality than the size or characteristics of the surface watershed.

7.0 EVALUATING ECOREGIONS

As with any new tool, the usefulness of ecoregions must be evaluated. However, the evaluation of a framework intended to depict patterns in the aggregate of ecosystem components is not an easy task. Although commonly done, an appropriate test is *not* how well patterns of a single ecosystem component, such as fish species richness or total phosphorus in streams, match ecoregions. The work of Larsen et al. (1988) in Ohio showed that the patterns of any one chemical parameter often do not demonstrate the effectiveness of ecoregions in that state. However, when the chemistry portion of water quality was illustrated for the Ohio reference sites using a principal components analysis, with a combination of components comprising nutrient richness on one axis and a combination of components comprising ionic strength on the other axis, the ecoregion patterns became quite clear. Similarly, methods of grouping biotic characteristics to express biotic integrity, such as the index of biotic integrity (IBI) (Karr et al. 1986), have effectively shown ecoregion patterns (Larsen et al. 1986). Because of the nature of ecoregions, the ideal way of evaluating them would be through use of an ecological index of integrity. Such an index has yet to be developed and would need to be regionally calibrated. Hence, there is necessarily some circularity in the evaluation process.

It must be recognized that the concept and definition of ecoregions are in a relatively early stage of development. The USEPA Science Advisory Board (SAB 1991), in their evaluation and subsequent endorsement of the ecoregion concept, strongly recommended further development of the framework, including collaboration with states regarding the subdivision of ecoregions, definition of boundary characteristics, and evaluation of the framework for specific applications. They saw the need for research to better understand the process by which the regions are defined, and how quantitative procedures could be incorporated with the currently used, mostly qualitative methods to increase replicability. To date, qualitative methods, although used for many applications where the usefulness of the results is more important than the scientific rigor of the technique used, have not been widely accepted. Research must be conducted to demonstrate how the two approaches are complementary. We need to examine the use of art in science, rather than assuming an either/or scenario. As we increase our awareness that a holistic ecosystem approach to environmental resource assessment and management is necessary, we must also develop a clearer understanding of ecosystems and their regional patterns. Essential to this is the development of ecological regions and indices of ecosystem integrity.

ACKNOWLEDGMENTS

The ecoregions/reference site concept has been developed through the efforts of many people, too many to acknowledge individually here. The strength and usefulness of the framework lies in the collaborative way it has been and is being developed — closely tied to its applications. State and federal resource managers and scientists who helped generate, carry out, evaluate, and otherwise contribute to the various "ecoregions" projects should receive much of the credit for the success of those projects. While not wishing to slight my fellow geographers, and the soils scientists and geologists who have all made invaluable contributions, I feel a special sense of gratitude to the biologists and ecologists for their ideas, mental maps, and understanding. Especially deserving of acknowledgement are the people my geographer colleagues and I have learned from and worked with at the USEPA laboratory in Corvallis.

## CHAPTER 6

Multimetric Approach for Establishing Biocriteria and Measuring Biological Condition*

Michael T. Barbour, James B. Stribling and James R. Karr

1.0 INTRODUCTION

The accurate assessment of biological condition requires a method that integrates biotic responses through an examination of patterns and processes from individual to ecosystem levels (Karr et al. 1986). Classical approaches select some biological attribute that refers to a narrow range of perturbations or conditions (Gray 1989; Karr 1991) such as reduced dissolved oxygen, biological oxygen demand, or selected toxicants (Karr 1991). Karr also pointed out that some techniques were used solely to detect fecal contamination, while others were limited to causal effects of chemical stress on organismal or population levels. The indicator species concept has dominated biological evaluations (Kremen 1992). Ecological studies typically focus on a limited number of parameters that might include one or more of the following: species distributions, abundance trends, standing crop, and production estimates. Invariably, parameters are interpreted separately with a summary statement about overall health. Just as single species toxicity testing is low on "environmental realism" because it ignores system-level responses (Buikema and Voshell 1993), these narrow ecological approaches may not be reflective of overall ecological health. Although valuable for measurement of selected anthropogenic effects, these approaches are not successful in screening for all types of degradation, including complex cumulative impacts (Karr 1991).

In contrast, an alternate approach is to define an array of measures or metrics that individually provide information on diverse biological attributes, and when integrated, provide an overall indication of biological condition. Gray (1989) states that the three best-documented responses to environmental stressors are (1) reduction in species richness, (2) change in species composition to dominance by opportunistic species, and (3) reduction in mean size of organisms. However, because each feature responds differently to different stressors (Gray 1989; Kelly and Harwell 1990), the best approach is to incorporate many attributes (for the multimetric approach) into the assessment process. The strength of a multimetric approach is its ability to integrate information from individual, population, community, and ecosystem levels and to allow evaluation with reference to biogeography as a biologically based indicator of water resource quality (Karr et al. 1986; Plafkin et al. 1989; Karr 1991; Karr and Kerans 1991). In combination, strengths of individual metrics, when integrated, minimize weaknesses they may have individually (Ohio EPA 1987a,b).

For the broad range of human impacts, a comprehensive, multiple metric approach is more appropriate. A metric is a calculated term or enumeration representing some aspect of biological assemblage

* This paper is, in large part, taken from Biological Criteria: *Technical Guidance for Streams and Small Rivers* (Gibson 1994). Support for the technical guidance is from the U.S. Environmental Protection Agency, Office of Science and Technology, Contract 68-C0-0093.

structure, function, or other measurable characteristic that changes in some predictable way with increased human influence (Fausch et al. 1990; Gibson 1994). Similarly, each assemblage (e.g., fish, macroinvertebrates, and periphyton) composing the aquatic community would be expected to have a response range to perturbation events or degraded conditions. Thus, biosurveys targeting multiple species and assemblages are more likely to provide improved detection capability over a broader range as well as protection to a larger segment of the ecosystem (Kremen 1992) than a single species or assemblage approach.

The multimetric concept was first successfully used with the fish Index of Biotic Integrity (IBI) (Karr 1981). The IBI aggregates various elements and surrogate measures of process into a meaningful assessment of biological condition. Karr (1981) and Karr et al. (1986) demonstrated that combinations of these attributes or metrics provide valuable synthetic assessments of the status of water resources.

Consistent routines for normalizing individual metric values provide a means of combining scores across metrics despite their initially dissimilar values. The aggregated number, or index, is used to judge condition of the biota. However, final decisions on management actions are not made on the single, aggregated number alone. Rather, if comparisons to established reference values indicate an impairment in biological condition, component parameters (or metrics) are examined for their individual effects on the aggregated value. This approach lends itself to a better understanding of the nature of the impairment, including which elements or processes of the community are most affected. This often provides insight about the factor(s) responsible for degradation and offers a diagnostic capability (see Yoder and Rankin, Chapter 17).

The focus of this discussion is primarily on stream systems, because that is the waterbody type where most of the developmental work has been done, to date. However, the concept and theoretical application to other waterbody types should be valid. Current research is being conducted to test the efficacy of this application to lakes, reservoirs, estuaries, and large rivers (Ohio EPA 1987a; Karr and Dionne 1991; Masters 1992; Gerritsen et al. 1994a; Gerritsen and Bowman 1994).

2.0 CONCEPTUAL FRAMEWORK OF METRICS

The validity of an integrated assessment using multiple metrics is supported by the use of measurements of biological attributes firmly rooted in sound ecological principles (Karr et al. 1986; Fausch et al. 1990; Karr 1991; Lyons 1992). The status of the biota as indicated by a number of appropriate attributes provides an accurate reflection of the biological condition at a study site. Large numbers of attributes have been used (e.g., see Karr et al. 1986; Fausch et al. 1990; Kay 1990; Noss 1990; Karr 1991). The use of each is essentially based on an hypothesis about the relationship between instream condition and human influence (Fausch et al. 1990).

A broad approach for development of metrics might be modeled after Holland (1990), Fausch et al. (1990), Paulsen et al. (1990), Karr and Kerans (1991) or Barbour et al. (1992). Candidate metrics are selected based on knowledge of aquatic systems, flora and fauna, literature reviews, and historical data (Figure 1). Candidate metrics are then evaluated. Less robust metrics, or those not well-founded in ecological principles, are excluded as a result of this research process. Metrics with little or no relationship to stressors are rejected. Core metrics provide useful information in discriminating among sites exhibiting either good or poor quality ecological conditions. Metrics based on the relative sensitivity of the monitored populations to specific pollutants, where these relationships are well characterized, can be useful diagnostics tools (see Yoder and Rankin, Chapter 17). Core metrics should be selected to represent diverse aspects of structure, composition, individual health, or processes of the aquatic biota (Figure 2). Together they form the foundation for a sound, integrated analysis of the biotic condition.

For a metric to be useful, it must be (1) relevant to the biological community under study and to the specified program objectives; (2) sensitive to stressors; (3) able to provide a response that can be discriminated from natural variation; (4) environmentally benign to measure in the aquatic environment; and (5) cost-effective to sample. Thus, metrics reflecting biological characteristics may be considered as appropriate in biocriteria programs if their relevance can be demonstrated, response range is verified and documented, and the potential for application in water resource programs exists. Regional variation in metric details are expected but the general principles used in defining metrics seem consistent over wide geographic areas (Miller et al. 1988).

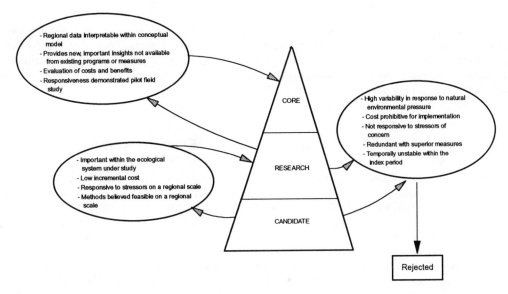

Figure 1. Tiered metric development process. (Modified from Holland 1990.)

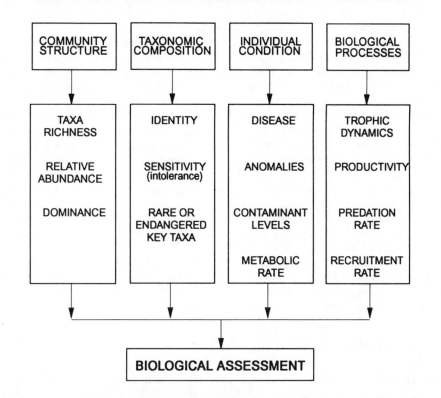

Figure 2. Organizational structure of the types of attributes that should be incorporated into biological assessment.

Pilot studies or small-scale research may be needed to define, evaluate, and calibrate metrics. Past efforts to evaluate the utility of individual metrics illustrate procedural approaches for evaluating the validity of a set of metrics (Angermeier and Karr 1986; Karr et al. 1986; Boyle et al. 1990; Davis and Lubin 1989; Karr and Kerans 1991; Barbour et al. 1992; Kerans et al. 1992; Kremen 1992; Lyons 1992; Resh and Jackson 1993; Fore et al., 1994; Kerans and Karr, 1994).

Perpetual evaluation of metrics and indices is an essential feature of the use of biocriteria; however, existing multimetric approaches are robust in their ability to measure biological condition because they: (1) rely on information about multiple attributes of biological systems; (2) relate expectations defined by minimally disturbed reference condition; and (3) incorporate biological and ecological principles that enable an interpretation of exposure/response relations.

3.0 POTENTIAL METRICS FOR DIFFERENT TARGET ASSEMBLAGES

A number of metrics have been developed and subsequently tested in field surveys of benthic macroinvertebrate and fish assemblages (Karr 1991). Because metrics have been recommended for fish assemblages (Karr 1981; Karr et al. 1986; Simon and Lyons, Chapter 16) and for benthic macroinvertebrates (Plafkin et al. 1989; Karr and Kerans 1991; Barbour et al. 1992; Resh and Jackson 1993; DeShon, Chapter 15; Kerans and Karr, 1994), they will not be reviewed extensively here. Rosen (Chapter 14) has reviewed proposed metrics for the periphyton assemblage.

Fish assemblage metrics (Table 1) are generally grouped into three classes (Karr 1981, 1991): (1) species richness and composition, (2) trophic composition, and (3) abundance and condition. These classes encompass and extend the characteristics identified by Gray (1989). Since the development of the IBI, which is the earliest of the multimetric indices, fish metrics have been extensively tested and evaluated in various regions of the country. One of the unique features of the IBI presently lacking in benthic indices is the ability to incorporate metrics on individual condition, although chironomid larvae deformities have recently been advocated (Lenat 1993).

Benthic metrics have undergone similar evolutionary developments and are documented in the Invertebrate Community Index (ICI) (DeShon, Chapter 15), Rapid Bioassessment Protocols (RBPs) (Shackelford 1988; Plafkin et al. 1989; Barbour et al. 1992; Hayslip 1993), and the benthic IBI (Kerans and Karr, 1994). Metrics used in these indices evaluate aspects of both elements and processes within the macroinvertebrate assemblage. Although these indices have been regionally developed, they are typically appropriate over wide geographic areas with minor modification. Selected metrics are listed in broad classes for each index (Table 2). Resh and Jackson (1993) tested the ability of 20 metrics used in 30 different assessment protocols for the benthic assemblage to discriminate between impaired and unim- paired sites in California. The most effective measures, from their study, were the richness measures, two community indices (Margalef's and Hilsenhoff's family biotic index), and a functional feeding group metric (percent scrapers). Resh and Jackson emphasized that both the measures (metrics) and protocols need to be calibrated for different regions of the country, and, perhaps, for different impact types (stressors). In a study of 28 invertebrate metrics, Kerans and Karr (1994) demonstrated significant patterns for 18 metrics and used 13 in their final B-IBI. Richness measures were useful as were selected trophic and dominance metrics.

Figure 2 illustrates a conceptual structure for attributes of biotic assemblages, which in an integrated assessment reflect overall biological condition. Comparison to reference conditions is essential to evaluate the extent to which study sites are influenced by human actions. A number of these attributes can be characterized by metrics within four general classes.

3.1 Community Structure

Taxa richness, or the number of distinct taxa, represents the diversity within a sample. Use of taxa richness as a key metric in a multimetric index include the ICI (DeShon, Chapter 15), the fish IBI (Karr et al. 1986), the benthic IBI (Kerans et al. 1992; Kerans and Karr, 1994), and RBP's (Plafkin et al. 1989). Taxa richness is also recommended as critical information in assays of natural phytoplankton assem- blages (Schelske 1984) and periphyton assemblages (Bahls 1993). Taxa richness usually consists of species level identifications but can also be evaluated as designated groupings of taxa, often as higher taxonomic groups (i.e., genera, families, orders, etc.) in assessment of invertebrate assemblages.

Table 1.   Fish IBI Metrics Used in Various Regions of North America

Alternative IBI metrics	A	B	C	D	E	F	G	H
1. Total number of species	X	X		X	X		X	
# Native fish species	X		X			X		X
# Salmonid age classes[a]					X	X		
2. Number of darter species	X			X	X			
# Sculpin species						X		
# Benthic insectivore species								
# Darter and sculpin species	X	X						
# Salmonid yearlings (individuals)[a]		X				X	X	
% Round-bodied suckers	X							
# Sculpins (individuals)							X	
3. Number of sunfish species	X				X			X
# Cyprinid species						X		
# Water column species		X						
# Sunfish and trout species				X				
# Salmonid species							X	
# Headwater species	X							
4. Number of sucker species	X	X				X		X
# Adult trout species[a]						X	X	
# Minnow species	X				X			
# Sucker and catfish species			X					
5. Number of intolerant species	X	X			X	X		X
# Sensitive species	X							
# Amphibian species							X	
Presence of brook trout				X				
6. % Green sunfish	X							
% Common carp						X		
% White sucker		X			X			
% Tolerant species	X							X
% Creek chub				X				
% Dace species			X					
7. % Omnivores	X	X	X	X	X			X
% Yearling salmonids[a]					X	X		
8. % Insectivorous cyprinids	X							
% Insectivores		X				X		X
% Specialized insectivores				X	X			
# Juvenile trout							X	
% Insectivorous species	X							
9. % Top Carnivores	X	X	X					X
% Catchable salmonids						X		
% Catchable trout							X	
% Pioneering species	X							
Density catchable wild trout							X	
10. Number of individuals	X		X	X	X	X	X	X[b]
Density of individuals		X						
11. % Hybrids	X	X						
% Introduced species					X	X		
% Simple lithophills	X							X
# Simple lithophills species	X							
% Native species							X	
% Native wild individuals							X	
12. % Diseased individuals	X	X	X	X	X	X		X

Note: X = metric used in region. Many of these variations are applicable elsewhere.

Key: A, Midwest; B, New England; C, Ontario; D, Central Appalachia; E, Colorado Front Range; F, Western Oregon; G, Sacramento-San Joaquin; H, Wisconsin.

[a] Metric suggested by Moyle or Hughes as a provisional replacement metric in small western salmonid streams.
[b] Excluding individuals of tolerant species.

Taken from Karr et al. (1986), Hughes and Gammon (1987), Ohio EPA (1987a,b), Miller et al. (1988), Steedman (1988), and Lyons(1992).

Table 2. Examples of Metric Suites Used for Analysis of Macroinvertebrate Assemblages

Alternative benthic metrics	ICI[a]	RBP[b]	RBP[c]	RBP[d] ID	RBP[d] OR	RBP[d] WA	B-IBI[e]
1. Total No. Taxa	X	X	X	X	X	X	X
% Change in Total Taxa Richness				X	X	X	
2. No. EPT Taxa	X	X		X	X	X	
No. Mayfly Taxa	X						X
No. Caddisfly Taxa	X						X
No. Stonefly Taxa							X
Missing Taxa (EPT)			X				
3. No. Diptera Taxa	X						
No. of Chironomidae Taxa				X		X	
4. No. Intolerant Snail and Mussel Species							X
5. Ratio EPT/Chironomidae Abund.				X	X	X	
Indicator Assemblage Index			X	X	X		
% EPT Taxa				X			
% Mayfly Composition	X						
% Caddisfly Composition	X						
6. % Tribe Tanytarsini	X						
7. % Other Diptera and Noninsect Composition	X						
8. % Tolerant Organisms	X						
% Corbicula Composition							X
% Oligochaete Composition							X
Ratio Hydropsychidae/Tricoptera		X			X		
9. % Ind. Dominant Taxon		X		X	X	X	
% Ind. Two Dominant Taxa							X
Five Dominant Taxa in Common		X	X		X		
Common Taxa Index			X				
10. Indicator Groups				X		X	
11. % Ind. Omnivores and Scavengers							X
12. % Ind. Collector Gatherers and Filterers							X
% Ind. Filterers				X		X	
13. % Ind. Grazers and Scrapers				X			X
Ratio Scrapers/Filterer Collectors				X	X	X	
Ratio Scrapers/(Scrapers + Filterer Collectors)		X					
14. % Ind. Strict Predators							X
15. Ratio Shredders/Total Ind. (% shredders)		X		X		X	
16. % Similarity Functional Feeding Groups (QSI)		X	X				
17. Total Abundance				X			
18. Pinkham-Pearson Community Similarity Index	X						
Community Loss Index					X	X	
Jaccard Similarity Index				X			
19. Quantitative Similarity Index (Taxa)		X	X				
20. Hilsenhoff Biotic Index		X		X	X	X	
Chandler Biotic Score				X			
21. Shannon-Weiner Diversity Index					X		
Equitability				X			
Index of Community Integrity				X			

[a] Invertebrate Community Index, Ohio EPA (1987b, DeShon (Chapter 15)).
[b] Rapid Bioassessment Protocols, Barbour et al. (1992) revised from Plafkin et al. (1989).
[c] Rapid Bioassessment Protocols, Shackelford (1988).
[d] Rapid Bioassessment Protocols, Hayslip (1993); ID = Idaho, OR = Oregon, WA = Washington.
  (Note: these metrics in ID, OR, and WA are currently under evaluation).
[e] Benthic Index of Biotic Integrity, Kerans et al. (1992).

Relative abundance of taxa refers to the number of individuals of one taxon as compared to that of the whole community. The proportional representation of taxa is a surrogate measure for community balance that can relate to both contaminant and enrichment problems. **Dominance** (e.g., measured as "percent composition of dominant taxon"; Barbour et al. 1992) or dominants-in-common (Shackelford 1988) is an indicator of community balance or lack thereof. Dominance is an important indicator when the most sensitive taxa are eliminated from the assemblages and/or the food source is altered, thus allowing the more tolerant taxa to become dominant.

3.2 Taxonomic Composition

Taxonomic composition can be characterized by several classes of information. **Identity** is the knowledge of individual taxa and associated ecological patterns and environmental requirements. Key taxa (i.e., those that are of special interest or ecologically important) provide information that is important to the condition of the targeted assemblage. The presence of exotics or nuisance species may be an important aspect of biotic interactions that relates to both identity and sensitivity. **Sensitivity** refers to the numbers of pollution-tolerant and -intolerant species in the sample. The ICI and RBPs each use a single metric based on species tolerance values (percent tolerant individuals and the HBI, respectively). A similar metric for fish assemblages is included in the IBI (number of intolerant species) (Table 1). Recognition of those taxa considered to be threatened and endangered provides additional legal support for remediation activities or recommendations. Species status for response guilds of bird assemblages as being threatened or endangered, their endemicity, or of some commercial or recreational value, also relates to the composition class of metrics (Brooks et al. 1991).

3.3 Individual Condition

Individual condition metrics focus on chronic exposure to chemical contamination. The condition of individuals can be rated by observation of either physical/morphological or behavioral characteristics. Physical characteristics of individuals that may be useful for assessing chemical contaminants would result from microbial or viral infection, some sort of teratogenic or carcinogenic effects during development of that individual. These would be categorized as **diseases, anomalies,** or **metabolic processes (biomarkers)**. Metrics of this nature have been implemented successfully in fish multimetric indices. A metric of individual condition is used for fish in the IBI as "percent diseased individuals" (Table 1). Possibilities exist for benthic macroinvertebrates, such as insect larval head capsule abnormalities or aberrant net-spinning activities of certain caddisflies, but these metrics are currently cost-prohibitive.

The underlying concept of the biomarkers approach in biomonitoring is that contaminant effects occur at the lower levels of biological organization (i.e., at the genetic, cell, and tissue level) before more severe disturbances are manifested at the population or ecosystem level (Adams et al. 1990). However, biomarkers may provide a valuable complement to ecological metrics if they are of pollutant-specific nature, responsive to sublethal effects, and the time and financial costs for measurement are consistent with available resources. Unusual behavior regarding locomotory, reproductive, or feeding activities is often an indication of physiological or biochemical stress. Oftentimes, behavior measures are difficult to assess in the field. McCarthy (1990) briefly discussed several studies that have shown biomarker responses correlate with predicted levels of contamination and with site rankings based on community level measures of ecological integrity. Approaches for assessing the response of stress proteins in organisms when exposed to chemicals are currently under development, but they must be cost-effective before they are likely to be widely used in biological assessment and criteria programs.

3.4 Biological Processes

Biological processes can be divided into several categories for consideration as potential metrics. **Trophic dynamics** encompasses functional feeding groups, and measures the condition of the food web for the system. Examples involve the relative abundance of herbivores, carnivores, and detritivores. Without relatively stable food dynamics, populations of the top carnivore, for example, reflect stressed conditions. Likewise, if **production** of a site is considered high based on organism abundance and/or biomass, and high production is natural for the habitat type under study (as per reference conditions), biological condition would be considered good.

Process metrics are available for several assemblages. For example, Table 1 indicates at least seven fish IBI metrics dealing with trophic status or feeding behavior in fish, focusing on insectivores, omnivores, predators, or herbivores. Also, number or density of individuals of fish in a sample (or an estimate of standing crop) might be considered a measure of production, and thus, considered in the process class of metrics. Additional information is gained from density measures when considered relative to size or age distribution. Three RBP metrics for benthic macroinvertebrates focus on functional feeding groups as surrogate measures of trophic status (Table 2) (Plafkin et al. 1989; Barbour et al. 1992). Brooks et al. (1991) use trophic level as one category for rating avian assemblages.

It is not necessary to establish metrics for every attribute of the targeted assemblage. However, the integration of information from several metrics, especially a grouping of metrics representative of the four major classes of attributes (Figure 2), improves and strengthens the bioassessment process. These metrics have a strong ecological foundation and enable the biologist to determine attainment or nonattainment of biological criteria.

4.0 MODEL FOR DEVELOPMENT AND AGGREGATION OF METRICS

The development of metrics for use in the biocriteria process can be partitioned into two phases. First, an **evaluation** of metrics is necessary to eliminate nonresponsive metrics and to address various technical issues (i.e., associated with methods, sampling habitat and frequency, etc.) relevant to the particular waterbody under study. Second, **calibration** of the metrics determines the discriminatory power of each metric and identifies thresholds for discriminating between "good" and "bad" sites. This process defines a suite of metrics that are optimal candidate for inclusion in bioassessments. Subsequently, a procedure for aggregating metrics to provide an integrative index is needed.

A conceptual model for processing biological data from the initial measurements through aggregation of metric values and scores into a composite index is provided by Paulsen et al. (1990) (Figure 3). This systematic process involves discrete steps, which are described as follows:

4.1 Step 1: Classification

A critical issue is to determine the regional extent over which a particular biological attribute is applicable. The procedure by which this is done can be viewed as part of the regionalization process. It is better to identify large areas over which calibrations will be performed rather than small areas. This allows regional, and perhaps, subregional, patterns to emerge if present. Streams are classified as described by Hughes (Chapter 4) and Omernik (Chapter 5). Stream classification provides relatively homogenous classes of streams for which biocriteria may differ among the classes. The best and most representative sites for each stream class are selected and represent the set of reference sites from which the reference condition is established. Regional scale modifications in management practices may be feasible, allowing for the significant recovery of impaired aquatic resources. Criteria for degraded regions may be set at different levels than those set for other areas, although we generally discourage that practice. By calibrating over large areas, decisions about level of protection can be made with a clear understanding of the differences among areas over which differing criteria might be established. The risk of error due to inappropriate classification across heterogeneous regions should be avoided.

4.2 Step 2: Survey of Biota and Habitat

Biological surveys of minimally impaired sites are conducted; these should be supplemented with habitat and some water quality data. Sampling from a gradient of conditions permits metric calibration and discrimination. Use of standardized methods is essential. Data are evaluated within the ecological context (waterbody type and size, season, geographic location, and other elements) that defines what is expected for similar waterbodies (Paulsen et al. 1990).

4.3 Step 3: Data Evaluation and Metric Calibration

4.3.1 Data Evaluation

Analysis of the biological data emphasizes the evaluation of biological attributes that represent the elements and processes of the "natural" community (Figure 2). Expected metric values vary as a function of species pools that form the colonizing potential for streams, regional characteristics (climate, geology,

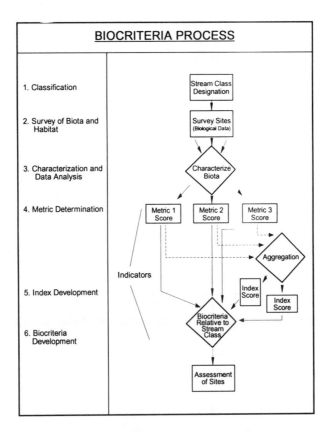

Figure 3. The conceptual process of proceeding from measurements to indicators of condition. (Modified from Paulsen et al. 1990.)

soils, land use, and regional scale barriers to colonization), and local site characteristics (habitat factors, including local barriers). By combining an examination of regional species pools, large watersheds, and important barriers to species movement, it should be feasible to identify calibration regions for biological metrics. Multi-state collaboration is encouraged in the development of these calibration regions; a benefit is that common methods and metrics can be established among states, and cross-state comparisons are enhanced.

Those metrics with a monotonic response, i.e., a linear change in value against a gradient of conditions (number of species in Figure 4), are often the best candidates for assessing impairment. Metrics that are not monotonic still may be informative metrics (intolerant species in Figure 4), but must be evaluated in the context of other metrics or known gradations of environmental contamination. Ambient sites other than reference sites should be surveyed as part of the database. Subsequent calibration of the metrics must address the ability to differentiate between impaired and nonimpaired sites.

4.3.2 Calibration of Metrics

Once sites are classified and the reference condition is established from a compiled set of reference sites, the expectations for each metric can be defined. Certain metrics may exhibit a continuum of expectations dependent on specific physical attributes of the reference streams. When different stream classes have different expectations in metric values a plot of survey data for each stream class may be useful to display the central tendency and selected percentiles for each metric (Figure 5).

For each metric, the sites are sorted by stream class (e.g., ecoregion and stream type) and plotted to define the spread in data and the ability to discriminate among classes. If such a representation of the data does not allow discrimination of the classes, then it will not be necessary to separate classes. That is, in such cases, the overlap of the metric values indicates that the same range will be applicable to a set of

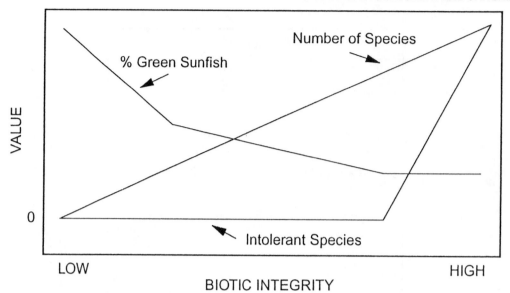

Figure 4. Conceptual depiction of the range of sensitivity of IBI metrics across the gradient from low to high biotic integrity. (From Karr, J. R. 1993. *Ecological Integrity and the Management of Ecosystems,* Woodley, Kay, and Francis, Eds., St. Lucie Press, 83–104. With permission.)

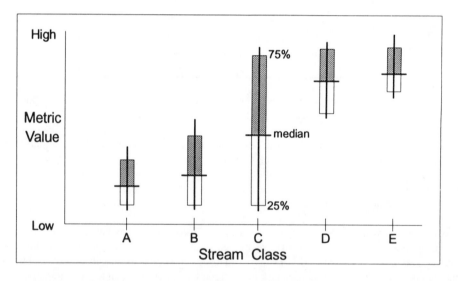

Figure 5. Box and whisker plots of metric values from hypothetical stream classes.

sites that represent different physical classes. Conversely, if differences in the biological attribute are apparent and appear to correspond to the classification, then separate classes are necessary to minimize the variability in each metric. This technique is especially useful if the covariates are unknown or do not exist, but a difference in stream class is apparent (Figure 6). The evaluation and calibration of the individual metrics may depend on whether the classification process was based on the best available sites or randomly selected sites. Historical data may present a problem because the site selection generally is not conducive to either of these processes.

4.4 Step 4: Metric Transformation

The use of a variety of metrics representing different biological responses to stress is a major advance in recent programs in biological monitoring. Combining unlike measurements is only possible when the

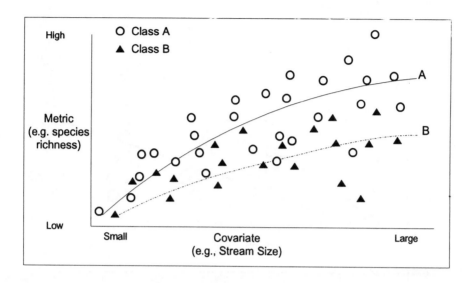

Figure 6. Metrics plotted with a continuous covariate (hypothetical example).

values have been "standardized" or normalized by a transformation through which measurements become unitless (Schuster and Zuuring 1986). Standardization of these measurements into a logical progression of scores is the typical means for comparing and interpreting unlike metric values. For each metric, the range of possible values is divided into scoring sections from some selected percentile. The divisions of this range correspond to increasing deviation from the expected value for a respective reference condition. Thus, the highest score would be assigned to the closest approximation of the reference values, and consecutive lower scores would be assigned to progressively lower metric values.

Within a class of streams, sorting biological data along an environmental gradient can be done for a composite of good and poor sites, thus providing a means for discrimination. For example, the total number of fish species changes as a function of stream size estimated by stream order or watershed area for a number of good and poor sites (Fausch et al. 1984). The authors showed that when these data are plotted, the points produce a distinct right triangle, the hypotenuse of which approximates the upper limit of species richness. Fausch et al. (1984) suggest that a line with a slope fit to include about 95% of the sites is an appropriate approximation of a maximum line of expectations for the metric in question and identifies the upper limit of the reference condition. The area on the graph beneath the maximum line can then be trisected or quadrisected to assign scores to a range of metric values as illustrated in Figure 7. The scores provide the transformation of values to a consistent measurement scale to group information from several metrics for analysis.

An alternative is when it is assumed that most, if not all, sites represent nonimpaired or minimally impaired conditions. The 50th percentile (values above the median) or a lower percentile such as the 25th (Ohio EPA 1987a,b) can be used as representative of the highest condition for each metric. This upper range would receive the maximum score and quartiles below the median would receive progressively lower scores. This approach is conducive to metrics that may have a modal response other than monotonic because upper bounds on the expected condition can be established. For example, taxa richness may be best in a region when the number of taxa are hypothetically between 25 and 35. However, in a nutrient enrichment situation, the number of taxa may increase to 37. In this approach, this particular condition would be assessed as nutrient enrichment.

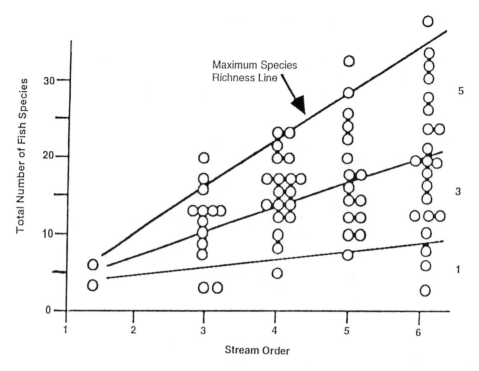

Figure 7.   Total number of fish species vs. stream order for 72 ambient sites along the Embarras River in Illinois. (From Fausch, D. O., J. R. Karr, and P. R. Yant. 1984. *Transactions of the American Fisheries Society* 113: 39–55. With permission.)

4.5 Step 5: Index Development

Following the transformation from values of various metrics to scores, an aggregation of metrics into an index can be accomplished. The index, such as the IBI, in turn, is calibrated by stream class to become part of the final assessment. Based on the IBI scores calculated for the surveyed reference sites, a range of expectations can be obtained. As an example of a regional calibration of the IBI, Ohio EPA determined the best expectations for each of its five ecoregions (Figure 8) for a class of stream sites, which ranged from 20 to 554 square miles drainage area. Using a "box and whisker" plot display, they illustrated that two of the ecoregions (i.e., Huron/Erie Lake Plains [HELP] and Western Allegheny Plains [WAP]) had IBI expectations distinctly different from the others. They then selected the 25th percentile of the IBI for their reference sites as the threshold for biocriteria for their warmwater aquatic life use (Yoder and Rankin, Chapter 9). Tiered aquatic life uses may best be addressed by various percentiles. For instance, Ohio EPA used the 75th percentile for their statewide database to establish a biocriteria threshold for their Exceptional Aquatic Life Use.

As described by Karr et al. (1986) and Plafkin et al. (1989), the range of pollution sensitivity exhibited by each metric differs among metrics; some are sensitive across a broad range of biological conditions, others only to part of the range (Figure 4). Those metrics that are sensitive (i.e., exhibit response) to changes in relatively unperturbed conditions are important indicators of the attainment of biological integrity. Metrics that are relatively insensitive at higher levels of biological condition, but have a detectable response in areas of lower biological condition, are important in the discrimination among impaired waters. This ability to discriminate among impaired sites provides management with important information to set priorities. Overlap in the ranges of sensitivity of the metrics helps to reinforce final conclusions regarding biological condition (Karr 1991), while metrics that are better able to differentiate responses at the extremes of the range of impairment enable a more complete bioassessment (Plafkin et al. 1989). The integrated multimetric analysis approach thus allows a broader assessment of condition than an analysis using a single metric. However, information from individual metrics is used to enhance the interpretative power of the integrated assessment.

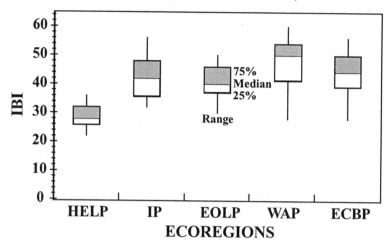

Figure 8.   Box and whisker plot of Ohio reference sites for the Index of Biotic Integrity.

Certain metrics are more responsive to specific stressors than to others. Unique combinations of biological community characteristics that identify one impact type over others are referred to as "response signatures" (Yoder 1991; Yoder and Rankin, Chapter 17). These have proved valuable in assigning causes and sources to water resource impairments noted by Ohio EPA. In some benthic protocols, the Hilsenhoff Biotic Index serves as a metric and uses a tolerance classification based on organic pollution effects; functional groups respond to either organics or toxicants and are useful for evaluating chemical contamination from nonpoint and point-source impacts (Plafkin et al. 1989).

Fish metrics offer similar information. Municipal effluents typically affect total abundance and trophic structure (Karr et al. 1986). Bottom-dwelling species (e.g., darters and sculpins) that depend on benthic habitats for feeding and reproduction are particularly sensitive to the effects of siltation and benthic oxygen depletion (Kuehne and Barber 1983; Ohio EPA 1987b).

4.6 Step 6: Biocriteria Development

Biocriteria are formulated from the indices and adjusted by stream classes and designated aquatic life uses. The biocriteria may be based on a single aggregated index (e.g., Ohio, Yoder and Rankin, Chapter 9) or established for several biological end points (e.g., New York, Bode and Novak, Chapter 8). The essential element of an effective biocriterion is a threshold from which a discrimination of impaired sites can be made. The confidence with which a judgment of biological condition can be made using biocriteria rests with the soundness and scientific validity of the metrics selected and tested throughout this process. Metrics that are poorly defined or based on a flawed conceptual basis provide erroneous judgments with the potential for erroneous management decisions.

5.0 DECISIONS FOR ESTABLISHING BIOCRITERIA THRESHOLDS

The general approach that seems to have evolved for the selection of a quantitative regional biocriterion is to choose a percentile along the distribution of indicator scores representative of minimally-impaired sites. To do this, a variety of decisions must be made, not all of which are technical or scientific in nature. One of the decisions is the selection of the measure of central tendency of the database as a numeric criterion, such as the median of the distribution of values, the 25th percentile, the 75th percentile, or some other percentile. As stated previously, the selection of the appropriate criterion depends on the content of the database. If only reference sites are included, a lower measure (e.g., 25th or 50th percentile) would be appropriate. However, if the database includes both impaired and unimpaired sites, a higher measure (e.g., 75th, 90th or 95th) would identify the best sites within the database. It should be noted that different relative "mixes" of impaired and unimpaired conditions could result in different biocriteria. This approach is best done when a relatively thorough sampling of the region can be accomplished. The criterion is best characterized by both a percentile and an expression of the variation among the scores of the reference sites. Incorporation of a measure of variation into the criterion allows distinction between criteria that have been established with low variation and criteria established with high variation. This could be important in determining how well real impairment can be reasonably expected to be detected. A recent article describes some approaches for defining how well various percentiles can be estimated and notes pitfalls in the use of extreme percentiles for these purposes (Berthouex and Hau 1991).

A second decision is to define the set of sites to which a biocriterion applies. Calibration procedures, as noted above, allow normalization of the effects of stream size, so index scores, such as the IBI, can be compared among streams of different sizes. It is important to define the regional extent over which a particular biocriterion is applicable. One may be tempted to establish biocriteria by ecoregion or subregion by choosing a particular percentile (such as the 75th percentile) that will be applied to the set of scores derived from minimally impaired sites in each ecoregion. An implicit assumption in this approach is that the relative condition of reference sites within each region is similar, and that a constant percentile is as protective in one region as it is in another. This assumption has not been examined in any detail. It might be more reasonable to argue for differing percentiles for different regions depending upon the nature of regional-scale degradation.

Examining the variance structure can give insight into the extent over which particular biocriteria might be applicable. If conditions are similar among regions, or if the differences can be associated with management practices that have a chance of being altered, it seems wise to combine the regions for the purposes of establishing biocriteria. If a strong case cannot be built for subregion-scale biocriteria, it is probably better to err on the side of overprotection rather than underprotection at this stage of biocriteria development by setting biocriteria over broad regions. Procedures should be developed that allow for both regional- and subregional-scale deviations from broadly established biocriteria. These procedures could be based upon building a weight of evidence case for modifications.

It will be necessary to describe some site-specific rules of exception to regional-scale biocriteria because of local natural limitations that prevent the achievement of the regional biocriterion. Certain natural channel configurations, such as through-bedrock or above natural barriers to dispersal, do not offer the habitat diversity of other channel configurations, and therefore cannot support the richness and diversity of other channel types nearby. Need for establishment of modified biocriteria due to limitations in habitat quality will become evident in application of habitat assessment routines (e.g., Barbour and Stribling 1991, 1994; Rankin, Chapter 13). Many other natural restrictions to achievement can be identified, and care must be taken that actual degraded conditions are not included as evidence for regional scale biocriteria modification.

6.0 SUMMARY

The strength of a multimetric approach is its ability to integrate information from individual, population, community, and ecosystem levels into a biologically based indicator of water resource quality. Metrics individually provide information on diverse biological attributes and represent some aspect of biological assemblage structure, function, or other measurable characteristics that change in some predictable way with increased human influence. In combination, strengths of individual metrics, when integrated, minimize weaknesses they may have individually. For the broad range of human impacts, a comprehensive, multiple metric approach is most appropriate.

For a metric to be useful, it must be (1) relevant to the biological community under study and to the specified program objectives; (2) sensitive to stressors; (3) able to provide a response that can be discriminated from natural variation; (4) environmentally benign to measure in the aquatic environment; and (5) cost-effective to sample. A number of metrics have been developed and subsequently tested in field surveys of benthic macroinvertebrate and fish assemblages. Metrics for the periphyton assemblage are currently under development. Metrics are categorized in broad classes of community structure, taxonomic composition, individual condition, and biological processes.

The model for development and aggregation of metrics follows a stepwise process, which includes: (1) classification of reference sites to partition natural variability and to identify homogeneous regions, (2) survey of the biota to characterize the biological attributes that will serve as the basis for the metrics, (3) selection and evaluation of candidate metrics, then calibration of the core metrics for discrimination of impairment, (4) transformation of the metrics through calibration to normalized scores, (5) aggregation of the core metrics into an index, and (6) development of biocriteria thresholds for assessment. Several decisions are inherent in developing biocriteria thresholds. First is to define the set of sites to which a biocriterion applies. The site classification and metric calibration steps are important. Equally important is the designation of aquatic life uses for the waterbodies under investigation. Second, the selection of the measure of central tendency of the database, such as the median of the distribution of values, the 25th percentile, the 75th percentile, or some other value is important to delineate the numeric criterion. A third decision is to describe or identify site-specific rules of exception to regional scale biocriteria because of local natural limitations that prevent the achievement of the regional biocriterion.

The assessment of biological condition using a suite of metrics to define biocriteria is rapidly becoming the method of choice among state water resource agencies (Southerland and Stribling, Chapter 7). The development of appropriate metrics for these programs follows an adequate description of the taxa and assemblages to be sampled, the biological characteristics of relatively undisturbed sites in the region (the "reference condition"), and, to a certain extent, the anthropogenic influences being assessed.

ACKNOWLEDGMENTS

First and foremost, we thank Dr. George Gibson, USEPA for the lengthy discussions and his insights on the biocriteria process and appropriate bioassessment approaches. Dr. Jeroen Gerritsen, Tetra Tech, also provided valuable input to our discussions and helped formulate an organized framework for presenting the multimetric concept for the biocriteria process. Our thanks go out to the many state agency biologists who have applied and refined the multimetric approach and have strengthened the concept through their interactions and testing. Ms. Catherine Deli provided typing and patience throughout the entire preparation. Mr. Steve Lipham and Ms. Christiana Gerardi assisted the process with the graphics and various aspects of the biocriteria document that became a basis for this chapter.

SECTION II:
WATER RESOURCE
PLANNING AND DECISION-MAKING

## CHAPTER 7

Status of Biological Criteria Development and Implementation

Mark T. Southerland and James B. Stribling

1.0 INTRODUCTION

Current efforts to develop and implement biological criteria are closely related to independent state efforts to apply biological monitoring information to water resource quality assessments. The diversity of approaches and programs pursued by the states range from exploring the utility of biological criteria; to using the concepts of biological assessment and biological criteria to enhance water quality programs; to developing sophisticated biological assessment methods and incorporating numeric biological criteria into water quality standards. This diversity is due to: (1) the length of time each state has been involved in biological assessment and criteria, and (2) the specific approach to biological criteria chosen by each state to meet the constraints imposed by their natural resources and their regulatory programs. In each state, the development of effective bioassessment methods provides the basis for establishment of biological criteria. These criteria are either narrative or numeric thresholds that are used to determine whether bioassessment results indicate impairment of the aquatic community. In some states, these biological criteria are formalized into water quality standards regulations, while in others informal biological criteria are used to support general standards or other water resource management activities. Although substantial national efforts are underway to provide consistency in the implementation of biological criteria (USEPA 1990a, 1991h, 1991g, 1991e; Gibson 1994), the actual development and implementation of biological criteria remains an activity of state water quality programs.

The U.S. Environmental Protection Agency (USEPA) has recently prepared national guidance on developing and implementing biological criteria. In September 1987, USEPA published a management study entitled, *Surface Water Monitoring: A Framework for Change,* that strongly emphasized the need to accelerate the development and application of promising biological monitoring techniques in state and USEPA monitoring programs (USEPA 1987a). In December 1987, the National Workshop on Instream Biological Monitoring and Criteria advocated the same measures, but also stressed the importance of combining new biological criteria and assessment methods with traditional chemical and physical procedures (USEPA 1987b). Both recommendations were presented at the June 1988 National Sympo-sium on Water Quality Assessment, where a workgroup of representatives from several state and federal agencies unanimously agreed that a national bioassessment policy should be developed to encourage the expanded use of new biological tools and to direct their rational implementation across water quality programs. In the years preceding these recommendations, Karr and Dudley (1981) and Karr et al. (1986) developed an operational definition of biological integrity. They also provided a practical approach for assessing ecological condition using fish assemblage structure and function based on this definition. Using this conceptual framework, Ohio EPA (1987a,b; 1989a,b) developed specific field methods and data analysis procedures for biological criteria based on surveys of macroinvertebrates and fish. In 1989, USEPA developed and published *Rapid Bioassessment Protocols for Use in Streams and Rivers: Benthic*

0-87371-894-1/95/$0.00+$.50
© 1995 by CRC Press, Inc.

Macroinvertebrates and Fish (Plafkin et al. 1989). These documents provide practical bioassessment methods that are being used as the foundation for biological criteria programs in many states.

In April 1990, USEPA's Office of Water Regulations and Standards issued a policy statement encouraging states to develop biological criteria (USEPA 1991c) and simultaneously published *Biological Criteria: National Program Guidance for Surface Waters* (USEPA 1990a). Further guidance was provided by the Office of Water in March 1992 with the publication of *Procedures for Initiating Narrative Biological Criteria* (USEPA 1992c). Currently, an implementation strategy is being developed to provide additional guidance to the states (USEPA 1993a). USEPA has also focused considerable effort on developing specific technical guidance. Two supporting technical documents are now available: *Biological Criteria: Guide to the Technical Literature* (USEPA 1991g) and *Biological Criteria: Research and Regulation-Proceedings of a Symposium* (USEPA 1991e). A third document, *Biological Criteria: State Development and Implementation Efforts* (USEPA 1991h), is also available and provides more detailed case studies than are given in this chapter. The ultimate objective of these USEPA activities is to produce a detailed technical guidance document for each waterbody type (streams, reservoirs and lakes, large rivers, estuaries and near-coastal waters, and wetlands). The first of these documents, *Biological Criteria: Technical Guidance for Streams and Small Rivers* (Gibson 1994), has been reviewed by the USEPA's Science Advisory Board (SAB 1993) and is now available.

As USEPA has been preparing formal guidance on biological criteria, the agency has also been sponsoring many regional and national workshops to facilitate state efforts to develop biological assessment methods and biological criteria programs. The first national workshop on biological criteria was held in Chicago in 1987 (USEPA 1987b; Simon et al. 1988). Since that time, USEPA has sponsored over 40 workshops on approaches to biological assessments and frameworks for developing assessment approaches. Every state in the nation has been represented at these workshops. Currently, routine training programs in bioassessment methods are being replaced by more frequent cooperative efforts in which states with common interests share experiences and work together to develop new approaches (e.g., the Mid-Atlantic Coastal Plains Workgroup). The emphasis of both state and federal efforts has been on the development and implementation of bioassessment methods for streams and small rivers, as both the scientific and experiential databases are greatest for these waterbodies. At the same, however, the fundamental bioassessment approaches being developed are applicable to all waterbody types.

There are a number of other federal programs that involve research and monitoring activities relevant to biological criteria development. The Environmental Monitoring and Assessment Program (EMAP) of the USEPA Office of Research and Development and cooperating units in the U.S. Forest Service and Bureau of Land Management have made substantial progress toward developing biological indicators for surface water assessment (Hunsaker and Carpenter 1990). EMAP also promises to provide monitoring data that may be useful for characterizing regional trends in the biological condition of surface waters. The U.S. Geological Survey has recently initiated a major program of integrated water quality assessment that includes ecological community components (Gurtz 1994). The National Water Quality Assessment (NAWQA) program is in the first of three years of intensive data collection that include measures of fish, invertebrate, and algal communities. This program will focus its survey and assessment activities on a total of 60 study units, each study unit a targeted watershed. On a rotating basis, groups of 20 study units will be sampled over a three year period; year four will be the beginning of the second group of 20 units, and so on. Thus, the total of 60 study units will have been sampled after nine years. This systematic NAWQA network of sampling sites will gather biological, chemical, and physical data that could prove invaluable to the development of biological criteria on a regional, if not local, level. The newly established National Biological Survey (NBS) will consolidate biological research within the U.S. Department of Interior (DOI) and promises to provide the focal point for an inventory of the nation's biological diversity (Larson 1993; NRC 1993). Reportedly, part of the NBS mission is the coordination of federal monitoring activities, which should lead to better and more efficient use of biological data and assessments. In addition, assessments of the quality of physical habitat structure are increasingly being incorporated into the biological evaluation of water resource integrity (Gibson 1994).

It is current USEPA policy that all states incorporate biological criteria into their water quality standards (USEPA 1991c). At the same time, the agency is encouraging the use of biological criteria throughout water resource management activities required under the Clean Water Act. For example, the 1994 guidance for preparation of water quality assessments (305b reports) includes the expanded use of biological integrity reporting as a primary goal (USEPA 1993b). Potential uses of biological criteria in the total maximum daily loads (TMDL) process are also being explored at both the state and federal levels. The combination of early state efforts and increased efforts from USEPA has caused virtually every state to examine their current bioassessment activities and attempt to develop the best biological criteria programs for their natural resources and regulatory programs.

2.0 ELEMENTS OF BIOLOGICAL CRITERIA PROGRAMS

The goal of biological criteria is to provide additional support for the state's water quality standards. In particular, biological criteria provide a mechanism for assessing aquatic life attainment based on the actual biological conditions of waterbodies. To achieve this, biological criteria must be created in the form of narrative expressions or numeric values that describe the biological optimum or highest biological potential of aquatic communities inhabiting waters of a given aquatic life use. The development of biological criteria by state water resource agencies depends on bioassessment to evaluate the condition of the resident biota as compared to an appropriate reference. The surveying of ambient biota has a long history in many states and usually takes the form of either coordinated monitoring networks or a series of special studies.

Development and implementation of biological criteria consists of four primary steps (Gibson 1994):

- Planning the biological criteria development process
- Designating reference conditions
- Performing the biosurvey
- Establishing biological criteria

Planning the biological criteria development process requires classification of surface water types, definition of designated uses, and a clear articulation of program objectives and data requirements for decision making. The planning process can be strengthened by interaction with other biological monitoring and criteria programs, thus potentially increasing geographic coverage and allowing joint utilization of reference databases. Designating reference conditions for biosurvey sites necessitates consideration of several factors: the type of biological data to be used to describe water resource condition, the habitat type to be sampled, the type of reference database to be developed (e.g., regional, ecoregional, or site-specific), the geographic and temporal scale to which the biological criteria may be applied, the means of ensuring habitat comparability, and the procedures for data evaluation. Performing the biosurvey of potential reference sites follows completion of the study plan, initiation of field and laboratory QA/QC procedures, and documentation of data collection and summarization methods. Establishing biological criteria is accomplished through state legal processes and USEPA review following evaluation of the survey results.

To a large extent, the status of biological criteria programs across the nation can be described by determining the presence of these activities in each of the states. As the capabilities of states to carry out the steps for implementing biological criteria increase, the quality of water resource assessments across the nation will improve. In addition to more accurately characterizing water quality problems, the application of biological criteria can: (1) provide the basis for designating high quality waters, (2) provide a framework for assigning nonpoint source pollution controls, and (3) provide a means for demonstrating water quality improvements.

3.0 STATUS OF BIOLOGICAL CRITERIA PROGRAMS

The efforts that states invest in developing biological criteria begin with biological surveys conducted to characterize the condition of the state's water resources. Indeed, the first state biological criteria programs began with the efforts of state biologists to apply their bioassessment results within a regulatory framework. The time spent by states in developing biological criteria includes both the bioassessments conducted and the activities related to the implementation of biological criteria within water quality standards.

Even before the conception of biological criteria, biological monitoring information was being used for many purposes within state water quality programs. Problem identification and compliance monitoring for discharge permits have been the major uses. Under sections 303(d), 304(l), and 305(b) of the Clean Water Act, states are required to identify all waterbodies that do not support balanced populations of shellfish, fish, and wildlife because of thermal, toxic, or other discharges. Furthermore, states must rank waterbodies for priority action and develop pollution control measures for protecting aquatic life. Many states have used biological monitoring to meet these requirements for identification and reporting of waterbody condition; however, the extent and form of biological survey data vary widely among states. At present, few states can characterize even a small fraction of their water resources with biological survey data. Nonetheless, some states are instituting comprehensive biological monitoring networks based on a rotational basin approach, wherein waterbody assessments rotate among watersheds on regular intervals.

A 1990 report on the feasibility of different environmental indicators for surface water programs (USEPA 1990b) recommended measures of biological community structure as a means to better represent the quality of water resources and to assess problems associated with nonpoint sources of pollution. To this end, the states and USEPA have recently developed provisions for reporting and analyzing biological community data as part of the process for assessing use support and preparing biennial 305(b) reports (USEPA 1993b). Collection of this type of data will likely increase over the next few years as biological criteria are developed in each state and as USEPA refines the guidance on how to incorporate biological community parameters into state standards. Insofar as past state biological criteria efforts grew out of biological monitoring efforts, future efforts will likely build on current monitoring programs.

In developing biological criteria for water quality programs, states have undertaken a wide range of efforts to improve bioassessment methods, transform biological monitoring activities into programs using biological criteria, and incorporate biological criteria into water quality standards or water resource management programs. The legislative and administrative environments of a state program ultimately determine the most effective structure for particular biological criteria programs. The following section describes current state efforts in biological criteria development and illustrates both the various differences and the many similarities of existing and emerging biological criteria programs.

3.1 Biological Assessment in the States

Florida, Kentucky, Maine, New York, North Carolina, Ohio, and Wisconsin pioneered the development of bioassessment programs in the 1970s and 1980s. These states dedicated the resources and developed the expertise needed to incorporate bioassessment into their water resource monitoring programs. In contrast, many other states did not undertake substantial bioassessment programs because they felt the cost of bioassessment outweighed the benefits. Currently, the Rapid Bioassessment Protocols (RBPs) (Plafkin et al. 1989) provide state programs with a conceptual framework for development of a cost-effective and time-efficient bioassessment approach.

Given the broad physiographic and climatic diversity of the United States, it was expected that the original form of RBPs for streams and wadable rivers would be modified on a regional, subregional, or statewide basis. Specifically, there would be different physical and biological expectations in different areas of the country, and thus different components of habitat structure and biological community composition would provide the most appropriate information for assessing biological condition in these areas. It is with this knowledge that many states have tailored a specific bioassessment approach. In some cases, the design, sampling, and analysis and interpretation have been based on RBPs for benthos (as per Plafkin et al. 1989), while in others they have been based on the Index of Biotic Integrity (IBI) for fish (as per Karr et al. 1986). The purpose of this section is to present a description of some of these modifications and to illustrate the different directions many programs have taken.

This overview is developed from available documentation on state bioassessment and monitoring programs and an informal survey of state staff biologists. This chapter is not intended to serve as a comprehensive synopsis of state activities. Some of the state program documents used to develop information for this summary were in draft form at the time of our review and are representative of preliminary programs. As further testing and implementation occurs by the states, it is likely that those programs will be improved.

Preliminary results of the state survey show that 47 states and the District of Columbia and Puerto Rico employ some form of biological monitoring in their water resource quality programs (Table 1 and Figure 1). Of these 45 states, about half have documentation (mostly in draft form) supporting the methods and analyses and providing program rationale. Those states with documentation are considered to be in the implementation phase of their programs. The remaining states are in the development and testing phase. Three states have not yet begun to formulate their bioassessment approach, but have initiated discussions within their own agencies and with the USEPA.

Forty-six of these states focus on the benthic macroinvertebrate assemblage as all or part of their sampling program. Of these states, 25 also have fish assemblage monitoring programs (Table 1 and Figure 1). Periphyton has received recent attention related to monitoring programs (Bahls 1993; Rosen, Chapter 14)) and its use in bioassessment is being evaluated by three states. Unfortunately, biologists with expertise and training in this area of applied ecology are not widely available. Most states are also incorporating physical habitat structure data into their assessments of water resource quality. Most state biological monitoring and assessment programs, whether in the development or implementation phase, pattern their data analysis procedures after the multimetric approach advocated by Karr (1981), Karr et al. (1986), Ohio EPA (1987a,b), Plafkin et al. (1989), and Barbour et al. (Chapter 6). This survey has shown that a number of states are using not only the multimetric approach, but are often using multiple assemblages. Similar in rationale to the multimetric approach for analysis of assemblage-level data, simultaneous use of such data from multiple assemblages can provide a broader sensitivity to potential stressors (Karr 1991). Bioassessments performed on community-level data have also been shown to provide cost and resource efficiencies relative to other assessment approaches (see Yoder and Rankin, Chapter 9).

3.2 Biological Criteria in the States

As we have seen, most states are developing bioassessment methods in advance of implementing biological criteria programs. However, few states have fully integrated biological criteria into their water quality standards or water resource management activities (Figure 2). Three states (Ohio, Maine, and Florida) currently have incorporated numeric biological criteria into their water quality standards. In the case of Ohio, these regulatory biological criteria are supported by the most extensive sampling, assessment, and implementation program in the nation. Other states, most notably North Carolina, are applying comparably rigorous numeric biological criteria programs to water resource management activities without incorporating them directly into water quality standards. In the case of North Carolina, these criteria are referenced in the state's water quality standards regulation. States such as Arkansas and Texas have used biological criteria to assign aquatic life use classifications based on natural conditions appropriate to different regions. States such as Connecticut, Nebraska, New York, and Vermont are using biological criteria to evaluate impairment and support compliance based on existing permits and administrative rules. In all, at least 24 states can be described as currently using biological criteria to support their water resource management. An additional 23 states appear to be developing biological criteria for use either in water quality standards or in water resource management. Program activities include development of bioassessment methods (see previous section), participation in regionalization studies, and design of implementation procedures. Only five of the states have no known biological criteria programs. Nearly 90% of the states have, or are developing, biological monitoring programs (Figure 1). Many, if not most, of these states are focusing their programs on the development and maintenance of biological criteria.

3.2.1 Water Quality Standards

In 1978, Florida became the first state to incorporate a numeric biological criterion into their water quality standards regulations. The longstanding freshwater criterion and the new wetlands standard both mandate specific levels of invertebrate species diversity. The species diversity within a waterbody, as measured by the index, may not fall below 75% of reference measures. Although this criterion has been used in enforcement cases to obtain injunctions and monetary settlements, it plays a limited role within the water quality standards program. Similarly, Delaware has specific numeric criteria within their water quality standards regulations that are used only for marina siting and the evaluation of effects of marinas on estuaries.

Table 1.   Status of Biological Assessment Programs (Based on the Target Assemblage Used) and Biological Criteria Programs (Based on How Bioassessment Results Are Used) in the United States

| State | Bioassessment Target Assemblage | | | Use Of Bioassessment As Biocriteria | | | |
	Fish	Benthos	Periphyton	Used in Regulations	Used in Water Resource Management	Biocriteria In Development	None or Unknown
AL		✔			✔		
AK		✔				✔	
AR	✔	✔			✔		
AZ		✔				✔	
CA		✔				✔[2]	
CO		✔				✔	
CT		✔			✔		
DE		✔		✔[1]	✔		
DC		✔				✔	
FL		✔		✔		✔[2]	
GA		✔				✔	
HI							✔
ID	✔	✔			✔		
IL	✔	✔				✔	
IN	✔	✔				✔	
IA	✔	✔				✔	
KS		✔				✔	
KY	✔	✔	✔		✔		
LA	✔	✔				✔	
ME		✔		✔			
MD		✔				✔	
MA	✔	✔			✔		
MI	✔	✔			✔		
MN	✔	✔			✔		
MS		✔			✔		
MO		✔				✔	
MT		✔	✔		✔	✔[3]	
NE	✔	✔			✔		
NV							✔
NH		✔				✔	
NJ		✔			✔		
NM		✔				✔	
NY		✔			✔		
NC	✔	✔			✔		
ND	✔					✔	
OH	✔	✔		✔			
OK	✔	✔	✔			✔	
OR	✔	✔			✔		
PA	✔	✔			✔		
PR						✔[2]	
RI	✔	✔				✔	
SC	✔	✔				✔	

Table 1 (continued). Status of Biological Assessment Programs (Based on the Target Assemblage Used) and Biological Criteria Programs (Based on How Bioassessment Results Are Used) in the United States

State	Bioassessment Target Assemblage			Use Of Bioassessment As Biocriteria			
	Fish	Benthos	Periphyton	Used in Regulations	Used in Water Resource Management	In Development	None or Unknown
SD							✔
TN	✔	✔				✔	
TX	✔	✔			✔		
UT		✔				✔	
VT	✔	✔			✔		
VA		✔				✔	
VI							✔
WA	✔	✔			✔		
WV	✔	✔				✔	
WI	✔	✔			✔		
WY		✔				✔	

[1] Has numeric biocriteria (benthos) for estuaries, specifically for regulation of marinas and marina siting.
[2] Currently in the process of developing a multimetric approach for assessing community-level effects using benthic macroinvertebrate data.
[3] Currently developing methods for assessment of lakes using periphyton data.

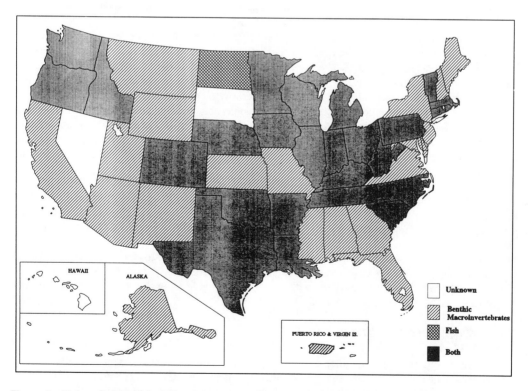

Figure 1. Status of state biological assessment programs based on benthic macroinvertebrates and fish as target assemblages (Kentucky, Montana, and Oklahoma are also evaluating the use of periphyton in assessments).

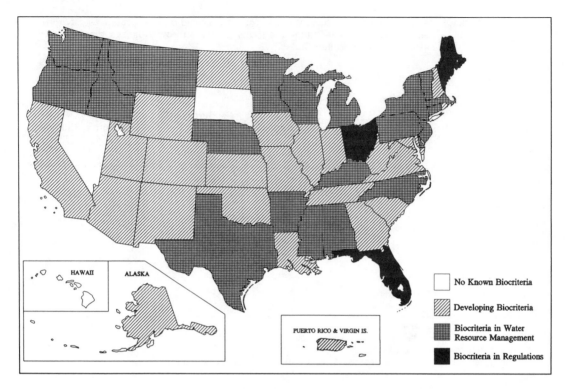

Figure 2.   Status of state biological criteria programs based on how biological assessment results are used.

In contrast, Ohio's biological criteria are applied extensively in both defining use classifications and in assessing water quality (Yoder and Rankin, Chapter 9). Use classifications based on biological criteria have been upheld in legal challenges, and in 1990, numeric biological criteria were directly incorporated into the state's water quality standards. The state uses an ecoregional reference site approach to determine least impacted expectations (biological criteria) for Ohio rivers and streams. Within each of the state's five ecoregions, criteria have been derived for three biological community indices (two for fish communities and one for macroinvertebrates). Ohio has used its biological criteria to demonstrate attainment of aquatic life uses and to detect environmental degradation. Twice as many impaired waters have been discovered by using biological criteria and chemistry assessments together than were discovered using chemistry assessments alone (Yoder 1991b).

The third state with biological criteria in water quality standards regulations is Maine (Davies et al. 1993; Courtemanch, Chapter 20). Maine developed specific aquatic life use classifications (based on representative natural aquatic communities) as a means of incorporating criteria based on statewide macroinvertebrate sampling into its water quality standards. In 1986, a revised state water quality classification law was enacted specifically to facilitate the use of bioassessments. The description of each of Maine's four water classes includes the aquatic life conditions necessary to attain that class. Maine has developed a set of decision rules to serve as numeric biological criteria based on a statewide database of macroinvertebrate samples. These criteria will insure consistency in water quality assessment, licensing and certification, and enforcement of water quality standards.

3.2.2 Water Resource Management

Among the states applying biological criteria in water resource management activities without codifying numeric criteria in water quality standards regulations, North Carolina has the most extensive program. Biological data and criteria are used extensively to identify waters of special concern or those with exceptional water quality. For example, the state employs biological criteria to assess High Quality

Waters (HQW), Outstanding Resource Waters (ORW), and Nutrient Sensitive Waters (NSW) that are at risk from eutrophication. Arkansas has rewritten aquatic life use classifications to reflect biological criteria developed for each of its ecoregions. An ecoregion project was conducted to define areas with naturally low dissolved oxygen and to develop different criteria for each region. Many Arkansas cities are designing wastewater treatment plants that meet the realistically attainable dissolved oxygen conditions determined by the new criteria. Texas has narrative biological criteria that describe aquatic life attributes on a sliding scale from limited to exceptional. Connecticut is developing qualitative bioassessment methods to complement narrative biological criteria for benthic macroinvertebrates. Vermont uses quantitative criteria to support two existing aquatic life narratives in its water quality standards. New York has developed numeric biological criteria to support enforcement actions and intends to incorporate these criteria into state water quality standards. Nebraska uses aquatic life bioassessments based on narrative biological criteria to support permit writing and pollution control. Oregon is developing water resource categories based on two aquatic life classes.

Many more states are in the early stages of developing the biological survey or reference site methods needed for biological criteria. Delaware (John Maxted, personal communication) and Minnesota (Patricia Bailey, personal communication) are attempting to develop reference conditions based on surveys of their unique coastal plain and lake-rich environments, respectively. Tennessee provides an example of a state that is upgrading its bioassessment activities with the goal of developing biological criteria. The state has conducted 20 rapid bioassessment surveys and intensive stream surveys and has established four ecoregional reference sites as part of its 319(h) Nonpoint Source Pollution Program. However, until more reference sites are established, the Division of Water Pollution Control (DPC) is using upstream reference sites to assess stream impacts on a case-by-case basis. Similarly, the state is using existing sampling metrics as it works toward developing a Tennessee Biotic Index (TBI) through refinements to the Hilsenhoff (Hilsenhoff 1987) and North Carolina (Lenat 1993) tolerance estimates (USEPA 1992a).

4.0 STATE EXAMPLES OF BIOLOGICAL CRITERIA DEVELOPMENT AND IMPLEMENTATION

As discussed in the previous section, the status of biological criteria programs across the United States can be thought of as the sum of all the states in various stages of developing and implementing biological criteria. Therefore, several examples of state biological criteria programs are provided in this section. Those states with no biological criteria activities represent the first point on the path to implementation of biological criteria. Developing biological criteria entails considerable research activities involving the accumulation of background data and the designation of expected least disturbed conditions. Implementation strategies must also be developed that may lead to incorporation of biological criteria into state water quality standards or their use in water resource management.

Opinions differ among states as to whether formal incorporation of biological criteria into state water quality standards should be the ultimate goal of all biological criteria programs. While the inclusion of detailed narrative and numeric biological criteria in state standards may be ideal, the regulatory environments of many states may require a more flexible approach to implementing biological criteria, i.e., as support to water resource management. A state usually chooses a biological criteria approach that enhances particular aspects of its existing water quality program. For example, some states focus on aquatic resource characterization through the use of biological criteria to refine aquatic life use classes. Other states use biological criteria for impact assessment, compliance monitoring, and program effectiveness monitoring. In general, biological criteria serve five fundamental purposes in state water quality programs (Gibson 1994):

- Aquatic resource characterization
- Refinement of aquatic life uses
- Arbiter of use impairment
- Identification of impact types
- Compliance (point source) and effectiveness (nonpoint source) monitoring

States may choose to concentrate on one or more of these uses of biological criteria when developing their programs. Although it can be argued that full protection of aquatic life requires the use of biological criteria for all five activities, the implementation of biological criteria in any one area serves to enhance the state's water quality program. The remainder of this section discusses the different approaches to the use of biological criteria by individual states. More detailed discussions of biological criteria development for the States of Maine, New York, Ohio, and Texas can be found elsewhere in this volume.

4.1 Biological Criteria in Water Quality Standards Regulations

The States of Ohio, Maine, and Florida have incorporated numeric biological criteria into their water quality standards regulations.

4.1.1 Ohio

Ohio has adopted a comprehensive program for applying biological criteria as a replacement for best professional judgment (BPJ) evaluations of surface water quality (see Yoder and Rankin, Chapter 9). To ensure that biological evaluations would be applicable to all of its surface waters, Ohio based biological criteria on ecoregions and regional reference sites. In February 1990, the Ohio EPA adopted both narrative and numeric biological criteria as part of the state water quality standards regulations.

Criteria for the Index of Biotic Integrity (IBI), Invertebrate Community Index (ICI), and Modified Index of well-being (MIwb) have been developed for different site types within each ecoregion. These numeric indices provide specific quantitative measures that must be met to attain the tiered aquatic life uses stipulated in Ohio's water quality standards. Full use attainment occurs if both the fish (IBI and MIwb) and macroinvertebrate (ICI) criteria are met. Partial use attainment occurs if the criteria of one organism group are met, but those of the other are not. Nonattainment occurs if none of the biological criteria are met, or if one organism group is in very poor condition. Ohio also uses the Area of Degradation Value (ADV) to indicate the severity and extent of biological impairment. The ADV provides a comprehensive view of the biological health of a stream reach and can be used to prioritize remedial treatments (Ohio EPA 1990a; Yoder and Rankin, Chapter 17).

Since 1984, the biological and water quality survey program and associated techniques have been used to evaluate nonpoint source impacts, toxicants, antidegradation issues, spills, combined sewer overflows, hazardous waste, posttreatment upgrades, and habitat modifications in the state. The data collected in these surveys have facilitated the discovery of impairments and an enhanced understanding of poorly defined problems. For example, Ohio has developed a conceptual model for a "gradient" of impact types based on the biological responses of several rivers. The information provided by Ohio's biological criteria program has been useful for virtually all regulatory, resource protection, and monitoring and reporting programs pertaining to surface waters.

This program has already played an important role in litigation and enforcement. Upgraded use designations based on biological criteria have been upheld in Ohio courts, and an appeal of these decisions was sustained in Ohio EPA's favor. Currently, Ohio EPA's resources cover approximately 75% of the National Pollution Discharge Elimination System (NPDES) issues that need at least one biosurvey evaluation. The agency has instituted a five-year rotational approach for NPDES permit reissuance and ambient monitoring support. This rotating system is designed to promote more efficient use of ambient monitoring resources and to ensure timely results.

New initiatives present the possibility of an expansion of the existing biological criteria program in Ohio. For example, the 1989 Ashtabula River Survey, made at the request of the Ohio Division of Emergency and Remedial Response, was the first official effort in support of a Natural Resource Damage Assessment (NRDA). The state's biological criteria and associated impairment quantification approaches are particularly useful for these natural resource projects.

4.1.2 Maine

In April 1986, after four years of negotiation with industry and environmental groups, the Maine legislature enacted the revised Water Quality Classification Law, which includes language specifically designed to facilitate bioassessments (Courtemanch, Chapter 20). Each waterbody class lists the descriptive aquatic life conditions necessary for attaining it. To implement the new classification system, the Maine Department of Environmental Protection has developed numeric biological criteria to support the statutory aquatic life uses in the Water Quality Classification Law.

The 1986 law was not designed to change existing water quality levels but to improve the Department of Environmental Protection's ability to monitor and manage surface waters. Under a previous law, a single aquatic life statement — "Discharges shall cause no harm to aquatic life" — was applied to the state's four water quality classes. Countless biological studies demonstrated that it was impossible to enforce this restrictive statement across all classes of effluent-receiving waters. Maine waters that were clearly attaining the minimum chemical and physical standards of the lowest class could not meet the "no harm to aquatic life" criterion because some sensitive indigenous species had been displaced. The 1986 law defines different levels of aquatic life use (ecological integrity) for each water quality classification and also specifies bacteria and dissolved oxygen criteria.

With its refined biological classification system and standard benthic macroinvertebrate database in place, Maine has identified sets of significant, measurable ecological attributes associated with each aquatic life standard. For example, the state's highest water quality class — AA — has a standard stating that "aquatic life shall be as naturally occurs." The ecological attributes identified for this standard are taxonomic equality (as compared to a nondegraded or minimally-impacted reference site), numerical equality (as naturally occurs), and the presence of pollution-intolerant indicator taxa. The identification of ecological attributes associated with each standard allows designation of indices and measures of macroinvertebrate community structure that are most sensitive to the evaluation of these sets of attributes.

4.1.3 Florida

Biological criteria have been used to evaluate water quality in Florida since 1950. However, they gained legal status only when a specific numeric biological criterion, based on the Shannon–Wiener diversity index ($\bar{d}$) criterion for macroinvertebrates, was incorporated into the Florida Administrative Code Water Quality Rules in 1978. The strict construction of this statute, in terms of sampling method and parameter computation, allow the criterion to be used to enforce the water quality standard. However, the criterion is not flexible enough to be used for many other water quality problems.

The statutory biological criterion's lack of flexibility can present problems, e.g., when drastic variations in $\bar{d}$ occur as a result of natural causes such as seasonal effects. Therefore, presentation of these $\bar{d}$ values to nonbiologists including administrators, lawyers, planners, and engineers requires substantial supporting explanation.

Biological criteria are also being developed in Florida for use in water resource management, e.g., in a two-phase (screening and definitive) approach that begins with qualitative sampling and analysis of benthic macroinvertebrate assemblages. This approach can evaluate point source discharges by detecting imbalances in aquatic flora and fauna. Recent research activities include creating a subregionalization of the state based on Omernik's ecoregions (Griffith et al. 1993a,b) and developing community assessment protocols based on USEPA's RBPs (Plafkin et al. 1989). Evaluation and optimization of specific benthic invertebrate indicators is currently underway. Revisions and additions to the standards are anticipated as Florida explores the use of ecoregions in setting biological criteria.

4.2 Biological Criteria in Water Resource Management

Although North Carolina has the most extensive nonregulatory approach to biological criteria, many other states use biological criteria in innovative ways to support water resource management.

4.2.1 North Carolina

North Carolina has used its extensive biological monitoring program as the basis for developing biological criteria to protect aquatic life in surface waters. The state uses peer-reviewed standard biological methods (Lenat 1988; Eaton and Lenat 1991), principally benthic macroinvertebrate sampling, to assess impairments of aquatic life as defined by narrative water quality criteria. Benthic criteria have been established for the mountain, piedmont, and coastal plain ecoregions based on long-term sampling at least-impaired reference sites. Sensitive indicator taxa (EPT) and tolerance-based metrics are used to assess the data (Lenat 1993). Biological classification criteria also define Outstanding Resource Waters and High Quality Waters. It is important to note that, in these classifications, excellent water quality must be identified from both biological and chemical monitoring data. Staff biologists conduct surveys to determine which watersheds and stream reaches should be given these new classifications.

Cumulative impacts associated with multiple discharges and nonpoint source inputs are the most difficult to identify through monitoring. North Carolina's Water Quality Program determined that measuring a second trophic level of organisms in free-flowing streams would aid assessments of such impacts; therefore, fish community structure surveys and (eventually) criteria will be developed to complement the macroinvertebrate program. This work should be especially helpful in addressing impacts from sedimentation, which is one of North Carolina's largest pollution problems.

Biological information has become integrated into every phase of operations within the state's Water Quality Section. Refined narrative biological standards in North Carolina's Water Quality Regulations (Overton 1991) support the use of bioassessments in evaluating point and nonpoint source pollution as well as in identifying and protecting the best uses of North Carolina's surface waters. Within North Carolina's program, bioassessments can identify temporal and spatial changes or trends in water quality including analysis of point source pollutant discharges, nonpoint source impacts, and cumulative impacts. These results, in turn, support 305(b) reporting and watershed management. They also can provide use attainability analyses for determining existing and appropriate uses, and identify watersheds with water quality higher than existing standards. North Carolina's bioassessments analyze trophic status for lake characterizations and assess existing or potential impacts relative to nutrient enrichment. They often provide data support for enforcement actions and complaint investigations, and can form the basis of water quality standards 401 certification for new facilities. Unlike other methods, bioassessments can be used to document instream improvements that result from wastewater facility upgrades and the implementation of best management practices.

4.2.2 Texas

The Texas Surface Water Quality Standards recognize the geologic and hydrologic diversity of the state by dividing major rivers, streams, reservoirs, estuaries, and bays into classified segments. The standards contain narrative biological criteria that describe aquatic life attributes (species richness and composition, diversity, trophic structure, and abundance) on a sliding scale from limited to exceptional (see Hornig et al., Chapter 10). Segment-specific uses such as aquatic life, contact or noncontact recreation, oyster waters, public water supply, aquifer protection, industrial water supply, and navigation may be assigned by the Texas Water Commission. Assignment of an appropriate aquatic life use to a waterbody is primarily driven by an assessment of biotic integrity. Narrative and numerical criteria are derived to ensure protection for some of these uses.

In 1987, a study was conducted to assess the applicability of the preliminary biological criteria and to determine when unclassified streams should be assigned aquatic life use designations. The study revealed that most of the streams possessed physical habitat heterogeneity that enhanced the development of communities of diverse aquatic fauna. In response to these findings, the Commission changed the manner in which it assigned aquatic life uses to unclassified waterbodies and initiated a three-year study to determine if the regional patterns would correspond to the ecoregions of Texas mapped by Omernik and Gallant (1987).

The resulting physical, chemical, and biological (macrobenthos and fish) data are being assessed to: (1) indicate the water quality, levels of habitat complexity, and biotic integrity that can be naturally attained within each region; (2) determine to what extent Texas ecoregions have distinctive fish and macrobenthic assemblages; and (3) regionally calibrate the existing quantitative biological criteria. The eventual goal of these studies is to develop water quality standards that are individually tailored to the different ecoregions of the state.

4.2.3 Arkansas

Arkansas addressed the specific problem of unattainable dissolved oxygen standards by restructuring its water quality program to include biological criteria associated with natural dissolved oxygen levels.

These biological criteria allowed Arkansas to reclassify streams and designate uses that would protect the existing fish communities observed in reference streams within the same ecoregion.

For the large rivers of Arkansas, secondary wastewater treatment was sufficient to meet the 5 ppm dissolved oxygen water quality standard. However, the majority of smaller towns were located on small headwater streams that never reached this level during low-flow periods — even under relatively unimpaired conditions. The state's wasteload allocation process had determined that these small towns had to meet effluent limits that were often overly stringent and cost prohibitive. Consequently, Arkansas determined that the water quality standard driving this process needed revision.

To address this problem, a three-year ecoregion project was conducted that characterized the streams within each ecoregion, developed a classification of streams, and provided a sound basis for developing realistic water quality standards and beneficial uses within ecoregions (Rohm et al. 1987). The ultimate result of this effort was the specific identification of the biological community to be protected and a methodology to ensure its protection. Arkansas has now implemented water quality standards for specific locations that are both higher and lower than the dissolved oxygen criteria. This has allowed small towns to begin building treatment plants that will attain the effluent limits specified in the new water quality standards.

4.2.4 Nebraska

Biological criteria in Nebraska's water quality standards are narrative and directed at preventing human activities that would significantly impact or displace identified key species. The key species listed in the standards are endangered, threatened, sensitive, and recreationally important aquatic species. Nebraska has recently developed additional narrative criteria for wetlands based on endangered and threatened species.

Nebraska has determined that enforcing water quality on the standards alone is difficult. Therefore, the Department of Environmental Control uses its standards and aquatic life evaluations to write permits. For example, although the water quality standards contain a "free of junk" provision, it is easier to establish the legal basis for a violation of a 404 permit to fill a wetland. Therefore, biological criteria derived from the ambient monitoring program in Nebraska are used to identify problem areas for enforcement by permit or for mitigation through increased nonpoint source prevention efforts.

Nebraska is currently expanding its evaluation approach by incorporating ecoregion- and resource-specific factors. Regional ICI and IBI values that indicate unimpaired conditions for various stream types have been developed. Nebraska hopes that this will lead to the establishment of numeric biological criteria in the future.

4.2.5 Vermont

Vermont uses biological criteria from ambient stream data to determine whether two different types of biological standards are being met. The following narratives are found in the state's water quality standards:

- No Significant Alteration of the Aquatic Biota (NSAAB). This is not a water use class standard, but is applied to all permitted indirect discharges from inground or sprayfield systems through a process that uses compliance monitoring data generated by the nonpoint discharger. The NSAAB criteria are designed to detect community-level changes that result from slight (benign) enrichment.
- No Undue Adverse Effect (NUAE). This standard is Vermont's use classification standard. At present, the same biological standard is applied to both Class B and C waters. Class C waters is set apart from Class B only because of human health concerns — as a bacterial standard.

Both of these biological standards are narrative statements within the state's water quality standards. The Vermont Department of Environmental Conservation has developed a set of administrative rules to define the NSAAB narrative standard. The department is negotiating with the state's Water Resources Board to develop a similar set of administrative rules to define NUAE. It believes that, by using the administrative rules process to define biological standards for a class of water, it can exercise the flexibility needed to sample and describe the different communities found in different ecotypes, e.g., lakes, rivers, and wadable streams.

4.2.6 Connecticut

Narrative biological criteria for benthic macroinvertebrates in lotic waters were incorporated into Connecticut's water quality standards in 1987. Connecticut routinely uses the bioassessment process to evaluate spill incidents, point-source impacts, and the effectiveness of waste treatment installations. Recent 305(b) assessments for the years 1988 and 1990 also included biomonitoring information as a measure of use attainment. In 1989, biological monitoring data were employed to assess use attainment and impairment at 22 sites in support of numeric criteria development for copper and zinc based on ambient water quality monitoring. Also in 1989, Connecticut initiated development of a numeric component to complement existing narrative biological criteria.

4.2.7 New York

The State of New York has developed a set of biological impairment criteria based on five measures of the benthic macroinvertebrate community (Bode and Novak, Chapter 8). These criteria are designed to measure significant biological impairment of the stream biota as determined by site-specific comparisons between locations upstream and downstream of given discharges. Using the paired-site comparison method (Green 1979), significant biological impairment in discharge sites can be determined relative to an upstream control or, if none are available, relative to a comparable nearby stream.

4.3 Cooperative Programs in Biological Criteria Development

An increasing number of multistate and multiagency cooperative efforts are focusing on the development of biological criteria. In particular, the regionalization efforts required to develop ecoregional reference conditions are most effectively conducted across state boundaries. Summaries of the more progressive of these interstate cooperative efforts are presented below.

4.3.1 Mid-Atlantic Highlands Assessment (MAHA)

In late 1992, the USEPA, Region 3, in cooperation with the Environmental Monitoring and Assessment Program (EMAP), the U.S. Geological Survey, the U.S. Fish and Wildlife Service, and the States of Virginia, West Virginia, Maryland, and Pennsylvania, began organizing an ecoregional monitoring effort called the Mid-Atlantic Highlands Assessment (MAHA). The original goals of this effort were as follows:

- To implement the EMAP probablistic site selection design within a regional framework in the Ridge and Valley ecoregion of the Appalachian Mountains
- To choose minimally impaired sampling areas (from the randomly selected sites) for establishment of ecoregional reference conditions and databases
- To work with the involved states (through calibration of methods and analysis of data) to determine regional expectations of biological conditions from which biological criteria could be established
- To evaluate and test the ability of employing standardized biological survey methods among the different states

In Year 1 of MAHA, 266 sites were sampled throughout the mid-Atlantic highlands stretching from Virginia, through West Virginia and Maryland, into Pennsylvania. Data collected include measures of both physical habitat and benthic macroinvertebrate, fish, and periphyton assemblages. These data will be analyzed to assess the status of biological condition and to address the goals as stated above.

4.3.2 U.S. Environmental Protection Agency, Region 10

The states of the USEPA, Region 10 (Oregon, Washington, Idaho, and Alaska) have cooperated to develop consistency among bioassessment procedures (Hayslip 1993). Through a series of meetings, workshops, and joint field trials, the states have combined their expertise and resources to resolve specific technical issues common to their needs. The states are now using similar methodologies based on the use of multiple metrics to generate comparable biological data that will be useful for understanding regional variability and characterizing expected biological conditions. By sharing data and expertise, the Pacific Northwest states expect to establish regional biological criteria with greater ecological realism and consistency across political boundaries.

4.3.3 Alabama/Mississippi Subregionalization Project

In a cooperative effort among the Alabama Department of Environmental Management, the Mississippi Bureau of Pollution Control, and the USEPA, Region 4, the States of Alabama and Mississippi have been able to develop subunits of ecological regions as an initial step towards establishment of a common bioassessment methodology and biological criteria (Griffith and Omernik 1991). Griffith and Omernik (1991) have delineated six southeastern plains subecoregions that are in common between these two states: (1) Blackland Prairie, (2) Flatwoods/Alluvial Prairie Margins, (3) Sand Hills, (4) Piedmont, (5) Southeastern Plains and Hills, and (6) Southern Pine Plains and Hills. Reference sites representative of the six subregions have been selected and, following testing of sampling results derived from different methods, the states will develop taxonomic lists of benthic macroinvertebrates, sorted by habitat types within each subregion. Subregional reference expectations will be established after comparing calculated metrics and indices.

4.3.4 U.S. Environmental Protection Agency, Region 1

Sponsored by the USEPA, Office of Science and Technology, Health and Ecological Criteria Division, the states of USEPA, Region 1 (Massachusetts, Connecticut, Rhode Island, Vermont, New Hampshire, and Maine) held an initial workgroup meeting focused on developing common reference conditions and a biological criteria framework for those states that have yet to develop them (Maine has recently promulgated their biological criteria). Griffith et al. (1993a) completed draft subecoregional delineations for Massachusetts, Connecticut, and Rhode Island, providing the geographic basis necessary for such considerations. The states are currently proceeding with a pilot study to test methodologies within a limited number of subregions. Their intent is to use standardized protocols to sample candidate reference sites and to determine the best mix of reference conditions for use in establishing biological criteria.

5.0 FUTURE DIRECTIONS AND CONCLUSIONS

This chapter has attempted to describe the status of biological criteria efforts in the United States by indicating the diversity of programs and activities in the states as of 1994. To date, a few states have developed comprehensive and sophisticated biological criteria programs that play a critical role in protecting water resource quality. Other states use biological criteria in more limited ways to enhance their water resource quality programs. Many more states are in research and development stages, increasing their bioassessment capabilities and incorporating the biological criteria concept into water quality standards activities. It is envisioned that by the end of the decade, nearly all of the state water resource agencies will have some form of biological monitoring, well established, and dedicated to monitoring the effectiveness of Best Management Practices (BMPs) and other mitigation measures.

The development and dissemination of biological criteria guidance by USEPA likely will further increase the pace of biological criteria development and implementation. Perhaps most important to the enhancement of biological criteria efforts, however, are the lessons being learned from state experiences.

Cooperative efforts to share experiences among states are occurring in USEPA Regions 10, 5, 4, 3, and 1. These cooperative programs are intended to characterize region-wide reference conditions and to develop standardized methods. Cooperative efforts among states offer several benefits, e.g., the sharing of resources and expertise and the development of a stronger database. These cooperative efforts will undoubtedly facilitate the development of biological criteria through better assessment methods and consistency across states.

Now that there is impetus from USEPA to develop and implement biological water quality standards, the primary limits to biological criteria development and implementation will be imposed by state legislatures, state environmental boards, and the regulatory environments of water quality programs. Arguments over the merits of narrative vs. numeric biological criteria will likely remain, and the key to future success will continue to be the dedication and ingenuity of state biologists.

ACKNOWLEDGMENTS

Some of the information in this chapter was drawn from research conducted during the preparation of the USEPA document entitled *Biological Criteria: State Development and Implementation Efforts* (USEPA 1991h). We are indebted to the state representatives who provided information for this document, as well as to Mike Barbour, two anonymous reviewers, and the editors of this book who commented on this chapter.

## CHAPTER 8

Development and Application of Biological Impairment Criteria for Rivers and Streams in New York State

Robert W. Bode and Margaret A. Novak

1.0 BACKGROUND OF BIOLOGICAL CRITERIA

1.1 Rationale and Benefits of Macroinvertebrate Surveys

Advantages of using macroinvertebrates as water quality indicators include: sensitivity to chemical and physical perturbations, limited mobility to avoid discharges, ability to detect intermittent discharges, abundance and ease of collection, and vital position in the food chain (Plafkin et al. 1989). In recent years these projected benefits have become realizations for many water quality programs, including programs in New York State. Macroinvertebrate surveys have been used in water quality monitoring and assessment by the New York State Department of Environmental Conservation since 1972, based on the mandate of the 1972 Clean Water Act (Section 101[a]) to "restore and maintain the chemical, physical, and biological integrity of the Nation's waters". The status of macroinvertebrate surveys in trend monitoring over the past two decades grew slowly, being advanced by: (1) assessing many streams with limited manpower, (2) reaching unambiguous water quality conclusions, and demarcating the zone of impact (Bode et al. 1986a), (3) detecting problems that were not detected by chemical sampling (Bode et al. 1986b), (4) tracking problems to previously unknown sources (Bode et al. 1990a), (5) determining if current permits are sufficiently protecting the biota (Bode 1988b), (6) demonstrating the ambient effects of discharges on stream life (Bode et al. 1990b), and (7) using faunal changes to exemplify "success story" improvements (Bode et al. 1992). The success of biosurveys using detection criteria in trend monitoring has led to the recommendation to use them in compliance monitoring. The potential value of macroinvertebrate monitoring in the regulatory process was the main impetus for developing impairment criteria for New York State rivers and streams. The ideal circumstance for applying criteria relating to aquatic life damage is one in which a pollution discharge did not exceed permit limitations or fail toxicity tests, but was thought to damage fish populations. The result of developing these criteria would be the transforming of the macroinvertebrate survey into a compliance monitoring tool for New York State streams.

1.2 Guidelines for Developing Criteria

The guidelines we used in developing impairment criteria considered the concerns of the biologist, the permittee, and the public. The biologist needs methods that are scientifically sound and defensible, resulting in conclusions that would assign significant impairment in cases where it was justified. The permittee wants methods that are standardized and objective, and criteria that are fair and reasonable, not assigning impairment where the biological change was minor or questionable. The public wants assurance that the environment is being protected, and should be kept informed of issues of environmental concern; therefore, the criteria should be based on indices that are simple and understandable in terms of biological health of the stream.

1.3 Detection Criteria and Impairment Criteria

Two types of criteria have been developed for protecting water quality in New York State streams, both using the assessment of macroinvertebrate communities. The first type, termed "biological detection criteria" establishes the expected biological condition at any site. The second type, termed "biological impairment criteria," establishes the amount of change that would be considered impairment to a biological community.

"Biological criteria" or "biocriteria" as defined by the U.S. Environmental Protection Agency (USEPA 1990a) are used to define the expected biological condition: "Biological criteria, or biocriteria, are numerical values or narrative expressions that describe the reference biological integrity of aquatic communities inhabiting waters at a given designated aquatic life use". Using this method, the reference conditions "describe the characteristics of waterbody segments least impaired by human activities..." (USEPA 1990a). The biological criteria approach of defining the reference condition most often defines criteria within an ecoregion, in which conditions in regions of ecological similarity are assumed to be homogeneous for a given stream size.

Biological impairment criteria are used to define significant impairment from a specific source, which distinguishes them from biological criteria. The impairment criteria use the paired-site method of comparing the downstream condition to an upstream control condition. Biological impairment criteria were derived from biological detection criteria, and may be seen as an extension of them. The original biological detection criteria consisted of expected index values for four levels of water quality, ranging from nonimpacted to severely impacted (Bode 1988a). Used for trend monitoring, they are a generalized set of values that are applied statewide to detect water quality problems by placing each site into one of the four categories of water quality. They were derived from data sets obtained by sampling macroinvertebrates from a wide variety of stream conditions, ranging from excellent to very poor water quality. The biological impairment criteria, derived from changes between categories of the detection criteria, were designed to diagnose impairment by confirming the severity and significance, and determining the source, using replicated upstream/downstream sampling.

Our decision to approach criteria from the standpoint of measuring impairment from a pollution source was influenced by two major factors. First, there was the infeasibility of establishing an expected biological condition for each of the waterbody use classes as defined in the state. In New York State, Classes AA and A fresh surface waters are designated for drinking, Class B waters are designated for swimming, Class C waters are designated for fish propagation and fishing, and Class D waters are designated for fishing. Each class also includes all uses for the classes below it. From a biological standpoint, the range of any given class is so great that a criterion established for an expected biological community would be so broad as to be ineffectual. As an example, Class A includes waters ranging from a small trout stream to parts of the Lower Hudson River; these segments are so classified because they are used for drinking water. It would be futile to attempt to establish a reference biological condition based upon such a grouping of stream segments.

The second important factor was the importance of the control, or reference site. Two methods of establishing the reference condition are currently used: site-specific and regional (USEPA 1990a; Gibson 1994). The site-specific approach usually uses the upstream–downstream method, while the regional approach establishes a condition based on a similar macrohabitat type (e.g., ecoregion). Since the comparability of the reference site to the test site in large part determines the strength of the conclusions about the test site, we believe that the site-specific approach is preferable in a regulatory setting, one in which a discharger may be fined and/or required to provide upgraded treatment.

The use of biological detection criteria to detect impairments and biological impairment criteria to diagnose them is in agreement with the objectives contained in the EPA biocriteria guidance, which states, "When water quality impairments are detected using biological criteria, they can only be applied in a regulatory setting if the cause for impairment can be identified" (USEPA 1990a). Our upstream/ downstream sampling regime allows identification of the source of impairment, and also provides a statistical significance that we felt could not be met by using generalized ecoregional criteria. When used together, biological detection criteria for trend monitoring can indicate when biological impairment criteria should be used for compliance monitoring. In application, the decision to implement impairment criteria sampling can be made in the field, based on field assessment of a problem.

2.0 SPECIFICATIONS OF THE CRITERIA

2.1 Methods

Two primary steps formed the foundation for development of biological detection criteria and biological impairment criteria in New York State, and these would seem to be basic to any development of criteria: standardization of methods, and using these methods to establish a database.

The components of biological monitoring methods that particularly need standardization are sampling, sample processing, and data evaluation. Two recent EPA publications have greatly advanced the standardization of field and laboratory methods for macroinvertebrates (Plafkin et al. 1989, Klemm et al. 1990). However, each state traditionally has conducted biomonitoring in a slightly different manner, and writing one's own standard operating procedure or quality assurance document provides a solid foundation for a database. For New York State biomonitoring, macroinvertebrate sampling since 1972 has mostly consisted of multiplate artificial substrates in deeper streams and rivers, and kick samples in streams with wadeable riffles. Since these were used to build the database, and are both EPA-recommended methods, they are used in our criteria sampling methods.

Riffle kick sampling in our impairment criteria protocol is standardized as much as possible to eliminate variability between collectors. The kick sample is taken by positioning an aquatic net (mesh size 0.9 mm) 0.5 m downstream, disturbing the bottom vigorously by foot for 2 min, while proceeding downstream on a 5-m diagonal course (Figure 1). This yields a semiquantitative sample. Four such samples are taken at each site, and 100 specimens are randomly sorted from each in the laboratory. Organisms are identified from three samples, and tested for similarity using percent similarity (Whittaker and Fairbanks 1958). If 50% similarity is not achieved, the most dissimilar sample is replaced by the fourth sample. This procedure safeguards against nontypical samples or outliers, since samples from a site may be taken by more than one collector. It is used as a quality assurance measure. Fifty percent similarity was selected as the criterion since it was found to exclude about 5% of all replicates.

Multiplate sampling is used for nonwadeable rivers, and streams without riffles. Our multiplate sampler consists of three hardboard plates, 6 in.^{2}, mounted on a turnbuckle and separated by $^1/_8$-in. and $^3/_8$-in. spacers. The sampler is usually suspended 1 m below the water surface, and exposed for 5 weeks. For impairment criteria testing, three multiplate samples are used from a sequence of three 5-week exposures, usually taken between May and October.

Building a database of macroinvertebrate data for New York State streams was accomplished through many years of monitoring. Sites sampled included a wide range of conditions including stream size, location, and water quality. The database used in the development of the criteria consisted of 214 kick sample data sets and 324 multiplate data sets; upstream discharges affecting water quality were known for most data sets. Although this database was accumulated through many years of monitoring rather than a deliberate database-building process, the fact that continuity in methods was maintained throughout is an important factor contributing to its value.

The biological impairment criteria developed for New York State streams and rivers are based on species-level identification of most taxonomic groups, including midges and worms, since the criteria are derived from a database of samples identified in this manner. The application of the criteria therefore also requires this level of identification.

2.2 Site Selection and Habitat Comparability

Sampling sites are usually located upstream and downstream of a suspected discharge or change in water quality. The primary concern in selecting the two sites is to assure that the physical characteristics are as similar as possible. To achieve this, four criteria were compared: (1) substrate particle size, (2) substrate embeddedness, (3) current speed, and (4) canopy cover. Embeddedness (as in Plafkin et al. 1989) is measured by examining several individual rocks and estimating the proportion of the rock below

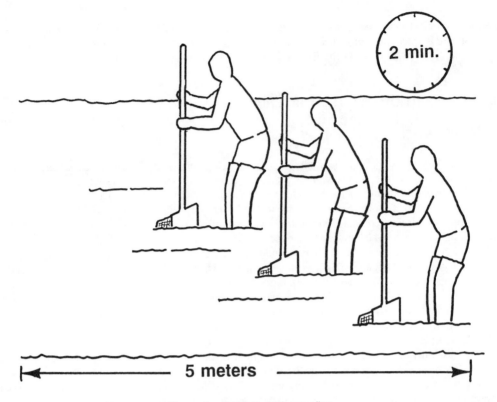

Figure 1. Replicate kick sampling.

the fine sediment line, usually evidenced by a color change. The substrate parameters are not applicable to multi-plate sampling. For kick sample sites, substrate particle size is obtained by making observational estimates to designate percentage of each of the seven EPA size categories, as listed in Weber (1973). These are then converted to phi values as in Cummins (1962), and mean particle size is calculated. The amount of allowable difference between the upstream and downstream site is 3 phi units for substrate particle size, and 50% change for embeddedness, current speed, and canopy cover. The four habitat parameters are used to minimize physical differences between the two sites, so that any differences between the macroinvertebrate communities can be assigned to water quality changes.

2.3 Indices

Five indices are used as criteria: biotic index, EPT richness, species richness, species dominance, and percent model affinity. These indices were selected among many others on the basis of accuracy, low variability, and simplicity (understandability by public). Additionally, these indices were chosen to represent different aspects of the community. For example, species richness measures the number of different species, and species dominance was selected to measure community balance. It is not uncommon for one attribute of the community, e.g., the biotic index, to change in response to an impact, while other attributes remain unchanged. Therefore it was necessary to institute several indices, each representing a different attribute of the community, and each independently applied. This means that an exceedance of criterion for any index is considered an impairment.

2.3.1 Biotic Index

The biotic index of Hilsenhoff (1987) is calculated by multiplying the number of individuals of each species by its assigned tolerance value, summing these products, and dividing by the total number of individuals. Tolerance values are assigned on a scale of 0 to 10, with 0 being the least tolerant and 10 being the most tolerant.

2.3.2 EPT Richness

EPT refers to the orders Ephemeroptera (mayflies), Plecoptera (stoneflies), and Trichoptera (caddisflies) three orders that are generally considered intolerant of poor water quality (Lenat 1987). The EPT value is the total number of species in these three orders.

2.3.3 Species Richness

Species richness is probably the most basic community index, being the total number of species found in the sample or subsample.

2.3.4 Species Dominance

This is the percent contribution of individuals of the most numerous species in the sample. It is used as a simple measure of community balance. Note that dominance may be compared between two sites regardless of whether the dominant species is the same.

2.3.5 Percent Model Affinity

This is a measure of similarity to a model nonimpacted community based on percent abundance in seven major groups (Novak and Bode 1992). Percentage similarity as defined by Whittaker and Fairbanks (1958) is used to measure similarity to a community of 40% Ephemeroptera, 5% Plecoptera, 10% Trichoptera, 10% Coleoptera, 20% Chironomidae, 5% Oligochaeta, and 10% other. This index is not used with multiplate samples.

3.0 DEVELOPMENT OF CRITERIA

3.1 Setting the Criteria

A criterion was developed for each index, based on the level of change that would be considered significant enough to indicate impairment of the community. Nonexceedance of the criteria for all indices would be considered protective of the community. The criterion for each index was derived from the biological detection criteria for the four levels of water quality impact: nonimpacted, slightly impacted, moderately impacted, and severely impacted (Figure 2). The detection criteria were derived early in our development of the Rapid Bioassessment procedure through the recognition and description of nonimpacted and severely impacted sites, and subsequently slightly and moderately impacted sites (Bode 1988a). The original detection criteria were modified slightly following additional sampling, and also were applied to multiplate samples. It is important to note that the four-tiered system has been widely used in biomonitoring because it recognizes the limitations of refinement and accuracy that can be ascribed to this science.

The numerical changes between biological detection criteria for each level of impact were used to develop biological impairment criteria for each index. The following indices and the criterion for each constitute the biological impairment criteria: biotic index (+1.5), EPT richness (–4), species richness (–8), species dominance (+15), and percent model affinity (–20) (Figure 3). The derivation of the impairment criterion for species richness is shown as an example in Table 1. The criteria were also applied to multiplate samples, except that percent model affinity was not used, because variability among percent contribution of major groups in multiplate samples from nonimpacted sites was found to be too great to establish a model community.

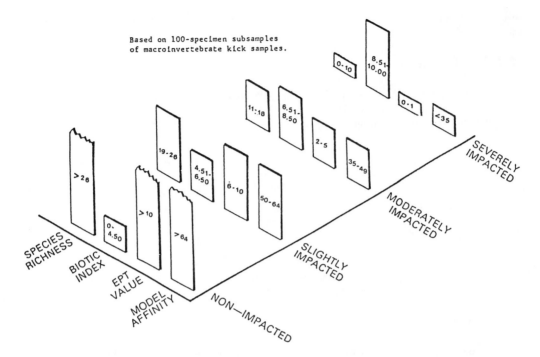

Figure 2. Biological detection criteria for flowing waters in New York State. Based on 100-specimen subsamples of macroinvertebrate kick samples.

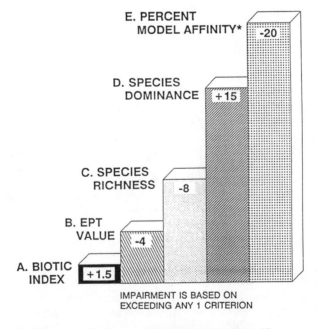

Figure 3. Biological impairment criteria for flowing waters in New York State. *Percent model affinity is not used with multiplate samples.

Table 1. Development of Impairment Criteria Δ Values from Detection Criteria Impact Categories for Species Richness

Detection criteria impact category	Range for impact category	Range average	Change (Δ)
Nonimpacted	27–35	31.0	
			31.0–22.5 = 8.5
Slightly impacted	19–26	22.5	
			22.5–14.5 = 8.0
Moderately impacted	11–18	14.5	
			14.5–5.0 = 9.5
Severely impacted	0–10	5.0	

3.2 Testing the Criteria

Over a 2-year period, the criteria were tested, focusing on the following questions: (1) is a 2-min kick sample as representative of the macroinvertebrate community as a 5-min kick sample, (2) is the variability between kick sample replicates at the same site within acceptable limits, (3) are the habitat criteria acceptable for ensuring comparability between upstream and downstream sites, (4) should seasonal variability place any seasonal restrictions on the application of the biological impairment criteria procedure, and (5) are the proposed impairment criteria sensitive enough to detect significant impairment, yet accurate enough to exclude false positives.

Testing of these questions (Bode et al. 1990c) yielded the follow results: (1) when using 100-organism subsamples and the five indices, the 2-min kick sample results were comparable to the 5-min kick sample results; (2) based on four replicates each at 16 sites, variability among replicates was acceptable, with an average of 62% similarity between replicates and a minimum requirement of 50% similarity between replicates; (3) using nine site pairs with differing habitats but similar water quality, the habitat criteria were adequate for excluding false indication of impairment, which occurred at three site pairs; (4) seasonal variability was tested by monthly sampling year-round at two streams, and showed that between-month comparisons should not be done, but upstream–downstream sampling on the same date is valid year-round; (5) based on 105 multiplate site pairs and 185 kick sample (unreplicated) site pairs, the biological impairment criteria were found to be sensitive and fair; based on field application of the full field methods for replicated kick sampling at eight site pairs, the methods were judged to be workable.

3.3 Modifying the Criteria

During the course of testing, several modifications were made to the criteria. Additionally, a t-test was added to provide statistical strength to a determination of significant biological impairment. This was a response to an upstream–downstream comparison that exceeded a criterion, but exhibited high variability at one of the sites. The t-test does not form the basis of the determination of impairment, but tightens the conditions for a genuine exceedance.

The following steps summarize the application of the biological impairment criteria: determine appropriate sampling method, determine habitat comparability and select sampling sites, conduct replicate sampling, sort and identify samples, calculate indices and means, apply criteria to determine exceedances, and calculate t-test to determine significance (Figure 4).

4.0 IMPLEMENTATION OF THE CRITERIA

4.1 Current Status

The biological impairment criteria developed have not yet gone through the review and comment process towards implementation into the state regulations. However, they are currently being applied in the determination of significant biological impairment from point sources, and are being used as *de facto* standards. Case histories are presented below showing the outcome of impairment criteria testing in a variety of situations.

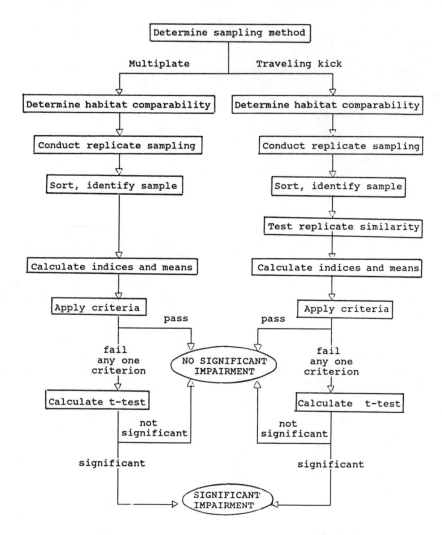

Figure 4. Biological impairment criteria flow chart.

4.2. Case Histories

4.2.1 Case History 1: Slight Change, No Exceedances

This small stream exhibited slight change downstream of a tributary that contained runoff and leachate from suspected landfill materials, resulting in high levels of iron. Both upstream and downstream sites were assessed as slightly impacted using the detection criteria, but a decline in most indices was seen at the downstream site. The impairment criteria methods of replicated kick sampling found that although decreases were present in species richness, EPT, and model affinity, none of these exceeded the criteria, and no significant impairment was indicated (Table 2).

4.2.2 Case History 2: Significant Impairment

This stream received improperly treated sewage effluent from a large municipality. The upstream site was assessed as nonimpacted using the detection criteria, and the downstream site was assessed as moderately impacted, with ammonia as the probable cause of impairment. The application of the

Table 2. Application of New York State Biological Impairment Criteria to a Stream with Slight Community Changes at the Downstream Site — No Significant Biological Impairment Demonstrated

	Species richness	EPT value	Biotic index	Species dominance	Model affinity
Upstream					
Replicate A	33	10	5.20	14	58
Replicate B	33	9	5.25	13	59
Replicate C	24	8	5.44	24	47
Mean	30	9	5.30	17	55
Downstream					
Replicate A	23	6	5.23	22	46
Replicate B	22	7	4.59	10	53
Replicate C	31	8	5.46	20	51
Mean	25	7	5.09	17	50
Net change	−5	−2	−0.21	0	−5
Criteria	−8	−4	+1.50	+15	−20
Exceedance (Y or N)	N	N	N	N	N

Table 3. Application of New York State Biological Impairment Criteria to a Stream Receiving Improperly Treated Sewage Effluent at the Downstream Site — Demonstration of a Significant Biological Impairment

	Species richness	EPT value	Biotic index	Species dominance	Model affinity
Upstream					
Replicate A	31	7	6.79	31	59
Replicate B	23	8	5.73	23	74
Replicate C	29	9	5.52	29	76
Mean	28	8	6.01	28	70
Downstream					
Replicate A	18	2	8.16	32	32
Replicate B	21	3	7.39	28	28
Replicate C	16	2	7.84	27	27
Mean	18	2	7.80	29	29
Net Change	−10	−6	+1.79	+1	−41
Criteria	−8	−4	+1.50	+15	−20
Exceedance (Y or N)	Y	Y	Y	N	Y

Table 4. Application of New York State Biological Impairment Criteria to a Stream with High Metals Levels Below an Industrial Discharge — Exceedance of One of the Criteria, but No Significant Biological Impairment Demonstrated

	Species richness	EPT value	Biotic index	Species dominance	Model affinity
Upstream					
Replicate A	19	3	6.01	41	50
Replicate B	22	6	5.63	18	63
Replicate C	25	4	5.73	31	55
Mean	22	4	5.79	30	56
Downstream					
Replicate A	19	5	5.47	25	61
Replicate B	12	3	5.61	31	44
Replicate C	12	2	5.94	44	41
Mean	14	3	5.67	33	49
Net Change	−8	−1	−0.12	+3	−7
Criteria	−8	−4	+1.50	+15	−20
Exceedance (Y or N)	Y	N	N	N	N

Table 5. Application of New York State Biological Impairment Criteria to a Large River Sampled with MultiPlate Samplers Upstream and Downstream of a Major Industrial Discharge — No Significant Biological Impairment Demonstrated

	Species richness	EPT value	Biotic index	Species dominance
Upstream				
Exposure A	20	3	7.72	34
Exposure B	17	6	7.99	33
Exposure C	23	7	8.14	20
Mean	20	5	7.95	29
Downstream				
Exposure A	19	2	7.62	19
Exposure B	15	3	9.00	48
Exposure C	17	1	9.15	56
Mean	17	2	8.59	41
Net Change	−3	−3	+0.64	+12
Criteria	−8	−4	+1.50	+15
Exceedance (Y or N)	N	N	N	N

Table 6. Application of New York State Biological Impairment Criteria to a Stream With an Impaired Upstream Control Site — Demonstration of a Significant Biological Impairment

	Species richness	EPT value	Biotic index	Species dominance	Model affinity
Upstream					
Replicate A	20	5	6.28	33	57
Replicate B	17	3	6.31	29	51
Replicate C	20	4	6.36	27	54
Mean	19	4	6.32	30	54
Downstream					
Replicate A	14	0	8.49	30	28
Replicate B	15	0	8.49	35	26
Replicate C	11	0	9.14	25	28
Mean	13	0	8.71	30	27
Net Change	−6	−4	+2.39	0	−27
Criteria	−8	−4	+1.50	+15	−20
Exceedance (Y or N)	N	Y	Y	N	Y

impairment criteria protocol resulted in exceedances for four out of five criteria, a strong confirmation of impairment (Table 3). This designation of impairment was supported by toxicity testing results, and subsequently resulted in a reevaluation of the treatment process of this facility.

4.2.3 Case History 3: Exceedance, but Not Significant

A long-standing heavy metals problem in this stream occurred below an industrial effluent discharge. Improvements in the treatment process were made in the year before this sampling, and macroinvertebrate changes were not as great as in previous years. However, a decrease of 8 in species richness was an exceedance, since the criterion for that index is 8 (Table 4). The calculation of the t-test showed that this change was not statistically significant at the $p = 0.05$ level, due to high variability in species richness at the downstream site. Therefore, significant impairment was not confirmed.

4.2.4 Case History 4: Multiplate Changes, but No Exceedances

This large, slow-moving river was sampled with multi-plate samplers upstream and downstream of a major industrial discharge. Exposures A–C represent three sequential 5-week exposures. Although all indices showed a worsening trend, none exceeded the criteria, and no significant impairment was indicated (Table 5).

4.2.5 Case History 5: Upstream Impairment

The upstream site of this stream was previously assessed as moderately impacted by unknown and nonpoint sources. The downstream site received untreated wastes from a major municipal sewage treatment plant, resulting in a downstream macroinvertebrate community of tolerant worms and midges. Despite the impaired condition of the receiving water, the impact of the sewage discharge resulted in criteria exceedances in EPT, biotic index, and model affinity (Table 6). Calculation of the t-test confirmed that the impairment was significant. As a follow-up note, this sewage treatment plant subsequently underwent a major upgrade, and later macroinvertebrate sampling revealed an improved community downstream.

4.3 Advantages of the Upstream–Downstream Method

During the application of the impairment criteria protocol, it became apparent there were several advantages over an ecoregional method. The first was the ability to assign an impairment to a specific discharge. In every case where an impairment was indicated, we would not have been able to assign the source without bracketing the discharge with upstream–downstream sites. The second advantage was precision. The difference between an allowable amount of change and an exceedance may depend on a few species or organisms. Upstream–downstream comparisons allow the precision to detect differences that would not be detected by ecoregional criteria. A third advantage to the protocol was defensibility of the results. A determination of significant biological impairment by a source is a conclusion that is statistically very defensible. A significant difference between two communities has been demonstrated, and since physical variables are accounted for by the habitat comparability criteria, the change must be assigned to water quality. A fourth advantage is that a change can be measured in receiving water that is already considered impaired. This was demonstrated in Case History 5, in which the upstream site was moderately impacted. This allows multiple determinations of impairment to be made along the course of a stream in which there are many discharges.

4.4 Future Role

Biological impairment criteria are intended to supplement existing chemical standards and toxicity testing requirements. In some cases, biological impairment criteria may detect water quality problems that are undetected or underestimated by other methods. In these cases, biological impairment criteria may be used as the sole basis for regulatory action.

There exist three possible levels at which the biological impairment criteria could be instituted in New York State: as informal guidance criteria (current status), as part of the Technical and Operational Guidance Series (a formalized set of state guidance documents), or within the State Water Quality Regulations. It has not yet been determined which level would be most appropriate for the criteria. In practice, the criteria as informal guidance are moving towards achieving the *de facto* status of standards, and are able to be effectively applied in their current form.

The impairment criteria have been applied in several cases since the protocol was completed and disseminated. In these cases we were requested to provide a definitive yes/no answer about whether significant impairment to the macroinvertebrate community existed as the result of a specific discharge. The protocol provided the means for obtaining such an answer in a defensible way, and the results have been used to make regulation decisions. Thus, the criteria have been successful in aiding the regulatory process, allowing the macroinvertebrate survey to be used as a compliance tool as well as a trend monitoring tool.

## CHAPTER 9

Biological Criteria Program Development and Implementation in Ohio

Chris O. Yoder and Edward T. Rankin

1.0 INTRODUCTION

The State of Ohio Environmental Protection Agency has been intensively monitoring the condition of Ohio's surface waters since the late 1970s. A major change in the program occurred in 1978 when Ohio EPA adopted a multiple-use system of water quality standards. This was a shift from the previous standards that specified one general aquatic life use for **all** waters of the state. The 1978 standards attempted to recognize the chemical, physical, and biological variability inherent in natural aquatic ecosystems by having a tiered classification scheme for different aquatic life uses (e.g., warmwater, exceptional, and coldwater habitats). While these classifications were based on ecological attributes, the criteria associated with each were entirely chemical/physical. These classifications formed the initial basis of the early narrative biological criteria developed in 1980 and were the forerunners of the framework from which numerical biological criteria were developed.

The manner in which biological data was analyzed also underwent changes. A qualitative/narrative system of evaluating biological data in the 1970s and early 1980s shifted to a more quantitative/numerical framework in the mid-1980s. New multimetric evaluation mechanisms such as the Index of Biotic Integrity (IBI; Karr 1981) and the regional reference site concept (Hughes et al. 1986) are the key components of this latter framework. However, the manner in which data were collected remained essentially unchanged. This permitted later comparisons of how these changes affected decisions about use attainment/nonattainment. Biological assessments were formalized in Ohio's programs with the adoption of numerical biological criteria in 1990. These efforts have resulted in a shift away from a sole reliance on regulatory and administrative activities as the principal measures of success to the inclusion of measures based upon environmental results.

1.1 Aquatic Life Use Classifications

Biological data has always played a central role in the Ohio water quality standards, particularly for the determination of appropriate and attainable aquatic life use designations. Aquatic life use designations are assigned to individual waterbody segments based on the potential to support that use according to the narrative and numerical criteria. Thus, it is not necessary to observe actual attainment of the criteria in order to designate a particular use. If this were so there would be little if any impetus to improve and rehabilitate degraded aquatic systems. The Warmwater Habitat (WWH) use replaced the general aquatic life use of the previous water quality standards (Regulation EP-1) in 1978. The narrative definition of this use specified the general aquatic community components that are characteristic of the majority of Ohio rivers and streams (i.e., warmwater fishery and associated biological and chemical components). The Exceptional Warmwater Habitat (EWH) use was intended for waters with unique and unusual assemblages of aquatic life (i.e., waters of unusually good chemical quality, significant populations of endangered and declining species, etc.). Other aquatic life uses include specialized categories (e.g., Coldwater Habitat, Seasonal Salmonid Habitat, Limited Warmwater Habitat). Although each use designation was defined based on ecological components, attainment of each was based solely on chemical water quality criteria prior to 1990.

A standardized classification hierarchy is used by the Ohio EPA to set measurable goals for specific surface waterbodies. As long as we are requested to make assessments of the integrity of surface water ecosystems, a framework to rank and partition natural variability is essential. Simply referring to a waterbody as a high-quality or significant resource is inadequate given the penchant for these characterizations to have unique attributes according to the individual making the pronouncements. Wilhelm and Ladd (1988) expressed frustration at the undisciplined usage of the terms *significant*, *exceptional*, and *high-quality* as unnecessarily vague and arbitrary. A structured and disciplined framework of classification that embodies the attributes of a desired state of well-being (i.e., biological integrity) is essential.

1.2 Narrative Biological Assessment Criteria

The usefulness of the biological data was initially limited by a lack of standardized and quantitative decision criteria. Criteria developed in 1980 consisted of narrative expressions and numerical biological index guidelines which reflected more directly the ecological components of the narrative aquatic life use designations. The narrative classification system consisted of assigning performance categories such as exceptional (meets the EWH use), good (meets the WWH use), fair, poor, and very poor, the latter three failing to attain the water quality standard goal uses. The purpose of the narrative classification system was essentially twofold: (1) to provide a systematic basis for assigning aquatic life uses to surface waters, and (2) to provide a standardized approach for determining the magnitude and severity of impairments to the aquatic biota. Considerable judgment was necessary to use these early narrative biocriteria. These played a major role in assigning and evaluating use designations, water quality management plans, and advanced treatment justifications. The narrative criteria also provided a partial basis for designating stream and river segments as attaining, partially attaining, or not attaining their designated aquatic life uses in the 1982, 1984, and 1986 Ohio 305(b) reports.

The numeric indices used to help define the narrative classification system were comprised of single-dimension measures such as taxa richness, the Shannon diversity index, and the Index of Well-Being (Iwb; Gammon 1976). Attainable expectations for a suite of narrative community attributes were based on our experience with sampling approximately 200 sites statewide. Some minor revisions were made in 1985, but the approach remained essentially the same through 1987. No effort was made to partition background variability by using ecoregions since the technology was not ready for use at that time. Ohio EPA now uses three biological indices as part of the numeric biological criteria: Index of Biotic Integrity (IBI), Invertebrate Community Index (ICI), and the Modified Index of Well-Being (MIwb).

2.0 BIOLOGICAL INTEGRITY: CONCEPTUAL BASIS AND PRACTICAL ISSUES

Without a sufficient theoretical basis it would be very difficult, if not impossible, to develop meaningful measures and criteria for determining the condition of aquatic communities. While it was the perception of biological degradation (i.e., fish kills, spills, odors, colors, gross pollution episodes, etc.) that stimulated the landmark environmental legislation of the past two decades, this biological focus was lost in the quest for easily measured surrogates (Karr 1991). The biological integrity provisions of the Clean Water Act (CWA), which initially proved difficult to determine in practical terms, were eventually defined (Karr and Dudley 1981). It was this latter definition that provided the theoretical underpinnings for developing a framework from within which numerical biological criteria could be derived. The essential elements of biological performance, natural habitats, and regions were dealt with in the early 1980s and together provided the framework and tools essential for biological numerical criteria derivation.

2.1 Reference Condition and Regional Expectations

Insufficient knowledge about regional expectations can result in misinterpretations about the severity of impacts in streams. The regional reference site framework offers a substantial advantage for the interpretation of community responses in addition to the derivation of biocriteria. By offering a more robust framework based on multiple and regionally calibrated reference sites, the chance for deriving an inappropriate biocriterion is greatly reduced. An approach that can use the same framework and information provides valuable consistency among the many different programs in which the protection of aquatic life is a goal.

2.1.1 Reference Sites and Reference Condition

The selection of reference sites from which attainable biological performance can be defined is a key component in deriving numerical biological criteria. Regional reference sites can fulfill a dual role as the arbiter of regionally attainable biological performance (which is the basis for numeric biological criteria) and as an upstream reference (more commonly referred to as a control) for determining the significance of any longitudinal changes. It is important to realize this duality and the differences between each role.

Reference sites should not be viewed in the same context as control sites since there are some important differences. Control sites are applied in a longitudinal upstream/downstream design characteristic of most water quality studies in lotic systems. While it is possible for reference sites to double as upstream control sites, the reverse is not always true. Ideally, reference sites for estimating attainable biological performance should be as undisturbed as possible and be representative of the watersheds for which they serve as models. Such sites can serve as references for a large number of streams if the range of physical characteristics within a particular geographical region are included (Hughes et al. 1986). This is one reason that the selection of only the most pristine sites as references is inadvisable. To do so would artificially restrict reference conditions to only rarely occurring benchmarks for evaluating progress or deterioration (Hughes, Chapter 4). While it is recognized that individual waterbodies differ to varying degrees, the basis for having regional reference sites is the similarity of watersheds within defined geographical regions. Generally, less variability is expected among surface waters within the same region than between different regions. This, is because surface waters, particularly streams, derive their basic characteristics from their parent watersheds. Thus streams draining comparable watersheds within the same region are more likely to have similar biological, chemical, and physical attributes than from those located in different regions.

2.1.2 Selecting Reference Sites

The selection of reference sites is another cornerstone issue in biocriteria derivation. Should reference sites be selected without prior detailed knowledge of the reference site sampling results? Or, should the sampling results be used to assist in the selection of reference sites? We believe the latter approach may introduce some unintentional bias into the biological criteria calibration and derivation process because of the inherent tendency to select the best sites instead of a more representative, balanced cross section of sites that reflect both typical and exceptional communities. In extensively disturbed regions and uniquely undisturbed regions, the method of reference site selection will likely be less of an issue because of the relatively homogenous conditions. However, in regions that have a gradient of disturbances, such as the more heterogeneous corn belt regions of the agricultural Midwest, the method of selection becomes more critical.

The mechanics for depicting reference conditions may vary according to the prevailing background conditions, which can range from relatively undisturbed, wilderness areas to highly disturbed, heavily used, and populated areas such as the corn belt ecoregions of the agricultural Midwest. For headwater and wadable streams, sufficient "least impacted" reference analogs usually exist; thus, the results of reference site sampling drive both the calibration of multimetric indices and the derivation of minimum, acceptable performance levels for aquatic communities. However, in extensively disturbed regions that have few or no suitable reference analogs, such as the larger waterbodies of the Great Lakes, the Ohio River, Chesapeake Bay, etc. an alternative approach must be used. Increased reliance on historical data and expert judgement is required in these situations.

2.2 Considerations for Biological Monitoring Programs

Management decisions based on biological criteria must be made under the direction of an aquatic biologist expert with the specific methods, organism group, indices, and criteria being used (Karr et al.

1986). Careful and effective sampling is another necessity which requires trained personnel who are able to contend with the site-specific characteristics of different waterbodies. Finally, taxonomic expertise must be adequate to accomplish organism identifications to the required level (Ohio EPA 1987a). Karr et al. (1986) provide additional cautions associated with using and interpreting biological data.

Six criteria that biological monitoring programs should be judged against have been defined (Herricks and Schaeffer 1985). These requirements, and how state programs that use multimetric evaluation mechanisms satisfy them, are

1. The measure(s) used must be biological. Ohio EPA uses the Index of Biotic Integrity (IBI), modified Index of Well-Being (MIwb), and Invertebrate Community Index (ICI) that are based solely on biological community attributes.

2. The measure(s) must be interpretable at several trophic levels or provide a connection to other organisms not directly involved in the monitoring. The ecological diversity of each of the three indices and the inclusion of two organism groups that have species which function at different trophic levels satisfies this requirement.

3. The measure(s) must be sensitive to the environmental conditions being monitored. The inherently broad ability of fish and macroinvertebrates to reflect and integrate a wide variety of environmental stresses and the redundancy of the IBI and ICI metrics themselves satisfy this requirement.

4. The response range (i.e., sensitivity) of the measure(s) must be suitable for the intended application. The biological indices and organism groups used by Ohio EPA have been demonstrated to have a high degree of sensitivity to even subtle changes in the environment and to a wide variety of environmental disturbance types. One example is the ability to discern community differences between streams of the same use designation.

5. The measure(s) must be reproducible and precise within defined and acceptable limits for data collected over space and time. Both the fish and macroinvertebrate sampling methods and evaluation indices have been shown to have consistent, reproducible expectations within acceptable limits (Rankin and Yoder 1990; Yoder 1991). Carefully following prescribed field and laboratory methods is a prerequisite to meeting this requirement.

6. The variability of the measure(s) must be low. The variability inherent to each of the three biological indices used by Ohio EPA has been shown to be quite low and within acceptable limits at relatively undisturbed sites (Rankin and Yoder 1990; Fore et al. 1994). Variation between samples clearly increases with environmental disturbance (Rankin and Yoder 1990; Yoder 1991a). Satisfying this requirement involves understanding the nature of variability that originates with sampling frequency and/ or seasonal influences.

Karr et al. (1986) found the application of the IBI (based on fish) satisfied these six criteria. The use of two additional indices (MIwb and ICI) and one additional organism group (macroinvertebrates) by Ohio EPA further satisfies these demands. These new-generation evaluation mechanisms, which are based on the recent improvements in ecological theory (Karr and Dudley 1981), provide a more comprehensive analysis of community information than do single-dimension measures such as diversity indices, species richness, indicator species, numbers, biomass, etc. Furthermore the IBI-type measures extract ecologically relevant information and provide a synthesized, numerical result that can be comprehended by nonbiologists.

2.3 Indicator Assemblages

Our experience has shown that at least two assemblages should be monitored (see Section 4.1.5 later in this chapter for more details). Fish and macroinvertebrates were chosen as the routine organism groups to monitor because each met the above criteria, have been widely used in environmental assessment, and there is an abundance of information about their life history, distribution, and environmental tolerances. The need to use two assemblages is apparent in the ecological differences between them, differences that tend to be complementary in an environmental evaluation. The different recovery rates between these two groups can provide insights about the degree to which a pollution problem has been abated. The value of having two assemblages showing the same result cannot be overstated and strengthens the assessment. The differing sensitivities of the two groups is not the same to all substances or in every situation. For example, representatives of one group may be able to tolerate and metabolize toxic substances that are highly detrimental to representatives of the other group. Such information can influence decisions to control certain substances or processes that might have been overlooked or underrated in an evaluation based on only one group. The use of these two groups is somewhat analogous to the use of a fish species and an invertebrate species as standard bioassay test organisms.

The use of macroinvertebrates is well established in state programs and the advantages are well known (see DeShon, Chapter 15). Fish, however, are a comparatively recent addition to state biological assessment programs. Fausch et al. (1990) and Simon and Lyons (Chapter 16) provide a comprehensive summary of the beneficial attributes of using fish to assess environmental degradation. The notion that fish are too mobile to use effectively as an indicator group (e.g., Waters 1992) is perhaps the most often raised liability of using this group. While the mobility of fishes compared to the more immotile groups (e.g., Unionidae and other macroinvertebrates) is obvious, this alone does not disqualify fish as a valid indicator. Stream and most riverine fish species are not excessively mobile to the point where they are unusable as indicators. The obvious need for every organism group to have a dispersal mechanism is critical if that species is to be sustained. Fish obviously accomplish this through swimming, the winged insect component of macroinvertebrates by drifting and flying, and even the highly immotile Unionidae via fish (Waters 1992). The majority of fish species encountered in warmwater rivers and streams are essentially sedentary during the summer and fall months (Funk 1955; White et al. 1975). Gerking (1953, 1959) showed that most stream fish moved less than 0.5 km in their lifetime with each species having a fraction of the population that moved longer distances.

3.0 BIOLOGICAL CRITERIA DEVELOPMENT AND DERIVATION

3.1 Framework for Deriving Numerical Biological Criteria

The framework within which biological criteria were established and used to evaluate Ohio rivers and streams includes the following major steps:

- Selection of indicator organism groups
- Establishment of standardized field sampling, laboratory, and analytical methods
- Selection and sampling of least impacted reference sites
- Calibration of multimetric indices (e.g., IBI, ICI)
- Setting of Numeric biocriteria based on the attributes of tiered aquatic life use designations
- Reference site resampling (10% of sites sampled each year)
- Periodic adjustments to the indices, biocriteria, or both as determined by reference site resampling results

The key steps in this process are summarized in Figure 1 and presume that narrative statements of biological community condition (i.e., designated aquatic life uses) already exist in the water quality standards and that a regionalization scheme (e.g., ecoregions) is also included. Once reference sites are selected and sampled (step I) the biological data is first used to calibrate the multimetric evaluation mechanisms (IBI and ICI; steps II and III). Three modifications of the IBI were developed following the guidance of Karr et al. (1986): one each for headwaters, wading, and boat sites. The ICI applies to all sites with no adjustments other than the drainage area calibrations for each metric that are accomplished during step II. The reference site results for the IBI, modified Iwb, and ICI are analyzed and then used to establish numerical criteria for each index (steps IV and V). A notched box-and-whisker plot method was used to portray the reference site results for each biological index by ecoregion (step IV in Figure 1; Figures 2 and 3). These plots contain sample size, medians, ranges with outliers, and 25th and 75th percentiles. Box plots have an important advantage over the use of means and standard deviations (or standard errors) because a particular distribution of the data is not assumed. Furthermore, outliers (i.e., data points that are two interquartile ranges beyond the median) do not exert an undue influence as they do on means and standard errors. In establishing biological criteria for a particular area or ecoregion we are attempting to represent the typical biological community performance, not the outliers. The latter can be dealt with on a case-by-case or site-specific basis if necessary.

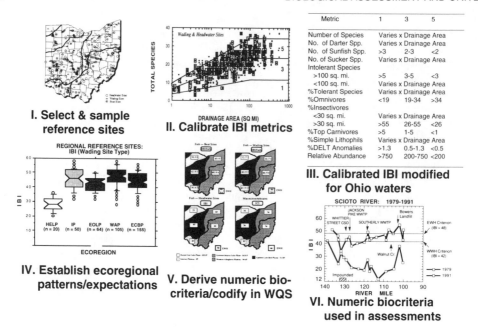

I. Select & sample reference sites

II. Calibrate IBI metrics

Metric	1	3	5
Number of Species	Varies x Drainage Area		
No. of Darter Spp.	Varies x Drainage Area		
No. of Sunfish Spp.	>3	2-3	<2
No. of Sucker Spp.	Varies x Drainage Area		
Intolerant Species			
>100 sq. mi.	>5	3-5	<3
<100 sq. mi.	Varies x Drainage Area		
%Tolerant Species	Varies x Drainage Area		
%Omnivores	<19	19-34	>34
%Insectivores			
<30 sq. mi.	Varies x Drainage Area		
>30 sq. mi.	>55	26-55	<26
%Top Carnivores	>5	1-5	<1
%Simple Lithophils	Varies x Drainage Area		
%DELT Anomalies	>1.3	0.5-1.3	<0.5
Relative Abundance	>750	200-750	<200

III. Calibrated IBI modified for Ohio waters

IV. Establish ecoregional patterns/expectations

V. Derive numeric bio-criteria/codify in WQS

VI. Numeric biocriteria used in assessments

Figure 1. Six of the key steps in the derivation and use of numerical biological criteria for the IBI (wading site type).

3.2 Calibration of Multimetric Indices

In order to establish biological criteria that are reflective of the legislative goal of attaining and restoring biological integrity in surface waters a calibration of the multimetric indices is needed. The practical definition of biological integrity as the biological performance exhibited by the natural or least impacted habitats of a region provides the underlying basis for designing a reference site sampling network. This is **not** an attempt to characterize pristine or totally undisturbed, pre-Columbian environmental conditions; such conditions exist in only a very few places, if any, in the conterminous United States (Hughes et al. 1982, Chapter 4). The landscape and aquatic ecosystems of Ohio have been significantly altered during the past 150 to 200 years (Trautman 1981). This includes massive deforestation and conversion to agricultural and urban land use, extensive use of rivers and streams for wastewater discharges, extensive drainage and elimination of more than 90% of the wetlands, and extensive modification of stream and river habitats through channelization, impoundment, and encroachment on the riparian zone. Together these activities have radically altered the lotic ecosystems of Ohio, much of which is essentially irreversible. Thus, expectations of how a biological community should perform are determined by the demonstrated make-up of natural communities at least impacted reference sites within a particular biogeographical region.

The initial selection of reference sites occurred during the Stream Regionalization Project (SRP) of 1983–84. The results of this effort are reported in Larsen et al. (1986) and Whittier et al. (1987). While the 1983–84 SRP focused on watersheds with drainage areas of 10 to 300 square miles the reference site network was consequently supplemented with data from additional locations with drainage areas of 1 to 6000 square miles sampled during 1981 to 1989 (Ohio EPA 1987b, 1989a). These included reference sites on larger streams, mainstem rivers, and headwaters streams throughout the state. The estuary sections of Lake Erie tributaries, the Ohio River, and inland lakes and reservoirs were not included in this analysis. However, work is underway to address biocriteria for these areas within the next 3 to 5 years.

The reference site results were pooled on a statewide basis prior to constructing the drainage area scatter plots (Figure 1; step II). Calibrating on a statewide (or other large area basis) as opposed to an ecoregion by ecoregion basis gives the resultant index important resolution between ecoregions. For example, it is useful to know that an index value of 30 means something different in the HELP (Huron–Erie Lake Plain) ecoregion as compared to the WAP (Western Allegheny Plateau) ecoregion while retaining comparability on a statewide basis. Having to deal with multiple, ecoregion-specific indices and resultant biocriteria values on a statewide and interregional basis would make communication and

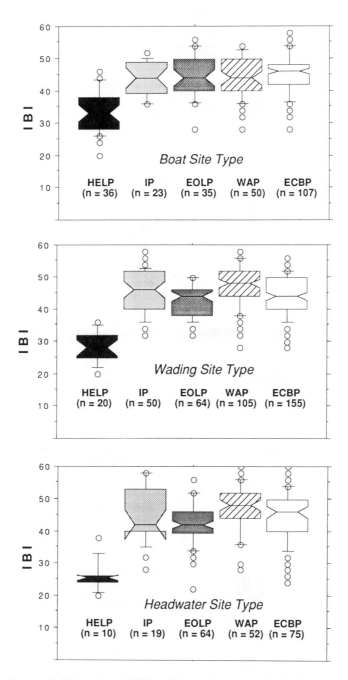

Figure 2. Notched box-and-whisker plots of Ohio reference site results for the Index of Biotic Integrity (IBI) showing minimum, maximum, median, and interquartile ranges (25th and 75th percentiles) and outliers for boat (upper), wading (middle), and headwaters (lower) site types. Notch overlap between ecoregions indicates that the median values are not significantly different at P < 0.05.

comparison much more difficult. Ideally, index calibration should occur on a broad spatial basis other than that defined by political boundaries. This is an area for further research and an opportunity for interstate cooperation.

Ohio EPA uses three indices as part of the numerical biological criteria: the Index of Biotic Integrity (IBI), the Invertebrate Community Index (ICI), and modified Index of Well-Being (MIwb). The IBI and ICI require calibration in order to establish individual metric scoring criteria tailored to the reference

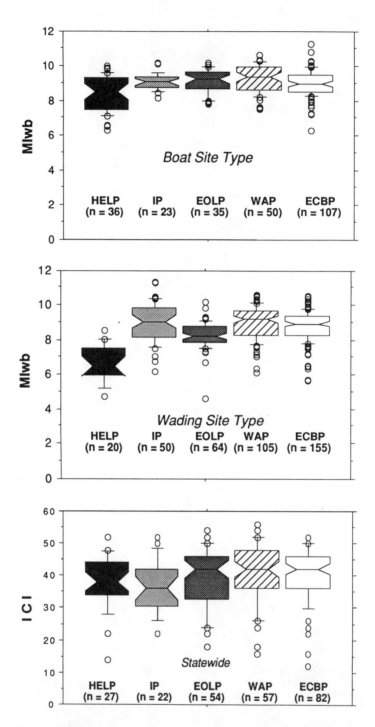

Figure 3. Notched box-and-whisker plots of Ohio reference site results for the Modified Index of Well-Being (MIwb) showing minimum, maximum, median, and interquartile ranges and outliers for boat (upper), and wading (middle) site types, and for the Invertebrate Community Index (ICI) in all streams (lower). Notch overlap between region indicates that the median values are not significantly different at P < 0.05.

conditions of the jurisdictional area, which in this case is Ohio. The MIwb does not require a spatial calibration prior to use. The sample value (i.e., number of species, percent tolerant species, etc.) of each of the 12 IBI metrics is compared to the range of values from the least impacted reference sites located within the same ecoregion. Each IBI metric then receives a score of 5, 3, or 1 based on whether the sample value approximates (5), deviates somewhat from (3), or strongly deviates from (1) the range of reference site values. The maximum IBI score possible is 60 (i.e., all 12 metrics receive scores of 5) and the minimum is 12 (i.e., all metrics receive scores of 1). We are contemplating a method to extend the lower range to 0, which would increase resolution in extremely degraded areas (see Simon and Lyons, Chapter 16).

To determine the 5, 3, and 1 values for each IBI metric the reference site database was first plotted against a log transformation of drainage area, the latter serving as an indicator of stream size (Figure 4). Other measures that have been used as an indicator of stream size include stream order (Fausch et al. 1984) and stream width (Lyons 1991). The decision to use drainage area was based on the availability and ease of calculation and greater relevance to stream size in Ohio. Stream order was viewed as being too coarse (Hughes and Omernik 1981) and stream width is simply not representative of stream size given the widespread historical modification of streams throughout Ohio. In other regions of the United States these and other parameters may be appropriate for use in the calibration process. Additional stratifying dimensions could include temperature, gradient, elevation, and lake acres or shoreline distance. The one concept that continues to surface throughout this process is that these are decisions that can only be made reliably by regional experts.

The plots for each metric (Figure 1, step II; Figure 4) were examined to determine if any visual relationship with drainage area existed. If a relationship was observed, a 95% line of best fit was determined and the area beneath trisected following the method recommended by Fausch et al. (1984). Wading and headwaters site data were combined for common metrics to determine the slope of the 95% line even though scoring for these metrics was performed separately; all boat site IBI metrics were calibrated separately. The IBI metric scores (i.e., 5, 3, or 1) for a sample are determined by comparing the site value to the trisected scatter plots constructed from the reference site database for each applicable metric (Figure 4). Certain metrics that showed no positive relationship with drainage area required the use of an alternate trisection method. Horizontal 5% and 95% lines were determined and the area between trisected. A bisection method was used only for the number-of-individuals metric. For two others (top carnivores and anomalies) the reference site database was examined and scoring criteria established following Karr et al. (1986) and Ohio EPA (1987b). The resultant 5, 3, and 1 values for these metrics are the same across drainage areas. A similar method of trisection was used by Hughes and Gammon (1987) for a modified IBI used in the lower 280 km of the Willamette River, Oregon.

Osborne et al. (1992) raised the issue of the potential confounding effect of the proximity of a site to larger waterbodies, particularly for smaller streams. They demonstrated that the location in a drainage network can significantly affect the number of fish species, particularly those which drain into higher-order main channel streams. Their recommendation was for IBI applications to consider the location of tributaries relative to the larger streams in the drainage network and that any spatial influences on IBI metrics be incorporated into the calibration process. While this work postdated the derivation of the Ohio EPA IBI biocriteria, we believe that this phenomenon is largely accounted for by two adjustments: (1) the modification of the IBI by Ohio EPA to reflect the unique faunal attributes of headwaters streams (which was not done by Osborne et al. 1992), and (2) the use of a tiered use classification system which additionally stratifies the variability induced by this phenomenon.

3.3 Deriving Numerical Biocriteria

Once the task of calibrating the biological indices is completed, the task of deriving the numerical biological criteria can proceed. For Ohio, this process is outlined in Figures 1 to 3. However, on what basis were the decisions to select a baseline numerical criterion value for each index made?

As was previously mentioned, Ohio EPA has employed a system of tiered aquatic life uses in the state water quality standards since 1978. These use designations are essentially narrative goal statements about the type of aquatic community attributes that are envisioned to represent each use. For the purposes of

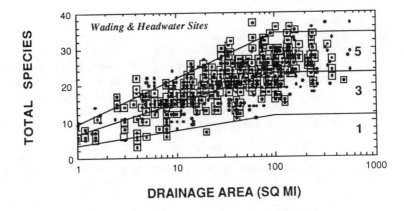

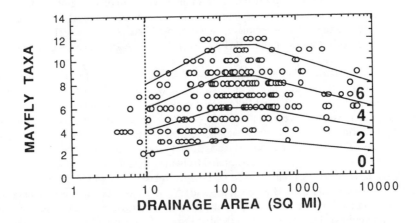

Figure 4. Examples of the technique used to calibrate the Index of Biotic Integrity (IBI) and Invertebrate Community Index (ICI) for the drainage area-dependent metrics of each index. The number of fish species (upper) and the number of mayfly taxa (lower) vs. drainage area demonstrate the use of the 95th percentile maximum species richness line (MSRL) and the trisection and quadrisection methods used to establish the IBI and ICI scoring criteria.

establishing numerical biocriteria the two most important uses are Warmwater Habitat (WWH) and Exceptional Warmwater Habitat (EWH). These use designations contain a narrative goal statement that is consistent with the definition of biological integrity (Karr and Dudley 1981) and specifies the numeric index thresholds which serve as the numeric biocriteria for each use. Numerical biological criteria for the Warmwater Habitat (WWH) use designation, which is the most commonly applied aquatic life use in Ohio, were established as the 25th percentile value of the reference site scores by index, site type (fish sampling only), and ecoregion (Figure 5; Table 1). The resultant numeric biocriteria for the WWH use vary by ecoregion in accordance with the narrative definition and the reference site results for each site type. It was felt that most of the least impacted reference results should be encompassed by the baseline WWH use designation for Ohio's inland rivers and streams. The selection of the 25th percentile value is analogous to the use of safety factors, which is commonplace in chemical water quality criteria applications. There is precedent for using quartiles to establish minimum safe values for aquatic life protection such as the 75th percentile pH, temperature, and hardness used to derive unionized ammonia-nitrogen and heavy metals design criteria for wasteload allocations, using >20% mortality for determining significance in bioassay results, or even the 10^{-6} risk factor for human exposure to carcinogens. In this sense the 25th percentile acts as a safety factor in the derivation process. Also, the difficulties in accurately estimating extreme percentiles (Berthoux and Hau 1991) argue against using lower percentiles. The inclusion of a 4-unit (IBI and ICI) and 0.5-unit (MIwb) insignificant departure below the 25th

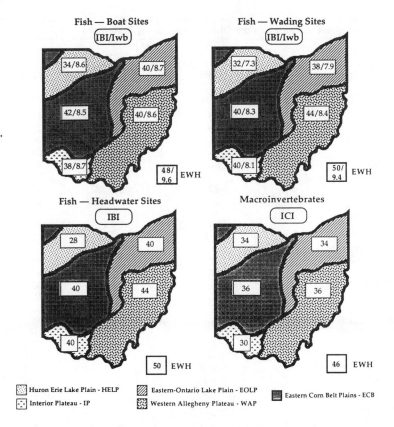

Figure 5. Biological criteria in the Ohio WQS for the Warmwater Habitat (WWH) use and Exceptional Warmwater Habitat (EWH) use designations arranged by biological index, site type for fish, and ecoregion. The EWH criteria for each index and site type is located in the boxes located outside of each map.

percentile (Ohio EPA 1987b) creates a band of acceptable departure from these criteria. Choosing the 25th percentile as the minimum WWH criterion is also conservative and reduces the influence of any unintentional bias induced by including potentially marginal sites.

A modified approach was necessary for determining the HELP ecoregion biocriteria. The HELP ecoregion is affected by significant and widespread historical land use and stream channel modifications dating from the 19th century. Setting the WWH criteria for the IBI and MIwb in this ecoregion involved detailed consideration of the extensive and essentially irretrievable physical stream habitat and watershed modifications. This distinction is made necessary by the widespread degree to which macrohabitats have been altered among the headwater and wadable streams in the HELP ecoregion. Intensive row-crop agriculture and attendant subsurface drainage practices (i.e., channel maintenance and tiling) have left few if any streams that match the intended definition of least impacted. As a result IBI and Iwb values from the wading and headwaters reference sites of this ecoregion reflect these environmentally degrading influences. Deriving the WWH wading and headwaters sites biocriteria involved an examination of IBI and MIwb results from **all** sites sampled during 1981–89 (Ohio EPA 1989b). IBI and MIwb values that marked the upper 10% (90th percentile) of **all** sites sampled were selected as an alternative to the 25th percentile of the HELP reference sites, which yielded lower values. Even with these adjustments, the resulting IBI and MIwb criteria are the lowest in the state. Establishing biocriteria for the HELP ecoregion is an example of the dilemma posed by extensively disturbed areas — maintaining a balance between setting a goal for watershed restoration efforts and the pragmatic implications of maintaining present-day socioeconomic activities.

Exceptional Warmwater Habitat (EWH) criteria were based on a combination of the entire statewide reference site data set (by organism group and site type) at the 75th percentile value statewide. EWH criteria (**upper** 25% of **statewide** reference sites) appropriately reflect the narrative definition in the Ohio

Table 1. Format for Numerical Biological Criteria that Apply to Three Major Aquatic Life Use Categories in the Ohio Water Quality Standards

Index Site Type Ecoregion	MWH[1] Channel modified	MWH[1] Mine affected	MWH[1] Impounded	WWH[1]	EWH[1]
Index of Biotic Integrity (Fish)					
Wading sites					
HELP	22			32	50
IP	24			40	50
EOLP	24			38	50
WAP	24	24		44	50
ECBP	24			40	50
Boat sites					
HELP	20		22	34	48
IP	24		30	38	48
EOLP	24	24	30	40	48
WAP	24	24	30	40	48
ECBP	24		30	42	48
Headwaters sites					
HELP	20			28	50
IP	24			40	50
EOLP	24			40	50
WAP	24	24		44	50
ECBP	24			40	50
Modified Index of Well-Being (Fish)[2]					
Wading Sites					
HELP	5.6			7.3	9.4
IP	6.2			8.1	9.4
EOLP	6.2			7.9	9.4
WAP	6.2	5.5		8.4	9.4
ECBP	6.2			8.3	9.4
Boat sites					
HELP	5.7		5.7	8.6	9.6
IP	5.8		6.6	8.7	9.6
EOLP	5.8		6.6	8.7	9.6
WAP	5.8	5.4	6.6	8.6	9.6
ECBP	5.8		6.6	8.5	9.6
Invertebrate Community Index (Macroinvertebrates)					
Artificial Substrates/Qual. Dipnet					
HELP	22			34	46
IP	22			30	46
EOLP	22			34	46
WAP	22	30		36	46
ECBP	22			36	46

[1] MWH — Modified Warmwater Habitat; WWH — Warmwater Habitat; EWH — Exceptional Warmwater Habitat.

[2] Does not apply to sites with drainage areas less than 20 square miles (e.g., headwater sites).

From OAC 3745-1-07, Table 12.

Water Quality Standards and apply uniformly across the state (Figure 5; Table 2). Streams and rivers designated EWH are characterized by an above average abundance of sensitive macroinvertebrate taxa and fish species, unique species assemblages (i.e., species designated as rare, endangered, threatened, and declining), and in larger streams, above-average populations of top carnivores.

The Modified Warmwater Habitat (MWH) use designation also requires the derivation of numeric biocriteria. However, while the same calibrated indices used to derive the WWH and EWH numeric biocriteria are employed, the MWH biocriteria are derived from a *different* set of reference sites. Aquatic communities in such streams are characterized by a predominance of tolerant species, a predominance of functional feeding guilds such as omnivores and generalists, and moderately reduced diversity. Abundance as reflected by fish numbers and biomass can be very high as the result of the increased productivity of tolerant species, omnivores, and generalists. Such communities are also tolerant of low dissolved

Table 2.   Ranges of Exceptional, Good, Fair, Poor, and Very Poor Narrative Ratings of Biological Performance for the IBI, Iwb, and ICI

Index/Site type	Exceptional	Good	Fair	Poor	Very Poor
Index of Biotic Integrity (IBI)					
Boat sites	48-60[a]	[b]47	[b]26	16-25	<16
Wading sites	50-60[a]	[b]49	[b]28	18-27	<18
Headwater sites	50-60[a]	[b]49	[b]28	18-27	<18
Modified Index of Well-Being (MIwb)					
Boat sites	>9.6[a]	[b]9.5	[b]6.4	5.0-6.3	<5.0
Wading sites	>9.4[a]	[b]9.3	[b]5.9	4.5-5.8	<4.5
Invertebrate Community Index (ICI)					
All sites	48-60[a]	[b]46	[b]14	2-12	<2

a — Values that fall within the area of insignificant departure (4 IBI and ICI units; 0.5 MIwb units) for the statewide EWH criteria are referred to as "very good."

b — This value varies by ecoregion and is the ecoregional criteria for the Warmwater Habitat use designation except in the HELP ecoregion for fish, where the value for the next highest ecoregional WWH criterion is used in lieu of the HELP WWH criterion (fish only). Values that fall within the insignificant departure of the ecoregional WWH criterion for each index (4 IBI and ICI units; 0.5 MIwb units) are designated as "marginally good."

Modified from Ohio EPA 1987b.

oxygen, elevated ammonia, and nutrient enrichment. As a result the numerical biocriteria are lower than the WWH biocriteria (Figure 6; Table 1).

Since this is not a CWA goal use, a use attainability analysis is required for each stream segment recommended for this use designation. MWH use designation as defined in the water quality standards applies to extensively modified habitats that are capable of supporting the semblance of a warmwater biological community, but where that community falls short of attaining the WWH biological criteria because of functional and structural deficiencies due primarily to alterations of the macrohabitat. Examples include most of the headwater and wadable streams in the HELP ecoregion, which have been extensively channelized and straightened. These activities must have been sanctioned by the Ohio Drainage Law, the Federal Flood Protection Act, or approved under Section 404 of the CWA.

Numeric biocriteria vary by ecoregion for the WWH and MWH uses because the levels of attainability differ within each ecoregion. In this way the structure of the biocriteria/tiered uses/ecoregions serve a dual function: (1) as an absolute measure of biological community performance, and (2) as attainable levels of biological community performance. In terms of the tiered use designations the reflection of the lowest degree of biological integrity is consistent with the Limited Resource Water (LRW) use designation and biological index values reflecting poor and very poor community performance. The WWH use represents the minimum goal of the CWA and the biocriteria are consistent with a good level of community performance (Figure 7) with the exception of the HELP ecoregion where it reflects fair performance. Again, the numeric biocriteria vary by ecoregion because of the varying levels of attainability within each ecoregion. The EWH use represents the highest level of community performance (Figure 7) and is applied statewide and irrespective of ecoregion. These biocriteria are consistent with an exceptional level of community performance and represent the highest level of biological integrity attainable given present-day conditions. The scales of each of the three biological indices at the EWH criteria still leave ample room for upward movement, which allows us to distinguish truly outstanding quality communities and the more widespread measurement of this type of change should it occur in the future.

The numerical biological criteria are limited in their application to inland rivers and streams and do not apply to the Ohio River mainstem, Lake Erie river mouth and harbor areas, or inland lakes and reservoirs. However, work is underway to develop methods and metrics for Lake Erie river mouth and nearshore areas, and the Ohio River mainstem. Developing biological criteria for these areas, while requiring different specific methods and metrics, will include the application of many of the same fundamental concepts that were employed in the development of numerical biocriteria for inland streams and rivers.

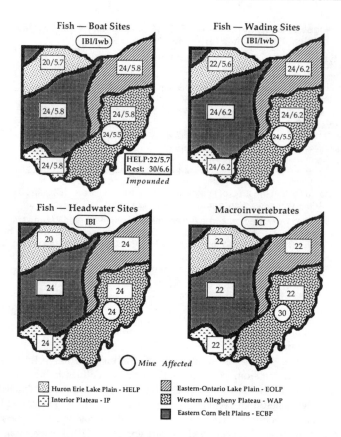

Figure 6. Biological criteria in the Ohio WQS for the Modified Warmwater Habitat Use (MWH) use designation arranged by biological index, site type for fish, modification type, and ecoregion. The MWH criteria for the impounded modification type is located in the box just outside the Boat Sites map. The biocriteria for the mine-affected modification type is represented by the circled value located in the WAP ecoregion on each map.

4.0 ROLE OF BIOLOGICAL CRITERIA IN STATE WATER RESOURCE PROGRAMS

4.1 Evaluating Aquatic Life Use Attainment/Nonattainment

The most common use of biological monitoring results is to determine the aquatic life use attainment status of the surface waterbody segment being evaluated. Ohio EPA (1987b) defines the procedure by which the biocriteria are used to determine whether or not the designated aquatic life use is being attained for Ohio rivers and streams. This process was used in the 1988, 1990, and 1992 Ohio Water Resource Inventory (305b report; Ohio EPA 1988, 1990b, 1992b), and is used routinely for site-specific applications of the biological criteria. Biological criteria are the principal arbiters of aquatic life use attainment and are used to determine if a site is fully attaining, partially attaining, or not attaining the designated aquatic life use. Use designations are listed by individual waterbody in the Ohio Water Quality Standards (Ohio Administrative Code 3745-1-08 through 3745-1-31) and determine which biocriteria values should be used when evaluating use attainment status.

Use attainment status is determined at each site using the three principal biological indices for which biocriteria have been derived. The following guidelines are used:

Full attainment — A use is considered to be fully attained when all of the applicable biological indices meet the biocriteria value for the applicable use designation, ecoregion, and site type; values that are within the nonsignificant departure range (4 IBI or ICI units; 0.5 Iwb units) are considered to meet the biocriteria.

Ohio's Biocriteria and Tiered Aquatic Life Use Designations

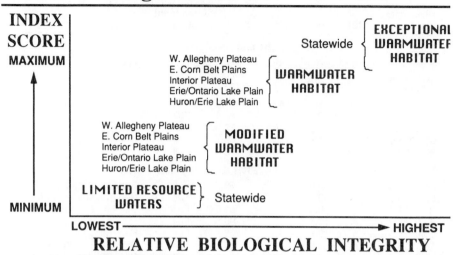

Figure 7. Relationship of relative biological integrity with the numerical biological criteria and tiered use designations. The additional stratification contributed by ecoregions is also shown.

Partial attainment — A use is considered to be partially attained if one or two biological indices indicate attainment, but others do not; for the WWH and EWH use designations the biological indices that fail to meet the applicable biocriteria must be at least within the fair range of performance (Table 2 lists the biological index values which reflect the different narrative levels of biological performance, i.e., good, fair, etc).

Nonattainment — A use is not attained if all of the biological indices fail to meet the biocriteria, or if either organism group reflects poor or very poor performance (Table 2), even if the other organism group meets the biocriteria.

4.1.1 Use Attainment Tables

To demonstrate the implementation of the preceding guidelines the attainment tables from the 1987 biological survey of the Kokosing River subbasin and the 1982 and 1990 surveys of the Hocking River, both located in central Ohio, are used here as examples (Tables 3 and 4). Each river represents two different extremes of aquatic life use attainment status. A use attainment table includes the sampling location for fish and macroinvertebrates by river mile (upstream to downstream), the IBI, MIwb, and ICI scores, the Qualitative Habitat Evaluation Index (QHEI; Rankin 1989, Chapter 13) score, the use attainment status (i.e., full, partial, or non), and any pertinent comments with regard to the proximity of the site to pollution sources and other impacts. This table is the official designation of use attainment status using the numerical biological criteria and forms the basis for all other uses including reporting (e.g., basin reports, 305b report) and assessment (e.g., Water Quality-Based Effluent Limit reports).

The Kokosing River example demonstrates longitudinal changes in use designation, different site types, and a change in ecoregion. As a result of the 1987 survey, the original WWH use designation was revised to EWH for a portion of the mainstem and some of the tributaries (i.e., designated as recommended segments in Table 3). In this situation it was the demonstrated attainment of the EWH biological criteria for both organism groups in the recommended segments that prompted the change. The original WWH designation was assigned by default in 1985 without the benefit of site-specific biological data. In addition to the longitudinal changes in use designation, there are also changes in site type, with the transition from headwaters and wading sites in the upper mainstem and tributaries to boat sites in the lower mainstem. The decision about where the change in ecoregions takes place involves more than merely following the boundaries on the ecoregion maps. Ecoregion boundaries are more transitional than

Table 3.   Aquatic Life Use Attainment Status for the Warmwater Habitat (WWH) and Recommended Exceptional Warmwater Habitat (EWH) Use Designations in the Kokosing River and Tributaries Based on Data Collected during June to September 1987

River mile (fish/invert.)	IBI	Mlwb	ICI[a]	QHEI[b]	Attainment status[c]	Comment
Kokosing River (1987)						
Erie/Ontario Lake Plain — WWH Use Designation (Existing)						
49.8[d]/49.8	56	N/A	38	85	Full	
—/48.4	—	—	46	—	(Full)	
46.4[d]/46.3	50	8.4	42	87.5	Full	
45.3[d]/45.2	50	7.7[ns]	44	36.5	Full	
40.6[d]/40.5	53	8.8	48	68.5	Full	
36.6[d]/35.0	46	7.8[ns]	48	65	Full	
30.7[d]/30.6	43	7.9	48	72	Full	
Erie/Ontario Lake Plain — EWH Use Designation (Recommended)						
28.7[e]/28.6	50	9.6	48	79	Full	Reg. ref. site; dst. N. Br.
25.5[e]/25.2	51	9.7	46	78	Full	Reg. ref. site
24.0[e]/24.2	50	9.6	46	91.5	Full	Dst. Mt. Vernon WWTP
23.0[e]/22.9	48	10.1	48	90.5	Full	
20.9[e]/18.0	53	9.8	46	76	Full	Reg. ref. site
16.1[e]/16.2	47*	9.8	46	80	Full	Dst. Gambier WWTP
11.7[e]/11.6	48	9.5[ns]	54	97.5	Full	Reg. ref. site
8.8[e]/8.7	51	9.9	38*	93.5	Partial	
W. Allegheny Plateau — EWH Use Designation (Recommended)						
6.3[e]/6.2	54	10.0	52	92.5	Full	
0.5[e]/1.5	46[ns]	10.1	48	87	Full	Reg. ref. site
N. Branch Kokosing River (1987)						
Erie/Ontario Lake Plain — WWH Use Designation (Existing)						
11.8[f]/11.6	43	8.8	46	57	Full	
6.3[f]/6.2	47	8.6	50	87.5	Full	Reg. ref. site
5.5[f]/4.8	49	8.8	52	73	Full	Dst. Fredricktown WWTP
2.2[f]/2.1	45	8.0	46	81	Full	Jelloway Creek (1987)
Erie/Ontario Lake Plain — EWH Use Designation (Recommended)						
4.4[f]/—	50	9.1	—	78.5	(Full)	Reg. ref. site
3.0[f]/1.6	49[ns]	8.4*	E	59	(Partial)	Dst. E. Br. Jelloway Cr.
Little Jelloway Creek (1987)						
Erie/Ontario Lake Plain — EWH Use Designation (Recommended)						
2.0d/6.3	54	N/A	E	84.5	Full	Reg. ref. site
E. Branch Jelloway Creek (1985)						
Erie/Ontario Lake Plain — EWH Use Designation (Recommended)						
2.3[d]/—	52	N/A	—	77	(Full)	Reg. ref. site
1.1[d]/—	42*	N/A	—	—	(Non)	Dst. Danville WWTP
0.3[d]/—	50	N/A	—	—	(Full)	Schenck Creek (1987)
Erie/Ontario Lake Plain — WWH Use Designation (Existing)						
2.8f/2.6	48[ns]	9.1	E	86	Full	Reg. Reference Site

* Significant departure from ecoregional biocriteria.
[ns] Nonsignificant departure from ecoregional biocriteria for WWH or EWH (4 IBI or ICI units; 0.5 Mlwb units).
[a] Narrative evaluation used in lieu of ICI (E, exceptional; G, good; MG, marginally good).
[b] Qualitative Habitat Evaluation Index (QHEI) values based on the most recent version (Rankin 1989).
[c] Attainment status based on one organism group is parenthetically expressed.
[d] Headwaters site type.
[e] Boat site type.
[f] Wading site type.

Table 3 (continued). Aquatic Life Use Attainment Status for the Warmwater Habitat (WWH) and Recommended Exceptional Warmwater Habitat (EWH) Use Designations in the Kokosing River and Tributaries Based on Data Collected during June to September 1987

Ecoregion Biocriteria: Erie/Ontario Lake Plain (EOLP)

Index — Site type	WWH	EWH	MWH[g]
IBI — Headwaters/Wading	40	50	24
IBI — Boat	40	48	24
MIwb — Wading	7.9	9.4	5.8
MIwb — Boat	8.7	9.6	5.8
ICI	34	46	22

[g] Modified Warmwater Habitat for channel modified areas.

Ecoregion Biocriteria: W. Allegheny Plateau (WAP)

Index — Site type	WWH	EWH	MWH[h]
IBI — Boat	40	48	24
MIwb — Boat	8.6	9.6	5.5
ICI	36	46	22

[h] Modified Warmwater Habitat for channel modified areas.

discrete, and defining where a change takes place is important for cross-boundary rivers such as the Kokosing. This involves examining the base maps of surficial geology (i.e., glacial geology in Ohio), soils, climax vegetation potential, and land use. Since rivers tend to export the characteristics of the parent ecoregion into the next this phenomenon is taken into account as well. The areas defined as most typical and generally typical by Omernik and Gallant (1988) in the receiving ecoregion are also used to assist in making this determination. In some cases the site-specific habitat attributes are used to help separate where the transition from one ecoregion to the other takes place.

The Hocking River presents a stark contrast in attainment status compared to the Kokosing (Table 4). The extensive nonattainment observed in 1982 was due to a combination of factors related to point source discharges, combined sewers, urbanization, and habitat impacts. Improvements made primarily in municipal wastewater treatment and industrial pretreatment were revealed in increased partial and full attainment in 1990 (Table 4). This example additionally demonstrates the use of the biological criteria to serve as a feedback tool for determining the success of pollution control programs.

4.1.2 Interpreting Results on a Longitudinal Reach or Subbasin Basis

The longitudinal examination of biological sampling results is also performed in an attempt to interpret and describe the magnitude and severity of departures from the numerical biological criteria. This is done by plotting the biological index results (IBI, MIwb, or ICI) by river mile for the subject survey area. Major sources of potential impact and the applicable numerical biological criteria are indicated on each graph. These graphs are also a standard reporting feature in basin specific reports. The results of fish and macroinvertebrate community sampling in the Scioto River during 1980 and 1991 are used as an example (Figure 8).

The interpretation of the overall meaning of the biological results and resulting use attainment status is best done on a longitudinal basis. For example, a site exhibiting marginal attainment in the midst of nonattaining sites which extend for some distance both upstream and downstream would be viewed as an exception with the assessment of widespread nonattainment prevailing. Conversely, the interpretation would have been quite different if the attainment status of the other sites was full. Thus, the significance, severity, and spatial "extensiveness" of the attainment status at the sampling sites is also important in determining the significance of any observed impairment.

4.1.3 Demonstrating Changes Through Time

One of the best proven uses for biosurvey data is for trend analysis. Biological assessments integrate chemical, biological, and physical impacts to aquatic systems and portray use attainment/nonattainment

Table 4. Aquatic Life Use Attainment Status for the Warmwater Habitat (WWH) Use Designation in the Upper Hocking River Mainstem Based on Data Collected During July to October 1982 And 1990

River mile (fish/invert.)	IBI	Mlwb	ICI	QHEI[a]	WWH Attainment status[b]	Comment
Hocking River 1982						
95.2/—	27*	6.1*	—	46	(Non)	Old channelization
93.2/—	23*	5.5*	—	—	(Non)	Channelization
92.0/92.0	17*	4.5*	44	48	Non	Urban development
—/91.2	—	—	8*	—	(Non)	Raw sewage evident
90.7/89.3	17*	4.0*	2*	40	Non	dst. Wheeling lift sta.
88.8/88.5	12*	0.6*	0*	48	Non	dst. Lancaster WWTP
—/87.3	—	—	0*		(Non)	
85.7/85.4	12*	1.8*	0*	62	Non	
83.1/82.9	20*	4.0*	0*	67	Non	ust. Rush Cr./Sugar Gr.
—/81.8	—	—	0*		(Non)	ust. Rush Cr./Sugar Gr.
81.4/81.3	17*	2.4*	12*	84	Non	dst. Sugar Gr, Rush Cr.
77.2/77.3	29*	6.8*	22*	63	Non	dst. Clear Cr.
73.3/73.5	31*	7.3*	18*	66	Non	at Enterprise
Hocking River 1990						
95.2/95.1	35ns	8.2	50	66.5	Full	Background
92.2/91.9	27*	6.5*	52	44	Non	Channelization
90.8/90.7	28*	6.9*	46	37	Partial	dst. CSOs, urban
89.4/89.4	24*	5.9*	38	42	Non	dst. Wheeling Lift Sta.
89.1/89.1c	25	4.7	26	49.5	N/A	Lancaster WWTP Mix.
89.0/88.9	30*	6.5*	32ns	38	Partial	dst. Lancaster WWTP
87.1/87.2	25*	4.9*	MGd	59	Non	
82.0/82.9	33*	6.5*	46	62	Partial	
81.2/81.3	39ns	8.1ns	44	57.5	Full	dst. Rush Creek
77.2/77.1	34*	7.5*	G	63.5	Partial	dst. Clear Cr.
73.2/73.4	43	8.1ns	G	63.5	Full	at Enterprise

* significant departure from ecoregional biocriteria; poor and very poor values are underlined.
ns Nonsignificant departure from ecoregional biocriteria (4 IBI or ICI units; 0.5 Mlwb units).
a All Qualitative Habitat Evaluation Index (QHEI) values are based on the most recent version (Rankin 1989).
b Use attainment status based on one organism group is parenthetically expressed.
c Biocriteria do not apply in mixing zones.
d Narrative rating used in lieu of ICI (G, good; MG, marginally good).

Ecoregion Biocriteria: Erie/Ontario Lake Plain (EOLP; RM 84.1-93.2)

Index — Site type	WWH	EWH	MWHe
IBI — Wading (RM96.2-93.2)	40	50	24
IBI — Boat (RM92.0-84.1)	40	48	24
Mlwb — Wading	7.9	9.4	6.2
Mlwb — Boat	8.6	9.6	5.5
ICI (RM92.0-85.4)	36	48	30

Western Allegheny Plateau (WAP; RM 73.5-82.9)

Index — Site type	WWH	EWH	MWHe
IBI — Boat (RM82.4-74.3)	40	48	24
Mod. lwb — Boat	8.6	9.6	5.5
ICI (RM82.9-73.5)	36	48	30

e Modified Warmwater Habitat for channel modified habitats.

in aggregate and direct terms. Frequency and duration considerations, which are difficult to adequately account for with most chemical monitoring approaches, are integrated by the resident aquatic life in the receiving waterbody. Figure 8 portrays the biosurvey results from the Scioto River downstream from Columbus, Ohio. A 40-mile segment has been sampled repeatedly over multiple years, which is a

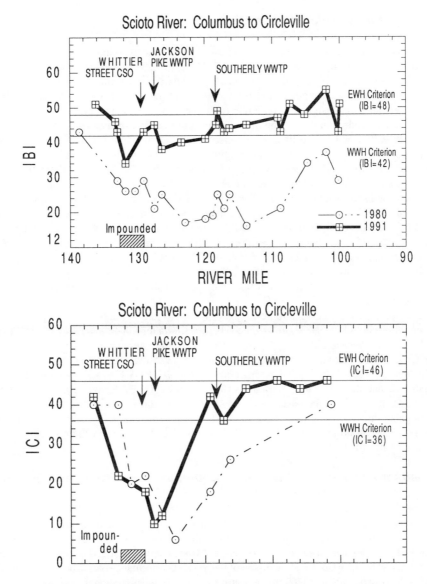

Figure 8. Longitudinal profile of the Index of Biotic Integrity (IBI; upper) and the Invertebrate Community Index (ICI; lower) for the Scioto River from upstream of Columbus to Circleville, Ohio based on artificial substrate and electrofishing samples collected during July to October 1980 and 1991.

prerequisite for trend analysis. As with many of our biosurveys, data is available before and after the July 1, 1988 compliance deadline for municipal wastewater treatment plants (WWTPs) to meet water quality-based effluent limitations specified by USEPA's National Municipal Policy. For most WWTPs throughout Ohio this resulted in significant reductions in the loadings of pollutants such as oxygen demanding wastes, suspended solids, ammonia, some nutrients, and in some cases heavy metals and other toxics. As a result, the overall loadings of conventional pollutants have markedly declined at both of the major treatment plants that serve the Columbus metropolitan area. This has resulted in improvements in both the fish and macroinvertebrate indices (Figure 8), which is a solid indication of overall water quality improvement. However, nonattainment remains in some reaches, particularly those in closest proximity to the immediate urban area in Columbus, which is impacted simultaneously by combined sewers, flow regulation, and habitat impacts.

4.1.4 Changes in Aquatic Life Use Designations

Since the initial adoption of tiered aquatic life uses in 1978, the assessment of the appropriateness of existing aquatic life use designations has continued. There were 394 changes to segment- and stream-specific aquatic life uses in six different water quality standard rulemaking changes between 1985 and 1992. The majority of these changes included the deletion of the State Resource Waters (SRW) classi-fication (116 segments), redesignation of EWH to WWH (95), the designation of previously unlisted streams (84), and the redesignation of the now defunct Limited Warmwater Habitat (LWH) use desig-nation to either WWH or LRW (90). Reevaluation of this number of stream and river segment use classifications was needed because most were originally designated for aquatic life uses in the 1978 and 1985 Ohio water quality standards mostly on a default basis. The techniques and decision criteria used then did not include standardized biological data or numerical biological criteria. Therefore, because the basin, mainstem, and subbasin biosurveys subsequently initiated in the early 1980s represented a first use of standardized biological data to evaluate and establish aquatic life use designations, many revisions were necessary and will continue to be made in the future until the backlog of waters needing evaluation is eliminated. Certain of these changes may appear to constitute downgrades (e.g., EWH to WWH, WWH to MWH, etc.) or upgrades (e.g., LWH to WWH, WWH to EWH, etc.). However, it is inappropriate to consider these changes as such because the 1985 through 1992 water quality standards revisions constituted the first use of an objective and robust biological evaluation system and database. The 1985–1992 changes are summarized in Figure 9.

Two of the proposed use designation changes were challenged in court because upgrades in desig-nated use could possibly result in more stringent permit limitations. The appeal of the revision of the existing LWH use assigned to the lower Cuyahoga River in northeastern Ohio to the WWH use was heard before the Ohio Environmental Board of Review (EBR) in 1988 (*NEORSD v Shank* 1991). The proce-dures outlined in Ohio EPA's biological criteria documents (Ohio EPA 1987a, b) including the numerical biological criteria and the Qualitative Habitat Evaluation Index (QHEI; Rankin 1989, Chapter 13) were under scrutiny as they formed the basis for defending the use change. The redesignation to WWH was upheld throughout the appeals process including the Ohio Supreme Court. The revision of the LWH use for the Ottawa River in northwestern Ohio to WWH was settled in favor of the revised use prior to a decision by the Ohio EBR.

4.1.5 Need for Multiple Assemblages to Determine Compliance with Biocriteria

The response of the fish and macroinvertebrate communities at more than 1300 sites were compared in terms of compliance with the IBI and MIwb for fish and the ICI for macroinvertebrates with the ecoregional biocriteria. The need to use at least two organism groups has been stated as being necessary for the Ohio EPA biological monitoring effort largely from a theoretical and philosophical viewpoint. This analysis provides a quantitative comparison of the relative performance of fish and macroinvertebrates as indicators of aquatic life use attainment status. Levels of full agreement, partial agreement, and disagreement were defined as follows:

Full agreement — This occurs when all three indices were in agreement about meeting or not meeting the respective ecoregional biocriteria; this was subdivided into attainment and nonattainment of the applicable use designation according to criteria specified in Ohio EPA (1987b).

Partial agreement — This occurs when one organism group indicates attainment of the ecoregional biocriteria and the other indicates nonattainment; this includes an indication of nonattainment by only one of the fish community indices in combination with macroinvertebrate community attainment or nonattainment and was subdivided into two groups, fish nonattainment and macroinvertebrate nonattainment.

Nonagreement — This occurs when one group and all attendant indices indicate attainment and the other group and all attendant indices indicate nonattainment; this was subdivided into fish nonattainment and macroinvertebrate nonattainment.

The comparisons were stratified into three stream and river sizes based on drainage area: (1) small streams <50 sq. mi. drainage area, (2) larger streams and small rivers (50 to 500 sq.mi.), and (3) large

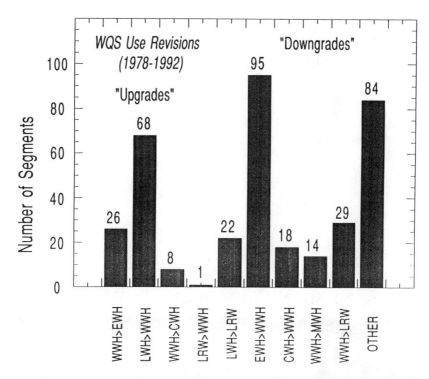

Figure 9. The number of Ohio river and stream segments where revisions to the aquatic life use designations have been made in the Ohio WQS as a result of the application of biological criteria, 1985–1992.

rivers >500 sq. mi. (Figure 10). This was done to determine if the level of agreement/disagreement was correlated with stream and river size.

The level of full agreement ranged from 43.5% for the large rivers (>500 sq. mi.) to 74.8% for the small streams (<50 sq. mi.). The streams and small rivers were intermediate at 65% agreement. Conversely, the level of nonagreement was highest for the larger rivers (33%), but least for the streams and small rivers (21.2%). The level of partial agreement followed the size order with large rivers at 23.5%, streams and small rivers at 13.8%, and small streams at 6.8%. These results are generally consistent with our previous perception that the highest agreement was in the small streams as evidenced by our original biocriteria protocol (Ohio EPA 1987b), which permits the use of only one organism group in these areas. Based on the comparison analysis (Figure 10) using only one group will be from 80.4% (macroinvertebrates) to 84.4% (fish) effective. Thus, knowing where the potential liabilities of having an incomplete evaluation are likely to arise, an informed decision can be made about whether or not one group alone will suffice. In the larger streams and rivers the effectiveness of the assessment decreases (74.9 to 88.1% for larger streams and 57.6 to 84.9% for larger rivers) to the point where both groups should be used in most situations. It may be possible to further refine this analysis to determine if there are specific situations where one group alone would reliably indicate attainment or nonattainment of designated aquatic life uses.

4.2 Water Resource Management Applications

There are several program areas within which biological criteria have been applied. Most include as a basic goal the maintenance and restoration of biological integrity and reporting on the success towards meeting that goal.

4.2.1 The Ohio Water Resource Inventory

Biological data and biological criteria are the principal arbiters of aquatic life use attainment status for the biennial Ohio Water Resource Inventory (CWA Section 305b report) the principal purpose of

SMALL STREAMS:
<50 SQ. MI. (n = 378)

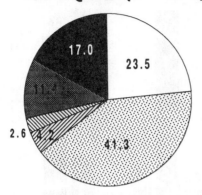

WADABLE STREAMS/RIVERS:
50-500 SQ. MI. (n = 700)

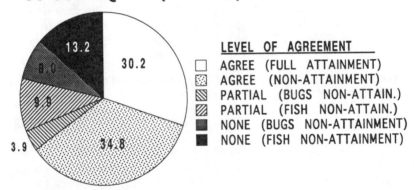

LARGE STREAMS & RIVERS:
>500 SQ. MI. (N = 269)

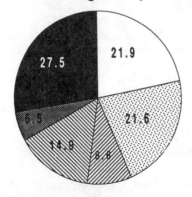

Figure 10. Results of the comparison of the frequency of agreement between the fish and macroinvertebrate communities to indicate attainment or nonattainment of the IBI, MIwb, and ICI biological criteria. The comparison was made for small streams <50 sq. mi. drainage area (upper), streams and small rivers 50 to 500 sq. mi. (middle), and larger rivers >500 sq. mi. (lower).

which is to report on the status of the state's waters. Perhaps the most important question that we endeavor to answer with the 305b report follows: Is water quality improving or worsening?

In Ohio we have attempted to develop a long-term database that will meet several previously identified shortfalls (USGAO 1986; Nichols 1992). One analysis of the statewide database performed as part of the 1992 Ohio Water Resource Inventory (Ohio EPA 1992b) was the use of a cumulative frequency diagram to show changes in the IBI and ICI at paired sites (Figure 11). In this analysis, for sites with more than two years of data, trends represent the difference between the earliest and latest results, most of which are approximately 10 years apart (Figure 11). Table 5 summarizes pertinent percentile shifts in the biological indices between the earliest and latest time periods and the results of a paired t test (using a t statistic and Wilcoxon's Z test) between these periods as well. The comparison of the two different time periods showed that the increased index scores for the later period were highly significant ($p < 0.0001$). For each index there has been a significant **positive** change over time at most sites. Another way to visualize these trends is to examine changes in the cumulative frequency distribution (CFD) of biological index scores between each time period (Figure 11).

Unlike the USEPA Environmental Monitoring and Assessment Program (EMAP) sampling design (Larsen, Chapter 18), the Ohio EPA database was not collected under a statistically random design for the location of sampling sites. The purpose of our effort was and remains to provide river- and subbasin-specific information for the evaluation of aquatic life use designations and attainment/nonattainment of aquatic life goals on a **local** scale. Although the aggregate design is biased, the sheer number of sites sampled (approximately 4500 locations) and thorough coverage of the streams and rivers with drainage areas greater than 100 sq. mi. (67% coverage statewide) makes statewide comparisons possible. It should be noted that the Ohio EPA basin monitoring efforts generally employ a sampling site density of at least ten times that proposed by either the EMAP national grid or U.S. Geological Survey National Water Quality Assessment (NAWQA) basin design.

4.2.2 Nonpoint Source Assessment and Management

Biological criteria can play an especially important role in nonpoint source assessment and management since they represent an important environmental end point. Gammon et al. (1983) documented a gradient of compositional and functional shifts in the fish and macroinvertebrate communities of small, agricultural watersheds in central Indiana. Community responses ranged from an increase in biomass with mild enrichment to complete shifts in community function with increasing enrichment. Impacts from animal feedlots had the most pronounced effects. In the latter case, the condition of the immediate riparian zone was correlated with the degree of impairment. Later work by Gammon et al. (1990) suggests that nonpoint sources are now impeding further biological improvements observed in larger rivers that resulted from reduced point source impacts. This is similar to observations that we have made in the Scioto River downstream from Columbus, Ohio, and elsewhere. Steedman (1988) used this approach to develop a relationship between these factors for watersheds tributary to Lake Ontario. The advantage of this approach is that it would allow land-use planners and managers to establish criteria for the conservation of high-quality watersheds and the restoration of degraded areas. However, before this approach can be widely employed, important information about land use, riparian zones, and subecoregional components must be more fully developed.

4.2.3 Dredge-and-Fill (401 Certifications)

Activities requiring a permit under Section 404 of the CWA must be certified as meeting provisions of the water quality standards by the state water quality agency. These are referred to as 401 certifications and largely pertain to wetlands and stream habitat impacting activities. The latter is termed hydromodification by the Ohio Nonpoint Source Management Plan (Ohio DNR 1990) and is another type of nonpoint source impact that has undoubtedly resulted in some of the most irretrievable impairments to aquatic life uses in Ohio. Stream habitat modification was identified as the third leading cause of aquatic life impairment in the 1992 Ohio Water Resource Inventory (Rankin et al. 1992), although the database was oriented primarily to the evaluation of point sources.

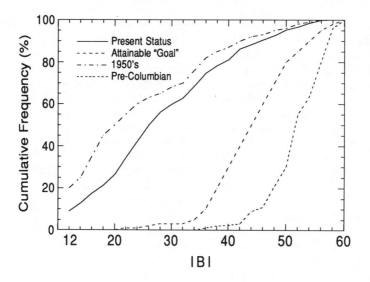

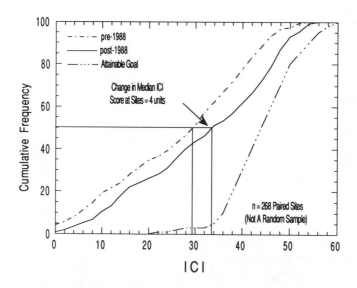

Figure 11. Cumulative frequency diagram (CFD) for the IBI (upper) comparing the present status with estimates of pre-Columbian conditions, the 1950s, and what we estimate as an attainable level of future performance, and the ICI (lower) in which the pre-1988 and post-1988 scores for 268 paired samples are compared on a statewide basis.

Streams channelized under the auspices of the Ohio Drainage Law (Ohio Revised Code 6131) are subject to routine maintenance activities, which include herbicide application, tree removal, sand bar removal, and the snagging and clearing of accumulated woody debris. In much of the HELP ecoregion productive row crop agriculture is not possible unless subsurface drainage systems are maintained. Ohio EPA has recognized that state-sanctioned channel maintenance will keep these streams in a permanently altered condition and will effectively prevent the attainment of the WWH biocriteria. The Modified Warmwater Habitat (MWH) use designation was developed as a middle ground in response to constraints imposed by Ohio drainage laws, the need for subsurface drainage to support existing agricultural land uses, and the unlikely prospect that attempts to restore the original habitat in these waterbodies will be

Table 5. Summary of "Paired" Data Sites with At Least Two Years of Biological Data from Streams in Ohio.

| | Biological Index Scores | | | | | |
| | IBI | | Mlwb | | ICI | |
Category	Earliest	Latest	Earliest	Latest	Earliest	Latest
10th percentile	14	19	3.1	4.3	6	10
25th percentile	21	24	5.1	6.1	16	20
Median	28	31	6.9	7.9	30	34
75th percentile	38	40	8.3	8.8	40	46
90th percentile	46.2	48	9.2	9.5	46	50
Mean	29.6	32.6	6.4	7.3	27.7	32.6
Paired t-test	442 df		403 df		267 df	
t Value	−8.558		−10.678		−6.689	
Mean difference	2.8		0.8		4.8	
(L minus E)	$p <0.0001$		$p <0.0001$		$p <0.0001$	
Wilcoxon (Z)	−8.039		−9.794		−6.403	
	$p <0.0001$		$p <0.0001$		$p <0.0001$	

Note: Data pairs represent earliest and latest index results for the ICI, Mlwb, or IBI at a site.

successful in the near future. Biocriteria and the attendant habitat assessment tools (e.g., QHEI, Rankin, Chapter 13) are essential to making these distinctions in an environmentally sound manner.

4.2.4 NPDES Permit Program

The usefulness of biocriteria in the NPDES permit program has been difficult for some to comprehend simply because direct limitations cannot be written in terms of the biocriteria nor can the biocriteria values be directly translated into effluent limitations. The confusion of the term *biocriteria* with the traditional *chemical criteria* is understandable, but unfortunate because this has led to an erroneous conclusion that the former must be used in permits. However, our experience demonstrates that the implementation and enforcement of NPDES permits is enhanced by the site-specific information provided by biosurveys within a biocriteria framework.

An example of the application of biocriteria to NPDES permits is with the Ottawa River in northwestern Ohio. In 1978 the Ottawa River, like many other heavily polluted stream and river segments in Ohio, was designated as a Limited Warmwater Habitat (LWH), which essentially established varied criteria for specific chemical parameters discharged by major point sources. In 1979 the first comprehensive biosurvey report was completed (Martin et al. 1979), which documented severe biological and chemical degradation. NPDES limitations during that time were based on technology guidelines or best engineering judgement. In 1986, the LWH use was replaced by the WWH use based on the results of the 1985 biosurvey, which had the benefit of employing some new assessment tools (e.g., habitat assessments) that the 1979 effort lacked. This designation was contested by the major permittees and the WWH use designation was upheld via a settlement agreement between the NPDES permittees and Ohio EPA. A cooperative study on the attainability of the WWH dissolved oxygen criterion was part of the agreement. In 1991, water quality-based effluent limitations including chronic WET limits were included in the permit re-issuance. The biosurvey results strongly indicated a toxic impact and were instrumental in justifying the water quality-based permits.

In 1992, the latest biological report was released and it indicated the persistence of toxic impacts. The degree of degradation identified by the 1991 biosurvey (Ohio EPA 1992a) prompted some unconventional detective work into identifying possible causes for severe nonattainment (the permittees claimed to be largely in compliance with their NPDES permits). One finding was that the sequence of minor permit violations for each of the three major point sources was such that the cumulative pattern of noncompliance would have placed all three **combined** in significant noncompliance status. None of the entities individually appear on the significant noncompliance list at this time. Given that the three discharges are within 0.8 miles of each other perhaps the determination of the degree of NPDES noncompliance should be viewed in the aggregate rather than singly in such situations. Spills and other releases were also documented as frequently occurring and leachaete from an industrial landfill was also discovered during the 1991 biosurvey. Later in 1992 and into 1993 discussions were initiated under the auspices of a local interest group comprised of citizens, local agencies, the NPDES permittees, and the Ohio EPA, in order to determine the feasibility of and possible ways to restore WWH use attainment. While agreement on the issues is not complete among the group members, the existence of a definitive set of data on instream conditions has been instrumental in maintaining the high interest level among the public and local government agencies. In this situation, and others like it, biocriteria alone are not used to develop permit limitations. However, they have been an indispensable part of the process because of the credibility and information they bring to the process to improve the Ottawa River and the establishment of specific biological (CWA) goals. Without the robust database generated within the underpinnings of the regional reference site approach, the effort to upgrade the aquatic life use to WWH might have failed in 1985 or become seriously restricted. This would have hampered any hope of meaningful restoration efforts and voided much of the current interest in attempting to restore the river.

Another example of the use of biocriteria is with the review of Permits to Install that are required by Ohio law (ORC 6111) prior to the installation of new, or expansion of existing, wastewater treatment facilities and processes including accessories such as sewer systems. The alignment of interceptor sewers for a new regional wastewater treatment facility was changed substantially based on the results of an intensive biological and habitat investigation of the effects of interceptor sewer construction on streams in southwestern Ohio (Ohio EPA 1992b). Essentially, the construction of the interceptor sewers as originally designed would have caused irretrievable impairments to the aquatic life uses due to permanent damage to stream beds in the area. In fact, the findings of this study not only prompted a reevaluation of the policies on regionalizing sewage flows and interceptor sewer alignments, but also prompted the development of a comprehensive stream protection policy as well.

Biosurvey information and biocriteria contribute more to the NPDES process than merely producing permit limitations. Key information about the environmental setting, characteristics of the receiving waterbody, and insights into the chemical/physical dynamics of the discharge(s) are either directly or indirectly reflected by the biota. This is extremely useful in enforcement and litigation proceedings since the extent and severity of degradation and information about the character of the aquatic resource are revealed to the arbiters of each case. In fact, our attorneys have requested biosurvey results for each major enforcement case because this type of information provides a basis from which the legal action can be considered reasonable.

4.3 Potential Non-Clean Water Act Uses

Biocriteria, because they measure the overall condition of aquatic communities and, hence, reflect the condition of the entire aquatic resource, are potentially useful outside the traditional purview of CWA programs. One of these areas is with nongame species, particularly the rare, endangered, threatened, and special status species listed by the Ohio Division of Wildlife. Presently, 25 species are listed as endangered, 8 species as threatened, 13 species as special interest, 5 as extirpated, and 2 as extinct; this represents more than 30% of the Ohio fauna. Of the 41 species listed by Ohio EPA as extremely intolerant, intolerant, and sensitive (Ohio EPA 1987b), 25 are listed as endangered, threatened, or special status. An **additional** 16 species are in the process of significant declines, some more rapidly than others (Ohio EPA 1992b). This increases to more than 40% the fraction of the Ohio fish fauna that is potentially imperiled. These trends are potentially symptomatic of other environmental problems that could eventually emerge to affect attributes of surface waters on which humans depend directly. This information was provided by the biosurveys conducted by Ohio EPA over the past 14 years, demonstrating the multiple uses of the same data. It also demonstrates the opportunity to utilize the dimensions of the data in ways that would otherwise become collapsed in the IBI evaluations. Nongame aquatic communities are not only indicators of acceptable environmental conditions for themselves, but also indicate that the water resource is of an acceptable quality for wildlife and human uses since they have the ability to integrate and reflect the sum total of disturbances in watersheds. While individual, site-specific watershed and stream disturbances may singly seem trivial, the aggregate result of these individual impacts emerges in the form of a degraded and declining fauna on a regional or watershed scale. We will have a very difficult time demonstrating this problem if we do not employ monitoring and assessment efforts that generate this type of information in a scientifically credible manner that the public will accept.

Another potential non-Clean Water Act use for biocriteria is in the management and assessment of lotic fisheries. The smallmouth bass (*Micropterus dolomieui*) is one of the most important game species in midwestern rivers and streams. Furthermore, this is a species that requires little or no external support in the way of supplemental stocking. However, like any other valued fish species, it does have specific habitat and water quality requirements. We examined the relationship between the occurrence and abundance of smallmouth bass with the IBI throughout the state (Figure 12). The overall pattern is that this species reaches the highest abundance and occurs most frequently at sites with IBI scores at least in the fair range and preferably in the good and exceptional range. As expected, the species declines sharply as the IBI indicates increasingly degraded conditions (i.e., poor or very poor). This analysis demonstrates the relevance of the IBI to and correlation with tangible resource benefits of direct importance to resource users specifically and the public in general.

5.0 EMERGING ISSUES

5.1 Initial Decisions and Other Considerations

There are a number of fundamental decisions that need to be made early in the development of biocriteria. This is a critical juncture in the process since these initial decisions will determine the overall effectiveness of the effort well into the future. Decisions about which sampling methods and gear to use, seasonal considerations, which organism groups to monitor, which parameters to measure, which level of taxonomy to use, etc. will need to be made. The axiom follows "When in doubt choose to take more measurements than seem necessary at the time since information *not* collected is impossible to retrieve at a later date." This does not apply equally to all parameters. For example, seasonality is a well-understood concept, therefore, it is not necessary to sample in multiple seasons for the sake of data redundancy. However, parameters that require little extra effort to acquire should be included until enough evidence is amassed to evaluate their relative worth. One example of this in Ohio is external anomalies on fish. A decision was made to record this information even though the eventual importance of its use was not immediately apparent. This one parameter has proven over time to be one of our most valuable assessment tools. For macroinvertebrates the issue of identifying midges to the genus/species level (as opposed to the family level) proved likewise to be a farsighted decision given the value of this group in diagnosing impairments. Samples could have been archived for later processing, but the logistical burdens that this would entail later on are even more undesirable.

5.1.1 *Costs of Developing Numeric Biocriteria*

Concern is frequently expressed not only about the practical utility of biological field data, but also about the resources needed to implement such programs (Loftis et al. 1983; USEPA 1985). Whole effluent toxicity evaluation has been advocated partly because it is viewed as more cost-effective than biological field evaluations (USEPA 1985). Our experience with using a standardized and systematic application of biological field monitoring techniques integrated with the traditional chemical/physical and bioassay assessment techniques allows a detailed comparison of the costs involved with each component.

The approach used by Ohio EPA to collect macroinvertebrate and fish community data is intended to secure an adequate sample, but not necessarily an exhaustive inventory. Fish relative abundance data is collected using standardized, pulsed DC electrofishing techniques. In an analysis of resources expended during federal fiscal year (FFY; October 1 to September 30) 1987 and 1988 the following were revealed:

- 8.44 work year equivalents (WYE) were used to collect 1277 samples at 617 sites.
- An average of 0.014 WYE or 29.1 hours per site were expended to plan, collect, analyze, interpret data, and produce reports at an average cost of $740 per site.
- This translates into 1 to 3 hours per sample with a field crew sampling 3 to 6 sites per day by working 10 to 14 hours per day.
- Postfield-season laboratory effort ranges from 1 to 3 weeks.
- A field crew consists of one full-time biologist and two interns.

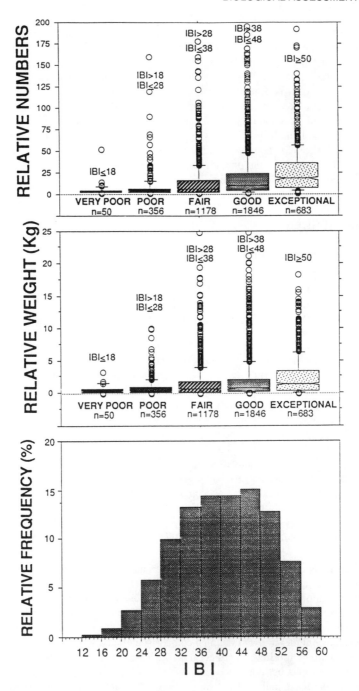

Figure 12. The relative frequency of occurrence (lower), relative biomass (middle), and relative number (upper) of smallmouth bass (*Micropterus dolomieui*) compared to the range of IBI scores observed in the Ohio EPA statewide database, 1979–1992.

Macroinvertebrate community data is intended to secure an adequate sample, but not an exhaustive inventory of all taxa possible. Relative abundance data is collected using a standardized approach (artificial substrates; DeShon, Chapter 15). In an analysis of resources expended during FFY 1987 and 1988 the following were revealed:

- 5.02 WYE were used to collect 323 samples at 323 sampling sites, setting or retrieving 4 to 6 sites per day.
- An average of 0.015 WYE or 33.2 hours per site were expended to plan, collect, analyze, interpret data, and to produce reports at an average cost of $824 per site.
- Laboratory effort is 12 to 20 hours per sample for artificial substrates and 2 to 6 hours for qualitative samples only.
- A field crew consists of one full-time biologist and one intern.

If the Ohio universe of named and perennial stream and river miles is used, approximately 1 WYE is needed for each 1000 miles. This translates to approximately 29 WYE to cover the biological monitoring needs across Ohio via a basin approach to water resource management. Out of nearly 100 WYE that were devoted to surface water monitoring and laboratory activities within the Division of Water Quality Planning and Assessment in FFY 1987 and 1988, 19.34 WYE or just over 19% of the total was devoted to ambient monitoring. When considered on the basis of agencywide water programs (e.g., NPDES permitting, enforcement, etc.) this percentage is approximately 6%.

Table 6 gives the unit cost of the four monitoring components in Ohio's program. Costs are broken down by sample collection, laboratory analysis, individual test, and evaluation as appropriate for each component. Included in the cost figures are all equipment, supplies, logistical, administrative, data analysis, and interpretation activities. States that do not operate extensive ambient bioassessment networks will need to be prepared for some rather sizeable start-up costs. While the cost analysis incorporated start-up costs for equipment and supplies, these were amortized over 5 or 10 years depending on the expected life of an item. Start-up equipment and supplies, for most states, could total from $200,000 to $500,000 depending on the number of field crews involved.

5.1.2 Data Management and Information Processing

The principal Ohio EPA data management system for fish, macroinvertebrate, and habitat data (Ohio ECOS) includes storage, processing, and analysis routines. Once field data are collected, processed, and finalized, the next step is to reduce the data to scientifically and managerially useful information. Data are tabulated in the field (fish and habitat) and laboratory (macroinvertebrates), and documented via chain-of-custody procedures; the data is entered directly into the electronic database. Basic information includes the field crew, waterbody name, date, and time. Site location is indicated by river mile (distance upstream from mouth) and latitude/longitude both of which are determined from USGS 7.5-minute topographic maps. A basin-river code system is used to electronically identify individual streams, rivers, and lakes. Sampling information includes method or gear type and other information relevant to the use of each. Ohio ECOS generates data summaries and reports for a variety of community measures, community composition, or individual species/taxon analyses.

5.2 Maintenance of the Reference Site Network

The adoption of numerical biological criteria includes the task of maintaining the reference database, which includes a planned resampling of all sites within a prescribed time frame. A concern that is frequently expressed is that by basing aquatic community performance expectations on contemporary conditions defined by present-day reference sites, aquatic life goals are somehow being frozen in time. This is why the concept of continual maintenance monitoring must be included as a part of the overall regional reference site approach. In Ohio, we have chosen to sample approximately 10% of the reference sites each year within the organization of the Five-Year Basin Approach. This will provide an opportunity to examine regional background aquatic community performance at periodic intervals (e.g., once every ten years) and make appropriate adjustments to the calibration of the multimetric indices, the numerical biological criteria, or both.

5.3 Biological Index Variability

A frequent criticism of ambient biological data is that it is too variable to function as a reliable component of surface water resource assessment. Natural biological systems are indeed variable and

Table 6. Cost Comparison of Macroinvertebrate Community and Fish Community Evaluations with Chemical/Physical Grab Sampling and Acute and Acute/Chronic Bioassay Tests

Sample collection	Analytical cost (Laboratory)	Cost per test/sample	Cost per evaluation
	Macroinvertebrate Community		
	Artificial Substrates (includes qualitative sample)		
N/A	$397	$824	$824
	Qualitative Sample Only		
N/A	$150	$275	$275
	Fish Community		
	Cost per sample		
N/A	N/A	$340	$340
	Cost per site		
N/A	N/A	$340	$740
	Chemical/Physical Water Quality (4.6 samples per site)		
$1124[1]	$529[2]	$359	$1,653
	Bioassay		
	Screening[3]		
$261	N/A	$1,191	$3,573
	Definitive[4]		
$261	N/A	$1,848	$5,544
	Seven-day[5]		
$261	N/A	$3,052	$9,156
	Seven-day[6]		
$1,973	N/A	$6,106	$18,318

[1] Includes cost of sample collection and data analysis only; based on an average frequency of 4.6 samples collected per site in 1987 and 1988.
[2] Analytical costs based on each sample being analyzed for five heavy metals ($7.00 ea.), four nutrients ($10.00 ea.), COD or BOD ($20.00 ea.), and two additional parameters ($20.00 for both); $115 per sample.
[3] 48-hour exposure to determine acute toxicity.
[4] 48- and 96-hour exposure to determine LC50 and EC50.
[5] Seven-day exposure to determine acute and chronic effects using a single 24-hour sample; cost based on analysis of one pipe only; costs for chemical analyses in sole support of the test are not included.
[6] Seven-day exposure using a composite sample collected daily (renewal); other factors in footnote 5 apply.

seemingly noisy, but no more so than the chemical and physical components that also exist within aquatic ecosystems. Certain dimensions of ambient biological data are quite variable, particularly population- or subpopulation-level parameters. Single-dimension community measures can also be quite variable. The new generation community evaluation mechanisms such as the IBI and ICI are sufficiently redundant so as to compress and dampen some of the aforementioned variability. Rankin and Yoder (1990) examined replicate variability of the IBI from nearly 1000 sites throughout Ohio and found it to be quite low at least impacted sites (Figure 13). Coefficient of variation (CV) values were less than 10% at IBI ranges indicative of exceptional biological performance and generally less than 15% for the good performance range. This is lower than the variability reported for chemical laboratory analyses and interlaboratory bioassays (Mount 1987). Variability as portrayed by CV values increased at IBI ranges indicative of increasingly impaired biological performance. Low variability was also found for the ICI with a CV of 10.8% for 19 replicate samples at a relatively unimpacted test site (DeShon, Chapter 15). The variability of the MIwb was determined to be on the order of ±0.5 MIwb units (Ohio EPA 1987b). Other investigators have reported similarly low variability with other biological indices (Davis and Lubin 1989;

IBI VARIABILITY: REPLICATE SAMPLES

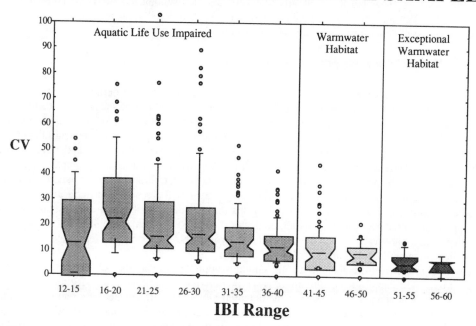

Figure 13. Coefficient of variation (CV) for a range of IBI scores at sites with three sampling passes per year. Boxes show median, 25th and 75th percentiles, minimum, maximum, and outlier values.

Stevens and Szczytko 1990). Fore et al. (1994) used differential statistical techniques and determined a variability of ±3 IBI units using the Ohio database. Cairns (1986) suggested that differences in variability rather than differences in averages or means might be the best measure of stress in natural systems. Variability must begin to be recognized as a part of the signal rather than noise alone (Karr 1991). Not only is the variability of the measures used as biological criteria low, the degree of variability encountered can also be a useful assessment and interpretation tool.

Ohio EPA has addressed the variability inherent to biological measures in three general ways:

1. Variability is **compressed** through the use of multimetric evaluation mechanisms such as the IBI and ICI.
2. Variability is **stratified** by the tiered use classification system, ecoregions, biological index calibration, and site type.
3. Variability is **controlled** through standardized sampling procedures that address seasonality, effort, replication, gear selectivity, and spatial concerns.

Lenat (1990) also described similar approaches to controlling and reducing the variability of ambient biological monitoring.

5.4 Concerns About Potentially Underprotective Criteria

Concern is frequently expressed about the potential for biological criteria to be underprotective. One specific point is that the selection of reference sites could be biased to impaired conditions. At least two factors used in setting the WWH and EWH criteria offer substantial protection against the potential influence of substandard or marginal reference sites. The IBI and ICI are based on a trisection and quadrisection calibration procedure, respectively, which focuses on a line of maximum value (i.e., 95% line; Figure 4). Thus the influence of sites with low metric values is negligible because the calibration procedure is controlled by the sites with the highest values. Secondly, choosing the 25th percentile of the reference site results as the biocriterion for each index eliminates values that were low because of factors that the resident biota could discern, but to which the initial reference site selection procedure was not sufficiently sensitive. Together these safeguards ensure that the resultant biocriteria are consistent with the goals of the CWA and protective of their designated uses. Additionally, the reference site resampling previously discussed is another important safeguard in the process.

5.5 How Many Reference Sites Are Enough?

We have frequently been asked this question as most are interested in deriving technically valid biological criteria at the lowest cost. Logically, enough reference sites must be selected to account for the range of natural variability among the least impacted reference sites within a region. Increased variability among reference sites, if it originates from natural sources and not sampling error, indicates the need to employ a stratification scheme among the reference sites for the purpose of biocriteria derivation. High variability among reference sites without obvious natural causes could be a result of sampling problems, which an increase in the number of reference sites would not correct. Stratification of natural variability is an essential component of biological criteria development if the resultant criteria are to become managerially useful. Our approach accomplished this through the use of tiered use designations, site types, and ecoregional stratification. Additional stratification variables could include mean annual temperature (e.g., warmwater vs. coolwater streams; Lyons 1992) and gradient (e.g., low-gradient vs. high-gradient streams; Leonard and Orth 1986).

Assuming proper stratification and a valid sampling approach we can then determine the minimum number of reference samples needed to arrive at a biocriterion (e.g., 25th percentile for the Ohio WWH use designation) that adequately represents the potential biological performance of a region. The range of natural variability will not be encompassed with an insufficient reference database on which stratified expectations are to be based. This could result in biocriteria that are either under- or overprotective of the biological performance defined by the designated aquatic life uses.

To illustrate the effect of reference site sample size on the Ohio EPA IBI biocriteria, we randomly selected sites from our reference database for each ecoregion and site type combination and, without replacement, recalculated the 25th percentile WWH biocriterion after samples were added in increments of five. The procedure was performed for 50 trials over 15 different sets of reference sites (5 ecoregions and 3 site types per ecoregion). The results were plotted on a three dimensional bar chart with the frequency at which a 25th percentile biocriteria value was randomly selected vs. sample size. The analog of an asymptotic relationship of a 25th percentile IBI value with increasing sample size defined the minimum number of reference sites that are needed to achieve a biocriterion value that encompasses the inherent background variability.

Our criterion to determine when the analog to an asymptotic relationship was reached was the point at which the variation in the 25th percentile value narrowed to one predominant index value in terms of the number of observations per aggregation category. Of the 15 sets of reference samples tested (5 ecoregions and 3 site types per ecoregion) this point ranged from a low of 10 to 15 samples for headwater sites in the Interior Plateau ecoregion to 75 to 80 samples for boat sites in the Eastern Corn Belt Plain (ECBP) ecoregion. The Huron–Erie Lake Plain (HELP) ecoregion appeared to require the fewest reference samples to reach the point of diminishing return (Figure 14) and the ECBP ecoregion appeared to require the most reference samples (Figure 15). The other ecoregions tended to be intermediate between the HELP and ECBP.

Ecoregions with widespread and uniform land disturbance, such as the HELP, require fewer samples to characterize the **present** reference condition while those with a greater degree of natural heterogeneity (i.e., ECBP) require the most samples. Most of the reference sites were sampled at least two times, which makes the safe minimum number of sites for the Ohio ecoregions from as few as 5 to 8 sites per ecoregion per site type/stream size strata for the more homogeneous ecoregions such as the HELP and IP to as many as 38 to 40 sites per ecoregion per site type/stream size strata for the more heterogeneous ecoregions (i.e., ECBP), which could also illustrate the need for further landscape stratification via subecoregions. We believe that if uncertainty exists about the variability within an ecoregion **more** sites should be used than too few. In our experience this would be approximately 35 to 40 sites per ecoregion per site type. This may vary across the nation as these figures are most representative of the agricultural midwestern United States.

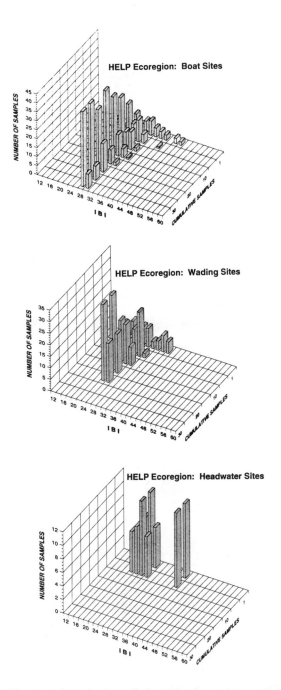

Figure 14.   Results of a random sampling selection technique of an increasing number of reference site samples (sampled without replacement) on the possible 25th percentile IBI values for the boat (upper), wading (middle), headwater (lower) site types in the Huron/Erie Lake Plain (HELP) ecoregion.

A failure to stratify variability where the clear need for a stratification scheme exists risks inaccurate biocriteria that may be underprotective of sites with greater biotic potential and overprotective of sites with lower biotic potential that otherwise would have been adequately protected by lower criteria. In contrast, attempts to stratify regions where little difference exists may lead to unnecessary regulatory complexity and an unsound and arbitrary scientific basis for biocriteria development.

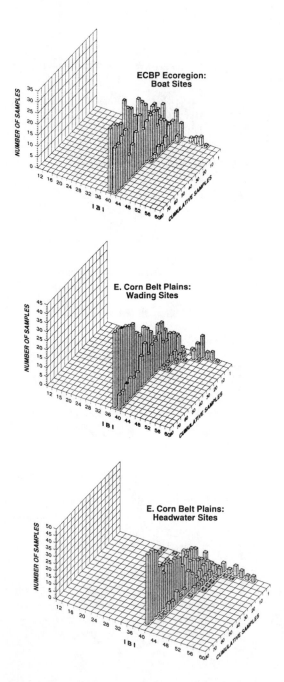

Figure 15. Results of a random sampling selection technique of an increasing number of reference site samples (sampled without replacement) on the possible 25th percentile IBI values for the boat (upper), wading (middle), headwater (lower) site types in the Eastern Corn Belt Plain (ECBP) ecoregion.

The minimum number of reference sites also depends on the statistics upon which the criteria will be based. Extreme percentiles (e.g., 5th, and 95th), because they represent the tails of distribution functions, are characterized by wider confidence bounds around the threshold statistic, and will require a larger number of sites before a stable asymptote is reached, whereas the median of the same distribution will reach an asymptote with fewer samples (Berthouex and Hau 1991). For example, if a state wants to

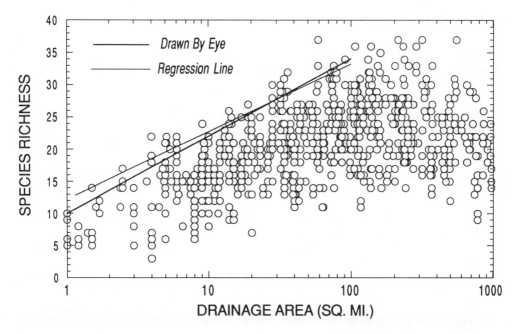

Figure 16. Comparison of the determination of the 95th percentile maximum species richness line (MSRL) using the traditional line-of-best-fit-by-eye method vs. a statistical derivation using the technique of Blackburn et al. (1992).

derive a biocriterion from a database of both impaired and unimpaired sites, which calls for a higher measure as a criterion such as the 90th percentile (Barbour et al., Chapter 6; Ohio EPA 1987b), then a larger number of sites will be required to achieve a representative biocriterion.

5.6 Index Calibration

The determination of the 95% line is one of the most important parts of the calibration process. While the line-of-best-fit method is presently accepted (Fausch et al. 1984), it is not a strict statistical derivation. As an experimental approach to possibly improve the objectivity of the 95% line determination we applied the technique described by Blackburn et al. (1992) in which a series of regression lines are determined across the upper surface of the wedge of points that result from the scatter plots of drainage area-dependent IBI metrics. Thus far we have determined this for the fish species richness metric. The results indicate a line that is not substantially different from the line-of-best-fit method (Figure 16). While this seems to initially confirm the line-of-best-fit method it appears to offer some important advantages, the most obvious of which is a statistically objective method for determining the 95% line. One important drawback, however, is the inability of the statistic to determine when and where the slope of the line should change. This was done by visual interpretation for several of the IBI and most of the ICI metrics (Figure 4).

Calibration issues that need further examination include determining the degree of convergence between the 5, 3, and 1 lines at the lower drainage areas, the nonlinear vertical distribution of the scatter plots for the percent of metrics, and how to determine scoring for metrics that have no apparent relationship with stream size. Lyons (1992) uses a calibration procedure where 10 divisions are made resulting in individual metric scores ranging from 1 to 10 and a 12- to 100-point scoring range. Other considerations include the consistent designation of trophic guilds, tolerance rankings, refined metrics, refined metric scoring, and regional calibration. For example, differences exist in the designation of feeding and tolerance guilds between states that share similar faunas. In addition, criticism has been leveled at intolerant species designations as reflecting rare, threatened, and endangered status more so than true environmental tolerance. While we have dealt with most of these issues in Ohio, these and other issues will arise elsewhere, thus regional consistency in achieving a resolution of these issues will be needed.

6.0 RECOMMENDATIONS

Although we have presented here a framework from within which numerical biocriteria can be developed and implemented by states, there remain important areas for future development and research. Some of these follow:

1. The continued development of more detailed regionalization frameworks (e.g., subecoregions) needs to proceed nationally. This is important not only for biocriteria development in terms of reference sites and calibration regions, but also in the application of biocriteria in watershed management and nonpoint source assessment.
2. Important issues such as the minimum number of reference sites, calibration regions, tiered use designations, etc. need to continue to be examined, particularly in different regions of the United States.
3. In keeping with recommendation 2, consideration should be given to establishing regional technical panels that could deal with the methods, regionalization, and biocriteria derivation issues on a regional basis. This could be accomplished, in part, within the existing administrative framework of the relevant federal agencies.
4. Biocriteria need to be formally incorporated into all relevant aspects of water quality management. The discovery of unknown or poorly understood problems is one of the strengths of bioassessment and biocriteria. Yet when problems are discovered they are not always acted on, especially if they do not fit into an existing regulatory framework. This could be accomplished via a redirection of existing state agency resources.
5. Biocriteria should be used to validate chemical-specific criteria and site-specific criteria modifications.
6. Based on our experience in Ohio, staffing in state programs should be a minimum of 1 WYE for every 1200 miles of perennial streams and rivers. This estimate may vary in other regions and should additionally incorporate lake acres in states with a predominance of this waterbody type.
7. The role of ambient bioassessments and biocriteria in determining compliance with the provisions of the CWA should be defined and formalized. This would include modifying the present capital- and resource-intensive system of tracking compliance at the source and on a pollutant-specific basis by supplementing and even replacing some measures with the more holistic and resource-focused biological measures. This may well prove to be a more cost- and information-effective approach.

ACKNOWLEDGMENTS

This chapter would not have been possible without the many years of field work, laboratory analysis, and data assessment and interpretation by members (past and present) of the Ohio EPA, Ecological Assessment Section. Extensive contributions were made by Dave Altfater, Randy Sanders, Marc Smith, and Roger Thoma (fish methods, MIwb, and IBI metrics development) and Mike Bolton, Jeff DeShon, Jack Freda, Marty Knapp, and Chuck McKnight (macroinvertebrate methods and ICI metrics development). None of this would have been possible without the excellent data management and processing skills of Dennis Mishne. Other staff who also made contributions to the process include Paul Albeit, Ray Beaumier, Chuck Boucher, Bernie Counts, Beth Lenoble, and Paul Vandermeer. Dan Dudley and Jim Luey contributed extensively to the early development and review of important concepts of biological integrity, ecoregions, reference sites, and biological monitoring in general. Charlie Staudt provided many hours of support in the development of the basic computer programs. Finally, Gary Martin and Pat Abrams (deceased) are credited for their solid management support for the concept of biological criteria and monitoring at the Ohio EPA. Marc Smith provided comments on earlier versions of the manuscript.

## CHAPTER 10

Development of Regionally Based Biological Criteria in Texas

C. Evan Hornig, Charles W. Bayer, Steve R. Twidwell, Jack R. Davis, Roy J. Kleinsasser, Gordon W. Linam, and Kevin B. Mayes

1.0 INTRODUCTION

The Texas Natural Resource Conservation Commission (TNRCC) is the state agency responsible for water quality regulations. TNRCC partitioned the major state waters into classified segments, each with a designated aquatic life use (exceptional, high, intermediate, or limited). Classifying all the minor waters (including approximately 3700 small tributaries) was, however, not practical. Therefore, the agency grouped together these minor waters under the same preliminary aquatic life designation. When a major point source control decision (such as an application for a new discharge permit) to one of the minor waters is anticipated, a biological and habitat assessment may be requested. This assessment is used to confirm or change this water's preliminary aquatic life designation. As a result, biological assessments play a critical role in deciding control requirements of many Texas dischargers.

Historically, biological assessments have often involved comparisons between upstream and downstream sites. Over time, however, TNRCC has identified several shortcomings to the approach of using upstream sites as references for evaluation of downstream sites. These shortcomings involve waters with multiple point or nonpoint source discharges and streams where the habitat or flow regimes differ between control and downstream sites. With many Texas streams, most or all of the stream flow originates as effluent.

More recently, TNRCC has used a set of statewide habitat and biological community metrics in aquatic life use evaluations (Twidwell and Davis 1989). As recommended by the U.S. Environmental Protection Agency's (USEPA) Rapid Bioassessment Protocols (Plafkin et al. 1989), metrics are combined into multimetric indices for fish, macroinvertebrate, and habitat evaluations. Metrics were developed from direct experience, professional judgement of state biologists, and evaluations of the literature. The multimetric fish index is based on the Index of Biotic Integrity (IBI) (Karr et al. 1986). A modification of this index better reflects Texas fish communities (Linam and Kleinsasser 1987). The modified IBI incorporates abundance, species richness (of entire collection and of select groups), intolerance, trophic composition, and fish condition. The multimetric index for invertebrates, (Texas Mean Point Score or MPS) is an average of scores (ranging from one to four) that rate species diversity (including diversity of the generally intolerant mayflies, stoneflies, and caddisflies), standing crop, and community trophic structure. Habitat Quality Index (HQI) sums the observation scores for instream cover, prevalence of riffles and depth of pools, bottom substrate, degree of flow fluctuations, channel sinuosity, bank stability, riparian cover, and aesthetics (Table 1). The HQI was adapted from several sources (Binns 1982; Platts et al. 1983; USEPA 1984b; Maret 1986; Ohio EPA 1989a). Aquatic life use designation is based on an evaluation of all three composite metrics.

Table 1. Habitat Quality Index Attributes and Scores Used by Texas Natural Resource Conservation Commission

Rating parameter	Attributes of subcategories (rating scores)			
Instream cover	Common (4)	Occasional (3)	Rare (2)	None (0)
Riffle/runs	Common (4)	Occasional (3)	Rare (2)	None (0)
Pool depth	>4 ft (4)	2 — 4 ft. (3)	< 2 ft. (2)	No Pools (1)
Bank stability	Eroded banks <10%; side slopes <30° (3)	Eroded Banks 10–30%; side slopes up to 40° (2)	Eroded Banks 31–50%; side slopes up to 60° (1)	Eroded Banks >50%; side slopes >60° (0)
Riparian width	>350 ft (3)	150–350 ft. (2)	15–150 ft. (1)	<15 ft. (0)
Flow fluctuations	Minor Little or None from base flow (3)	Moderate Debris along middle portion of banks (2)	Severe Debris high on banks (0)	Severe Intermittent stream (0)
Channel sinuosity	High (3)	Moderate (2)	Low (1)	None (0)
Bottom substrate	>50% Gravel or larger (3)	31–50% Gravel or larger (2)	10–30% Gravel or larger (1)	<10% Gravel or larger (0)
Aesthetics	Wooded or unpastured natural area; water exceptional (3)	Some development (fields, pastures, dwellings); water discolored (2)	Developed, but uncluttered; water discolored or turbid (1)	Offensive; cluttered, water discolored or turbid (0)

Total score: 26–30 exceptional; 21–25 high; 15–20 intermediate; <15 limited.

Texas recognized the need to improve on these metrics by basing them on the conditions found in the healthiest streams of the state's regions. In 1986, TNRCC and the Texas Parks and Wildlife Department (TPWD) initiated the Texas Aquatic Ecoregion Project. The project's goal was to locate and characterize many of the state's minimally impacted streams. Besides the establishment of a regionally tailored reference database from which to derive designated-use assignments, results from this project would also provide biological benchmarks for measuring improvement or degradation of environmental quality by TNRCC and others.

Field collection for the project ended in 1990, with sample processing completed in 1992. This chapter describes the methods employed, provides results from preliminary data analyses, and discusses expected approaches for developing regionally based criteria in Texas.

2.0 METHODS

2.1 Selection of Reference Sites

Candidate reference streams are bordered by mature riparian zones and flow through watersheds most closely resembling that of the natural vegetation for the associated ecoregion; typically forests, brush-lands, or grasslands. The streams' watersheds lacked urban development, obvious point sources of pollution, channelization, or atypical nonpoint sources of pollution. Streams selected represent a range of watershed sizes in each ecological region of Texas. Ecoregions are based on Omernik's ecoregion map for the south-central states (Omernik and Gallant 1987). These ecoregions are developed through analyses of existing maps of regional patterns in land surface form, land use, potential natural vegetation, and soil type. Variables important to determining aquatic ecosystem attributes include water chemistry, flow regime, habitat structure, and food source (Omernik 1987). The ecoregion map also depicts those areas most typical of each ecoregion.

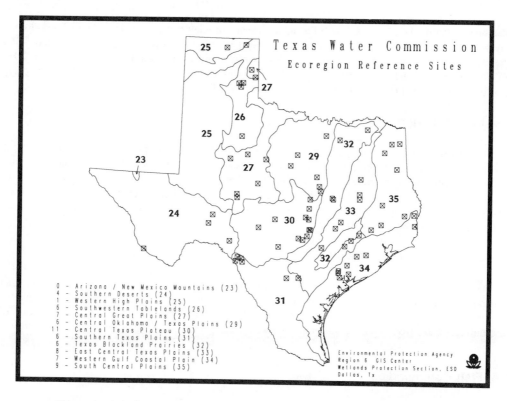

Figure 1. Location of reference sites used in the Texas Aquatic Ecoregion Project.

Streams, whose entire watersheds are within these most typical portions of ecoregions, are chosen as the primary candidates for the study. However, because of the scarcity of minimally impaired perennial streams in the more arid ecoregions of Texas, some streams with watersheds extending beyond the most typical areas of these ecoregions are also included. Eleven of the twelve Texas ecoregions were sampled during the study. The Arizona/New Mexico Mountains Ecoregion was not included since it extends only slightly into Texas, at Guadalupe Mountains National Park.

A minimum of two perennial streams from each of three watershed size groups (<50 mi², 100 to 200 mi², and >300 mi²) were selected from each ecoregion. Several intermittent streams were also selected for sampling. Figure 1 shows the location of the sites sampled on the candidate reference streams.

2.2 TNRCC Pilot Study

Texas Natural Resources Conservation Commission conducted a pilot study of six candidate ecoregion reference streams during 1986 to 1987. The pilot study helped to refine field and analytical procedures and determine the need to sample over multiple seasons and multiple stream sites (Twidwell and Davis 1989). Results from this study showed more information would be obtained by sampling more individual streams than more sites on the same stream. During 1988 to 1990, TPWD joined TNRCC to sample an additional 66 streams. These streams were all sampled during the summer period (June through September), to correspond to critical low flow and elevated temperature conditions.

2.3 Physicochemical and Bacterial Parameters

Dissolved oxygen, pH, temperature, and specific conductance were measured in the field hourly over a 24-h period with either a Hydrolab® Surveyor II or Data Sonde. Surface water grab samples obtained at each site were analyzed at the Texas Department of Health and Environmental Chemistry Laboratory by Standard Methods (APHA 1985). Fecal coliform samples were analyzed within 6 h using the membrane filter method (APHA 1985). Physicochemical parameters measured include: carbonaceous 5-day biochemical oxygen demand (CBOD$_5$), total suspended solids (TSS), volatile suspended solids (VSS), orthophosphate (OP), total phosphate (TP), kjeldahl nitrogen (KjelN), ammonia nitrogen (NH$_3$-N), nitrite nitrogen (NO$_2$-N), nitrate nitrogen (NO$_3$-N), chlorophylla (Chl$a$), chloride (Cl), sulfate (SO$_4$) total dissolved solids (TDS), turbidity, total alkalinity (T-Alk), and hardness.

2.4 Habitat Parameters

Watershed areas and stream gradients were determined from county highway maps and USGS 7.5 topographic maps, respectively. Stream flow was recorded with a Marsh-McBirney® Model 201 flow meter. Instream habitat and riparian characteristics were normally measured at five transects along a 0.5 to 1.0 km stream reach. Habitat parameters included stream width and depth, instream cover (large woody debris, boulders, undercut banks, vegetation) substrate composition, tree canopy, and stream bank features (slope, stability, vegetative cover).

2.5 Benthic Macroinvertebrates

Benthic macroinvertebrates were collected from riffles at most sites with a 1-ft^2 Surber sampler (Klemm et al. 1990). At sites where no riffles were found (primarily in eastern Texas), grab samples of the soft bottom substrate were collected with a 0.25-ft^2 Ekman sampler (Klemm et al. 1990). All macroinvertebrates from three replicates per site were field-preserved and identified in the laboratory to the lowest possible taxonomic level.

2.6 Fish Communities

Fish were collected from all habitats (riffle, run, and pool) in proportion to their availability. The various types of cover and substrates were sampled until no new species were found. Sampling methods and durations were dictated by available habitats, flow regime, and water chemistry (conductivity). Seines, backpack electrofishing, and boat electrofishing were the gear types employed, respectively, at 98%, 73%, and 10% of the sites. Seining was used at most sites to complement electrofishing; and, in streams where high conductivity precluded use of electrofishing gear, seining became the primary sampling method. At sites where a combination of gear was used, the mean number of seine hauls was 6.5, with a total length of 63.6 m of stream sampled. The average duration of backpack shocking was 14 min with an average length of 90 m of stream sampled. Additional effort was required at sites where only one collection method was used.

The primary sampling method was electrofishing with a battery powered backpack unit (Smith-Root® Type VII). Sampling was conducted in an upstream direction to eliminate turbidity effects caused by bottom sediment disturbance. An attempt was made to net all stunned fish observed. Typically, one pass was adequate to acquire the data needed for IBI calculation. Larger fish were identified to species, enumerated in the field, and usually released. All other specimens were preserved in formalin and returned to the lab, where they were identified to species and enumerated. All fish were examined for external deformities, skeletal anomalies, eroded fins, tumors, and lesions.

3.0 ANALYSIS OF RESULTS

3.1 Physical and Chemical Characteristics — Regional Patterns

Physical, chemical, and habitat data were compiled, summarized, and tested (using ANOVA) for differences among ecoregions (Bayer et al. 1992). Several habitat parameters exhibit geographical patterns. Canopy cover increases from west to east, corresponding to the moisture gradient across the state. Bottom substrates of streams in the hill country of central Texas have higher percentages of gravel and cobble than substrates of northern or eastern plain streams. Other habitat parameters, such as instream cover and maximum pool depth, display no clear geographical trends.

Regional trends were demonstrated for several chemical parameters. Total dissolved solids in far eastern Texas streams are approximately 100 mg/l, severalfold less than in the south, and an order of magnitude less than found in western Texas. Phosphate levels are highest in the Gulf Coastal Plains Ecoregion. Several water quality factors, most notably total suspended solids, and fecal coliforms show no apparent regional patterns; differences within ecoregions are much greater than differences among ecoregions.

Dissolved oxygen (24-h average concentrations) tend to be lowest in the eastern and coastal ecoregions. Because of its importance to treatment control requirements and its suspected relations with other, easily measured, environmental features, dissolved oxygen in these regions was further investigated. Using a stepwise multiple regression, variations in three parameters, bedslope, stream flow, and canopy cover, were found to explain 72% (coefficient of determination or $R^2 = .72$) of the variation in dissolved oxygen. Using the data from the ecoregion study, TNRCC is drafting new dissolved oxygen criteria for nontidal streams of eastern and southern Texas. These proposed criteria will consider bedslope and stream flow conditions, using the following multiple regression equation developed from the study's data:

$$DO = 9.054 + 0.551*\ln(Q + 0.0003) + 0.686*\ln Bd - k$$

where, Q is stream flow in m^3/s; Bd is bedslope expressed as meters per kilometer of stream length, and k is a conservatively applied correction factor (1.64) for percent stream canopy.

As demonstrated in Figure 2, this equation predicts the actual dissolved oxygen concentrations from a majority of these ecoregions' 42 reference sites (including nine Arkansas sites). Application of this modification provides a more realistic match of criteria with attainable standards when compared to statewide standards.

3.2 Aquatic Communities — Regional Patterns in Taxa Composition

A total of 518 macroinvertebrate taxa were identified from 81 collections. As many as 85 taxa were collected from a 3-ft^2 area. These taxa included many chironomids and oligochaetes, identified to the lowest possible level. An ordination technique, detrended correspondence analysis (DCA) was used to compare between-site similarities in community composition of the biological collections. Using this method, sites clustering closely together on an ordination plot were more similar in their composition. The analyses were performed on a PC using Version 3.10 of CANOCO (ter Braak 1988, 1990). The DCA performed best when (1) samples from soft substrates were excluded, (2) inadequate samples (<100 individuals) were excluded, (3) rare species (found in less than four samples) were excluded, and (4) the counts were transformed (octave or log base 2 scale put to percent data). In addition, all collections with less than 20 taxa or a Texas Mean Point Score of less than 2.5 were eliminated from consideration as invertebrate reference collections. Using the above criteria, a total of 193 invertebrate taxa from 58 collections were subjected to the community ordination.

The resulting ordination showed that invertebrate community composition varied geographically, but not strictly along the ecoregion lines (Figure 3). Invertebrate collections from the three eastern ecoregions clustered separately from the other Texas ecoregions. Sites from a northern ecoregion (Southwestern Tablelands) grouped with sites located within the northern areas of neighboring ecoregions. The third major cluster consisted of sites located in the central and west-central parts of Texas, including sites from the southern areas of northern Texas ecoregions (Central Great Plains and Oklahoma/Texas Plains). Interestingly, there is a part of the Blackland Prairies Ecoregion that clustered separately, east of the main body of this ecoregion, being surrounded by one of the eastern Texas ecoregions (East Central Texas Plains). The two reference sites located in this disjunct part of the Blackland Prairies Ecoregion clustered between the two major groupings of eastern and central ecoregions. Possibly, both ecoregional (soils, etc.) and zoogeographical (dispersal) factors decide distribution of macroinvertebrates.

The results suggest that faunal distribution is partially a function of nonecoregion variables. To construct new regional boundaries based on this limited data set al.one, however, may be problematic. Further analyses of the data are planned that will employ additional existing geographical classifications; testing which one best describes spatial trends among the biological (fish and benthos), chemical, and habitat data. These geographical classifiers include aquatic ecoregions (Omernik and Gallant 1987),

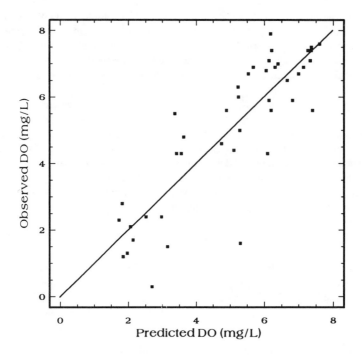

Figure 2. Comparison of observed dissolved oxygen at 42 sites located in the ecoregions of eastern Texas and southeastern Arkansas to the DO predicted from bedslope, stream flow, and canopy cover data at these sites. See text for mutiple regression equation used to predict DO.

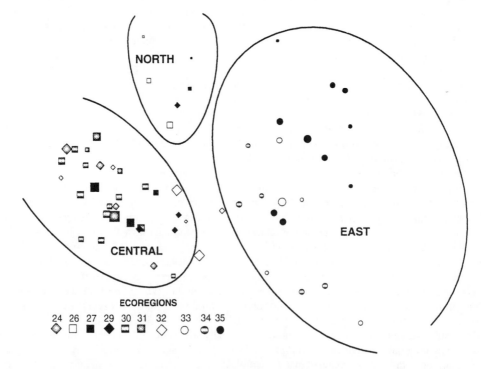

Figure 3. Reference site ordination using detrended correspondence analysis of macroinvertebrate taxonomic composition (size of symbol is proportional to taxa richness).

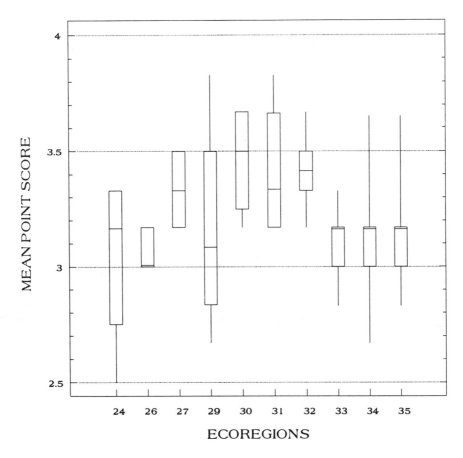

Figure 4. Texas mean point score (MPS) summaries by ecoregion. Boxplots depict ranges, 25th and 75th percentiles, and medians.

aggregated ecoregions (Omernik and Gallant 1989), biotic provinces (Blair 1950), game areas (Hubbs 1982), and river basins. Physical and chemical data will be subjected to principal components analysis (PCA), while DCA will be applied to the fish and benthos data. Site scores from the PCA and DCA axes will then be analyzed separately using analysis of variance (ANOVA). The regional classification schemes listed above will be evaluated by the relative fit (ANOVA variance ratio) of their respective categories (regions or provinces).

3.3 Aquatic Communities — Regional Patterns in Biometric Scores

Currently, the Water Quality Standards implementation procedures uses an invertebrate mean point score (MPS) value of 3.0 statewide to classify a stream supporting high aquatic life uses. Although results from the Texas Aquatic Ecoregion Project indicate that this MPS value reasonably reflects conditions in the best streams of Texas, differences among ecoregions are apparent (Figure 4). Regional modifications using this data set may raise this MPS expectation by approximately one third of one point for the spring dominated streams in the central part of the state (ecoregions 30, 31, and 32) and one sixth of one point for the eastern ecoregions (33, 34, and 35).

Preliminary analyses of fish data also reveal some apparent patterns in metrics (Bayer et al. 1992). For example, the greatest number of species (generally greater than 20) were found in eastern Texas. Lower diversity (often 10 or less species) was observed in the Texas panhandle and in western Texas. Modified forms of the Index of Biotic Integrity have previously been used in Texas to evaluate fish communities. Recently, a modified IBI was used to establish numeric criteria to support the state water quality standards for the Trinity River (Kleinsasser and Linam 1990). Components of this IBI will be further modified on a regional basis to reflect the geographic differences found in this study. Besides species richness, modifications in IBI components are expected to better reflect regional patterns in fish families sensitive to perturbations. One possibility may be the substitution of cyprinids in areas where darters are depauperate. Differences were also observed between runoff-dominated and spring-dominated springs in the Trans Pecos area, and criteria modifications will additionally incorporate these differences. Minimal modifications are expected in eastern Texas, with greater changes expected in the Panhandle and western Texas. Because the habitats (pool/riffle/run) are sampled in proportion to their availability, the metrics reflect the combination of habitats typical for a region. For example, the criteria in eastern Texas will reflect the absence of riffle habitat. The fish community criteria will also reflect the sampling method common for the region; typically electrofishing, complemented by seining.

4.0 SUMMARY

This study's biological (particularly macroinvertebrate) collections cannot be expected to represent the entire range and variety of conditions in the state's many reference-quality streams. However, the study (1) enables the state to tie their metrics to actual stream data, and (2) results in regional differences in criteria, a substantial improvement over statewide standards. The state considers this initial database and resulting metric modifications to be an initial step in the regionalization of standards. TNRCC and TPWD are continually revisiting reference sites and will periodically reevaluate the results as needed to further refine biological metrics. Texas will then use these modifications to update its Water Quality Standards implementation procedures for biological criteria.

## CHAPTER 11

Biocriteria: A Regulated Industry Perspective

Robin J. Reash

1.0 BIOCRITERIA: NEW CHALLENGES FOR REGULATED INDUSTRY

Biocriteria (biological criteria) are being developed and implemented with increasing frequency as water quality management tools for state agencies responsible for administering the National Pollutant Discharge Elimination System (NPDES) program. Relying on U.S. Environmental Protection Agency's (USEPA) national guidance on biocriteria (USEPA 1990a) and national policy on biosurveys/biocriteria (USEPA 1991c), states are initiating programs that will result in the adoption of narrative and/or numeric biocriteria within water quality standards. The statutory mandate of required biocriteria development is somewhat controversial. Though USEPA's position is that the development and adoption of biocriteria is required by states within the water resource management program, it should be noted that some industry groups have questioned USEPA's authority to require biocriteria under the Clean Water Act. Nonetheless, USEPA has reaffirmed its position that states must initially adopt narrative biocriteria to comply with statutory requirements under Sections 303 and 304 of the Clean Water Act. Specifically, USEPA has advocated that states are to adopt narrative biological criteria into state water quality standards during the FY 1991–1993 triennium and then adopt numeric biocriteria by the end of FY 1996 (USEPA 1990). Regulated industry must become aware of this implementation "clock" and closely follow the development of biocriteria in individual states, whatever the implementation schedule happens to be. Because so few states have progressed to the point of proposing legally binding narrative or numeric biocriteria at this time, industry involvement has been relatively limited on a national scale.

The State of Ohio has led all other states in the derivation and regulatory adoption of numeric biocriteria, currently being the only state with approved biocriteria for tiered aquatic life use designations within ecoregions (see Southerland and Stribling, Chapter 7; Yoder and Rankin, Chapter 9). With the exception of facilities on specific waterbodies (e.g., acid mine drainage streams, the Ohio River, and Lake Erie) all regulated dischargers in Ohio must consider their compliance with numeric biocriteria during the NPDES permit renewal period.

American Electric Power Company (AEP) subsidiaries have several coal-fired electric generating facilities that are permitted by the Ohio EPA. Ohio Power Company's Muskingum River Plant and Columbus Southern Power Company's Conesville Generating Station are located on the Muskingum River (Figure 1) and are subject to Ohio EPA's inland biocriteria. Both Ohio EPA and AEP have conducted numerous biosurveys near these plants and each has assessed compliance with ecoregion-based biocriteria using these biosurvey results. A detailed discussion of these results is given in Section 2.3.

This chapter summarizes the technical challenges of biocriteria compliance from a regulated industry perspective. As previously mentioned, industry has not been forced to address issues related to numeric biocriteria on a national scale. At this time, only regulated industries in Ohio must comply with ecoregion-based numeric biocriteria. The facilities operated by AEP subsidiaries represent only a few of the many

0-87371-894-1/95/$0.00+$.50
© 1995 by CRC Press, Inc.

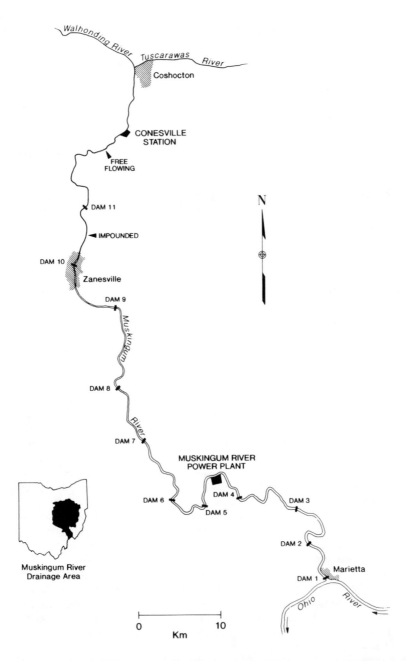

Figure 1. Location of two coal-fired power plants on the Muskingum River: Conesville Station (located on free-flowing portion) and Muskingum River Plant (located on impounded portion). Navigation dams are indicated relative to power plants and major cities.

facilities that are regulated under Ohio EPA's NPDES program. Therefore, permitting experiences for AEP facilities are unique and cannot be considered representative of Ohio's regulated industry as a whole. Biosurvey data are unique for individual facilities and Ohio EPA assesses compliance with all applicable criteria by site. The following discussion on technical advantages and deficiencies of biocriteria implementation in Ohio is intended to stimulate discussion among industry and regulatory agencies to address unresolved issues, and to encourage other industries to work with state agency staff at all stages in biocriteria development.

1.1 Biocriteria as a New Regulatory Tool

The development and adoption of biocriteria into state water quality standards poses challenges as well as opportunities. As a new regulatory tool, industry must determine the effects that biocriteria compliance will have on their individual facilities. Such assessments are familiar to industry. Projections of compliance have been, and are continually being conducted, for chemical-specific and whole effluent toxicity criteria in Ohio and all other states. The real challenge for industry and state agencies is that biocriteria, being an integrated biologically based regulatory tool, will vary tremendously both among states and within states. There can be no "national" biocriteria that all states or even several States may adopt as identical, in contrast to national water quality criteria. Even the adoption of "generic" narrative biocriteria within nearby states may not be valid because biological expectations may differ tremendously. It seems reasonable that states must assess at least some state-specific biosurvey data and existing land-use impacts before even a valid narrative biocriteria can be developed. During this early stage, it will be prudent for industry to closely follow agency developments. Some facilities may want to assume a proactive role by gathering data using state-approved methodologies, or methodologies that agencies are currently investigating. The risks and benefits of this proactive strategy are discussed in more detail below.

There are technical challenges for industry during the development of state-specific biocriteria (Van Hassel et al. 1992). Understanding the underlying concepts and theoretical assumptions of biocriteria may be a challenge to environmental managers that have little or no experience with biological assessments. Clearly, the regulation of complex wastewaters through biocriteria will be different compared with the traditional chemical-specific/end-of-pipe approaches. Ohio EPA (1987a, 1990a) has published comprehensive overviews of biocriteria rationale, whereas the overview by Karr et al. (1986) is generally regarded as the primary publication advocating using biological integrity within water quality management programs. Environmental managers within industry can become well acquainted with basic biocriteria concepts by reviewing the four documents cited above. Further discussion of specific challenges to industry are summarized below.

1.1.1 Differences With Chemical-Specific Criteria

Chemical-specific criteria are discrete values that should not be exceeded to protect a designated use. By using design flows and standard mass-balance equations, water quality-based effluent limits (WQBELs) can be reasonably predicted using chemical-specific criteria. Furthermore, the analysis of wastewaters for specific constituents is relatively rapid and a sizeable database can be generated within a short period. Biocriteria typically represent a composite of independent metrics or calculations that are summed to produce a value without units (e.g., the Index of Biotic Integrity and Modified Index of Well-Being). Predicting compliance with biocriteria will typically require more data, be more costly, and require staff biologists (or consultant biologists) who are familiar with data analysis techniques and interpretation. Instream biological data can vary considerably both temporally and spatially: thus, replicate samples are necessary to define the biological integrity near a particular facility. Though Ohio EPA has determined that instream biological assessments are cost-effective when compared with chemical-specific and toxicity testing methods (Ohio EPA 1990a), the agency has spent considerable years refining its field protocols and data analysis procedures (see Yoder and Rankin, Chapter 9).

Although there are fundamental differences in predicting compliance with chemical-specific criteria vs. biocriteria, industry should perform *both* chemical and biological analyses simultaneously. The collection of paired data allows testing the hypothesis that community-level measures respond predictably with varying concentrations of measured pollutants. Some regulatory agencies rely on a limited number of biosurveys often conducted in one year only and without measured water quality data to assess potential wastewater impacts. Long-term monitoring data, although costly to obtain, are invaluable for industries who need to refute or confirm perceived impacts based on small sample sizes.

Like chemical-specific criteria, assessment of whole effluent toxicity (WET) compliance for a particular facility may or may not be comparable with biocriteria compliance. Ohio EPA (1990b) studied the level of agreement between bioassay and instream biosurvey results using data collected near 43 separate facilities. The agency found good agreement between results in about 20% of cases when effluent (end-of-pipe) toxicity was assessed, and in about 30% of cases when mixing zone toxicity was measured. There is some empirical evidence that instream biological response and instream ambient toxicity are closely associated (Dickson et al. 1992). Marcus and McDonald (1992) caution against wide-scale extrapolation of these study results, citing biases in statistical design and a lowered association between instream measures and ambient toxicity when more subtle, or borderline, toxicity is occurring. Based upon a review of relevant studies, the author believes that WET results, especially instream toxicity assessments, will more closely parallel biocriteria compliance compared to using a chemical-specific approach. The reason is that WET and instream toxicity tests are a true biological-response assessment. Nonetheless, industry will likely be forced to utilize facility-specific strategies and data requirements when assessing biocriteria compliance. These strategies must rely on assessing various combinations of chemical-specific and WET criteria in the absence of instream biosurvey data.

1.1.2 Understanding How They Are Derived

In general, numeric biocriteria will be derived using one of two approaches: the site-specific reference condition, or regional reference condition (USEPA 1990a). Industry personnel should understand the difference between the two approaches and establish a dialogue with regulatory agencies in order to offer their insight on the most valid approach. Industry is already familiar with the site-specific reference approach, where a traditional upstream vs. downstream design is used to assess potential discharge impact. The regional reference design can result in biocriteria based on a paired watershed approach (one watershed impaired, one watershed unimpaired), or an ecoregion-based approach (USEPA 1990a). Evidently, some states are developing numeric biocriteria with inadequate sample sizes, e.g., one field season of sampling at ecoregion-specific reference sites. This "one year, one pass" sampling design has basic ecological flaws, and competent agency biologists should readily understand the pitfalls of such a meager database. Industry personnel should become aware of the process that states are using to derive biocriteria and communicate their concerns when technical flaws may result in unrealistic, or invalid, biological expectations outside of reference sites.

1.1.3 The Statutory Basis of Biocriteria

The statutory basis of biocriteria has been summarized by USEPA (1990a, 1991c). Industry should be aware of the principal statutory provisions pertaining to biocriteria. Sections 303, 304, and 308 of the Clean Water Act contain pertinent language for development of protective criteria and biological assessment methods. (Please see Adler, Chapter 22 for additional information on statutory bases for biological criteria.)

1.2 Compliance With Biocriteria

Once narrative or numeric biocriteria are developed, proposed, and adopted into state water quality standards (and subsequently approved by USEPA), industry must choose strategies that will ensure compliance with biocriteria, as well as any applicable chemical-specific or WET criteria. As previously discussed, projecting compliance with biocriteria may be difficult, especially if little or no instream data are available. Industries that choose a proactive strategy by performing biosurveys that are not required will obtain results that allow a direct assessment of biocriteria compliance. Moreover, if a regulatory agency performs biosurveys to assess attainment of applicable biocriteria, data collected by a permittee can be used to expand the site-specific database, and more importantly, validate or contest results obtained through agency studies.

1.2.1 Biocriteria as Part of "Independent Application"

USEPA has issued a policy of "independent application" for all applicable criteria to determine whether designated uses are being attained (USEPA 1991f). This policy affirms that an exceedance of any applicable chemical-specific, WET, or biocriteria will result in the nonattainment of a designated use despite evidence indicating that the other two criteria are, or will not be, exceeded. This policy not only integrates all three types of assessments but requires that the most stringent (i.e., limiting) criteria be applied in water quality-based toxics control programs. This position is summarized in USEPA's (1991c) guidance on biocriteria as follows:

> The failure of one method to confirm an impact identified by another method would not negate the results of the initial assessment. This policy, therefore, states that appropriate action should be taken when any one of the three types of assessments determine that the standard is not attained.

There are some significant technical flaws with this policy if implemented on a wide-scale basis, with no exceptions. Miner and Borton (1991) provide arguments against the independent application approach, emphasizing the underlying statistical requirements for such an approach. There may be instances when states are justified using this policy in the effluent characterization and hazard assessment process. When there is little or no information on potential receiving stream impacts and/or whole effluent toxicity, states should use a conservative approach in permitting the particular facility.

In contrast, there are some instances where the independent application approach is not justified because it may result in overprotective effluent limitations. In general, the independent application approach assumes that both the amount and quality (i.e., relevance) of data used to assess all three criteria types are equivalent (USEPA 1991f). In most instances, however, this will not be true. Many facilities will likely have considerable data for one assessment but relatively little for another. In such cases a risk assessment approach (not an independent application approach) should be taken and appropriate effluent limitations be required that (1) will protect, but not overprotect, the receiving stream, (2) are commensurate with the results of existing data, and (3) utilize a weight-of-evidence, best professional judgement process. Considerations of statistical power should be integrated into the risk assessment process. Thus, a weight-of-evidence approach can become increasingly obvious when hypothesis testing (statistical power) demonstrates the unnecessary usage of independent application.

Chemical-specific and whole effluent approaches are tools that only predict a level of protection for instream aquatic life whereas the biocriteria approach measures the actual level of protection. Logically, there will be instances when demonstrated compliance of biocriteria should take precedence over one or both of the other criteria. This is especially relevant when a chemical-specific criterion is being exceeded but the biological community indicates no impairment. Many water quality criteria were derived from toxicity tests that exposed test organisms to highly bioavailable fractions of toxicants. Unless site-specific adjustments of protective chemical criteria are made where toxicity mitigation is demonstrated, many facilities could have effluent limitations that are overly protective. Overly protective effluent limitations may require costly alternate treatment technologies that cannot reasonably be expected to provide real environmental benefits.

In summary, industries must understand the decision-making policies that govern the water quality-based toxics control program in each state. If biocriteria have not been adopted then the issue of independent application is absent, or less complicated. In states where numeric biocriteria have been adopted, the issue of independent application may be complicated. In general, regulatory agencies *and* industry are favored when states have some flexibility in administering the NPDES program. USEPA is correct in stating that required compliance with all criteria results in a powerful regulatory tool (USEPA 1991f). The independent application policy, however, seems to dismiss the professional judgment that states often utilize when a differing array of site-specific data are available.

1.2.2 Interaction With Regulatory Agencies

There are many opportunities where industry can interact with agencies during the biocriteria development process. Interaction during the early formulation stage is critical if regulated industry desires a voice in strategy and timetable. Besides attendance at agency-sponsored workshops and public meetings, industry biologists can inspect field-sampling and sample-processing techniques. Regulatory agencies should have well-established quality assurance procedures for field and laboratory protocols. For biocriteria development, expertise in fish and macroinvertebrate identification is crucial, and a standardized chain of custody procedure should be verified by industry. Ohio EPA (1987) has published a peer-reviewed standard operating procedures manual for fish and macroinvertebrate assessments. Because these standardized methods are used to gather data for biocriteria index calculations, regulated industry in Ohio has learned that close adherence to these methods is beneficial in resolving use attainment issues.

There are many advantages in establishing a constructive dialogue between regulated industry and agencies during the strategy planning phase. Resolving technical issues at the formulation stage is unmistakably far more profitable than trying to resolve disputes after strategy and methodologies are set.

2.0 THE OHIO EXPERIENCE

2.1 Overview of Ohio EPA Biocriteria Requirements

By adopting numeric biocriteria into Ohio's water quality standards, Ohio EPA explicitly requires one (and only one) approach in demonstrating attainment of the aquatic life use: compliance with biocriteria. In other words, Ohio EPA considers biocriteria indices a direct gauge of attainment or nonattainment of the aquatic life use. Unless it can be demonstrated that the designated aquatic life use of a waterbody cannot be attained due to habitat limitations or long-term irretrievable conditions, effluent limitations will be modified as appropriate when noncompliance of biocriteria has been demonstrated. Logically, there must be some empirical evidence indicating a relationship between a facility discharge and nonattainment of applicable biocriteria. This is a crucial question when *reasons* for biocriteria nonattainment are elucidated. Often, debates between regulated industry and Ohio EPA center on the interpretation of biosurvey data, especially regarding the dynamic relationship between instream pollutant levels and measured biological response. A robust statistical treatment of these data often can help determine whether or not the biological response is attributable to the measured pollutant.

Ohio's biocriteria vary among ecoregions and size of the waterbody. With the establishment of numeric biocriteria, however, industry should have no confusion as to what target biological community must be attained to comply with biocriteria. In Ohio, industry cannot change the regulatory implementation of biocriteria, but they can collect data which either confirms or challenges Ohio EPA's assessment of use nonattainment.

Ohio EPA has established a rule that attainment of biocriteria should be granted disproportionate weight in demonstrating overall use attainment, due to the fact that chemical-specific and WET criteria are only surrogate measures of biological integrity. This precedence for biological criteria is embedded in Ohio's statutory water quality standards:

> Demonstrated attainment of the applicable biological criteria in a waterbody will take precedence over the application of selected chemical-specific or whole-effluent criteria associated with these uses. (Ohio Revised Code 3745–1-07)

Industry should be aware that (1) demonstrated nonattainment of the biological criteria may result in more stringent chemical-specific limitations, and (2) USEPA has not yet approved this departure from its policy of independent application. Until the issue of independent application is resolved, Ohio's regulated industry must develop a strong technical case that biocriteria are met near a given facility or that nonattainment is due to other factors independent of a facility's operations.

2.2 Technical Validity of Ohio EPA Biocriteria

From an industry perspective, the most attractive feature of Ohio EPA's biocriteria is the empirical foundation. In general, Ohio EPA derived numeric biocriteria using a systematic standardized sampling of least impacted reference sites in each of the five ecoregions (see Yoder and Rankin, Chapter 9). From the standpoint of defining *regional* expectations of biological performance, the validity of actual Index of Biotic Integrity, Modified Index of Well-Being, and Invertebrate Community Index values has been demonstrated. Ohio EPA's Users Manual (Ohio EPA 1989) provides a good overview of how reference site data were translated into regional biocriteria. Ohio EPA has conducted biosurveys throughout the state for nearly 20 years. Replicate samples were taken at most sites, including ecoregion reference sites. The extensive database compiled by Ohio EPA has allowed the derivation of defensible biocriteria values.

Fore et al. (1994) analyzed the statistical properties of Index of Biotic Integrity values among several Ohio streams and concluded that, although IBI scores at a site are not distributed normally, the usage of a two-sample t-test and ANOVA model could be cautiously applied for hypothesis testing.

Ohio EPA's statewide database should serve as a model to other states who are developing numeric biocriteria. Though some states may be tempted to gather data quickly without an adequate number of replicates at several reference sites, this strategy is not sound and will only result in poorly validated metric cutoffs. Some states, in addition, tend to ignore data variability rather than let the variability determine the resultant criteria. The inherent variability of biological community data demands that adequate replicates be taken. More importantly, this variability must be defined empirically and incorporated into resulting biocriteria. Ohio EPA correctly analyzed the biological data variability and incorporated this variability into the numeric biocriteria.

There are technical aspects of Ohio EPA's biocriteria that had no precedent and thus required best professional judgement. Fausch et al. (1990) state that one disadvantage of using the IBI is the subjectiveness of defining metric criteria (e.g., maximum species richness lines). One notable item is the 25th and 75th percentile cutoff for ecoregion-specific biocriteria, which Ohio EPA uses. The specific percentile cutoff for biocriteria compliance or noncompliance varies with use designation. The 25th percentile of each biocriteria index, based on ecoregion-specific reference site data, is considered the minimum criterion for the Warmwater Habitat use designation. For waters that support highly diverse communities (the Exceptional Warmwater Habitat use), the 75th percentile value of the combined statewide reference site data is used as the minimum criterion. The designation of these percentile values to judge attainment of designated values seems reasonable. From an industry perspective, the percentile cutoffs are acceptable because they recognize the variation in index scores within a given ecoregion. For example, the designation of 25th percentile for the minimum attainment value acknowledges that 25% of the reference sites in a given ecoregion do not attain this minimum value, for a given index. Thus, even if a discharger has a facility located on a "least impacted" waterbody, there is a 1 in 4 chance that a given index score will fall below the ecoregion minimum due to factors independent of the facility's operations.

Even though sites that are not influenced by point-source discharges should not have an expectation of 100% biocriteria compliance, Ohio EPA still must interpret results carefully when upstream reference sites show a low compliance frequency. Ohio EPA uses an averaging of index scores to compare with the minimum ecoregion criterion. Thus, a discharger does not have to demonstrate 100% compliance with minimum criteria; rather, the average value is compared to the percentile cutoff value. When a relatively small database (e.g., three replicates or passes in a given field season) must be analyzed, comparisons of site-specific average scores to established biocriteria or upstream scores may be appropriate. When a larger database is available, however, the agency should use true hypothesis testing in accordance with USEPA policy (USEPA 1991c).

2.3 Technical Problems With Ohio EPA Biocriteria

Using standardized methodologies, Ohio EPA has developed three biocriteria indices that are implemented on an ecoregion-specific basis. With numeric biocriteria, the agency can assess numerous waterbodies for attainment of the aquatic life use based on regional expectations. Temporal trends in biological performance, especially useful for waterbodies that had varying degrees of historical impact, can be tracked with greater resolution when numerical community-based indices are used as a benchmark.

There are some technical problems with Ohio EPA's biocriteria. These flaws do not undermine the fundamental concepts of Ohio EPA's biocriteria, but they will require some rethinking of certain implementation steps. These problems are applicable to any state that develops ecoregion-based biocriteria, and are not unique to Ohio. Three problem areas discussed below are (1) site-specific modification of biological expectation, (2) biocriteria for regulated vs. unregulated waterbodies, and (3) derivation of large river biocriteria, with emphasis on the Ohio River.

2.3.1 Site-Specific Modification of Biological Expectation

Ohio EPA's biocriteria were developed to define *regional* expectations of biological performance. Several reference sites were sampled in each ecoregion to provide numerical estimates of "least impacted"

Figure 2. Location of lower Muskingum River and lower Scioto River relative to the Ohio River. Geographic locations of Ohio ecoregions are indicated. WAP = Western Allegheny Plateau ecoregion. (From Ohio EPA. 1992.)

conditions. Ohio EPA acknowledges, albeit briefly, that site-specific modifications may be appropriate in some circumstances:

> In situations where the biological criteria are not met because of the natural attributes of the surface water and/or watershed, a site-specific modification of the criteria may be performed. This procedure recognizes that there may be habitats that do not meet the ecoregional criteria due to unique, site and/or watershed specific characteristics. (Ohio EPA 1989, pp. 7–5)

Site-specific habitat constraints are a possible reason for nonattainment, but these can be addressed using an intensive habitat assessment or Ohio EPA's Qualitative Habitat Evaluation Index (see Rankin, Chapter 13). The extrapolation of minimum criteria to sites having drainage areas much larger than ecoregion-specific reference sites may also cause nonattainment. In such cases, the calibration of indices is exceeded due to a lack of comparable reference sites. An example of this technical problem is the experience at one of American Electric Power Company's coal-fired generating facilities, Conesville Station. Conesville Station is located on the Muskingum River at River Mile 102 (distance from Ohio River confluence), near Coshocton, Ohio (Figure 1). The drainage area of the Muskingum River at Conesville Station is 4882 mi². The mainstem Muskingum River (and most of the entire drainage area) lies within the Western Allegheny Plateau (WAP) ecoregion (Figure 2). Ohio EPA has adopted minimum biological criteria for the WAP ecoregion (pertaining to Warmwater Habitat aquatic life use) as 40 (IBI), 8.6 (Modified Index of Well-Being, or MIwb) and 36 (Invertebrate Community Index, or ICI). These criteria were derived from collections at least impacted reference sites, none of which were on the Muskingum River mainstem.

One relevant question is: how comparable are reference sites to the mainstem Muskingum River near Conesville Station? Ohio EPA does list the location and drainage area of all ecoregion reference sites (September, 1989 addendum to the Users Manual), as well as summary statistics for drainage area, species richness, and biocriteria index scores. Among least impacted reference sites within the WAP ecoregion, the sites with the highest drainage areas are those on the Tuscarawas River. These sites have drainage areas that are no more than 53% of the area at Conesville Station. Also, these sites were recently upgraded to the higher use of Exceptional Warmwater Habitat, further illustrating that they are less comparable to conditions at Conesville Station.

The only other reference sites within the WAP ecoregion that have similar drainage areas to that at Conesville Station are those on the lower Scioto River. The Scioto River is a tributary to the Ohio River, about 185 river miles downstream of the Muskingum River/Ohio River confluence (Figure 2). Conesville Station is an additional 102 river miles upstream from this point. Thus, a considerable distance (approximately 300 river miles) separates Conesville Station from the nearest ecoregion reference site having a similar drainage area. The comparability between the two areas is even more speculative considering that the lower Muskingum River is impounded, being regulated by navigation dams. In contrast, all sites on the lower Scioto River are free-flowing. Thus, even though both sites have similar drainage areas, some differences in biotic communities would be expected due to zoogeographic and habitat factors. For example, faunal similarity between midwestern river systems is dependent on river mile distance, along with other factors (Robison 1986).

A site-specific modification of biological expectation, or performance, would be reasonable if biocriteria index scores on the mainstem Muskingum River just upstream of Conesville Station have a different distribution than those for ecoregion reference sites. Biosurveys were conducted near Conesville Station from 1988 to 1991 using methods that conformed to Ohio EPA protocols or with minor deviations. Ohio EPA conducted fish sampling along the entire Muskingum River mainstem in 1988; AEP conducted subsequent studies from 1989 to 1991. Biocriteria scores (IBI, MIwb) were compiled for four sites just upstream of Conesville Station, using both Ohio EPA and AEP data. Samples collected in 1990 were not used due to high flow conditions. A total of 27 individual samples were compiled for sites just upstream of Conesville Station; 51 samples were taken within the entire WAP ecoregion. Thus, the site-specific sample size is slightly more than 1/2 of the entire ecoregion sample size.

Table 1 indicates the statistical parameters of IBI and MIwb data for WAP ecoregion reference sites and reference sites just upstream of Conesville Station. Table 1 provides evidence that the population of pooled ecoregion reference sites has distinct statistical parameters compared to sites in the mainstem Muskingum River near Coshocton, Ohio. The median values of the IBI and MIwb are consistently lower at site-specific reference sites. The lowered 75th and 25th percentile values is notable because Ohio EPA regards the 25th percentile in a given ecoregion as the minimum biological expectation. The above data suggests that the WAP ecoregion criteria of 40 IBI units and 8.6 MIwb units represent unrealistic biological performance expectations when compared to sampled reference sites near Conesville Station.

Though both point source and nonpoint source impacts have been documented upstream of Conesville Station, these impacts are intermittent (i.e., flow dependent) and thus there is no reason to believe that these have caused a continual, systematic decline in index values. Nonpoint sources probably have some effect on biological performance at many ecoregion reference sites (Whittier et al. 1987). Actually, from a biological perspective, the differences in index score parameters are not unexpected, because: (1) no sites on the mainstem Muskingum River were used for ecoregion criteria, (2) the relatively large sample size upstream of Conesville Station is much higher than at any individual ecoregion site, and (3) the only ecoregion reference site with a comparable drainage area is separated by about 300 river miles. From a regulatory perspective, these data provide good technical justification for a modified biological expectation. In this case, a discharger has met the burden of proof by providing an adequate database to evaluate the reasonableness of an ecoregion-based expectation.

2.3.2 Biocriteria for Regulated vs. Unregulated Waterbodies

Ohio EPA's biocriteria for the WAP ecoregion are identical for both free-flowing and impounded sections of the Muskingum River (Figure 2). As previously indicated, the agency did not use sampling results for the mainstem Muskingum River in the derivation of WAP ecoregion biocriteria. Thus, sampling results for least impacted reference sites on impounded river segments were never used to derive minimum biocriteria. This lack of reference stream data on impounded river segments presents problems for dischargers located on the impounded Muskingum River. The assumption of Ohio EPA's WAP biocriteria is that biological performance should be similar between sites on free-flowing reaches and sites on impounded reaches, assuming the drainage areas are similar. This assumption is not realistic, as discussed below.

The most significant problem with Ohio EPA's assumption is that the agency has adopted identical biocriteria for free-flowing and impounded reaches without testing the hypothesis of no differences in

Table 1.   Statistical Parameters of Fish Collection Sites and Biocriteria Index
           Scores for Ohio EPA's Western Allegheny Plateau Reference Sites
           and Site-Specific Locations on the Mainstem Muskingum River Near
           Conesville Station.

Parameter	WAP ecoregion reference sites	Muskingum River reference sites (RM 103.5 to RM 105.8)
No. of samples	51	27
Drainage area of sites		
Mean (mi^2)	1860	4875
Minimum (mi^2)	90	4870
Maximum (mi^2)	6471	4880

Index of Biotic Integrity (IBI)

Mean	44	38
(SE)	0.9	1.3
Median	44	36
Range	28–54	24–52
Quartile		
Lower	40	34
Upper	50	42

Modified Index of Well-Being (MIwb)

Mean	9.3	8.1
(SE)	0.1	0.2
Median	9.4	8.1
Range	7.5–10.7	6.7–9.7
Quartile		
Lower	8.6	7.6
Upper	10.0	8.7

Table 2.   Summary of Fish Community Differences Found at Sites on
           the Free-Flowing and Impounded Portions of the Muskingum
           River During Electrofishing and Seining Studies in 1989

	Study Location	
Parameter	Conesville Sta.[1] (free-flowing)	Muskingum River[2] (impounded)
Total no. species	54	41
Total fishes collected	18,572	6,152

Relative abundances (%); number collected in parentheses

Sand shiner	15.8 (2,904)	0.0
Bluntnose minnow	11.3 (2,078)	0.18 (11)
Quillback	0.04 (8)	1.7 (104)
Highfin carpsucker	0.17 (32)	0.02 (1)
Smallmouth buffalo	0.0	1.7 (104)
Silver redhorse	1.3 (231)	1.0 (59)
Northern hog sucker	0.35 (65)	0.0
Flathead catfish	0.10 (19)	1.7 (104)
Rock bass	0.71 (131)	0.1 (7)
Orangespotted sunfish	0.0	0.7 (40)
Longear sunfish	0.0	0.8 (51)
Smallmouth bass	4.1 (743)	0.3 (19)
Spotted bass	0.0	11.0 (679)
Largemouth bass	0.23 (43)	0.09 (6)
Greenside darter	0.92 (168)	0.0
Banded darter	0.88 (161)	0.0
Freshwater drum	0.02 (4)	2.2 (138)

[1] Results based on 32 electrofishing samples and 16 seine (riffle) samples.
[2] Results based on 48 electrofishing samples.

biological performance between the differing habitats. Recent fishery surveys conducted at Conesville Station (located in the free-flowing upper Muskingum River) and Muskingum River Plant (located at River Mile 28.0 in the impounded lower river) demonstrate the differences in both species composition and relative abundance among the sites. Results of the 1989 fishery surveys are given in Table 2.

As would be expected, the fish community found at each site reflects the predominant habitat features. A riverine/riffle community is present near Conesville Station whereas a large river/lentic community is present near Muskingum River Plant. Major compositional differences were observed in most of the dominant families. At Conesville Station, spotfin shiner dominated the cyprinid catch with sand shiner and bluntnose minnow being fairly common. At Muskingum River Plant, the family Cyprinidae was dominated by emerald shiner, but sand shiner and bluntnose minnow were absent or uncommon. Within the sucker family, the subfamily Ictiobinae was represented by highfin carpsucker at Conesville Station whereas quillback and smallmouth buffalo were fairly common at Muskingum River Plant. All redhorse species (subfamily Catostominae) and the northern hog sucker were more common near Conesville Station. Flathead catfish were considerably more common at Muskingum River Plant. The composition of sunfish species was different at each plant site. Rock bass, smallmouth bass, and largemouth bass were more abundant at free-flowing sites, whereas orangespotted sunfish, longear sunfish, and spotted bass were collected exclusively at Muskingum River Plant. Not surprisingly, darter species were either absent at Muskingum River Plant or were more common near Conesville Station. Freshwater drum were considerably more abundant in the impounded reaches of the river. The higher species richness and total fish abundance is largely due to the presence of riffle habitats near Conesville Station, but none near Muskingum River Plant.

These data clearly demonstrate the presence of two different fish communities on the same waterbody, yet Ohio EPA's biological criteria do not reflect these differences because all ecoregion reference sites were located on free-flowing streams and rivers. The fish community near Muskingum River Plant is actually more similar to assemblages in the nearby Ohio River compared to assemblages in inland rivers. Because Ohio EPA is currently conducting studies on the Ohio River for the purpose of deriving biological criteria (see Section 2.3.3), there appears to be no sound biological reason why the lower Muskingum River, having unique habitat characteristics, should have identical biological criteria with free-flowing portions, or with other free-flowing rivers in the same ecoregion. Ohio EPA, as a minimum, should have tested the hypothesis of no differences in biological performance for reference sites on impounded and free-flowing sections.

2.3.3 Derivation of Large River Biocriteria (Emphasis on Ohio River)

Ohio EPA has initiated sampling at nearshore zones on the Ohio River for selection of reference sites and eventual derivation of biocriteria for upper and middle river sections that border the State of Ohio (Sanders 1991). In addition, the Ohio River Valley Sanitation Commission (ORSANCO) has initiated sampling of nearshore zones along the entire Ohio River using an electrofishing procedure that generally follows Ohio EPA's protocols. The results of these surveys will apparently be pooled to derive proposed biocriteria for the mainstem Ohio River.

The derivation of numeric biological criteria for large rivers presents technical challenges for regulatory agencies. First, the definition and delineation of reference sites is problematic. For the Ohio River, the identification of "least impacted" reference sites (i.e., similar to those selected for ecoregion-specific inland watersheds) is probably not possible. Because of the enlarged physical dimensions of larger rivers and ease of faunal transfer within navigation pools, nonimpacted reference sites (which represent a target biological expectation) likely do not exist on large rivers. This factor changes the benchmark, or biological performance gauge, that will be used to assess attainment or nonattainment of the aquatic life use in the Ohio River. As a surrogate, a near-field reference site is probably the most valid target for regulatory assessments on large rivers.

A second problem is habitat comparability. Large rivers, such as the Ohio River, have extensive reaches of relatively unproductive habitat (e.g., shallow sloping banks with sand/muck substrate) with little or no attractive habitat feature (e.g., no log piles, overhanging vegetation, or gravel/cobble substrate) within these reaches. Productive and heterogeneous habitats are highly patchy and obviously these constraints will directly influence the diversity and abundance of fishes.

Another problem with large river biocriteria development is data variability. Temporal variations in fish community parameters must be expected in large rivers. Where significant seasonal variation is documented, this variability must be accounted for in the derivation of numeric biocriteria if the biocriteria are, in fact, based on multiseason sampling. This variability was demonstrated during a 1991 fisheries study at six power plant sites along the Ohio River (EA 1993). The plant sites were located on the upper, middle, and lower Ohio River between River Miles 54 and 946. Fishes were sampled in June or July, August, and September or October. For electrofishing and gill net samples there were 12 of 24 (50%) cases where a significant ($P < 0.05$) temporal variation was observed for either catch per unit effort or total biomass, at a particular plant site. The effects of temporal variation (both seasonal and year-to-year, if applicable) obviously must be accounted for when deriving numeric biocriteria for large rivers.

In addition to temporal effects on fishery parameters in the 1991 study cited above, statistical associations of calculated IBI and MIwb values with independent variables indicated that river flow, water temperature, and forage fish abundance significantly affected score values at combined locations (Reash, in press). For combined samples at six power plant sites ($N = 108$), statistically significant inverse correlations with calculated IBI values (using Ohio EPA inland methodology) were as follows: percent gizzard shad ($r = -0.43$; $P<0.001$) and river flow at time of sampling ($r = -0.21$; $P<0.03$). Significant inverse correlations with the Modified Index of Well-Being were: percent gizzard shad ($r = -0.31$; $P<0.002$) and water temperature ($r = -0.21$; $P<0.04$). These results indicate that (1) data interpretation of Ohio River biosurveys may be problematic due to stochastic factors, and (2) highly standardized methodologies will be required to minimize the influence of confounding factors regarding a site-specific assessment.

Ohio EPA has analyzed electrofishing data for Ohio River nearshore zones using the Modified Index of Well-Being (MIwb) and Index of Biotic Integrity (IBI). The MIwb appears to be suitable for the Ohio River as this index has been used in a wide range of waterbodies to assess changes in structural attributes of fish communities (Fausch et al. 1990). The application of Ohio EPA's inland version of the IBI to Ohio River mainstem sites is questionable, however. The most obvious reason that the inland metrics should not be applied to the Ohio River is that Ohio EPA's published reference sites do not encompass large, impounded rivers. Although the agency is currently in the process of recalibrating the IBI metrics for Ohio River sites, sampling data should not be interpreted using the inland IBI until this recalibration and validation is completed.

AEP, along with other electric utility companies, has sponsored annual ecological studies near coal-fired power plants since the early 1970s. Based on results of this long-term study, recommended modifications to Ohio EPA's inland IBI metrics (for application to the Ohio River) are listed in Table 3. New Ohio River-specific metrics may need to be developed if modifications to the inland IBI metric are not sufficiently sensitive to detect significant shifts in structural and functional community parameters. These new metrics may require a more generic approach. For example, a metric such as total number of trophic guilds (relative to a nearby reference site) may be sensitive enough to detect significant water quality degradation, but generic enough to prevent a "false positive" finding of use impairment that is actually caused by temporal effects or habitat effects. Whether established or new metrics are proposed for large river biocriteria, these metrics must be validated regarding their empirical foundation and ability to detect impairment beyond a reference condition.

Because the Ohio River is a large biological system with species composition constantly changing, interpretation of sampling data will require considerations of zoogeography, historical abundance and distribution, and historical ranges of variability (both population-specific and community-based parameters). Pearson and Pearson (1989) discussed historical trends of faunal composition and abundance. Van Hassel et al. (1988) reported significant distributional and temporal trends of Ohio River fishes in the upper and middle river, along with a segregation of species based on reproductive guilds, habitat preferences, and feeding habits. There appears to be sufficient evidence that IBI metrics will differ for the upper and middle Ohio River due to zoogeographic factors. The influence of zoogeographic and physicochemical factors on fish distribution in the upper and middle river was discussed by Reash and Van Hassel (1988). Navigation locks and dams restrict faunal transfer between navigation pools, thus creating a continuum of community similarity that is somewhat predictable. Reash (1992) developed a regression equation that predicts the similarity of Ohio River fish assemblages along the Ohio River. Using a two-variable model of river distance and drainage area difference, community similarity could

Table 3. Suggested Modifications to Ohio EPA's Inland Index of Biotic Integrity (Boat Method), for Potential Application to the Ohio River, an Impounded Large River

IBI metric	Current inland[a] cut-offs	Suggested modification
No. of species	>20, 10–20, <10	Cut-offs may need revision; sites with sparse habitat often yield <20 species
Percent round-bodied suckers	>38, 19–38, <19	Acceptable for upper river, possibly acceptable for middle river. Cutoffs will need modification
Sunfish species	>3, 2–3, <2	May require a lower expectation in middle river, where rock bass and pumpkinseed are rare
Sucker species	>5, 3–5, <3	Scoring cutoffs will require modification; hog sucker, white sucker, and black redhorse are rare in middle river
Intolerant species	>3, 2–3, <2	A new list of "Ohio River intolerant species" will be needed. Many intolerant species for inland metric are small stream forms
Percent tolerant	<15, 15–27, >27	Same comment as above. Scoring cutoffs will need modification
Percent omnivores	<16, 16–28, >28	Scoring cutoffs will need modification. A greater number of omnivore species would be expected in large, impounded rivers
Percent insectivores	>54, 27–54, <27	This metric is questionable for the Ohio River. Fewer insectivorous species present due to lentic-like hydrology. Benthic production of food organisms much less than in free-flowing systems
Percent top carnivores	>10, 5–10, <5	May require modification for upper and middle reaches
Percent simple lithophils	Varies w/drainage area	Questionable for use in Ohio River due to limited area with hard substrate
Percent DELT anomalies	<0.5, 0.5–3.0, >3.0	Scoring cutoffs will need modification based on Ohio River samples. A greater abundance of carp, catfish, and Ictiobinae suckers in impounded rivers may inflate the prevalence of DELT anomalies
Fish numbers	<200, 200–450, >450	Scoring cutoffs will need modification. Dense clusters of forage species will cause wide variation in total numbers

[a] Indicated cutoffs correspond to metric scores of 5, 3, and 1, respectively.

be predicted with a high degree of statistical confidence ($r^2 = 0.88$). Such factors should be considered when regulatory agencies assess the aquatic life use in the Ohio River using numeric biocriteria.

3.0 SUMMARY

Biological criteria will be developed and adopted in state water quality standards as generic narrative criteria or numeric criteria. According to USEPA, biocriteria will have no less legal weight as chemical-specific and whole effluent toxicity criteria, and thus regulated industry must understand the conceptual foundations of biocriteria and the technical aspects of data collection, data interpretation, and biocriteria derivation. Regulated industry should work with water resource agencies to ensure that standardized methodologies are used and especially that numeric biocriteria are valid and have a sound empirical foundation. Regulatory agencies must avoid the temptation of collecting a sparse amount of data to derive biocriteria indices.

The State of Ohio has utilized an extensive database of statewide biological surveys to derive numeric biocriteria based on the ecoregion approach. Ohio EPA's three biocriteria indices have a sound empirical foundation regarding incorporation of the broad historical database. The establishment of minimum biological performance in Ohio's five ecoregions allows a concise compliance target that industry can readily assess using Ohio EPA methodologies. Industry should take the initiative to confirm or challenge a regulatory agency's assessment of aquatic life use nonattainment, due to the fact that causes of nonattainment may be independent of instream pollutant levels. Technical flaws to Ohio EPA's biocriteria are discussed. Ohio EPA, and all other agencies, will need a concise mechanism for site-specific biocriteria modification. Large impounded rivers present technical challenges to biocriteria derivation. The assumptions (hypotheses) of biological expectation in large rivers must be tested before data are analyzed using methodologies applied to small and medium-sized rivers.

ACKNOWLEDGMENTS

I thank Wanda Vestermark and Toni Nijssen, who dutifully typed the manuscript. Greg Seegert provided insightful comments on Ohio River biocriteria. I thank the following individuals for contributing useful comments on a draft version: William Patberg, Alan Gaulke, Scott Matchett, Timothy O'Shea, Timothy Lohner, and John Van Hassel. Two anonymous reviewers provided improvements on a draft version. The cited Ohio River fishery data are from reports of the Ohio River Ecological Research Program, an interutility study sponsored by American Electric Power Company, Ohio Edison Company, Ohio Valley Electric Corporation, Cincinnati Gas and Electric Company, and Tennessee Valley Authority.

## CHAPTER 12

Freshwater Benthic Macroinvertebrates and Rapid Assessment Procedures for Water Quality Monitoring in Developing and Newly Industrialized Countries

Vincent H. Resh

1.0 INTRODUCTION

The rapid assessment approach is the most recent trend in a century-old attempt to use biology in the assessment of water quality (Hynes 1960; Woodiwiss 1964; Sládeček 1979; Metcalfe 1989; Cairns and Pratt 1993; Resh and Jackson 1993; Resh and Norris 1994; Davis, Chapter 3). In North America and Western Europe, several rapid assessment procedures have been formulated, and some have even been codified through legislation. Because rapid assessment offers a cost-effective approach to water quality monitoring, the application of such an approach in developing or newly industrialized countries (hereafter referred to as developing countries) is as appealing (or even more appealing) than its use in developed and industrialized nations.

Water quality assessment programs in developing countries have generally been concerned with public health issues; safe drinking water has been the primary emphasis of these programs (Resh and Grodhaus 1983). Interest in environmental monitoring in some developing countries has increased in recent years (e.g., for habitat conservation programs) but, except for testing the effects of insecticidal treatments (such as effects of black fly control measures on nontarget organisms), macroinvertebrates have generally not been part of these assessments.

Macroinvertebrates are the most commonly used group of organisms in biological monitoring in industrialized countries, and Rosenberg and Resh (1993) have examined the advantages and difficulties to consider in using macroinvertebrates in water quality assessments (Table 1). In terms of developing countries, advantages 1 through 5 readily apply; however, difficulties 2, 3, 4, 5, and especially 6 must be considered and mitigated where possible (see below). Perhaps most importantly, the use of benthic macroinvertebrates and rapid assessment procedures can provide accurate information in surveys of pollution effects at a fraction of the cost and technical expertise than is required when using other assessment approaches (e.g., water chemistry).

The purpose of this paper is to examine: (1) measures that are being used in rapid assessment programs in industrialized countries that could be applied to the design of water-quality monitoring programs in developing countries, and (2) other considerations that arc necessary in designing rapid assessment procedures for use in such situations. The discussion will emphasize assessments of stream environments, but aspects of monitoring ponds and lakes also will be mentioned.

0-87371-894-1/95/$0.00+$.50
© 1995 by CRC Press, Inc.

Table 1.   Advantages and Difficulties to Consider in Using Benthic Macroinvertebrates for Biological Monitoring

Advantages	Difficulties to consider
1. Being ubiquitous, they are affected by perturbations in all types of waters and habitats	1. Quantitative sampling requires large numbers of samples, which can be costly
2. Large numbers of species offer a spectrum of responses to perturbations	2. Factors other than water quality can affect distribution and abundance of organisms
3. Their sedentary nature allows spatial analysis of disturbance effects	3. Seasonal variation may complicate interpretations of comparisons
4. Their long life cycles allows effects of regular or intermittent perturbations, variable concentrations, etc. to be examined temporally	4. Propensity of macroinvertebrates to drift may offset advantages of being sedentary
5. Qualitative sampling and analysis are well developed, and can be done using simple, inexpensive equipment	5. Perhaps too many methods of analyses are available
6. Taxonomy of many groups is well known and identification keys are available	6. Certain groups are not well known taxonomically
7. Many methods of data analysis for macroinvertebrate communities have been developed	7. Benthic macroinvertebrates are not sensitive to some perturbations, such as pathogens and trace amounts of some pollutants
8. Responses of many common species to different types of pollution have been established	
9. Macroinvertebrates are well suited to experimental studies of perturbation	
10. Biochemical and physiological measures of individual-organism stress to perturbations are being developed	

Summarized from Rosenberg and Resh (1993).

2.0 WHAT ARE RAPID ASSESSMENT APPROACHES?

Resh and Jackson (1993) described the use of rapid assessment approaches in water quality monitoring as somewhat analogous to the use of thermometers in assessing human health; easily obtained values are compared to a threshold that is considered normal, and large deviations indicate that further examination is necessary. The potential applicability and limits of rapid assessment approaches also can be seen in the human temperature analogy: What measures are biologically relevant (the thermometers)? What are the thresholds that these measures should be compared to (the normal body temperature)? And how much deviation from the threshold is indicative of illness?

The application of the term "rapid assessment" to water quality monitoring is confined largely to North America but programs in several European countries share common elements, many of which arose from the saprobien system (Metcalfe 1989). Moreover, some North American programs include concepts of evaluating community structure and/or function that are typical of quantitative studies. Rapid assessments differ from both the saprobien system and quantitative studies in that they are usually characterized by involving more than one type of measurement, and some summarization of these measurements is used to compare them with predetermined thresholds rather than relying on statistical comparisons. Rapid assessment approaches have been adopted (or programs are being developed) in water quality monitoring programs in the majority of the state water quality programs in the United States and, in many respects, the adoption of rapid assessment approaches has renewed interest in including biological appraisals in water quality assessments.

Rapid assessment approaches are intended to identify water quality problems and to classify aquatic habitats according to a variety of water resource criteria. Beside cost-effectiveness, rapid assessments offer standardization of analysis and accuracy of habitat classification. Their application to benthic macroinvertebrates follows their earlier, successful use with fish communities (Karr 1981; Karr et al. 1986; Plafkin et al. 1989; Karr 1991).

Rapid assessment procedures using benthic macroinvertebrates involve sampling and analysis techniques that are designed to fulfill two objectives: (1) effort (and consequently cost) is reduced in assessing

Table 2. Selected Examples of Measures Used in Rapid Assessment Protocols

I. Richness measures	II. Enumerations	III. Community Diversity and Similarity
Number of taxa	Number of individuals (or biomass)	Shannon's Index (Shannon 1948)
Number of Ephemeroptera, Plecoptera, Trichoptera (EPT) taxa	Percent EPT individuals	Margalef's Index (Margalef 1951)
Number of families	Percent Chironomidae individuals	Menhinick's Index (Menhinick 1964)
Niche occupancy forms (Mason 1979)	Ratio of EPT/Chironomidae individuals	Simpson's Index (Simpson 1949)
Number of Ephemeroptera, Plecoptera, Trichoptera, and Diptera taxa (considered individually)	Ratio of Hydropsychidae/Trichoptera	Coefficient of Community Loss (Courtemanch and Davies 1987)
	Percent individuals of numerically dominant taxa	Jaccard Coefficient (Jaccard 1912)
	Percent nondipterans	Pinkham–Pearson Community Similarity Index (Pinkham and Pearson 1976)
	Percent non-Chironomidae, Diptera, and noninsect individuals	Number of dominant taxa in common
	Percent individuals in major groups	Number of taxa in common
	Relative abundance of different groups	Quantitative Similarity Index
	Percent Ephemeroptera, Trichoptera, Tanytarsini Chironomidae, other Dipterans, and noninsects (considered individually)	Percent change in taxa richness
	Percent tolerant groups	Number of unique species per site
		Number of missing EPT taxa at study site (cf. reference site)

IV. Biotic indices	V. Functional measures	VI. Combination indices
Belgian Biotic Index (De Pauw and Vanhooren 1983)	Percent shredders (Cummins 1988)	Mean Biometric Score (Shackleford 1988)
Biotic Condition Index (Plafkin et al. 1989)	Percent scrapers (Cummins 1988)	Invertebrate Community Index (Ohio Environmental Protection Agency 1987b)
Biotic Index (Chutter 1972; Hilsenhoff 1982a, 1987, 1988)	Percent collector-filterers (Cummins 1988)	Biological Condition Score (Winget and Mangum 1979)
BMWP Score (Wright et al. 1988)	Ratio of scrapers/collector-filterers (Cummins 1988)	
Florida Index (Ross and Jones 1979)	Ratio of trophic specialists/generalists (Maine Department of Environmental Protection 1987)	
Indicator-organism presence	Types of functional-feeding groups	
ISO score (ISO 1984)	Functional group similarity	
Community Tolerance Quotient (Winget and Mangum 1979)		
Saprobic Index (Zelinka and Marvan 1961)		
Dominance of tolerant groups (Plafkin et al. 1989)		
Indicator Assemblage Index (Shackleford 1988)		

See Resh and Jackson (1993) and Gibson (1994) for examples of rapid assessment protocols that use these measures.

Based on Resh (in press) and Gibson (1994).

environmental conditions at a site, relative to what would be needed if quantitative approaches were used; this is done by reducing numbers of habitats sampled, the number of replicate samples taken, the time of sorting animals from substrate, etc.; and (2) results of site surveys are summarized, often by expressing the results of analyses from several impact-related measures as a single score and/or by using environmental-quality categories. The development and implementation of rapid assessment programs is discussed in detail by Southerland and Stribling (Chapter 7), Plafkin et al. (1989), Resh and Jackson (1993), and Resh and Norris (1994).

3.0 BACKGROUND INFORMATION AND BIOLOGICAL MEASURES USED IN RAPID ASSESSMENT APPROACHES

Dozens of different measures have been proposed or used in rapid assessment programs. Table 2 lists >50 population-, species assemblage-, and community-level measures that have been reviewed by Resh and Jackson (1993). The community measures include examination of both structural and functional aspects.

The measures in Table 2 require various levels of taxonomic discrimination or identification, and in some cases information about pollution tolerance or trophic status. In general, what levels of background information about macroinvertebrates are available in developing countries and how does this affect decisions about which of these measures would be appropriate to use?

3.1 Taxonomy

Taxonomic keys for use worldwide are available for order-level identification of aquatic macroinvertebrates. In addition, taxonomic keys for most regions have been constructed that can be used to identify macroinvertebrate families. However, there are exceptions; for example, no keys are available for identifying the families of Trichoptera present in southern India. Certainly, however, generic-level keys are lacking for most developing regions.

The lack of identification keys limits the ability to assign below-family-level taxonomic names to specimens collected; however, it does not inhibit sorting of at least some of these specimens into what are likely to be species-level groups. Clearly, this is easier for some taxa (e.g., species of Ephemeroptera, Plecoptera, Trichoptera) than others (e.g., various Diptera).

3.2 Pollution Tolerance

It has long been known that certain groups of macroinvertebrates are more pollution-tolerant than others (Cairns and Pratt 1993; Johnson et al. 1993). Resh and Unzicker (1975) concluded that species-level designations are required for the accurate determination of pollution status; study objectives ultimately determine whether this level is needed (Resh and McElravy 1993). However, it has been known for decades that certain groups of macroinvertebrates are, in general, more tolerant of pollution (e.g., tubificid worms, rat-tailed maggots of the family Syrphidae) than are others (e.g., ephemerellid mayflies and rhyacophilid caddisflies). Similar comparisons for all groups, however, are not as clear-cut; for example, all Oligochaeta, Chironomidae, or collector-filtering organisms could not be considered to be more tolerant of pollution than, say, all Crustacea, Ephemeroptera, or shredder organisms.

3.3 Trophic Status

Most measures of benthic macroinvertebrate communities examine their *structure*. An advantage of examining the trophic status of component populations is that it also provides information on *functional* aspects of the community as well. In North America (where functional feeding-group approaches are more commonly used as a measure of community function than elsewhere), trophic status is usually determined after a specimen has been identified by consulting the trophic-status designation for that genus in the appropriate tables of Merritt and Cummins (1984). However, Cummins (1988) stated that functional groups should be based on mouthpart morphology and the means of food acquisition. Given this consideration, the picture key approach of Cummins and Wilzbach (1985) for assigning specimens to functional groups without formal identifications is appropriate. This approach would allow functional measures to be used in assessments in developing countries.

An advantage of using the functional-group approach is that morphology and behavior are the basis for group assignment; thus, the lack of taxonomic keys is not a deterrent. Functional feeding-group approaches may also be used as surrogate measures of more complicated processes, such as secondary production, energy flow, etc. (see below).

4.0 CHOOSING APPROPRIATE RAPID ASSESSMENT MEASURES

Which of the >50 measures in Table 2 would be appropriate to use in developing countries, given the issues of taxonomy, pollution tolerance, and trophic status previously discussed?

4.1 Richness

Of the five richness measures that have been used routinely in rapid assessments (Table 2), only "number of families" actually requires below order-level identifications; as indicated above, family identifications can be done for most parts of the world. The picture-key approach to macroinvertebrate identification (as in McCafferty 1981 or Clifford 1992) may be a useful tool in this regard, although very usable conventional dichotomous keys to aquatic invertebrates are available in Lehmkuhl (1979), Merritt and Cummins (1984), Thorp and Covich (1991), and other sources listed in Merritt and Cummins (1984) and Resh and Rosenberg (1984). The other richness measures simply involve counting the number of taxa present. The sorting of some groups of macroinvertebrates into species-level categories is easily learned, especially for groups with distinctive (and often large) forms, and those that usually occur without closely related taxa. However, other groups usually have more than one taxon co-occurring, they require slide preparation and high microscopic magnification for differentiating among taxa, or they require considerable identification training. Illustrations of structures that could be used in determining species-level differences may aid in the discernment of taxa; although not currently available as compilations, they would not be difficult to prepare.

Arguably, larvae of many species of co-occurring Ephemeroptera, Plecoptera, and Trichoptera generally can be separated into species groups with a certain degree of training; also, this generally requires less expensive magnification equipment (i.e., a dissection microscope is usually sufficient) than needed for other groups, such as the Diptera. Estimations of number of families in these three groups could be determined with some training in the use of keys; however, different keys may use different family-level groupings (e.g., for Ephemeroptera, one key may list 3 families in the order and another 16; see Resh and Jackson 1993). Therefore, in terms of effort relative to information obtained, one can conclude that the number of Ephemeroptera, Plecoptera, and Trichoptera (EPT) taxa is probably the most reasonable choice among the richness measures in Table 2 for use in rapid assessment procedures in developing countries. It should also be remembered, though, that some areas of the world, such as oceanic islands or even vast areas such as South Australia, may have few representatives of one or more of these three groups.

4.2 Enumerations

All of the enumerations listed in Part II of Table 2, except for percent tolerant groups, require little taxonomic knowledge and can be used in rapid assessment programs in developing countries. The total number of individuals is the easiest measure to determine; the additional information obtained from estimation of biomass is probably not worth the technical problems of standardizing drying and weighing procedures. Because specimens of Ephemeroptera, Plecoptera, Trichoptera, Chironomidae, non-Diptera (determined by simply counting the number of Diptera and subtracting it from the total number of individuals), noninsects, and even Tanytarsini Chironomidae can be distinguished readily with some training, the proportions of these groups could also be determined. The percent individuals in major groups and the relative abundance of different groups could be obtained by sorting of specimens into groups. However, if the measure of percent tolerant groups is used, family- or order-tolerance characterizations would be required for all groups of organisms encountered, and this would involve the problems mentioned previously. If number of Ephemeroptera, Plecoptera, and Trichoptera (EPT) taxa is chosen as a richness measure, the proportion of individuals of these three groups relative to the total number of individuals collected may be the best choice of the enumerations listed in Table 2.

4.3 Community Diversity and Similarity

Calculation of the diversity measures given in Table 2 requires that taxa be enumerated at the species level. The taxonomic problems in doing this are the same as discussed for number of taxa above. However, if a community structure measure is desired, the Sequential Comparison Index (SCI, Cairns et al. 1968), which is a modification of Simpson's Index (Patil and Taillie 1976), may be appropriate to use. The SCI is calculated using the following steps:

1. Organisms are sorted from the substrate and debris, swirled in a jar or a vial for randomization, poured out in a flat tray, and then randomly arranged in rows.
2. An X is marked on a sheet of paper to represent the first organism. If the organism next to it is the same species, another X is marked on the paper; if it is a different species, an O is marked instead; the third specimen is then compared to the second, and the same symbol is used as before if it is the same species or the other symbol is used if it is a different species.
3. The procedure is continued through all the rows (Figure 1B), with the last specimen in a row being compared with the first specimen in the subsequent row; at least 100 to 200 organisms should be examined and compared.
4. The index is calculated by dividing the number of runs by the total number of individuals; when two organisms are compared they are part of the same run, so a sequence of XOOXOXOOXO would contain 8 runs and an SCI of 8/10 or 0.8 (Figure 1B).
5. Values close to one (e.g., 0.8 to 0.9) indicate a high-diversity benthic macroinvertebrate communities; low values (e.g., 0.1 or 0.2) indicate low diversity communities. Large differences between comparable sites (see discussion below) indicate perturbation at the site with the reduced value.

Gottfried and Resh (1979) developed a teaching module to demonstrate the use of the SCI; they found it appropriate for students with no prior training in aquatic insect identification. Given the many difficulties involved in using the community diversity and similarity measures listed in Table 2, the SCI may be the most appropriate diversity measure to use in developing countries.

4.4 Biotic Indices

Calculation of these indices (Part IV, Table 2) requires tolerance scores for macroinvertebrate taxa (i.e., values that are weighted in terms of the pollution sensitivity of each of the different taxa encountered). Tolerance scores are usually given at the generic level, although they are sometimes also assigned at the species or family levels. A family-level biotic index for organic pollution has been proposed (Hilsenhoff 1988), and it does work in many regions (Resh and Jackson 1993); however, even if family-level identifications can be accurately made, the extrapolation of family-level tolerances from North America to other continents is risky. Given tolerance and identification considerations, biotic indices may not be appropriate to use in developing countries at this time.

4.5 Functional Measures

Some of the functional feeding-group measures in Part V of Table 2 are also appropriate for use. A procedure for including them in assessments involves the following steps (from Cummins and Wilzbach 1985):

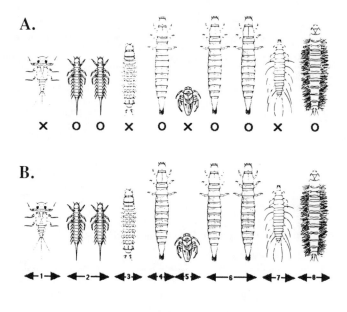

$$SCI = \frac{\text{Number of runs}}{\text{Number of individuals}} = \frac{8}{10}$$

Figure 1. Calculation of the Sequential Comparison Index of Cairns et al. (1968). (A) Comparison of specimens to determine number of runs; (B) calculation of index.

1. Take a sample (or replicate samples) from the following subhabitats at a site: (a) leaf-litter accumulations (which is the CPOM or coarse particulate organic matter fraction); (b) large branches and logs; (c) fine detritus (which is the FPOM or fine particulate organic matter fraction); (d) periphyton (which is the algae attached to some rock and wood surfaces); (e) vascular plants or moss, when available.
2. Sort macroinvertebrates from each subhabitat sample.
3. Assign macroinvertebrates to functional feeding groups (see Table 3, Part A or Cummins and Wilzbach's picture key) and count them. Estimates of relative biomass in each category can also be determined by using simple volume-displacement.
4. Calculate ratios of different functional groups, and compare these ratios with predicted ratios for the different types of streams (Table 3, Part B).
5. Predicted ratios may vary regionally but those presented in Part B of Table 3 may serve as a starting point for later modification. Large departures from predicted ratios may indicate functional impairment.

The absence of a functional feeding group when a resource base is present may indicate that instream impact has occurred; moreover, the ratios calculated using the above procedures also may be indicative of changes in the resource base. For example, percent scrapers or the ratio of scrapers to shredders and collectors may be indicative of productivity–respiration ratios or the relative proportion of autochthonous compared to allochthonous material in the resource base. These functional feeding-group ratios would be sensitive to changes in light and/or nutrient inputs. Percent shredders, or the ratio of shredders to collectors, are indicative of the ratio of CPOM to FPOM resource base and these ratios are sensitive to riparian alteration. The ratio of scrapers to filtering collectors or filtering collectors to gathering collectors is indicative of the relative importance of transported FPOM to sedimentary FPOM. These ratios are sensitive to increases of FPOM inputs from riparian alterations, effluents, outfalls from impoundments, etc.

Table 3.

A. Functional Feeding-Group Assignment of some Commonly Encountered Aquatic Taxa

Functional group	Molluscs	Crustaceans	Insects
Scraper	Limpets and snails	—	Caddisflies with mineral cases; flat-bodied mayflies
Shredder	—	Isopods and amphipods	Caddisflies with organic cases; dull, sluggish stoneflies; large craneflies
Filtering-collector	Clams or mussels	—	Net-spinning caddisflies; tube-inhabiting midges; black flies
Gathering-collector	—	Isopods and amphipods	90% of Chironomidae; cylindrically shaped mayflies
Predator	—	—	10% of Chironomidae; bright-colored active stoneflies; dragonflies; beetles; legless maggots with large jaws

B. Examples of Expected Functional Feeding-Group Ratios

Ratio	Shaded, small streams (0.5–10 m wide)	Open, small stream	Open, medium-sized stream (~10 m wide)	Large river (>30 m wide)
Shredders/total individuals (percent shredders)	>0.25	>0.10	<0.05	<0.01
Collectors/total individuals (percent collectors)	>0.50	>0.40	>0.50	>0.75
Scrapers/total individuals (percent scrapers)	<0.10	~0.25	>0.40	<0.10
Predators/total individual (percent predators)	~0.10	~0.10	~0.10	~0.10
Shredders/collectors	>0.30	>0.15	<0.10	<0.05
Scrapers/collectors	<0.25	>0.25	>0.25	<0.10
Filtering-collectors/gathering-collectors	<0.50	~0.40	~0.50	~0.50

This table is based on information in Cummins and Wilzbach (1985); consult that source for a pictorial key to functional-feeding groups.

Of the six functional feeding-group measures in Table 2, percent shredders, percent scrapers, and the ratio of scrapers/collector-filterers could be determined using Table 3, Part A or the picture key approach. Depending on whether an individual site is fully shaded or open, percent shredders or percent scrapers, respectively, may be the most appropriate to use (see discussion below).

4.6 Combination Indices

The indices in Part VI of Table 2 each require either species-level discernment or tolerance scores at some point in their calculation, and as such may not be currently appropriate for use in developing countries. However, as will be seen below, this does not preclude the development of other combination indices (e.g., multiple metrics, see Barbour et al., Chapter 6) that are appropriate for use in such situations.

5.0 DESIGNING RAPID ASSESSMENT PROGRAMS

The design of an accurate, cost-effective rapid assessment program involves several considerations, such as:

1. the use of reference sites or background data to determine whether impacts have occurred,
2. choice of habitats to be sampled within a site,
3. choice of sampling device(s) to be used,
4. choice of which taxonomic groups are to be used for analysis and the level(s) of identification,

5. choice of measures to be used,
6. and perhaps most importantly, how the results obtained are used in ascertaining whether or not impact has occurred, or the level of the impact present. The above considerations are discussed in detail in Resh and Jackson (1993), and discussed here as they could apply to developing countries.

5.1 Reference Sites

The choice of the number and location of reference sites in any water quality assessment is a function of the objectives of the study. For example, examination of the effects of a point (or direct) source of pollution, such as an effluent pipe, involves choosing sites that are varying distances above and below the point source (and which assumes no other sources of pollution above that source). However, in this type of design, the question of uncontrolled (or unaccounted for) variables is always an issue: Do these sites differ in some way (e.g., substrate type or amount of discharge) other than just the presence of the point source of pollution? Therefore, reference and control sites must be similar in as many characteristics as possible (e.g., flow, substrate type, depth, and seasons of sampling), otherwise the conclusions of an environmental assessment can be affected by this choice.

An approach called Before-After-Control-Impact (BACI) has been suggested as a way of dealing with the problem of control sites (Stewart-Oaten et al. 1986; Cooper and Barmuta 1993). This involves examining trends over time that occur both in the potentially impacted and in the control sites. Methods for choosing appropriate study sites are discussed at length in several general treatments (e.g., Green 1979; Norris et al. 1992; Cooper and Barmuta 1993; and references therein). In using rapid assessment procedures both in developed and developing countries, the choice of appropriate control sites is a critically important decision.

An alternative to selecting control sites is to compare information obtained at a site with generalized but region-specific background data (see Hughes, Chapter 4; and Omernik, Chapter 5). This requires that the fauna and habitats be well known regionally; this is not the current situation in most developing countries.

5.2 Choice of Habitats

There are basically two alternatives in choosing habitats for sampling: (1) limit sampling to a single habitat (e.g., riffles in streams) or zone (e.g., littoral or profundal zone, or according to depth in ponds or lakes); or (2) sample all subhabitats in proportion to their relative abundances. For most programs in developing countries, the first approach, coupled with standardization of activities at a site, will probably result in better comparisons among sites and will minimize the influence of uncontrolled variables.

5.3 Sampling Devices

The device used should provide both a representative collection of macroinvertebrate taxa and indicate the relative abundance of the different taxa present in a habitat. Dip- and kick-nets work fine in streams; dip-nets or grab (e.g., Ekman) samplers are appropriate in lakes. The most important consideration in using these devices is in standardizing their use at the different sites. For example, sampling a riffle habitat for a set amount of time (say 10 min — with the time chosen ultimately reflecting organism abundance) by turning over stones and collecting all the organisms encountered, or by disturbing substrate in 1-m^2 quadrats and letting all organisms flow into a kick-net, may be appropriate; likewise, taking a predetermined and consistent number of net sweeps through the water column may be appropriate for a pond or marsh. The important point is that the same technique should be used at each site that is to be compared.

5.4 Taxonomic Groups and Recommended Measures

The choice of groups to be identified and analyzed is related to the measures that will be used, as was discussed in detail above. For streams with riffle or cobble substrate, consideration of the following four measures is recommended: (1) for richness — number of Ephemeroptera, Plecoptera, and Trichoptera taxa (actually identified only at the order level but distinguished one from the other at the species level),

Table 4.   Variability and Accuracy of Various Rapid Assessment Measures as Applied to Various California Streams

Measure	Within-site CV	Interyear CV	Spatial or temporal similarity among unimpacted sites	Spatial similarity between impacted and control sites
Number of Ephemeroptera, Plecoptera, and Trichoptera taxa	18%	31%	≥70%	≤14%
Sequential Comparison Index (as Simpson's Index)	7%	19%	≥70%	≤12%
Percent Ephemeroptera, Plecoptera, and Trichoptera	14%	44%	≥40%	≤1%
Percent Scrapers	51%	42%	≥70%	≤8%
Percent Shredders	92%	160%	≥20%	≤2%

Note: Within-site coefficients of variation [CV, calculated as (standard deviation/mean) × 100] are based on data from Prosser Creek, California (Needham and Usinger 1956); interyear CVs are based on Big Sulpher Creek and Hunting Creek, California (Resh et al. 1990, Resh and Jackson 1993). Spatial or temporal similarity for nonimpacted sites and similarity with a thermally impacted site is based on the analysis in Resh and Jackson (1993).

(2) for enumerations — percent Ephemeroptera, Trichoptera, and Plecoptera individuals, (3) for diversity — the SCI, which only requires discrimination between two individuals at a time, and (4) for functional measures — shredders/total number of individuals or percent shredders (the proportion of this functional feeding group) in a FPOM collection of submerged leaves, and percent scrapers in the other subhabitat collections; both are determined using the picture key of Cummins and Wilzbach (1985) or the general outline in Table 3.

5.5 Ascertaining Impact

Finally, how do these measures indicate whether impact has or has not occurred? Basically, values for the four recommended measures described above are compared between the control (i.e., the unaffected) site and the other study sites that are potentially affected. The magnitude of differences in these values are used to determine if impact has occurred. This approach assumes that standardized sampling approaches have been used at all sites, and that the control and study sites differ only in the presence of pollution. In addition, it should be remembered that not all forms of pollution will produce differences in all these measures.

But how much should values differ in order to indicate impact? We can begin to examine this by considering the information in Table 4, which summarizes values of measures obtained at different sites from various California, USA, streams. For each of five measures, including the four mentioned above, four values are given in Table 4: (1) the within-site and (2) between-year Coefficients of Variation, (3) the percent similarity among non-affected sites (either between different sites or at the same site over different years), and (4) the percent similarity between a control site and an impacted site.

The variability of the measures at unimpacted sites (i.e., the within-site and interyear CVs in Table 4) influence how much similarity among unimpacted sites may be expected. In general, measures that have low variability will be more useful for intersite comparisons than those with high variability. For each of the measures in Table 4, the percent similarity at the impacted site (fourth column) was lower than the temporal or spatial similarity at the nonimpacted site or sites (third column). Plafkin et al. (1989) have proposed similarity values along a gradient of impact for some of these measures, which are summarized in Table 5; note that the similarity values they used are much higher than those found for the California streams reported in Table 4. Appropriate similarity values for unaffected sites should be determined for local situations.

6.0 MAKING RAPID ASSESSMENTS EFFECTIVE

In using rapid assessment procedures and benthic macroinvertebrates as a monitoring tool in developing countries, special training of personnel is needed in a few key areas, including: (1) how to choose

Table 5.   Similarity Among Impacted and Control Sites for Various Measures Expressed as Different Categories of Stream Impairment

Measure	Nonimpaired	Slightly impaired	Moderately impaired	Severely impaired
Number of Ephemeroptera, Plecoptera, and Trichoptera taxa	>90%	80–90%	70–80%	<70%
Ratio of numbers of Ephemeroptera, Plecoptera, and Trichoptera to numbers of Chironomidae	>75%	50–75%	25–50%	<25%
Percent shredders	>50%	35–50%	20–35%	<20%

Modified from Plafkin et al. 1989.

study sites and controls, (2) how to conduct standardized sampling, and (3) how to calculate measures correctly. Detailed, but clear, instructions that address these areas must be provided. For eventual use of all measures presented in Table 2 in developing countries, four types of information are needed:

1. Usable taxonomic tools to distinguish taxa at the species level for richness measures
2. Identification keys for designating organisms that are tolerant or intolerant to various forms of pollution (e.g., Lathrop 1989)
3. Establishment of pollution-tolerance values for commonly encountered taxa
4. Establishment of regionally based background information on the magnitude of differences that should be expected for the different measures when impacted and nonimpacted conditions occur

These goals eventually can be accomplished; until then, the procedures outlined in this paper can provide a basis for beginning effective water quality monitoring programs in developing countries.

What can scientists from industrialized countries do to assist in the establishment of effective water quality monitoring programs in developing countries? Clearly, they can assist in providing answers to the above-mentioned needs, and in fact many are doing so already. It seems that as many aquatic macroinvertebrate taxonomists from industrialized countries are doing research in developing countries as do research in their own countries! However, the research that they do often does not lead to identification keys that can be used locally (Costa Rica is an exception). This form of "scientific imperialism" must change.

Moreover, the experiences and techniques currently used in developed countries may also have applicability to nearby developing countries. For example, appropriate measures, tolerance scores, and similarity values for unimpacted and impacted conditions developed in the United States may have applicability to developing countries in Central and South America, those in Japan to East Asian countries, and those in South Africa to other African countries.

And what can scientists in developing countries do to assist both themselves and scientists in industrialized countries? By avoiding self-perpetuating or even codified "traditions," scientists in developing countries can explore new avenues for measuring biological effects of water pollution. With reciprocal exchanges of information, rapid assessment approaches in both developing and developed countries can then become even more accurate, cost-effective, and widely used.

ACKNOWLEDGMENTS

I thank J. K. Jackson (Stroud Water Research Center, Avondale, Pennsylvania), R. H. Norris (University of Canberra, Australia), and K. G. Sivaramakrishnan and K. Venkataraman (Madurai College, Madurai, India) for long and useful discussions on this topic, D. M. Rosenberg (Freshwater Institute, Winnipeg, Manitoba), and B. Statzner (University of Lyon) for their comments on this manuscript, and K. W. Cummins and M. A. Wilzbach for their comments on the use of their functional feeding-group key.

SECTION III:
METHODS ADVANCEMENT
AND TECHNICAL APPLICATIONS

## CHAPTER 13

Habitat Indices in Water Resource Quality Assessments

Edward T. Rankin

1.0 INTRODUCTION

A key concept of the Clean Water Act is the protection of biological integrity of the streams and rivers of the United States. Basic to maintaining diverse, functional aquatic communities in surface waters is the preservation of the natural physical habitat of these ecosystems. As obvious and basic as this concept seems, regulatory and protective efforts regarding habitat have been minimal (Hughes et al. 1990; Karr 1991). As a result many thousands of miles of United States streams have been and continue to be degraded each year (Benke 1990; NRC 1992). This loss of habitat quality has resulted in extinctions (Williams et al. 1989), local extirpations (Karr et al. 1985), and population reductions (Trautman 1981; Ohio EPA 1992) of fish species and other aquatic fauna (e.g., Williams et al. 1993) in the United States. In contrast to many other human impacts, habitat loss can be essentially irretrievable over a human time frame.

The lack of consistent habitat-protective efforts is reflected in the extreme inconsistencies in reporting aquatic habitat problems across the nation. Habitat, both instream and riparian, can be the factor most limiting aquatic community potential in streams and rivers. Observed habitat conditions are usually the result of complex interplay between hydrogeomorphological factors and anthropogenic landscape alterations (Gregory et al. 1991; Hill et al. 1991). Surprisingly, water chemistry is often the only one of five major factors that affect biological integrity (Figure 1) assessed to determine aquatic life use attainment. While all states and territories collect fecal coliform, dissolved oxygen, and other chemical specific data, few states effectively monitor for habitat destruction and alteration or effectively integrate existing habitat and biosurvey work into surface water monitoring programs. Often, habitat impacts are only considered in a framework of gamefish management (Osborne et al. 1991). Figure 2, derived from the National Water Quality Inventory (USEPA 1992), illustrates this problem. Of the 47 states or territories reporting impairment data on streams and rivers for the 1992 report, 25 did not report habitat as a cause of problems.

Much of the regulatory emphasis of the Clean Water Act as interpreted by the United States Environmental Protection Agency (USEPA) has focused on point sources of pollution (e.g., wastewater treatment plants [WWTPs], and industries) because of the obvious threats to human health and the relative ease, from a regulatory viewpoint, of dealing with a discrete source of pollution. Unfortunately, many of the most serious remaining problems and threats to the biological integrity of ecological systems (e.g., habitat destruction, urbanization and suburbanization, mining, grazing, and agricultural impacts) do not fit well into a point source control conceptual framework. This chemical-specific load reduction framework is insufficient by itself to address most habitat and development-related impairments to ecosystems. The development of the "River Continuum Concept" (Vannote et al. 1980; Minshall et al. 1983) resulted in an abundant literature that described the connections among the landscape, habitat quality and water quality, and that support holistic approaches to water resource management.

Efforts to reduce habitat destruction will require us to examine multiple scales of impacts, from landscape ecosystem to microhabitat scales and a move away from reductionist approaches (Karr 1991).

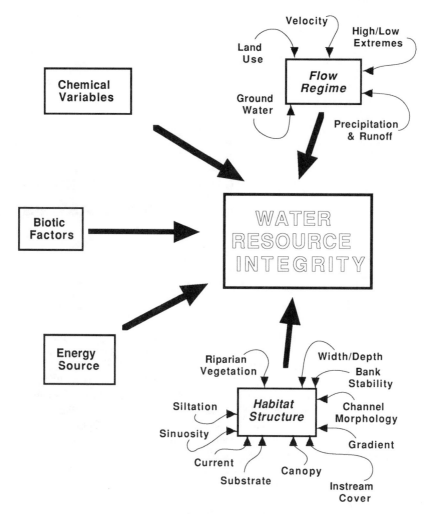

Figure 1. The five major factors that affect Water Resource Integrity with more detail on factors often used in habitat indices.

Such an effort will require states to improve their monitoring abilities to address and consider effects at each scale of impact. Monitoring data, including habitat assessments, can be used to rank physical and biological resource quality of streams, identify those streams and rivers threatened by anthropogenic changes, and to provide insight on possible remedies. Conversely, for areas that have had severe and essentially irretrievable (from a social and economic viewpoint) habitat losses, monitoring data can direct efforts towards areas where abatement efforts can be cost-effective and successful.

The objectives of this paper are to (1) review existing habitat assessment indices and methods, (2) explain how Ohio EPA incorporates its habitat index (the QHEI) in its integrated monitoring efforts, (3) delimit specific uses and limitations of habitat indices in state water resource quality management programs, and (4) encourage examination of water resource impacts at multiple spatial and temporal scales.

2.0 BACKGROUND

2.1 Approaches to Habitat Assessment

Various types of stream habitat indices and methodologies have been used in North America over the past 20 to 30 years (Table 1). The first, and most frequent use of habitat indices has been to relate the

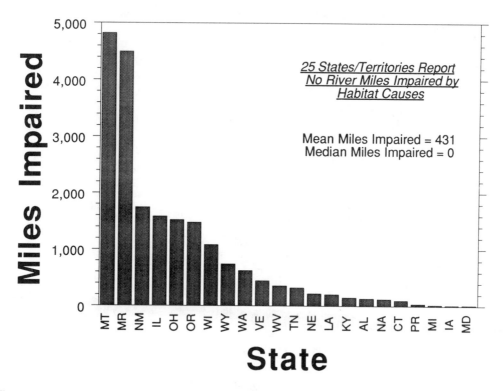

Figure 2. Miles of streams and rivers impaired by habitat causes, by state, summarized in the 1992 National Water Quality Inventory (USEPA 1992e). Data were reported by states to the USEPA through their 1990 305(b) reports.

standing crop or population of a target species, often a sport fish, to habitat characteristics in a stream. The usual goal of such work is to define limiting habitat factors to allow managers to manipulate stream habitat to enhance fishable populations. Most states use some type of transect or index method to accomplish such habitat assessments. Examples of this abound for salmonids (Binns and Eiserman 1979; Platts et al. 1983) as well as warmwater species (Layher and Maughan 1985; Layher and Brunson 1992). Another use of habitat indices, especially in the western United States is to determine the minimum or optimal stream flows that would protect habitat characteristics essential to the life history of one or more target species. The United States Fish and Wildlife Service (USFWS) has developed much of this work and it is summarized by Bovee (1982, 1986). Recent work by Hill et al. (1991) broadens this concept to include the importance of out-of-channel flow to habitat and riparian maintenance. This supports the call by Stalnaker (1990) for "progress beyond the minimal flow…, to focus on scientific principles in understanding riverine systems."

The third and most recent use of habitat indices, is as an integral part of water pollution control programs in states (Karr 1991). These states have recognized that the threats to biological integrity of streams are much more extensive than water quality threats alone. These monitoring programs use habitat indices to characterize the causes and sources of impacts and to help ascertain the potential for waters to support aquatic communities. Most of these habitat techniques focus on aquatic community responses rather than species-specific responses to changes in habitat quality, although the concepts are similar. The remainder of this paper will deal with the use of habitat indices in state water resource quality monitoring that focus on biological communities and biological integrity.

2.2 Habitat Indices in Use by States

Osborne et al. (1991) summarized habitat assessment programs in states of the American Fishery Society North Central Division (Midwest and Upper Midwest). Most of the habitat assessment efforts at

Table 1.   A Selected Listing of Habitat Indices and Their Design Purpose Used in North America Over the Past 30 Years

Index/methodology	Purpose	Ref.
HEP/HSI	Relate habitat quality to single species carrying capacity	Terrell (1984); Layher and Maughan (1985)
HQI	Habitat as predictor of trout standing crop	Binns and Eiserman (1979)
BSC	Habitat quality used with IBI to determine biotic potential of a stream reach	Illinois EPA (1989); Hite (1988)
Transect methods	Assesses various aspects of stream habitat by taking measurements along transects in a reach	Dunham and Colotzi (1975); Platts et al. (1983); Armour et al. (1983); Duff et al. (1989)
Habitat diversity/ complexity	Shannon index application using substrate, depth, and velocity	Gorman and Karr (1978); Schlosser (1982)
HI	Missouri's habitat quality index based on ten components relating present status to pristine condition	Fajen and Wehnes (1982)
HCI	Habitat condition indicator for streambank and instream components	Duff et al. (1989)
BCI/DAT	Species diversity using habitat, species dominance, and taxa	Winget and Mangum (1979); Mangum (1986)
QHEI	Visual habitat method correlated with fish community condition (e.g., IBI)	Rankin (1989, 1991); Ohio EPA (1989)
IFIM	Used to determine flow needs of stream fish species	Bovee (1982, 1986)
RBP habitat qual.	Habitat evaluation based on stream classification guidelines for Wisconsin	Plafkin et al. (1989); Barbour and Stribling (1991); Ball (1982); Platts et al. (1983)

the time of the survey were directed towards game/sport fish management rather than towards broader efforts of protecting biointegrity or biodiversity. Four of these states used habitat assessments as a component of biological assessment, aquatic life use designations, or to identify reference reaches (IL, NE, OH, and WI). A more recent survey (Abe et al. 1992) of states in USEPA Region 5 (i.e., IL, OH, MI, IN, MN, and WI) has found that most states have begun or have initiated programs to address biological integrity within the framework of overall surface water protection programs. Other states around the country have reported on the use of the Rapid Bioassessment and Habitat Assessment Protocols of the USEPA (Primrose et al. 1991; Plotnikoff 1992; Hayslip 1993) or other similar methods incorporating the work of Ball (1982), Mangum (1986), Winget and Mangum (1979), and Platts et al. (e.g., Maret 1988 in Nebraska; Simonson et al. 1993 in Wisconsin). The burgeoning effort expended on assessment of biological integrity prompts consideration of some of the uses, limitations, and needs for habitat assessment protocols and a review of the existing literature.

3.0 METHODS

3.1 QHEI derivation

Many of the topics discussed in this chapter will be illustrated with examples from Ohio's experience with integrating the "Qualitative Habitat Evaluation Index" (QHEI) into its surface water monitoring program. The following is a brief summary of the calculation and scoring of the QHEI. A more detailed explanation is found in Ohio EPA (1988) and Rankin (1989).

The use of the QHEI is dependent on visual estimates of habitat features. Each of the habitat attributes assessed is summarized in Table 2. A sample data sheet is illustrated in Figure 3. Definitions and procedures for scoring the attributes in Table 2 are detailed in Ohio EPA (1988) and Rankin (1989); all staff using this index in Ohio go through a yearly training program. Scores for each category of the QHEI were originally assigned on the basis of a literature review of the response of warmwater fish species and communities to various habitat characteristics. These original scores were adjusted by examining the

Table 2.   Physical Habitat Attributes Scored in Ohio EPA's Qualitative Habitat Evaluation Index (QHEI)

I. Substrate quality
   a. Two most predominate substrate types
   b. Number of substrate types
   c. Substrate origin (tills, limestone, etc.)
   d. Extensiveness of substrate embeddedness (entire reach)
   e. Extensiveness of silt cover (entire reach)
II. Instream cover
   a. Presence of each type in the reach
   b. Extensiveness of all cover in reach
III. Channel quality
   a. Functional sinuosity of channel
   b. Degree of pool/riffle development
   c. Age/effect of stream channel modifications
   d. Stability of stream channel
IV. Riparian quality/bank erosion
   a. Width of intact riparian vegetation
   b. Types of adjacent landuse
   c. Extensiveness of bank erosion/false banks
V. Pool/riffle quality
   a. Maximum pool/glide depth
   b. Pool/riffle morphology
   c. Presence of current types
   d. Average/maximum riffle/run depth
   e. Stability of riffle/run substrates
   f. Embeddedness of riffle/run substrates
VI. Local stream gradient (ft/mi) from 7.5' topographic map

Note: Habitat attributes are visually estimated over a 150 to 500-m reach that corresponds to a biological sampling reach.

response of the IBI, collected at a series of least impacted and habitat modified reference sites, to each of the QHEI habitat characteristics (Rankin 1989). These reference sites are the same as those used to derive Ohio's biological criteria, and are also discussed in Yoder and Rankin (Chapter 9) and DeShon (Chapter 15). This database, augmented with some newer data, will be used to illustrate many of the concepts discussed in this chapter.

Stream flow data (periodicities, peaks, minimums, etc.) are not an explicit part of the QHEI. The flow regimes to which a stream are subject, however, are a fundamental consideration when interpreting habitat and biological data. Incorporating the effects of flow on streams can be accomplished by stratifying streams according to their flow characteristics as has been proposed by Poff and Ward (1989). Extremely high, flashy flows can be limiting in certain small streams as can very low flows. However, in many situations in Ohio flow is not limiting and flow data are not always readily available; thus, it was excluded as part of the index. In addition, our habitat sampling is generally done in concert with biosurvey sampling which integrates and reflects the effects of past flow events. For some regions of the country measures of flow may be an essential component to include in an index.

4.0 RESULTS AND DISCUSSION

4.1 Regionalization of Habitat Approaches

A ecoregional approach to examining and managing surface waters has many advantages for organizing ecological data and interpreting man's impact on rivers and streams (Hughes et al. 1990). Ecologically pertinent stratification can simplify sampling approaches (Gallant et al. 1989) and provide a conceptual and operational framework for defining biotic potential or biotic limitations (Hughes et al. 1990). Advantages for considering regional differences in habitat assessments are especially convincing. Habitat features, which often affect or limit biological communities, are a consequence of geomorphologic and other natural factors that are the basis of regionalization efforts such as ecoregions (Omernik 1987). For some areas of the United States the application of ecoregions to water resource components has been successfully demonstrated (Larsen et al. 1986; Rohm et al. 1987; Whittier et al. 1988).

OhioEPA Qualitative Habitat Evaluation Index Field Sheet   QHEI Score: [____]

Stream_____ RM_____ Date_____ River Code_____
Location_____ Scorers Name:_____

1] SUBSTRATE (Check *ONLY* Two Substrate *TYPE BOXES; Estimate % or note every type present);

TYPE	POOL RIFFLE		POOL RIFFLE	SUBSTRATE ORIGIN		SUBSTRATE QUALITY	

TYPE   POOL RIFFLE   POOL RIFFLE   **SUBSTRATE ORIGIN**   **SUBSTRATE QUALITY**
☐ ☐-BLDR /SLABS[10] ___ ___   ☐ ☐-GRAVEL [7] ___ ___ Check ONE (OR 2 & AVERAGE)   Check ONE (OR 2 & AVERAGE)
☐ ☐-BOULDER [9] ___ ___   ☐ ☐-SAND [6]   ___ ___   ☐ -LIMESTONE [1] SILT: ☐ -SILT HEAVY [-2]
☐ ☐-COBBLE [8]   ___ ___   ☐ ☐-BEDROCK[5] ___ ___   ☐ -TILLS [1]   ☐ -SILT MODERATE [-1]
☐ ☐-HARDPAN [4] ___ ___   ☐ ☐-DETRITUS[3]___ ___   ☐ -WETLANDS[0]   ☐ -SILT NORMAL [0]   Substrate
☐ ☐-MUCK [2]   ___ ___   ☐ ☐-ARTIFICIAL[0]___ ___ ☐ -HARDPAN [0]   ☐ -SILT FREE [1] ___ ___
☐ ☐-SILT [2]   ☐ -SANDSTONE [0] EMBEDDED ☐ -EXTENSIVE [-2]
NOTE: (Ignore sludge that originates from point-sources;   ☐ -RIP/RAP [0]   NESS: ☐ -MODERATE [-1]   Max 20
score on natural substrates)   ☐-5 or More [2] ☐ -LACUSTRINE [0]   ☐ -NORMAL [0]
NUMBER OF SUBSTRATE TYPES: ☐-4 or Less [0]   ☐ -SHALE [-1]   ☐ -NONE [1]
COMMENTS_____ ☐-COAL FINES [-2]

2] INSTREAM COVER   AMOUNT: (Check *ONLY* One or
                  TYPE: (Check *All* That Apply)   check 2 and *AVERAGE*)
☐ -UNDERCUT BANKS [1]   ☐ -DEEP POOLS> 70 cm [2] ☐ -OXBOWS [1]   ☐ - EXTENSIVE > 75% [11]   Cover
☐ -OVERHANGING VEGETATION [1] ☐ -ROOTWADS [1]   ☐ -AQUATIC MACROPHYTES [1] ☐ - MODERATE 25-75% [7]
☐ -SHALLOWS (IN SLOW WATER) [1] ☐ -BOULDERS [1] ☐ -LOGS OR WOODY DEBRIS [1] ☐ - SPARSE 5-25% [3]
☐ -ROOTMATS [1]   COMMENTS: ☐ - NEARLY ABSENT < 5%[1] Max 20

3] CHANNEL MORPHOLOGY: (Check *ONLY* One PER Category OR check 2 and *AVERAGE*)

SINUOSITY	DEVELOPMENT	CHANNELIZATION	STABILITY	MODIFICATIONS/OTHER	
☐ - HIGH [4]	☐ - EXCELLENT [7]	☐ - NONE [6]	☐ - HIGH [?]	☐ - SNAGGING	☐ - IMPOUND.
☐ - MODERATE [3]	☐ - GOOD [5]	☐ - RECOVERED [4]	☐ - MODERATE [2]	☐ - RELOCATION	☐ - ISLANDS
☐ - LOW [2]	☐ - FAIR [3]	☐ - RECOVERING [3]	☐ - LOW [1]	☐ - CANOPY REMOVAL	☐ - LEVEED
☐ - NONE [1]	☐ - POOR [1]	☐ - RECENT OR NO		☐ - DREDGING	☐ - BANK SHAPING
		RECOVERY [1]		☐ - ONE SIDE CHANNEL MODIFICATIONS	

Channel Max 20

COMMENTS:_____

4]. RIPARIAN ZONE AND BANK EROSION - (check ONE box per bank or check 2 and AVERAGE per bank)   ★River Right Looking Downstream★

RIPARIAN WIDTH	FLOOD PLAIN QUALITY (*PAST 100 FOOT RIPARIAN*)		BANK EROSION	
L R (Per Bank)	L R (Most Predominant Per Bank)	L R	L R (Per Bank)	Riparian
☐ ☐-WIDE > 50m [4]	☐ ☐-FOREST, SWAMP [3]	☐ ☐-CONSERVATION TILLAGE [1]	☐ ☐-NONE/LITTLE [3]	
☐ ☐- MODERATE 10-50m [3]	☐ ☐-SHRUB OR OLD FIELD [2]	☐ ☐-URBAN OR INDUSTRIAL [0]	☐ ☐-MODERATE [2]	
☐ ☐-NARROW 5-10 m [2]	☐ ☐-RESIDENTIAL,PARK,NEW FIELD [1]	☐ ☐-OPEN PASTURE,ROWCROP [0]	☐ ☐-HEAVY/SEVERE[1]	Max 10
☐ ☐- VERY NARROW <5m[1]	☐ ☐-FENCED PASTURE [1]	☐ ☐-MINING/CONSTRUCTION [0]		
☐ ☐- NONE [0]				

COMMENTS:_____

5.]POOL/GLIDE AND RIFFLE/RUN QUALITY

MAX. DEPTH	MORPHOLOGY	CURRENT VELOCITY [POOL & RIFFLES!]		Pool/
(Check 1 ONLY!)	(Check 1 or 2 & AVERAGE)	(Check *All* That Apply)		Glide
☐ - >1m [6]	☐-POOL WIDTH > RIFFLE WIDTH [2]	☐-EDDIES[1]	☐-TORRENTIAL[-1]	
☐ - 0.7-1m [4]	☐-POOL WIDTH = RIFFLE WIDTH [1]	☐-FAST[1]	☐-INTERSTITIAL[-1]	
☐ - 0.4-0.7m [2]	☐-POOL WIDTH < RIFFLE W. [0]	☐-MODERATE [1]	☐-INTERMITTENT[-2]	Max 12
☐ - 0.2- 0.4m [1]		☐-SLOW [1]		
☐ - < 0.2m [POOL=0]	COMMENTS:_____			

                                   CHECK ONE OR CHECK 2 AND AVERAGE

RIFFLE /RUN DEPTH	RIFFLE/RUN SUBSTRATE	RIFFLE/RUN EMBEDDEDNESS	Riffle/Run
☐ - GENERALLY >10 cm;MAX > 50 [4]	☐-STABLE (e.g.,Cobble, Boulder) [2]	☐ - NONE [2]	
☐ - GENERALLY >10 cm; MAX < 50[3]	☐-MOD. STABLE (e.g.,Large Gravel) [1]	☐ - LOW [1]	
☐ - GENERALLY 5-10 cm[1]	☐-UNSTABLE (Fine Gravel,Sand) [0]	☐ - MODERATE [0]	Max 8
☐ - GENERALLY < 5 cm [RIFFLE=0]		☐ - EXTENSIVE [-1]	Gradient
COMMENTS:_____		☐ - NO RIFFLE [Metric=0]	

6] GRADIENT (ft/mi): _____   **DRAINAGE AREA (sq.mi.):**_____   %POOL: [____]   %GLIDE: [____]   Max 10
                                                                      %RIFFLE: [____] %RUN: [____]

EPA 4520   4/7/92

Figure 3. Qualitative Habitat Evaluation Index (QHEI) field sheet used by Ohio EPA.

It is obvious from differences in geomorphology and anthropogenic impacts around the country that states or agencies need to tailor habitat monitoring methodologies that (1) reflect factors that are likely to be limiting in their jurisdiction, (2) are sensitive to the range of habitat disturbance likely to be encountered, and (3) provide for an assessment of conditions necessary to evaluate and maintain genetic biodiversity and viability of aquatic biota. Adjusting indices for local geomorphology, land use, and biota will likely result in more accurate predictions or at least a more useful tool for examining responses of the aquatic biota. Hayslip (1993), for example, held workshops in USEPA Region 10 where water quality personnel adjusted and field tested biological and habitat parameters of the Rapid Bioassessment Protocols. National efforts, by USEPA and other federal agencies, should focus on developing guidelines for ecoregionalization of methods and they should promote training and proper quality assurance/quality

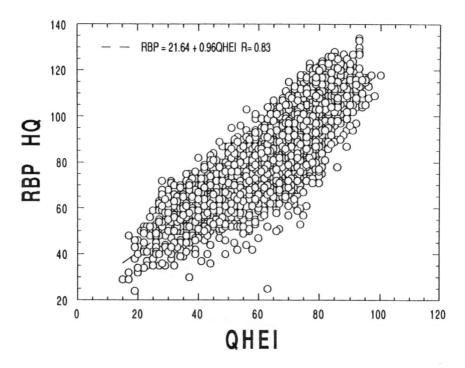

Figure 4. Qualitative Habitat Evaluation Index (QHEI) for all sites in Ohio EPA's database in relation to the Rapid Bioassessment Protocol Habitat Quality (RBP HQ) methodology derived from the same data.

control (QA/QC) procedures (e.g., as was done in Region 10 with regional modifications to the Rapid Bioassessment Protocols; Hayslip 1993). Common habitat attributes of streams, however, should have standard definitions as well as minimum standards for measurement (Armentrout 1981). A likely avenue for regionalizing habitat measures would be through the Environmental Monitoring and Assessment Program (EMAP; Paulsen et al. 1991; Kaufman 1993) in concert with national monitoring efforts of other agencies (USFWS Biomonitoring of Environmental Status and Trends [BEST] Program), USGS National Water-Quality Assessment [NAWQA] Program (Meador et al. 1993); and USFWS Aquatic Ecosystem Analysis Program (Mangum 1986a, 1986b).

To illustrate potential downfalls of accepting a national monitoring tool without local adjustments, I compared Ohio's Qualitative Habitat Evaluation Index (QHEI) was compared to USEPA's rapid bioassessment protocols for habitat quality assessment (RBP HQ). USEPA urges users of the protocols to tailor them to regional conditions (Barbour and Stribling 1991); thus, this is not an effort to "validate" this methodology, but, rather to illustrate a loss in the power of a habitat index when not adjusted to regional conditions.

The major difference between the QHEI and the RBP habitat tool is not the types of variables examined, but rather in the weighting of these factors as to their influence on biological integrity. To illustrate this each of the "metrics" or categories of the RBP HQ was generated from data components also collected for the QHEI. Because the RBP HQ results were derived from the QHEI results the actual degree of relationship could differ if the two were scored independently. However, the pattern of results related to differences in score weightings discussed below should still be valid.

The QHEI and the RBP HQ are significantly correlated (Figure 4). When compared statewide with the IBI at Ohio EPA reference sites (least impact and physically modified), however, the QHEI explained more of the variation in the IBI than did the RBP HQ (Figure 5). The use of other response variables such as the number of sensitive species at a site and the percent of individuals captured that were tolerant showed similar patterns (Figure 6). The better performance of the QHEI compared to the RBP HQ is not attributed to some inherent superiority in the QHEI. Rather, the improved explanatory power is likely related to: (1) the fact that the QHEI was calibrated, and metric scores were weighted, based on both literature reports *and* observed correlations of the IBI with habitat characteristics at a series of least

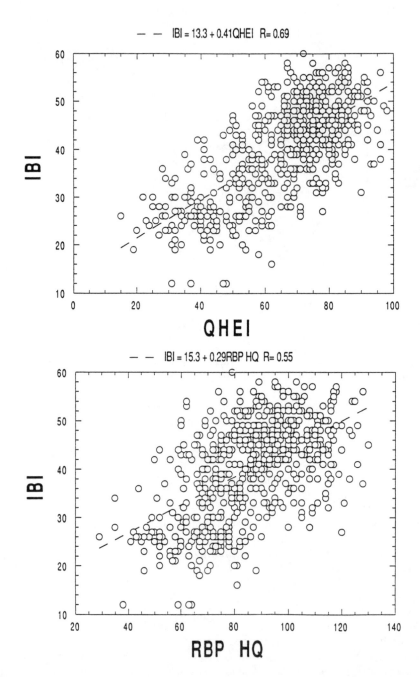

Figure 5. Top: Index of Biotic Integrity (IBI) values in relation to the QHEI at Ohio EPA's natural and modified reference sites (ecoregions differentiated by point type). Bottom: Index of Biotic Integrity (IBI) values in relation to the RBP HQ at Ohio EPA's natural and modified reference sites (ecoregions differentiated by point type).

impacted and physically modified references sites, (2) the QHEI is designed to measure those components most important to one organism group; fish, and (3) the relationships between QHEI and IBI most strongly reflect the anthropogenic disturbances common to Ohio (e.g., channelization and riparian destruction).

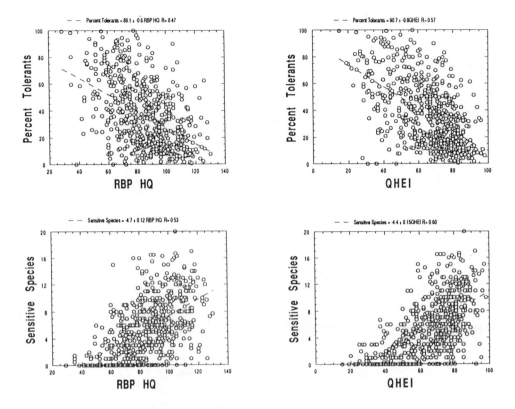

Figure 6. Top left: Percent tolerant individuals in relation to the RBP HQ at Ohio EPA's natural and modified reference sites. Top right: Percent tolerant individuals in relation to the QHEI at Ohio EPA's natural and modified reference sites. Bottom left: Number of sensitive species in relation to the RBP HQ at Ohio EPA's natural and modified reference sites. Bottom right: Number of sensitive species in relation to the QHEI at Ohio EPA's natural and modified reference sites.

4.2 Essential Components of any Habitat Index

Given that each habitat index needs to be calibrated regionally, preferably to a group of reference sites, are there components that should be common to any habitat quality index? The underlying effect of geomorphology on lotic ecosystems results in number of habitat characteristics that should be considered in any index. Calibrating habitat expectations upon reference conditions should account for the relative influence of geomorphology and stream energy (gradient) and important factors should become evident. Karr and Dudley (1981) summarize general characteristics of natural and modified streams in the eastern United States while Platts et al. (1983) provides a thorough analysis of many of the field techniques pertinent to all streams. The USEPA Rapid Bioassessment Protocols are largely based on the work of Platts et al. (1983) and Ball (1982). Bisson et al. (1981) provide excellent drawings and descriptions of common channel characteristics (pool/glide/riffle types). Other useful resources are Hynes' (1970) *The Ecology of Running Waters,* the various symposia that summarize many of the types of habitat assessment approaches used by workers across North America (Krumholz 1981; Armentrout 1982), and summaries of techniques and state programs by regional chapters of the American Fishery Society (e.g., Western Division, American Fishery Society 1985; Osborne et al. 1991). Thorough reading of some of the excellent geomorphology and hydrology texts that exist (e.g., Leopold 1964; Morisawa 1968; Gordon et al. 1992) will provide much insight to the types of forces that may be influential in various regions of the country. In addition, EMAP, which crosses political boundaries, should eventually provide regional data and useful field tests of habitat techniques in small streams and rivers. Preliminary discussions from an EMAP design workshop (Kaufman 1993) suggested potential stratifications for data analysis (ecoregion, size, and gradient) and eight categories of stream attributes (channel/riparian interaction, stream size, channel gradient, habitat type/distribution, channel substrate, riparian vegetation, and anthropogenic alterations) that are likely candidates for field data collection.

Barbour and Stribling (1991) describe four generic categories of stream types, mountain, piedmont, valley/plains, and coastal, for which the relative importance of habitat characters will differ. Much of the variability in habitat conditions among these stream types is related to the inherent energy in streams and the types of materials through which the streams flow (bedrock, sand, tills, alluvial deposits, etc.). Barbour and Stribling (1991) modified the RBP habitat procedures to provide separate methods for high gradient (riffle/run dominance) and low gradient (pool/glide dominance) streams. Mangum (1986a, b) describes the use of the Biotic Condition Index (BCI), which uses species diversity based on dominance and taxa to describe instream habitat conditions, and which can be used with or without RBP methodologies.

As mentioned earlier, most states have some group within their natural resource departments or universities that have some expertise in habitat assessment that should be involved in the development of procedures. For states beginning to incorporate habitat assessments into surface water protection programs the most conservative approach is to collect information on an array of habitat features and then refine an index on the basis of the desired response variable (e.g., IBI) from a series of reference sites (including physically impaired sites). The caveats of Platts (1981) to consider the accuracy and precision of assessment techniques should be addressed when choosing assessment tools. A sensible approach would be to include existing techniques where some idea of the variability of the measures has been assessed (e.g., Platts et al. 1983) and later modify these techniques if necessary rather than creating new methods.

Hawkins et al. (1993) suggest a hierarchical approach for classifying stream features, based on increasingly fine descriptions of stream morphology and hydrology within channel units. In such an approach, habitat measures may be made, for example, within a rapid (fine resolution), which is a subset of turbulent, fast water channel units (coarse resolution). Their categories for the classifications largely retain the nomenclature of Bisson et al. (1982) and Helm (1985). Use of a hierarchical approach such as this could be a catalyst towards some useful standardization of habitat type classification in streams (Hawkins et al. 1993).

Habitat altered sites, though they may be avoided by some biologists, should be sampled to provide a range of conditions under which to examine community responses to various impacts. The following section discusses habitat attributes that most indices should consider; useful (but not exhaustive) references for each attribute are included.

4.2.1 Substrate Type and Quality

All habitat indices should measure several characteristics of substrates. For most streams with all but the lowest gradient the type and quality of substrate conditions can be limiting. For most streams coarser substrates (gravels to boulders) are more likely characteristic of unaltered reference conditions. The addition of finer substrates via erosion is generally associated with land use changes and habitat modifications. As fines fill up the interstices of the larger substrates, the substrates are considered to be embedded. Some measure of the percent fines or degree or extent of embeddedness (Everest et al. 1981; Platts et al. 1983) is common to most habitat indices. Similarly, the degree to which substrates are covered by clayey-silts is also a common attribute of habitat indices. Sedimentation is widely held as responsible for degradation of fish communities in warmwater (Trautman 1981; Berkman 1987) and coldwater (Tappel and Bjornn 1983; Platts et al. 1989) streams, and many of the mechanisms of this degradation (loss of spawning habitat, lowering of interstitial dissolved oxygen, loss of habitat space, reduction in benthic production) have been well documented (Chapman 1988).

Management decisions should influence the choice of substrate assessment methods. Visual estimates of substrate types and conditions (Platts et al. 1983; Bain et al. 1985; Ohio EPA 1988; Plafkin et al. 1989) can be useful overall indicators of substrate quality; however, more specific objectives may call for more statistically rigorous methods (Everest et al. 1981; Platts et al. 1983) of substrate assessment. The sediment embeddedness criteria for salmonid spawning (Burton et al. 1991) discussed earlier provides a situation justifying very detailed field and lab measurements.

4.2.2 Instream Physical Structure/Cover

The presence of instream physical structure has a significant influence on aquatic organisms (Weshe 1980; Angermeier and Karr 1984; Weshe et al. 1987). Common attributes in habitat indices include the percent of a study reach with cover and recording the occurrence and extent of various types of cover. Types of physical structures or cover frequently measured or recorded include logs and woody debris, boulders, aquatic macrophytes, rootwads and rootmats, undercut banks, deep pools, and overhanging vegetation. Riparian forests are important contributors to instream cover (Murphy and Koski 1989). With the loss of riparian vegetation and the extensive dragging of woody debris from stream channels throughout much of the United States, streams are likely to have much less debris than has been historically present. For example, in old-growth streams in Oregon, Sedell et al. (1984) reported that woody debris "intervened" 16 to 18 times per 100 m of stream, a much higher rate than commonly found in the less-than-pristine streams of Ohio. Andrus et al. (1988) found that riparian trees must grow longer than 50 years to ensure an adequate, long-term supply of woody debris for stream channels. Similarly, in Delaware, analysis of stream recovery from modifications showed the most recovery at 30 years, roughly the time needed for trees to grow and start to fall into the stream channel (Maxted, unpublished data). In some low-gradient stream and river systems, physical structures such as logs and woody debris are the major source of invertebrate production (Benke et al. 1985; Benke 1990). The importance of instream structure has been documented for both coldwater and warmwater aquatic life. Physical structure can function to create pools and depth and velocity heterogeneity, to reduce export and increase processing of organic matter, as refuge from predation, as a substrate for prey organisms, as resting places from high velocity flows, and as spawning and nursery habitat (Angermeier and Karr 1984).

4.2.3 Channel Structure/Stability/Modifications

The natural channel morphology of stream and rivers is related to the geomorphology of an area, especially the energy of a stream (related to stream gradient) and the erodability of the material through which it flows. Streams in high-gradient areas generally flow "straighter" and erode less than streams in low-gradient areas that often meander through alluvial sediments that are more easily eroded. Unfortunately, hundreds of thousands of miles of streams and rivers have had their natural channel morphology altered significantly enough to impair both channel maintenance and the aquatic life in these systems. Alteration of the stream channel affects streamflow, the aquatic biota, and many of the characteristics measured by habitat indices (Emerson 1971; Trautman and Gartman 1974).

In Ohio most unaltered streams are sinuous or meandering. Some areas of the state were vast wooded wetlands (Black Swamp in northwestern Ohio) where stream channels were probably not well defined (Trautman 1981). The most common modifications in Ohio included channel straightening and deepening (channelization) for agricultural drainage and flood control. These activities destabilized streambanks and bottom substrates by increasing local gradients, increasing sedimentation, and exacerbating the peaks of storm flows. Such activities are also indicative of land use activities too close to stream channels that lead to increased sediment and pollutant runoff. Besides such physical changes, modified streams may alter recruitment of young fish to large rivers, resulting in fewer piscivores and insectivores and more omnivores and herbivore–detritivores. Thus, in Ohio we have habitat metrics that reflect ranges of sinuosity, channel modifications, and stream channel stability. A simple measure of sinuosity is the ratio of the stream path between two points and the straight line distance between these points (Leopold et al. 1964). The QHEI includes a functional correlate of sinuosity, pool formation on outside bends, to augment this habitat attribute.

Each state needs to determine the natural channel forms expected in a region and design its habitat indices so they detect important changes to the morphology and the other habitat attributes such changes may affect. It must be remembered that changes in channel morphology will export the effects downstream and sometimes upstream (e.g., head cutting). Other activities such as dams may reduce downstream deposition of alluvium and lead to bank erosion and changes in stream channel morphology (Johnson 1992). Thus, it is important to be able to detect the affects of upstream activities on channel morphology in the attributes of a habitat index.

4.2.4 Riparian Width/Quality

The quality and extent of the riparian vegetation is another critical component of a habitat index. More than other habitat components, however, the effects of removing or disturbing riparian vegetation often work at landscape scales (Gregory et al. 1991). While channel modifications have both immediate and downstream and upstream influences, the influence of riparian disturbance may be less evident in the immediate vicinity of a disturbance but become evident throughout a basin as riparian disturbances accumulate. In addition, the relative influence of riparian floodplain size to ecosystem function increases with stream size (Schlosser 1991). The importance of these large scale influences will be discussed later in this chapter.

Steedman (1988) examined the IBI in relation to land use and the existence of riparian zones near Toronto, Canada, and found significant correlations with both factors. He was able to generate a contour plot of qualitative IBI ratings as function of the percent urban land use and the proportion of upstream channels with intact riparian forest. Such studies can serve as models for the types of data needed to make habitat information a much more useful planning tool for preserving ecological integrity and riparian areas.

It is likely that no stream or river in Ohio has truly mature riparian forests that function as climax riparian forests functioned before European settlers arrived. In studies in "old-forest" areas of the Pacific Northwest, large logs were found to reside in a channel for a century or more (Sedell et al. 1984). Thus what could appear as a relatively undisturbed stream in Ohio could actually be an early successional stage with regard to riparian condition. In states with a great age variety in riparian forests, some estimate of forest maturity should be included in an index or accounted for in reference site stratification.

Habitat indexes often estimate the width of the riparian vegetation (i.e., trees, shrubs, and wetland) and the specific ages, stability, and species present. Because riparian tree species often require specific environmental conditions, such as out-of-bank flows, to germinate and grow (Hill et al. 1991), the composition of the vegetation can provide insight into previous environmental conditions. As riparian vegetation is degraded, lost functions include maintenance of narrow and deep channels (Platts and Rinne 1985), ineffective nutrient removal (Schlosser and Karr 1981; Lowrance et al. 1984; Peterjohn and Correll 1984), increased water temperatures (Karr and Schlosser 1977; Schlosser and Karr 1981), sedimentation and increased bank and bed erosion (Karr and Schlosser 1977, 1978), loss of terrestrial litter inputs and increased rate of organic export (Sedell et al. 1984), and loss of cover through woody debris "starvation" and loss of bank-related cover (e.g., undercut banks, rootwads, and rootmats) (Karr and Schlosser 1977). Work over the last decade by investigators, such as Bencala (1993), have emphasized the importance of riparian areas for maintaining the quality and function of the hyporheic zones of streams.

4.2.5 Bank Erosion

Bank erosion problems often occur hand-in-hand with riparian vegetation disturbance; however, bank erosion can occur in areas with "apparently" intact riparian vegetation. Stream channel alterations upstream in a watershed can drastically alter high flow characteristics making erosion problems more common downstream. Livestock grazing in riparian areas can also increase bank erosion and the formation of false banks. Typical modeling approaches to sediment runoff (e.g., Universal Soil Loss Equation, USLE) often do not account adequately for contributions from bank erosion or deposition (Schlosser and Karr 1981) nor the relative importance of particle types in runoff (e.g., clay vs. sand), which can have profound influences on aquatic community integrity (Ohio EPA 1992).

The streambank soil alteration rating and the streambank vegetative stability rating of Platts et al. (1983) and Duff et al. (1989) are widely used measures of bank erosion and the potential for bank erosion. The ability of a stream bank to erode will vary by region with the steepness of banks, bank materials (e.g., bedrock vs. alluvial soils) and stream gradient. As with the other measures, examination of reference sites will be useful in defining expectations for bank conditions.

4.2.6 Flow/Stream Gradient

Flow is not explicitly measured in Ohio's QHEI; however, as discussed earlier stream flow characteristics influence many of the habitat attributes of streams. Hill et al. (1991) examined four flow regimes that maintain physical and biological resources in stream ecosystems: (1) flood flows that form floodplain and valley features; (2) overbank flows that maintain surrounding riparian habitats, adjacent upland habitats, water tables, and soil saturation zones; (3) in-channel flows that keep immediate streambanks and channels functioning; and (4) in-channel flows that meet critical fish requirements. Streams in Ohio with the highest biological quality have natural flow regimes that include occasional flood flows that create and cleanse habitat, but which are not so frequent (as in urban streams) that they repeatedly scour bottom substrates and "reset" invertebrate communities (Matthews 1986). Work has shown that highly variable and unpredictable flow regimes (e.g., from urban runoff or controlled releases from dams) can have strong influences on fish assemblages (Bain et al. 1988). Extreme low flows are generally a problem in headwater areas (northwestern Ohio) where drainage activities have sped water off the landscape rather than slowly releasing water to stream channels. Small streams with such variable flows are dominated by tolerant and pioneering fish species that can withstand fluctuations in dissolved oxygen and temperature (Schlosser 1985; Matthews 1990; Schlosser 1990). Such broad landscape changes (e.g., draining most of NW Ohio) are responsible for the reductions in the distribution of many fish species across Ohio (Trautman 1981; Ohio EPA 1992; Yoder and Rankin, Chapter 9).

Each state will need to decide the advantages of explicitly including a measure of flow in a habitat index or using flow as an ancillary variable for interpreting symptoms of flow related problems that are observed in various habitat metrics and the biological communities from biosurvey data. One promising approach is to stratify streams by their flow characteristics. Poff and Ward (1989) examined streamflow characteristics across the United States, and on the basis of flow variability, flood regime patterns, and extent of intermittency distinguished nine stream types: harsh intermittent, intermittent flashy, intermittent runoff, perennial flashy, perennial runoff, snow melt, snow/rain, winter rain, and mesic groundwater. Examining biological performance and habitat conditions within such a conceptual framework, or for small regions a more finely divided framework, could improve explanations of patterns seen in aquatic assemblages and provide another useful form of stratification.

4.2.7 *Riffle-Run/Pool-Glide Quality/Characteristics*

Unaltered streams and rivers in the Midwest typically have fast, deep riffle/run complexes with large diameter substrates and deep pools with extensive physical structure. Even streams that, because of low gradient, lack riffles and runs often have a variety of flow regimes and depth heterogeneity associated with outside bends and meander patterns. Lobb and Orth (1991) examined a large warmwater stream in West Virginia and found five habitat-use guilds associated with the types of riffle/pool habitats: edge pool, middle pool, edge channel, riffle, and generalist. Degradation or loss of these types of habitats will eliminate or reduce abundance of the species in these guilds. Thus, it is important to measure the quality of these types of stream habitats.

Reference conditions can be used to determine the expected riffle/run and pool/glide types and their qualities for a region. In Ohio, stream channelization, siltation/sedimentation, and riparian destruction generally result in the loss of deep pools, the degradation of riffle habitat, and the predominance of shallow pool or glide habitat. The quality of riffles, runs, and pools is a direct result of the balance between erosion and deposition in natural systems. Many warmwater fish and macroinvertebrate species are habitat specialists and are eliminated as riffle and/or pool habitats are degraded. In Ohio, species associated with clean pool habitats appear to be especially vulnerable as relatively minor increases deposition of fine sediments has eliminated habitats and reduced distributions for many of the species over a wide area (e.g., sand darter, crystal darter, bigeye chub, and harelip sucker) (Trautman 1981). As sedimentation increases, even riffle habitats can become covered with fine substrates or more likely have large substrate interstices embedded with fine materials. Dunham and Collotzi (1975), Platts et al. (1983), and Duff et al. (1989) provide a rating for pool quality for small streams that incorporates pool morphology, stream depth, and instream physical cover and that could be modified for smaller or larger streams. Bisson et al. (1981) and Helm (1985), provide descriptions of various types of pools and riffles that are found in natural streams, and Hawkins et al. (1993) suggest a hierarchical framework for classifying such habitat types. As sediment delivery increases to a stream the morphology of the stream channel changes accordingly, often by becoming linear or convex in profile (Heede and Rinne 1990). Detailed measurements of stream channel morphology using cross-sectional transect techniques can provide statistics on changes in morphology, i.e., is it becoming wide and shallow or narrow and deep

(Olson-Rutz and Marlow 1992). Most index approaches, including the pool metric of the QHEI, which was derived from the Platts et al. (1983) and the (Habitat Condition Index HCI) from Duff et al. (1989), include some estimate of pool depth, morphology, instream cover, and sometimes velocity characteristics.

4.3 Importance of Reference Sites

A robust set of regional reference sites is critically important to accurately use biosurvey and habitat data. Single or multiple upstream control sites are important for interpreting longitudinal changes in the biota or habitat quality, but regional reference sites allow the quality of a stream to be placed in a broader perspective. In one sense, the use of single reference sites is a tie to the point source conceptual approach towards regulating water quality. For nonpoint or habitat problems, landscape-wide or broadscale land use problems that may be affecting habitat quality could likely be affecting the "control" condition as well. Anthropogenic changes may interact with the local geomorphology and have effects that may only be understood well when compared to regional patterns in biointegrity and habitat quality. Even when examining localized channel impacts such as bank erosion, the precipitating actions for such problems may originate from activities upstream in the basin.

The number of sites needed to accurately define baseline conditions will vary with the heterogeneity of the reference region and the variability inherent in the data. Yoder and Rankin (Chapter 9) consider this question for deriving biocriteria in Ohio (the 25th percentile of regional reference sites as a baseline for the Warmwater Habitat aquatic life use). To use habitat data effectively there also needs to be a suite of physically modified reference sites free from point source impacts. These modified reference sites should incorporate a broad range of habitat problems to allow sufficient resolution to document biological responses to multiple limiting factors. In Ohio, we have modified sites that reflect channel alterations, impounded streams, and nonacid, mine-related habitat impacts.

4.4 Need for Standardized Approaches and Quality Assurance Procedures

A call for "standardized" approaches to habitat assessment is not at odds with the call for regionalization of habitat assessment indices. Within a state or region, after one or more methodologies are selected, they need to be well defined, including specific purposes and objectives for each approach. For states or groups just beginning to develop or adopt habitat assessment procedures, effort to define QA/QC procedures is essential. The system, however, should be flexible enough to allow evolution in the specifics of each method. Sufficient regional reference sites should be sampled and habitat data compared to biosurvey results before a complete methodology is selected. As a result, it is advantageous to collect a broad spectrum of habitat data and examine multiple methodologies. Factors to consider when determining which individual habitat characters to measure include: (1) habitat characters with minimal between-user variation, (2) habitat characters that are clearly linked to biological responses, (3) the inclusion of habitat characters that measure each scale that can affect the biota (e.g., microhabitat to landscape), (4) characteristics likely to be affected by the major categories of stream alterations in a state, and (5) the time and effort required to measure/estimate the characteristic (i.e., cost-effectiveness).

Training is essential for reducing the between user error in habitat assessment methodologies. At Ohio EPA, we have yearly multiday training sessions consisting of classroom and hands-on field exercises. Data from these training sessions are used to refine the methods and to reduce user variation. In Ohio, we have found that it is extremely useful for "office" staff as well as field staff to be well versed in the methodologies and to know why examining habitat is important in protecting aquatic life uses. Incorporating these staff members into such training forces a clarification of the objectives of collecting such data.

Each state should develop a stream protection policy that clearly defines the need, mechanism, and technical justification for protecting stream habitat. In agencies still steeped in the point source conceptual framework of pollution control, it is important in making regulatory staff aware of the importance of habitat to ecological integrity. For example, it was our district water pollution control staff that, because of an awareness of habitat's significance to the biota, were able to identify that a permit-to-install (PTI) for a sewer line down the channel of a high-quality stream would have severe consequences for aquatic life and likely violate Ohio biological criteria (Ohio EPA 1992b).

4.5 Variability and Resolution of Habitat Indices

Data from habitat methods training sessions and special studies of data variation should be used to define the variability inherent in a habitat index. The degree of variability in an index will define the appropriate uses and limitations of an index and determine when more-detailed work is required. One way to examine between-user variability is to set up a set of test sites that users of the methods independently score for a given habitat technique. Data from these sites can be examined with quality control techniques and, potentially, users could be certified in the use of the tool.

Ideally, an index should minimize measurement error while maximizing the ability to distinguish important variation in habitat quality. For broad-based monitoring programs, cost-effectiveness must also be considered. An index that is cost-effective but does not provide the resolution needed to support decisions for an agency is useless, as is an index that provides resolution but is cost-prohibitive for general use.

Fortunately, indices such as the QHEI have sufficient resolution to support their desired uses under most circumstances, and are inexpensive to implement. Ohio uses the QHEI to explain changes in biological communities (as measured by the IBI); however, certain objectives may require more resource-intensive investigations, and it is important to recognize when an index like the QHEI is inappropriate. For example, in Idaho, salmonid spawning is a protected beneficial use for certain streams (Burton et al. 1991). The Idaho Department of Health and Welfare has developed protocols for assessment of dissolved oxygen and fine sediment in salmonid redds that affect salmonid embryo survival. Such a methodology is (1) sensitive to the types of impacts important in those streams, and (2) relates to a beneficial use of sufficient value to make a more intensive assessment justified.

Variation in the QHEI in Ohio is sufficiently low to make it useful for the objectives of our agency, which include: designating and protecting aquatic life uses, discerning causes and sources of impact in intensive surveys in watersheds, issuing PTIs for sewer lines and other construction, and supporting specific program activities such as 401/404 water quality certifications (i.e., certifying that dredge or fill operations in streams will not violate a state's water quality standards). Data from our training sessions, where QHEIs were generated independently by field staff and other trainees, found a strong, significant correlation between individual scoring and scoring of the instructor (Figure 7). Further confidence in decision making is provided by sampling multiple stations and incorporating other sources of data, especially fish community data. Both the habitat and biological data, including the biocriteria in our water quality standards, have been successfully court tested (*Northeast Ohio Regional Sewer District v. Shank* 1991).

4.6 Regional/Ecoregional Differences in Habitat

Habitat impacts that are likely to affect aquatic ecosystems around the country are often related to: (1) the geomorphology of a region and its effect on habitat diversity, and (2) land use activities typical to a region or ecoregion. For example, in Ohio agriculture is likely the most widespread activity that affects stream habitat and aquatic life (47% of Ohio's land surface is engaged in crop production, Hoorman et al. 1992). Hoorman et al. (1992) estimate that up to 55% of the agricultural area of Ohio needs "drainage improvements" to permit agricultural production. The effects and the need for such drainage vary regionally with relief and soil type. The Huron–Erie Lake Plain (HELP) ecoregion in Ohio, for example, suffers from both the need and effects of such drainage.

Figure 8 illustrates some of the differences between habitat characteristics related to ecoregions in Ohio streams. The most obvious difference between the HELP ecoregion and the other ecoregions is the overall low gradient of the HELP streams (Figure 8). Low-gradient streams are perhaps the most susceptible to sediment degradation and habitat destruction of all Ohio streams because the retention time of sediment is high and the stream energy, critical in developing and maintaining habitat diversity, is low. For all the habitat metrics shown here, the HELP had the most consistently poor habitat quality. For most of the other habitat parameters, the metric scores did not differ substantially with the exception of the interior plateau, which had more high-gradient streams (high relief and fewer large river sites) and more low-riffle scores. This small section of Ohio has many high-quality streams with good pools but riffles that may become intermittent during parts of the summer. Exploratory charting such as this is very useful for detecting regional patterns in data, especially when stratified by factors such as stream size, gradient, flow types, etc.

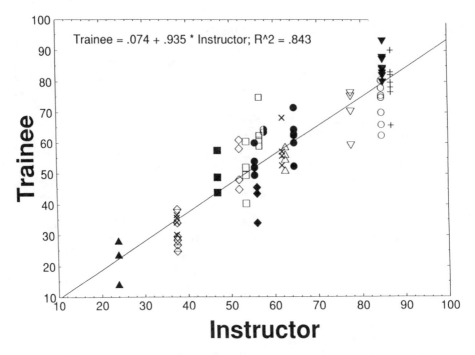

Figure 7. QHEI scores of individuals taking a QHEI training course in relation to QHEI scores generated by the instructor at small streams in Ohio. Data from five training sessions, one per area of Ohio (NW, NE, SE, SW, central). Common point symbols for a given instructor score indicate the same site.

4.7 Effect of Scale on Habitat Disturbance

As discussed earlier, habitat conditions are dependent on local geomorphology and anthropogenic influences on ecosystems. Much of the types of habitat assessments done in typical monitoring programs tend to focus on small scales of impact, usually at the microhabitat or, at most, the level of a several-hundred-meter reach. Assessments may focus on comparing a "site" (reach) to some reference "site" or sites. The "potential" of a study site to support aquatic life is then based on how close in quality this site is to a reference condition (Plafkin et al. 1989; Mangum 1990; Barbour and Stribling 1991). Such an approach, while useful in many cases, may not be sensitive to the effects of large-scale disturbances on stream ecosystems. In essence, impacts may not be totally predictable on the basis of site-specific habitat assessments alone because larger-scale disturbances are affecting the biota.

Frissel et al. (1986) and Gregory et al. (1991) advance a hierarchial conceptual framework of classifying stream habitat that incorporates various temporal (days to hundreds of years) and spatial (particle to stream network or subbasin) scales. The effects of anthropogenic changes on habitat should be considered at each of these scales (Schlosser 1991) when monitoring lotic systems. Such arguments are supported by Ohio EPA's monitoring data and in the theoretical ecological literature dealing with local extinctions and sources and sinks of individuals in ecosystems (Pulliam 1988).

Areas of Ohio that have had severe, large-scale landscape changes (HELP ecoregion) often have lower biological integrity and fewer species even in remaining areas of relatively good habitat. Presumably, local extinctions (e.g., 44% of original species in Maumee River drainage have declined or been extirpated; Karr et al. 1985) result from large, expansive areas of poor and modified habitat that act as "sinks" for production. In contrast, areas of Ohio with relatively intact landscapes and stream habitat often have high biological integrity. Here, short stretches of relatively poor habitat can often have much higher species richness and biological integrity than "predicted" from site-specific habitat assessments.

The effect of habitat disturbances at large scales can be seen in Ohio EPA fish community and QHEI databases. Subbasin-wide estimates of habitat quality in an area were estimated from average QHEI scores from each of 93 subbasins (all sites in our database; N = 2462) and plotted vs. average IBI scores

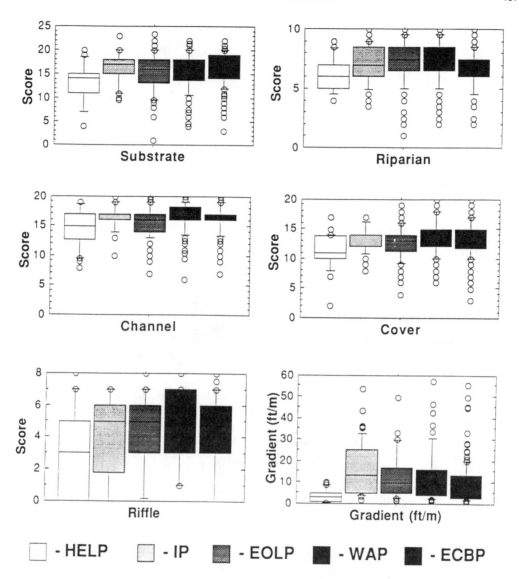

Figure 8. Box-and-whisker plots, by ecoregion, of QHEI metric scores for substrate score, riparian score, channel score, cover score, riffle score, and local gradient (ft/mi) for unmodified reference sites.

from our reference sites in these subbasins. There is a significant positive relationship between subbasin-wide estimates of habitat quality and reference site biological integrity (Figure 9). This pattern can also be seen when data from two individual subbasins (Little Auglaize River: habitat devastated; Twin Creek: habitat relatively intact) are examined (Figure 10). Although QHEI scores overlap (the highest QHEI scores in the Little Auglaize and the lowest in Twin Creek) the resulting IBI scores for a given QHEI differ substantially. Some evidence has suggested that high-quality oases of habitat could harbor sensitive species in areas that have been impacted by agriculture, urbanization, etc. (Luey and Adelman 1980). The Little Auglaize River in Ohio no longer has any oases of sensitive species. It is unknown how large such an oasis would need to be to remain viable and not subject to extirpations during "bottlenecks" of environmental stress (e.g., drought) in modified ecosystems. In most of Northwestern Ohio, headwater streams have been severely modified. Since headwater streams export problems downstream (e.g., sediment, flow, and tolerant species), species in a downstream "oasis" are likely to decline or be extirpated.

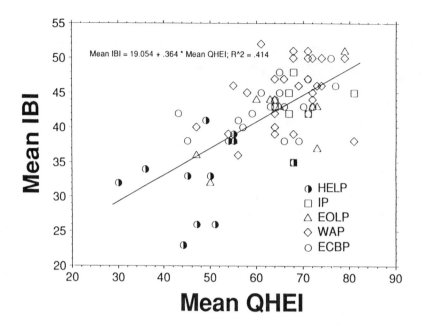

Figure 9. Average habitat quality in Ohio subbasins estimated by QHEI scores (all data from database) and average IBI scores at unmodified reference sites in Ohio subbasins. Ohio is divided into 93 subbasins.

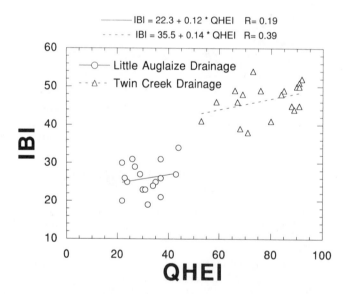

Figure 10. IBI in relation to the QHEI at two subbasins in Ohio, the Little Auglaize River in northwestern Ohio and Twin Creek in southwestern Ohio. The Little Auglaize River basin is a habitat poor, highly modified subbasin while Twin Creek has higher quality habitat with much less stream habitat disturbance.

The varying affects of different scales of habitat impacts on aquatic life and the importance of considering streams as open systems has important consequences for regulatory agencies charged with protecting streams. Too narrow a focus on specific sources of impacts (e.g., point sources) to the exclusion of other important factors (habitat and nonpoint) leads to the underprotection of streams. Similarly, there is often a focus on a study "site" and its impacts and aquatic potential rather than on a study reach or some larger scale. In day-to-day activities of a regulatory agency it is important to point out to clients (e.g., dischargers) that short stretches of modified stream do not preclude application of stringent water quality rules. Similarly, regulatory agencies need to protect against piecemeal degradation of stream habitats that would eventually result in large-scale devastation to aquatic life.

5.0 APPLICATION OF HABITAT INDICES IN WATER RESOURCE QUALITY ASSESSMENTS

In Ohio, habitat assessments are an integral part of our intensive survey program (Yoder 1991). Important uses of habitat assessment information include aquatic life use designations and as a tool in our intensive watershed surveys. The following sections provide examples of these uses in Ohio.

5.1 Habitat Indices in Stream and Basin-Intensive Surveys in Ohio

Habitat assessments (QHEIs) are done at all stream sites by the same field crew and during the same time period in which fish community data are collected. Besides its function in designating the proper aquatic life use, the QHEI assessment is used to explain causes and sources of impacts to the aquatic life. Although the final QHEI score is useful in interpreting habitat effects, we rely heavily on the component habitat characteristics to explain community impacts. Data from the QHEI and the fish communities at our reference sites and physically modified reference sites were used to derive habitat attributes that are characteristic of least-impacted or physically modified streams (Rankin 1989). Chi-square statistics were used to classify those attributes most often associated with low IBI or high IBI values (Rankin 1989). As the number of modified habitat attributes increase, the likelihood of having IBI scores similar to reference conditions decreases (Figure 11). These patterns of community response from our reference sites and the personal experience of our biologists are the basis of our interpretation of the patterns observed in intensive survey data. Patterns in biological response between the biota, water column chemistry, sediment chemistry, effluent characteristics, and land use patterns are all combined with the basic habitat condition to isolate the factors likely responsible for aquatic life impairment. Yoder (1994) provides some explanation on how biological responses ("signatures") can help in the interpretation of complex environmental data.

The following example will illustrate the use of the QHEI in intensive surveys. Data presented below have been summarized in an Ohio EPA biological water quality report (Ohio EPA 1991).

5.1.1 Hocking River

The Hocking River, located in southeastern Ohio, is a medium-sized river (1,197 mi^2 drainage area) of about 100 mi in length. The Hocking River headwaters are in glacial deposits of Fairfield County southeast of Columbus, Ohio, and it flows southeasterly through unglaciated, rugged topography to the Ohio River (Ohio EPA 1991). Since early in this century the river has been affected by industrial discharges and municipal and combined sewer discharges, especially in its headwaters near Lancaster, Ohio, and by acid mine drainage, nonacid mine effects (sediment), agricultural polluted runoff, severe bank erosion and sedimentation, and channelization (Shurrager 1932; Ohio EPA 1991). Shurrager (1932) surveyed the fish communities of the Hocking River in 1931-32 and the Ohio EPA surveyed the upper section of the river (RMs 73 to 95) in 1982 and the entire river in 1990 (Ohio EPA 1991).

In 1982, the Lancaster WWTP and industrial discharges to the WWTP were responsible for severe impacts to fish and macroinvertebrate assemblages throughout most of the study area (Figure 12). As a result of the 1982 survey, the City of Lancaster upgraded their WWTP in the late 1980s and an effective pretreatment plan was initiated. Improved treatment led to the elimination of exceedances of Ohio Water Quality criteria for ammonia-N, total zinc, and dissolved oxygen (Ohio EPA 1991). Macroinvertebrate data reflected the vastly improved water quality, recuperating from minimum Invertebrate Community Index values (ICI) near the WWTP during 1982 to exceptional levels of biotic condition in 1990 (Figure 12). Ohio EPA uses Hester-Dendy artificial substrates to collect macroinvertebrates (Ohio EPA 1988; DeShon, Chapter 15). Macroinvertebrate communities collected this way respond largely to changes in water quality constituents and do not usually reflect macrohabitat impacts to streams, especially if

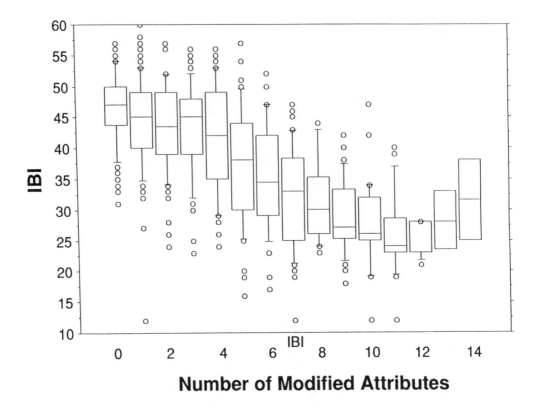

Number of Modified Attributes

Figure 11. Boxplots of the IBI in relation to the number of modified warmwater habitat attributes at natural and modified reference sites in Ohio.

macrohabitat impacts are localized. Fish communities, however, respond strongly to both water quality and habitat disturbances. Like the macroinvertebrates, the fish assemblages were severely impaired by water quality impacts in 1982. In the 1990 survey the extensive habitat impacts were "unmasked" by improving water quality and are now largely limiting the fish community in the Hocking River. The results of the survey of the entire river in 1990 indicated that throughout the length of Hocking River habitat impacts, including channelization, riparian removal and bank erosion, mine-related sedimentation, and severe substrate embeddedness, significantly impact fish assemblages. Table 3 summarizes for the Hocking River important habitat attributes that commonly affect fish assemblages; reference sites from the Hocking River are included for comparison. This table format is a useful tool for examining the cumulative affects of habitat destruction on streams.

The inclusion of biosurvey and habitat data in the assessment of the Hocking River was indispensable for determining impacts and setting priorities for future stream improvements (e.g., restoring riparian forests and reducing bank erosion). If habitat data were not collected or if only a single organism group was surveyed, the relative magnitude of the various impacts would likely have been distorted. Our survey experiences in Ohio have often shown us that our presurvey, study-plan-derived perceptions of the impacts in a basin may be false. The consequences of not using integrated, intensive monitoring approaches in water management programs are inaccurate diagnoses of the causes and sources of impairments and water resources left underprotected. Data compiled nationally (see Figure 2) strongly suggest habitat is overlooked as a major limiting factor to biological integrity.

5.2 Habitat Assessment Techniques and Use Designations

In Ohio, instream biocriteria are the arbiters of aquatic life use designations; however, habitat assessment data plays an integral role. The achievement of the ecoregional biocriteria for a stream assures it of *at least* the associated designated use, regardless of the score of the QHEI. However, in many

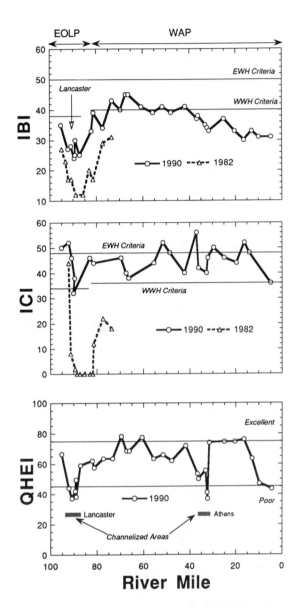

Figure 12. IBI, ICI, and QHEI in relation to River Mile in the Hocking River near Lancaster, Ohio in 1982 and 1990.

situations streams being designated or redesignated have water quality impacts that preclude the use of the biota alone to define the aquatic life potential of a stream. In these situations we rely heavily on the habitat assessment information.

Integral to Ohio's biocriteria is a "tiered" system of aquatic life uses which define the *baseline* expectations for the aquatic life of a stream. Yoder and Rankin (Chapter x) defines these aquatic life uses in more detail. About 10% of Ohio's streams are designated as Exceptional Warmwater Habitat (EWH), which encompasses most of the streams that harbor endangered or threatened species and the most diverse assemblages. The majority of streams are designated as Warmwater Habitat (WWH) and contain balanced, reproducing assemblages of fish and macroinvertebrate communities. Ohio also has two "non-fishable" aquatic life uses: Modified Warmwater Habitat (MWH) and Limited Resource Water (LRW). Limited Resource Waters are those extremely small (<3 mi²), ephemeral, and highly modified (often urban) streams that are not likely to support any semblance of a natural community. If any organisms are

Table 3. Warmwater and Modified Warmwater Habitat Attributes at Sites in Streams in the Hocking River Basin

(01-001)—Hocking River
Year: 90

River mile	QHEI	Gradient (ft/mi)	Ref. site	Total WWH attributes	Total (high influence) MWH attributes	MWH attributes	MWH (High)/WWH attributes	MWH (mod.)/WWH attributes
95.2	66.0	13.1		7	0	4	0.13	0.63
92.2	44.0	4.7	M	2	2	5	1.00	2.67
90.8	37.0	4.7		3	2	5	0.75	2.00
89.4	41.5	4.7		3	2	5	0.75	2.00
89.1	49.0	4.7		4	0	6	0.20	1.40
89.0	37.5	4.6		3	2	5	0.75	2.00
87.1	58.5	4.6		6	0	3	0.14	0.57
82.0	62.0	6.6		7	0	3	0.13	0.50
81.2	57.5	4.6		4	0	8	0.20	1.80
73.2	63.5	3.1		6	0	5	0.14	0.86
72.2	77.5	3.1		6	0	5	0.14	0.86
69.5	68.0	3.7		7	0	2	0.13	0.38
67.3	68.0	3.4		4	0	6	0.20	1.40
66.2	77.5	3.4		7	0	5	0.13	0.75
60.6	63.5	2.3		7	0	5	0.13	0.75
55.7	65.5	6.4		6	0	5	0.14	0.86
51.6	62.0	1.9		7	0	6	0.13	0.75
47.9	72.0	1.9		5	1	5	0.33	1.33
41.8	53.5	1.9		6	0	7	0.14	0.86
36.7	50.0	1.9		5	2	8	0.50	1.67
36.2	55.5	1.9	M	3	0	8	0.25	2.25
33.1		1.6		3	0	8	0.25	2.25

WWH attributes (■): No channelization or recovered; Boulder/cobble/gravel substrates; Silt-free substrates; Good/excellent development; Mod/high sinuosity; Extensive/moderate cover; Fast current/eddies; Low/normal embeddedness; Max depth >40 cm; Low/no riffle embeddedness.

High influence attributes (●): Channelized or no recovery; Silt/muck substrates; Low Sinuosity; Sparse/no cover; Max depth <40 cm (WD, HW).

Moderate influence attributes (◄): Recovering channel; Heavy/mod. silt cover; Sand substrates (BT); Hardpan origin; Fair/poor development; Low/no sinuosity; Only 1-2 cover types; Intermittent and poor pools; No fast current; High/mod. embeddedness; Ext./mod. riffle embeddedness; No riffle.

			Code					
32.5	41.0	1.6		2	2	8	1.00	3.67
32.3	37.0	1.6		1	2	9	1.50	6.00
31.5	74.0	1.6		7	0	5	0.13	0.75
25.0	74.0	1.6		7	0	4	0.13	0.63
20.4	74.0	2.2		8	1	3	0.22	0.56
16.3	75.5	2.2		8	0	3	0.11	0.44
12.9	63.0	0.1		6	0	6	0.14	1.00
9.8	47.0	0.1		3	2	7	0.75	2.50
4.6	43.0	0.1		3	2	6	0.75	2.25

(01-037)—Scotts Creek
Year: 78

			Code					
8.9	76.0	14.2	R	8	0	2	0.11	0.33
8.1	69.5	14.2	R	7	0	2	0.13	0.38

(01-100)—Federal Creek
Year: 90

			Code					
11.4	81.5	8.3	R	9	0	0	0.10	0.10

Year: 84

			Code					
1.3	74.0	2.3	R	9	0	1	0.10	0.20

Year: 90

			Code					
1.3	75.5	2.3	R	8	0	1	0.11	0.22

(01-400)—Clear Creek
Year: 82

			Code					
16.3	55.0	6.1	M	3	1	7	0.50	2.25
14.2	48.0	8.7	M	3	1	7	0.50	2.25
7.3	89.5	9.0	R	9	0	0	0.10	0.10

Year: 90

			Code					
2.0	77.5	4.1	R	9	0	1	0.10	0.20

(01-420)—Muddy Prairie Run
Year: 82

			Code					
0.7	85.0	15.3	R	10	0	0	0.09	0.09

(01-430)—Dunkle Run
Year: 82

			Code					
0.1	69.0	7.0	R	6	1	5	0.29	1.00

(01-510)—Durbin Run
Year: 82

			Code					
0.4	31.0	1.7	M	1	5	6	3.00	6.00

(01-520)—Turkey Run
Year: 82

			Code					
1.4	67.0	10.8	R	8	0	2	0.11	0.33

Reference site codes: R — unmodified reference site; M — reference site with modified physical habitat.

present they are generally pioneering species that are exceptionally tolerant to poor water quality, extremely modified habitat, and intermittent stream flow. The MWH use is a relatively new use that is designed to protect those streams that will not attain the WWH use because of extensive habitat modifications, but that have a significant permanent assemblage of tolerant organisms that would not be adequately protected by the LRW use. Presently, MWH streams have the same water quality criteria as WWH streams except for dissolved oxygen and ammonia-N, of which the typical assemblages in MWH stream are tolerant.

By examining the preponderance of various modified habitat attributes and unmodified habitat attributes at multiple sites in a stream we can determine the likelihood that a stream will or will not be able to achieve a particular aquatic life use (see Figure 11). The flowchart in Figure 13 summarizes and simplifies an example of assigning an aquatic life use to a stream that is a candidate for either a MWH aquatic life use or a WWH aquatic life use. The first fork in the flowchart diverges on whether biosurvey data are available (Figure 13). If only habitat data are available, aquatic life use designations will only be made in very simple and obvious situations, such as small HELP ecoregion streams. The HELP ecoregion of Ohio has been so extensively ditched and drained that many of the small streams in this region are incapable of supporting a WWH use. In most other cases, however, aquatic life use decisions are made with biosurvey and habitat data.

For the present example, illustrated in Figure 13, if biosurvey data are available we will examine them for attainment of the appropriate WWH ecoregion biocriteria (Ohio EPA 1988). If the stream achieves the criteria it will assigned the WWH use regardless of habitat scores. If the WWH use is not achieved then it is considered for the MWH use only if there have been extensive physical alterations. With no physical disturbances the stream is assigned an WWH use or, in some circumstances (e.g., a wetland stream), becomes a candidate for site-specific biocriteria modification.

A stream with extensive habitat modifications becomes a candidate for the MWH if the modifications are substantial enough to preclude a WWH use and the likelihood of recovery of habitat conditions to support a higher use is low. The habitat attributes considered are listed in Figure 13. It is important to note that numerous sites along a stream are examined before making a use decision. We will, under some situations, assign different uses to different segments of a river where the potential of a stream obviously differs between segments (e.g., WWH segment in rural area, MWH within urban area). The situation in Figure 13 largely applies to MWH streams related to channel activities, but we also have MWH criteria that apply to nonacid mine-affected streams and impounded streams.

A large and robust set of reference sites (both least impacted and physically modified) is indispensable when designating aquatic life uses. By linking our aquatic life uses directly to our biocriteria we have made a direct connection between uses and the methods for assessing use attainment. Our large data set gives us a knowledge base for interpreting the potential of streams where the proper use may not be obvious.

The importance of tiered aquatic life uses to appropriate designation of those uses cannot be overstated. In Ohio, the combination of stream size stratification (headwater, wading, and boat types), ecoregions, and tiered aquatic life uses provides a flexible and workable approach for defining aquatic life potential. The incorporation of the MWH use has provided a needed intermediate use between the WWH use and a RW. This additional "tier" provides increased protection to assemblages that may have otherwise been designated as LRW. As illustrated in the flowchart (Figure 13) ecoregions can play a significant part in determining which habitat influences are most important to aquatic life. The HELP ecoregion is unique in Ohio because its low gradient and high water table led to extensive drainage that allows clays, silts, and other sediments to accumulate instream.

QHEI scores alone do not always reflect the potential of a stream. Streams can have a single attribute limiting the biota yet have relatively high QHEI scores. The coal-bearing WAP ecoregion, for example, can be devastated by extreme sediment plumes from unreclaimed surface mining while still retaining relatively intact channel and riparian conditions and relatively high QHEI scores. In other ecoregions larger streams can have intact channel and riparian characteristics, but headwater streams with modified channels can deliver sediment to the mainstems at extremely high rates, which can accumulate in pools. In Figure 5 these are generally the sites that are furthest below the regression lines.

Figure 14 illustrates IBI and QHEI data for four streams in Ohio, one that has an EWH designation, one with a MWH designation, and two with WWH designations, one in the HELP and another in the ECBP ecoregion. The similarity between the biological and habitat scores for each of these streams, with

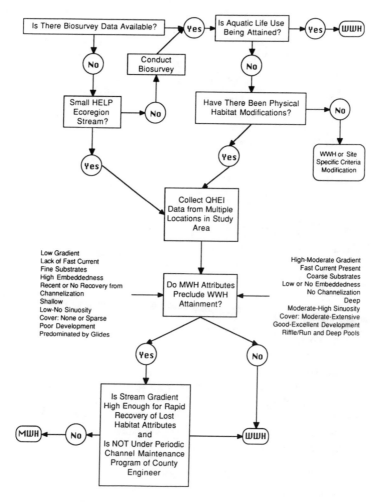

Figure 13. Flowchart summarizing the decision criteria used when assigning an aquatic life use to a stream that is a candidate for a Warmwater Habitat use (WWH) or Modified Warmwater Habitat use (MWH).

the exception of the IBI in Flatrock Creek (the WWH stream in the HELP ecoregion), is obvious. In Ohio the HELP ecoregion has much lower "expectations" for biological performance than other ecoregions in the state (Ohio EPA 1988b). Because this region of the state has been so extensively modified by stream drainage and channelization activities, reference sites meeting the "least impacted" definition for other regions of the state could not be found. Instead, the reference benchmark here was "best attainable" and derived from the 90th percentile of all sites in the region. Flatrock Creek, even though relatively unmodified and with a riparian zone comparable to other areas of the state, is severely affected by sediment and flow originating from headwater streams almost totally stripped of riparian vegetation, channelized, and drained. The Little Auglaize River, a MWH stream also in the HELP ecoregion, has had its mainstem and its headwater tributaries severely modified so that even the reduced expectations of the HELP WWH aquatic life use are unattainable.

The Kokosing River in the EOLP and WAP ecoregions is an obvious contrast to the Little Auglaize and Flatrock Creek drainages. Both the Kokosing River and its headwater streams are relatively intact and have sufficient gradient to continually flush excess sediment originating from agricultural land use practices. The proportion of land use in forest is much higher than in the HELP ecoregion. The increased relief also reduces the likelihood of riparian encroachment and the need for extensive drainage work.

Fourmile Creek is a WWH stream in the ECBP ecoregion in southwestern Ohio. The land use in this area is primarily row-crop agriculture; however, the relief is greater than in northwestern Ohio, especially in the lower part of the drainage. In contrast to Flatrock Creek, Fourmile Creek has had less of its

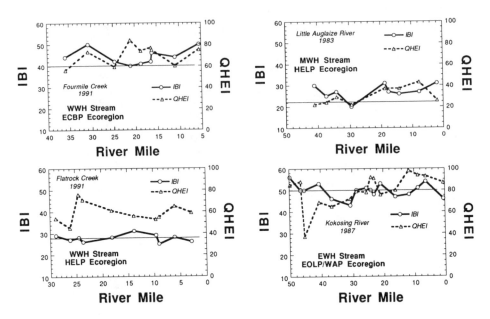

Figure 14. Longitudinal patterns in IBI and QHEI scores for a MWH stream (Little Auglaize River), an EWH stream (Kokosing River), and two WWH streams, one in the HELP (Flatrock Creek) and the other (Fourmile Creek) in the ECBP ecoregion.

headwaters modified by drainage but has been impacted from a WWTP. Riparian encroachment and extensive agriculture in the upper part of this drainage and a WWTP and an impoundment in the middle part of this stream may be masking the potential for this stream to reach the EWH aquatic life use. Improvements in the WWTP operation and recent plans to install nonpoint Best Management Practices (BMPs) in the upper basin may make this stream a candidate for a reassessment of its use when it is next assessed in Ohio's five-year basin monitoring approach.

5.3 Habitat Data in 401 Water Quality Certifications

The 401 water quality certification program is an important user of stream habitat data. The 401 program provides the states the ability to comment on and deny certain stream activities that deal with the fill of material into the nations waterways. States can deny or ask for modification of projects that could impair the state's water quality standards, including biocriteria. Thus, inclusion of biocriteria into water quality standards provides a direct link between habitat and water quality standards. In Ohio, numerous projects have been denied or modified because the activities would have degraded those habitat attributes upon which the biota is dependent. The process in these cases often follows those outlined for the intensive surveys. Some cases are complex and require biosurvey data and habitat data to arrive at a decision for denial or modification, while others are simple and can be denied or approved based upon what we have learned from our reference sites and the types of modifications that are planned.

Without biocriteria and without the collection of biosurvey and habitat data, there is a increased likelihood that destructive activities could occur because: (1) they do not usually violate chemically based water quality standards, or (2) biosurvey and habitat data are rarely integrated into regulatory decision making. Even with biocriteria, streams can still be degraded where the activities are covered by the nationwide 401 permit (e.g., in Ohio modifications on less than 1000 ft of stream) or are allowed by other, often conflicting regulations (e.g., state drainage laws). These concerns are similar to those of Schaeffer and Brown (1992) who reported that the "plethora of regulations" has not been successful in stopping the destruction of riparian habitats. They attributed this lack of success to federal statutes that protect water quality and quantity rather than wildlife habitat or biotic integrity. Federal rules and regulations (e.g., the Clean Water Act) need to promote ecosystem integrity rather than narrowly focusing on water quality in order to provide more practical, comprehensive, and cost-effective protection of water resources.

6.0 SUMMARY AND CONCLUSIONS

After Ohio instituted biosurvey and habitat assessment techniques in its water resource monitoring programs it became obvious that the state was not adequately protecting its streams and rivers and, as a result, biocriteria were incorporated into its water quality standards. In Ohio's 1988 305(b) report, its summary of the status and trends in the state's water quality for the U.S. Congress, the inclusion of biocriteria in addition to chemical criteria led to a significant increase in the number of miles reported as impaired (Ohio EPA 1988). The identification of habitat and other nonchemical impacts to aquatic life was responsible for much of this change in the assessment. In subsequent reports (Ohio EPA 1990, 1992) substantial improvements in water quality related to improved wastewater treatment, have been documented; however, we have also documented more habitat disturbance and little or no habitat restoration. Existing regulations, such as Section 404 of the Clean Water Act (33 U.S.C. § 1344), do not provide a broad enough approach to adequately protect streams from habitat degradation (Schaeffer and Brown 1992).

The impacts we observe in Ohio are not limited to Ohio or even the Midwest. Recent work by Benke (1990), Karr (1991), and Allan (1993) and even articles in the *New York Times* (Stevens 1993) have reported on the widespread degradation to the nation riverine resources. The opening paragraphs from William Steven's article in the *New York Times* (Stevens 1993) summarizes the problem well:

Two decades of Federal controls have sharply reduced the vast outflows of sewage and industrial chemicals into America's rivers and streams, yet the life they contain may be in deeper trouble than ever.

The main threat now comes not from pollution but from humans' physical and ecological transformation of rivers and the land through which they flow. The result, scientists say, is that the nation's running waters are getting biologically poorer all the time and that entire riverine systems have become highly imperiled.

The recent report by the National Research Council (NRC 1992) also documented the impoverished condition of the nation's rivers and recommended that: (1) erosion control programs should be accelerated for soil conservation and environmental restoration, (2) grazing practices should be changed to minimize damage to river–riparian ecosystems, (3) erosion control, where feasible, should favor "soft" engineering over "hard" engineering (e.g., channelization) approaches, (4) nonfunctional or non-cost-effective dikes and levees should be breached to reestablish hydrological connections between riparian habitats and rivers, and (5) riparian areas should be classified, in land-use and wetland systems, on the basis of their connections to rivers. This committee also set a goal of restoring 400,000 mi of riparian–river ecosystems (12% of total U.S. rivers and streams) within the next 20 years (NRC 1992).

Given that the destruction of habitat is of major importance nationwide it seems essential that states have the tools to assess the extent and magnitude of these impacts and the tools to eliminate and reverse habitat destruction. Recent reviews of game fish habitat restoration efforts have resulted in recommendations of integrated management of land use, particularly in riparian areas (Lyons et al. 1988; Lyons and Courtney 1990).

All biological sampling protocols require some form of habitat assessment to permit accurate interpretation of results. Since the USEPA requires states to have narrative biocriteria in their state water quality standards many states will be instituting biosurvey and habitat assessment programs (USEPA 1990). In this chapter we argue that, to be effective, states must have a program of sufficient size to allow repeated sampling of natural and physically modified reference sites. These reference sites will be the basis for biological criteria and for developing habitat assessment techniques tailored to specific regions of the country. Regional efforts to define reference conditions and develop region-specific habitat indices should be organized by USEPA regions in concert with states and other institutions that have a stake in such collaborative efforts (e.g., USFWS, USGS, and local agencies and groups). Such groups need to coordinate stream policies that will ensure adequate habitat protection across state and political boundaries.

As states begin to amass habitat data, many of the techniques reviewed here should be refined. States should remain open to advances in habitat assessment techniques to ensure a reversal of the present slide in habitat and ecological integrity of the nations rivers. Some work has been done in using more advanced statistical techniques to analyze biological and habitat data including expert systems and machine learning techniques (Anderson et al. 1991). However, there has been little support for ecological and habitat research by USEPA *relative* to water chemistry/toxicological research. Unfortunately, national efforts for habitat protection and nonpoint source control have been meager and expenditures are still dominated by research and management priorities skewed towards "toxic" chemicals and point sources of pollution. Point sources are still serious problems in the United States, but USEPA-sponsored efforts to rank relative risks suggest more emphasis must be placed on protecting ecological systems and reversing habitat destruction. Until spending is increased or, more likely, spending priorities changed, biological and physical evidence suggests the quality of the nations rivers will continue to deteriorate (Benke 1990; NRC 1992).

With the sad state of federal support for habitat and ecological monitoring, why should states add another "fiscal" burden by spending essential resources on such monitoring? Simply put, the amount of money spent on monitoring is dwarfed by the amounts states require cities and industry to spend on treatment of effluents. Analyses performed in Ohio suggest that without biosurvey and habitat data there is a high risk of missing nonchemical and chemical impacts to streams (Ohio EPA 1990). There is a smaller but still significant risk of "finding" a water quality impact where one really does not exist when monitoring data are insufficient. This could result in a regulatory action that might cost hundreds of thousands of dollars or more to an entity, with costs passed along to consumers. Biosurvey and habitat data allow states to rank areas on the basis of need for remediation or protection. The interpretation and use of monitoring data in an ecoregional framework rather than in a political or hydrological framework can also lead to more accurate estimations of problems on state or national scales (Omernik and Griffith 1991).

The specter of millions of dollars being misspent on environmental controls without strong evidence of the efficacy of the treatment, indicates that money spent on high-quality monitoring programs is money well spent. Initiatives under discussion, such as pollution trading, will likely fail, or at least fail to control many of the factors limiting river integrity, if these initiatives are not based on accurate environmental information, including an effort to quantify habitat quality.

ACKNOWLEDGMENTS

This work would not have been possible without the staff of the Ecological Assessment Section of the Ohio EPA. The development of Ohio's biocriteria and habitat assessment techniques are the result of a team effort by many individuals. Drafts of this paper were reviewed by Chris Yoder, Marc Smith, Roger Thoma, Dave Altfater, Bernie Counts, and three anonymous reviewers.

## CHAPTER 14

Use of Periphyton in the Development of Biocriteria

Barry H. Rosen

1.0 INTRODUCTION

Periphyton are usually dominated by algae attached to or living in proximity to substrates, and may include bacteria, microinvertebrates, and associated organic materials. Algae provide food for many primary consumers, such as macroinvertebrates and herbivorous fish. Periphyton are not the usual focus of public concern unless the accumulation of organisms causes unsightly growth, taste and odor problems in drinking water sources, interferes with fishing, or impedes water flow (Welch et al. 1988). In contrast to the relatively low public concern about periphyton, they have been used extensively in the analysis of water quality for several decades (Kolkwitz and Marrson 1908; Patrick 1968; Lange-Bertalot 1979; Stevenson and Lowe 1986; Watanabe et al. 1988).

Periphyton have proven useful for environmental assessments because they have rapid reproduction rates and short life cycles, and thus they respond quickly to perturbation (Stevenson et al. 1986). They also are in direct contact with the water and are directly affected by water quality (Round 1991; VanLandingham 1976). Periphyton are collected by sampling methods that are rapid and easily quantifiable and data can be standardized into indices (Leclercq and Depiereux 1987; Watanabe et al. 1990). In all aquatic habitats, some periphyton will always be present and typically there is greater taxonomic richness in the periphyton than in higher trophic levels (i.e., macroinvertebrates and fish). The available ecological information about each species, which includes physical, chemical, and pollution tolerances, collectively termed autecological information, allows periphyton to be used as a diagnostic tool. Periphyton may also be sensitive to pollutants that other organisms tolerate relatively well (Willemsen et al. 1990; Mundie et al. 1991) and they cannot avoid pollutants because they are immobile. Periphyton are particularly responsive to nutrient input and can be used as indicators of nonpoint sources of pollution (Stevenson 1984).

Periphyton are one of the best indicators of disturbances in the catchment as well as of instream chemical alterations. Several studies on streams have illustrated the usefulness of periphyton in detecting nutrient disturbance (VanLandingham 1976; Willemsen et al. 1990; Stevenson et al. 1991), acidity (Mulholland et al. 1986; Planas et al. 1989), turbidity (Chessman 1986), metal toxicity (Rushforth et al. 1981; Weitzel and Bates 1981; Lampkin and Sommerfeld 1982), urban storm water (Willemsen et al. 1990), and other environmental disturbances originating external to the stream itself. These studies indicate that periphyton respond to a great variety of pollutants and can be used to accurately diagnose the probable causes of degradation of a stream or river. Periphyton assemblages not only reflect the water quality of a stream, they can also be an indicator of the riparian cover, which limits light quantity and quality (DeNicola et al. 1992). In addition, simple field observations, such as bleaching or absence of periphyton, can be the first signal that a recent disturbance has occurred.

Three primary groups, or assemblages, of organisms can be found in streams: periphyton, macroinvertebrates, and fish. Using a minimum of two of these assemblages is important in stream assessments to provide a more comprehensive understanding of stream condition. Periphyton are particularly relevant in the assessment of streams because they are often the base of the food web. In addition, in small streams that do not support fish, the two assemblages readily available are periphyton and macroinvertebrates.

Certain states, such as Montana (Bahls 1993), have selected periphyton and macroinvertebrates as the most cost-effective assemblages to use for assessment of stream condition and have developed biocriteria based on periphyton community analysis. Montana (Bahls 1993) has two sets of numeric biocriteria for periphyton that are used as assessment tools, one for mountain streams and one for plains streams, which illustrates the versatility of this assemblage. In Kentucky (Kentucky Division of Water 1993), periphyton data are used to construct a diatom bioassessment index (DBI). The DBI is used to rank the algal communities as "excellent," with no impairment of "aquatic life use," (see Yoder and Rankin, Chapter 9, for a discussion of aquatic life designated uses); "good," with slight or no impairment of aquatic life use; "fair," with moderate impairment, which equals "partially supporting" aquatic life use; or "poor," with severe impairment and not supporting aquatic life use. Oklahoma (Oklahoma Conservation Commission 1993) is developing rapid bioassessment protocols using the diatom community.

2.0 CONSIDERATIONS IN USING THE PERIPHYTON ASSEMBLAGE

2.1 Stressor Responsiveness

Periphyton are very sensitive to a number of stressors and the response is displayed by the loss of sensitive species and replacement by, or increase in, tolerant species. Changes in species composition are due to differential performance of species in different environmental conditions. One of the main reasons measures of periphyton have good diagnostic capability is that the range and tolerance of a large number of taxa are known for important environmental variables.

Periphyton response to natural macrohabitat changes can be distinguished from response to anthropogenic impacts. For example, logging the riparian corridor of a stream, which allows more light and sediments to affect the periphyton assemblage, will cause a decrease in the shade-tolerant species and an increase in the sediment-tolerant species. In streams that have no natural canopy, the assemblage may have the same shade-tolerant forms, but the sediment-tolerant species will not be present.

The periphyton assemblage is mostly nonmotile and cannot avoid a particular stress. A chronic stressor will affect the community composition and can be detected. The community will integrate the effects of a stress over a relatively short period of time, e.g., since the last scouring event, and will do a fair job of indicating brief periods of stress (day to month time scale).

2.2 Ease of Quantitative Sampling

Periphyton exist in all macrohabitats in which light, water, and inorganic nutrients are available. Samples taken from riffles or runs are sufficient to characterize the stream. Sampling time can be as rapid as 20 min/site, as shown by the methods used in Montana (Bahls 1993). Only a few rocks need to be sampled, because each has a periphyton assemblage with hundreds of thousands of individuals. The appropriate techniques can easily be described and incorporated into a sampling protocol (Aloi 1990; Kentucky Division of Water 1993; Cattaneo and Roberge 1991; Bahls 1993; Oklahoma Conservation Commission 1993), and usually include scraping the entire surface of several rocks of different sizes collected at random (Bahls 1993). For diatoms alone, 10 to 30 taxa are commonly encountered in a 500-organism count. Additionally, 5 to 10 taxa present may be soft-body forms. Compositing samples from several part of a reach will increase the number of periphyton species.

Macroalgae may occupy the substrates and are more difficult to quantify. One method is to record percent cover of macroalgae over substrates and then take a subsample for laboratory identification. Sample processing and species proportional counts of a sample should take no more than 3 h. Periphyton are always present in streams, including intermittent or ephemeral streams, when water is present. The periphyton in dry stream beds form crusts that can also be analyzed. The assemblage information is preserved because the diatoms have silica cell walls that allow identification of dead cells.

2.3 Measurement Stability and Variation

The periphyton community redevelops after scouring of stream substrates by freshets (Peterson and Stevenson 1990). As temperature, nutrients, pH, and dissolved solids change seasonally, the periphyton also change. The use of relative abundance alleviates the variability associated with collecting a sample from the community that is recovering from a scour event because the same species will recolonize the substrates present if the other environmental variables have not changed. The variability associated with seasonal change can be alleviated using species autecologies. Although natural succession may occur, one set of organisms will be replaced by ecologically similar organisms that provide an equally useful indication of aquatic conditions.

Sample variability can result as a function of substrate angle and positioning, which creates several microhabitats that are colonized by specific species. If an individual rock is considered to be the entire sample, samples from the same macrohabitat, such as a riffle, may appear to have distinct communities. The most accurate way to decrease sample variability is to collect from only one type of habitat within a reach and to composite many samples within that habitat. Riffles and runs, with current velocities of 10 to 20 cm/sec, are best because biomass is least variable in these habitats. The edges of pools and substrates, such as logs and submerged vegetation, could also be sampled, but these habitats are less desirable because they are generally not important to periphyton production. If separate habitats are sampled, it is important to keep them separate for subsequent analysis.

The periphyton assemblage may change up to 30% in taxa present from year to year within the spring–summer period (Stevenson and Peterson 1990). Although this is a substantial change in species, the assessment of stream condition based on indices is much less variable. Differences among sites would be much greater than year-to-year variability at the same site (Stevenson and Hashim 1986). Data exist from several years of studies that support these conclusions (Evenson et al. 1981; Chessman 1986; Leclercq and Depiereux 1987; McCormick and Stevenson 1992).

2.4 Diagnostic Power and Guild Information

Algal physical microhabitat guilds are well understood, but algal guilds based on trophic level, such as those for fish and invertebrates, are not appropriate. Separate indices for diatoms present are sensitive to pH, flow rate, inorganic nutrients, organic enrichment, salinity, temperature, toxic organics, metals, and siltation. These indices are based on the assemblage information and existing autecological databases and are used to infer stream characteristics.

Using constrained ordination [canonical correspondence analysis (CCA)] and weighted averaging (WA) regression, the relationships between measured physical and chemical variables and the algal distribution can be determined (ter Braak and Prentice 1988). In addition, WA and CCA are used to produce more autecological information for future investigations and index development. When certain taxa have narrower tolerances, making them better as indicators of condition, WA can weight these taxa more heavily, which can improve diagnostic power (Charles and Smol 1994). The multivariate approach provides a powerful means of maximizing ecological information obtained from periphyton assemblages.

2.5 Information/Cost Ratio

The information derived from the identification and enumeration of the algal component of the periphyton provides a wealth of ecological information for the cost involved in producing it. Laboratory analysis of species composition is labor intensive. Diatoms must be identified to species since species within a genus often have different requirements and tolerance (Palmer 1969; VanLandingham 1976). It takes an average of about 2 h per sample to identify the 500 organisms to species. Sample processing time decreases as the processor gains taxonomic expertise.

Identification of organisms to species allows the use of indices. VanLandingham (1976) claims that over 3000 species have autecological information in the literature, which can be built into an autecological database appropriate for the periphyton in a region. The most detailed and current consensus on the tolerance of diatoms was compiled by Leclercq and Maquet (1987) and Descy and Coste (1990). The condition of the stream can be inferred from the taxa present and what is known of their requirements and tolerances (Archibald 1972; Schoeman 1976). Certain species or groups of species are key indicators of polluted condition, such as the 20 species of diatoms identified by Lange-Bertalot (1979) as the most pollution-tolerant taxa worldwide.

2.6 Field and Laboratory Requirements

Periphyton have traditionally been sampled by collecting them from artificial substrates that are placed in streams and colonized by periphyton over a period of time (i.e., 2 weeks), or by sampling natural substrates. Both techniques can be used to monitor and assess stream condition. The advantages and disadvantages of artificial and natural substrates have been discussed extensively (Pryfogle and Lowe 1979; Stevenson and Lowe 1986; Aloi 1990). Oklahoma (Oklahoma Conservation Commission 1993) is using artificial substrates while Montana (Bahls 1993) and Kentucky (Kentucky Division of Water 1993) recommend natural substrates. Artificial substrates can be made of a variety of materials (e.g., glass slides and wooden doweling) and can be floated just below the surface or anchored to the bottom of the stream. After collection of these substrates, the periphyton are scraped off and processed. For natural substrates, sampling gear consists of a small knife, a tooth brush, and aluminum foil (for quantification of surface area) for rocks that can be removed from the stream. For bedrock samples, gear usually includes a 3- to 4-in. diameter PVC pipe with a rubber seal that can be held in place while the rock is brushed clean and the sample removed by suction (Kentucky Division of Water 1993). Rocks can be placed in a small enamel tray for removal of the periphyton and then in 4-oz plastic collection jars and preserved (Lugol's or buffered formalin).

Laboratory technicians must be trained in identification of diatoms to species, and soft algae to the genus level, and must have the ability to use a microscope, prepare slides, and enter data into a personal computer. For the identification of samples, approximately 2 h/sample are needed initially and 1 h/sample additionally if difficult species identifications require examination of the literature. This is an estimate, and most routine analyses, from counts to final metric or index, are usually completed in a 3-h period. This does not include time set aside to accumulate the appropriate references and do simple laboratory preparation. Another aspect of quality assurance is photographic documentation of identifications. Visits to appropriate museums are also advisable, with periodic taxonomic workshops for laboratory personnel.

Spatial variability in population abundance is well known for periphyton (Jones 1978; Pryfogle and Lowe 1979), and is caused in part by substrate, rate of flow, and light intensity. The use of species composition and composite sampling from the same habitat may reduce this variability. In developing biocriteria for streams and rivers, riffles or runs are adequate; however, if separate habitats (e.g., riffle, run, or pool) are sampled, it is possible to determine the between-habitat variability.

3.0. CHARACTERISTICS OF PERIPHYTON THAT CAN BE MEASURED

3.1 Estimates of Taxonomic Composition

Assessments of water quality and biotic integrity can be produced by interpreting the taxonomic composition of a sample. Autecological information for many algae (VanLandingham 1976), particularly diatoms, has been recorded in the literature for over a century (Cleve 1899; Hanna 1933; Lowe 1974; Patrick and Roberts 1979). The great taxonomic richness and level of microhabitat specialization of algal species provide considerable material for quantitative analysis and metric and index development (Patrick 1968). Periphyton taxonomic data have been used to develop indices of pH, nutrient enrichment, salinity, organic enrichment, sedimentation, and toxic pollution (Descy 1979; Sumita and Watanabe 1983). The specific physiological tolerances of each species of algae suggest that many other aspects of water quality could be determined from periphyton taxonomic composition.

Montana (Bahls 1993), Kentucky (Kentucky Division of Water 1993), and Oklahoma (Oklahoma Conservation Commission 1993) all have used information from taxonomic composition of periphyton assemblages to evaluate the condition of streams. For the development of biocriteria and bioassessment protocols, Montana (Bahls 1993) used only diatoms, although soft algae may eventually be incorporated into the assessment process. The diatom metrics used in Montana include a diversity index, a pollution index, a siltation index, and a similarity index. In Kentucky (Kentucky Division of Water 1993), the DBI metrics include diatom species richness, species diversity, percent community similarity, a pollution tolerance index, and percent sensitive species. In general, one expects that streams fully supporting aquatic life use have high taxa richness, intolerant taxa are present, and the community similarity index approximates that of reference sites. In streams that do not support aquatic life use, taxa richness is low, pollution-tolerant taxa are dominant, and there is low community similarity compared to reference sites.

3.2 Estimates of Standing Crop

Another aspect of the periphyton that is commonly measured, and is easily quantifiable, is standing crop. For stream assessments, a combination of standing crop estimates and taxonomic analysis is ideal. Standing crop is particularly valuable as an indicator of nutrient concentrations or toxicity. Using standing crop alone for assessing and detecting changes in water quality, however, has some limitations. For example, storm (natural) disturbances can scour a stream and possibly lead to erroneous estimates of the number of organisms present. Standing crop is also influenced by grazing from invertebrates and fish, which can complicate interpretation. More than one measure of standing crop is recommended because each measure has limitations for use in a monitoring program, and ratios [e.g., chlorophyll-*a*/ash-free dry mass(AFDM)] are also useful for detecting impacts.

Chlorophyll-*a* is commonly measured and provides a rapid estimate of standing crop. Welch et al. (1988) indicated that nuisance levels of periphyton are present at 100 to 150 mg chlorophyll-*a* m^{-2}. Oklahoma (Oklahoma Conservation Commission 1993) uses chlorophyll-*a* as one metric in a multimetric assessment of stream condition. Interpretation of chlorophyll data can be difficult, though, because of the adaptability of organisms to various physical and chemical conditions. For example, the same organism may have an order of magnitude more chlorophyll under low-light and nutrient-rich conditions than under high-light and nutrient-poor conditions (Rosen and Lowe 1984). Other problems associated with chlorophyll-*a* are high temporal variability and difficulty in interpreting trends.

Ash-free dry mass (AFDM) of periphyton samples provides an estimate of standing crop similar to chlorophyll-*a*. A periphyton sample will include microscopic heterotrophs and organic detritus as well as algae. These can cause an overestimate of the autotrophic component of the standing crop. AFDM values are part of the assessment methodology developed by Kentucky (Kentucky Division of Water 1993). High and low values can be expected in streams that have poor water quality; a high value would indicate nutrient stimulation of growth and a low value could indicate toxicity. Relying exclusively on this measurement of standing crop is not recommended; however, in combination with other indicators, such as DBI, AFDM can be useful.

The algae in the periphyton are a diverse group of organisms from several phylogenetic classes. With this diversity comes a variety of cells of different sizes that may not be ecologically or functionally equivalent in the community. Instead of simply counting the number of cells per unit area, one can determine cell biovolume and use it to account for the differences in sizes of cells that are enumerated. Biovolume may be too elaborate and time consuming for routine monitoring programs because average dimensions of cells of each species are needed to estimating cell biovolume. Different classes of algae have different proportions of internal structure occupied by vacuoles, which are important sources of variation in measurements of biovolume.

4.0 POSSIBLE ASSESSMENT METRICS AND ANALYSES

A combination of several metrics, which use the relative abundance of species, can be used to evaluate periphyton assemblage characteristics and water-related changes (van Dam 1982). Several new approaches, including multivariate analysis and weighted average metrics, make periphyton analysis a better indicator than it was 5 or 10 years ago. The following discussion presents examples of methods presently used in evaluating the periphyton assemblages.

For diatoms, Montana (Bahls 1993) uses a diversity index, a pollution index (i.e., most tolerant, less tolerant, and sensitive groups), a similarity index, and a siltation index that examines the proportion of *Navicula* and *Nitzschia*. Another proposed index that is sensitive to siltation would add more epipelic diatoms (e.g., *Surirella*) to the current siltation index (a motile/nonmotile diatom ratio). For soft-bodied algae, used only to support the diatom assessment, Montana uses three metrics: dominant phylum, indicator taxa, and number of genera.

The Diatom Bioassessment Index (DBI) developed by the State of Kentucky (Kentucky Division of Water 1993) is currently used in water quality assessments. Metrics used to construct the DBI include diatom species richness, species diversity, percent community similarity to reference sites, a pollution tolerance index, and percent sensitive species. Scores for each metric range from 1 to 5 and these scores can then be translated into descriptive site bioassessments such as excellent, good, fair, or poor, which are used to determine fully, partially, or not supporting aquatic life use.

The Rapid Bioassessment Protocols (RBPs) developed for Oklahoma (Oklahoma Conservation Commission 1993) parallel the RBPs for fish and invertebrates described by Plafkin et al. (1989). Protocol I is a screening tool used to identify areas of a stream that exhibit impairment/nonimpairment of aquatic life use without reference conditions comparisons. Metrics for this protocol include the number of taxa present and a sequential comparison index, as well as some manipulations of these kinds of data (Cairns et al. 1968). Protocol II is based on comparisons with reference conditions from other parts of the stream that are not impaired, and classifies streams as unimpaired, moderately impaired, or severely impaired. Metrics for this protocol incorporate the Protocol I metrics and add a diversity index (Cairns et al. 1971), and a diatom index. Protocol III incorporates additional measures of community structure and a measure of community production. Protocol III is used to classify streams more finely as unimpaired, or as slightly, moderately, or severely impaired. Metrics for this protocol (Oklahoma Conservation Commission 1993) incorporate most of Protocol II, as well as a modified diversity index and a modified evenness index (Cairns et al. 1971), chlorophyll measurement, and a diatom index.

Eutrophication indices were developed by Smith (1966) and modified by Lowe (1974) for discerning nutrient concentrations and humic contributions for diatoms. Multivariate analyses, such as CCA and WA, are useful because they make use of the information in assemblages to quantitatively infer ecological characteristics (e.g., pH and biological oxygen demand). When used in a monitoring program, these methods provide an indication of what specific environmental characteristics are changing over time, and can be used to help diagnose the causes of changes in the aquatic system.

The Pollution Tolerance Index, developed by Lange-Bertalot (1979), highlights organisms sensitive to toxic substances. An index developed by Palmer (1969) ranks several genera, including soft-bodied forms, by their tolerance to organic pollution. The Halobien Index was developed by Kolbe (1927) for salinity.

5.0 SUMMARY AND CONCLUSIONS

Periphyton are very useful in developing biocriteria for streams. The metrics currently used by states for periphyton are well developed and are used to determine if a stream is supporting aquatic life use as designated under the Clean Water Act. Because periphyton are the major primary producers in most stream systems and are often the base of the food web, they are important indicators of biological integrity.

Periphyton have been used for over 50 years for monitoring streams and rivers, and several reports are available that describe the ecological characteristics of the taxa in these assemblages. In addition, there are several databases that describe the known ranges of taxa along chemical, physical, and pollution gradients. Analysis of variance components and historical assessments of stream and river quality are possible with the many samples that have been collected and archived in museums during the last 90 years (e.g., Philadelphia Academy of Natural Sciences, California Academy of Sciences).

Growth is directly dependent on nutrients, which make periphyton an excellent indicator of point and nonpoint sources of nitrogen and phosphorus enrichment. Excess growth is of public concern when it is unsightly and causes fish kills. A list of the taxa present and their proportionate abundance can be analyzed using several indices to determine biotic integrity and diagnose specific stressors. Because of their characteristics, and the work already conducted by states such as Kentucky, Montana, and Oklahoma, periphyton can be very useful in the development of biocriteria.

ACKNOWLEDGMENTS

I am grateful to the contribution from the Environmental Monitoring and Assessment Program (EMAP) participants in the Stream Design Workshop, held on February 26-28, 1992 in Cincinnati, Ohio. The contributions from Loren Bahls, Donald Charles, R. Jan Stevenson, Gary Collins, Ben McFarland, and Lythia Metzmeier at this workshop and their subsequent discussions were invaluable in putting together this chapter. I also would like to thank Robert Hughes and Susan Christie for helping with formatting and editorial content.

The research described in this article has been funded by the U.S. Environmental Protection Agency. This document has been prepared at the EPA Environmental Research Laboratory in Corvallis, Oregon, through contract #68-C8-0006, with ManTech Environmental Technology, Inc. It has been subject to the agency's peer and administrative review and approved for publication. Mention of trade names or commercial products does not constitute endorsement or recommendation for use.

## CHAPTER 15

Development and Application of the Invertebrate Community Index (ICI)

Jeffrey E. DeShon

1.0 INTRODUCTION

Aquatic macroinvertebrates have been used widely as an indicator group for many years in pollution studies involving flowing waters. Cairns and Pratt (1993) provide a detailed account of the current and historical use of macroinvertebrates in freshwater biomonitoring. At the Ohio EPA, macroinvertebrate communities have been collected and analyzed since the agency's inception in 1973 in an effort to provide biological data to be used in the water quality monitoring and assessment process. At least one collection has been made from over 2300 locations displaying a wide variety of water quality conditions within the state. Early assessments of macroinvertebrate data depended on the individual expertise of the biologists who collected the data. With the aid of tools such as the Shannon–Wiener Diversity Index (Shannon and Wiener 1949; Wilhm 1970) and a healthy dose of "best professional judgment," numerous narrative evaluations of water quality problems were made over the years. However, the inherently subjective nature of such evaluations was often considered a major liability, especially in complicated environmental issues involving permit holders and litigation. As a result, a more objective means to assess macroinvertebrate data was sought.

As an offshoot of the 1983–84 Ohio Stream Regionalization Project, a cooperative pilot venture with USEPA/ERL-Corvallis (Whittier et al. 1987), methods were researched to develop a multimetric macroinvertebrate index patterned after the concept of the Index of Biotic Integrity (IBI) developed for fish community assemblages by Karr (1981) and refined by Fausch et al. (1984). The result was the Invertebrate Community Index (ICI), which is now used as the principal assessment tool by Ohio EPA macroinvertebrate biologists for monitoring and assessment activities in all free-flowing rivers and streams in Ohio. In 1987, numeric ecoregional biological criteria were developed and codified in the Ohio Water Quality Standards by the Ohio EPA with the ICI being one of three biological indices applied (Yoder and Rankin, Chapter 9).

2.0 METHODS SUMMARY

The primary sampling gear used by the Ohio EPA for the quantitative collection of macroinvertebrates in streams and rivers is the modified multiple-plate artificial substrate sampler (Hester and Dendy 1962). The sampler is constructed of $1/8$ in. (3 mm) tempered hardboard cut into 3-in.^{2} (7.5-cm^2) plates and 1-in.^{2} (2.5-cm^2) spacers. A total of eight plates and twelve spacers are used for each sampler. The plates and spacers are placed on a 3 in. (7.5 cm) long, $1/4$ in. (6 mm) diameter eyebolt so that there are three single spaces, three double spaces, and one triple space between the plates. The total surface area of the sampler, excluding the eyebolt, approximates 1 ft^2 (roughly 0.1 m^2). A sampling unit consists of a composite cluster of five substrates tied to a construction block that is colonized in-stream for a 6-week period beginning no earlier than June 15 and ending no later than September 30. Detailed descriptions of the placement, collection, and processing of the artificial substrates are available (Ohio EPA 1989a). In addition to the artificial substrate sample, routine monitoring also includes a qualitative collection of macroinvertebrates that inhabit the natural substrates at the sampling location. All available habitat types are sampled and voucher specimens are retained for laboratory identification. More specific information for the collection of this sample has also been detailed (Ohio EPA 1989a). For the purpose of generating an ICI value, both a quantitative and qualitative sample must be collected at a sampling location.

The use of artificial substrates for monitoring purposes has a number of advantages. According to Rosenberg and Resh (1982), the major advantages in using artificial substrates in general are that they:

- allow collection of data from locations that cannot be sampled effectively by other means,
- permit standardized sampling,
- reduce variability compared with other types of sampling,
- require less operator skill than other methods,
- are convenient to use, and
- permit nondestructive sampling of an environment.

The authors also listed a number of disadvantages. These include:

- incompletely known colonization dynamics,
- long exposure times to obtain a sample,
- loss of fauna on retrieval,
- unforeseen losses of artificial substrates, and
- inconvenient to use and logistically awkward.

Generally, however, the authors concluded that these problems could be minimized by adhering to strict guidelines concerning sampler placement and collection and data analysis and interpretation.

Klemm et al. (1990) specifically focused on the use of modified Hester-Dendy multiple-plate artificial substrate samplers and listed the following advantages:

- are excellent for water quality monitoring,
- provide uniform substrate type,
- allow for a high level of precision,
- provide habitats of known area for a known time at a known depth.

The authors noted that colonization of macroinvertebrates should be relatively equal in similar habitats and should reflect the capacity of the water to support aquatic life. Although acknowledging that these samplers may exclude certain mollusks or worms, they concluded that a sufficient diversity of benthic fauna is collected to be useful in assessing water quality.

Thus, by exploiting the strengths and yet recognizing and controlling the liabilities, the use of multiple-plate artificial substrate samplers serves as an important component of macroinvertebrate sampling in Ohio and helps to ensure that a standardized approach to monitoring a wide variety of sites is maintained. When selecting any sampling method, it is imperative to have a clear definition of the objectives of the sampling as well as an understanding of the potential shortcomings of using that method for the collection of macroinvertebrates.

A composited set of five artificial substrate samplers of eight plates each has been used by the Ohio EPA in collecting macroinvertebrate samples since 1973. At this level of effort, it has been found that consistent, reproducible ICI values can be scored despite the collections of often highly variable numbers of individual organisms. The latter is a result of the tendency of macroinvertebrate populations to have naturally clumped (i.e., negative binomial) distributions in the environment. Results of analyzing replicate composites of five artificial substrates have shown that variability among calculated ICI values is at an acceptable level. Details of that analysis can be found elsewhere in this chapter. The reliability of the sampling unit not only depends on the fact that colonization surface areas are standard, but equally

Table 1.   Current Taxonomic Keys and the Level of Taxonomy Routinely Used by the Ohio EPA for Various Macroinvertebrate Taxonomic Classifications

Porifera: Species (Pennak 1989)
Coelenterata: Genus (Pennak 1989)
Platyhelminthes: Class (Pennak 1989)
Nematomorpha: Phylum/genus (Pennak 1989)
Ectoprocta: Genus/species (Thorp and Covich 1991)
Entoprocta: Species (Thorp and Covich 1991)
Annelida
  Oligochaeta: Class (Pennak 1989)
  Hirudinea: Species (Klemm 1982)
Arthropoda
  Crustacea
    Isopoda: Genus (Pennak 1989)
    Amphipoda: Genus (Pennak 1989)
      Gammarus: Species (Holsinger 1972)
    Decapoda
      Cambarus and *Fallicambarus*: Species
        (Jezerinac and Thoma 1984)
      Orconectes and *Procambarus*: Species
        (Jezerinac 1978)
      Palaemonetes: Species (Pennak 1989)
  Arachnoidea: Class (Pennak 1989)
  Insecta
    Ephemeroptera: Genus (Edmunds et al. 1976,
      Merritt and Cummins 1984)
      Baetidae: Genus/species (Morihara and
        McCafferty 1979, McCafferty and Waltz 1990)
      Heptageniidae
        Stenonema: Species (Bednarik and
          McCafferty 1979)
      Ephemerellidae
        Dannella: Species
          (Allen and Edmunds 1962)
        Ephemerella: Species
          (Allen and Edmunds 1965)
        Eurylophella: Species
          (Allen and Edmunds 1963a)
        Serratella: Species
          (Allen and Edmunds 1963b)
      Baetiscidae
        Baetisca: Species (Burks 1953)
      Ephemeroidea: Species (McCafferty 1975)
    Odonata: Family/genus
      (Merritt and Cummins 1984)
      Anisoptera: Genus/species (Needham and
        Westfall 1955, Walker 1958, Walker and
        Corbett 1975)
    Plecoptera: Genus (Stewart and Stark 1988)
      Perlidae
        Acroneuria: Species (Hitchcock 1974)
        Paragnetina: Species (Hitchcock 1974)
        Perlinella: Species (Kondratieff et al. 1988)
      Perlodidae: Species (Hitchcock 1974)
    Hemiptera: Genus (Hilsenhoff 1982b, Merritt and
      Cummins 1984)
    Megaloptera: Genus (Merritt and Cummins 1984)

Nigronia: Species (Neunzig 1966)
Neuroptera: Genus (Merritt and Cummins 1984)
Trichoptera: Genus (Wiggins 1977, Merritt and
  Cummins 1984)
  Philopotamidae: Species (Ross 1944)
  Hydropsychidae
    Hydropsyche: Species
      (Schuster and Etnier 1978)
  Rhyacophilidae
    Rhyacophila: Species (Flint 1962)
  Leptoceridae
    Ceraclea: Species (Resh 1976)
    Nectopsyche: Species (Haddock 1977)
Lepidoptera: Genus (Merritt and Cummins 1984)
Coleoptera: Genus (Hilsenhoff 1982b, Merritt
  and Cummins 1984)
  Dryopoidea: Genus/species (Brown 1972)
Diptera: Family/genus
  (Merritt and Cummins 1984)
  Ceratopogonidae
    Atricopogon: Species (Johannsen 1935)
  Chironomidae: Genus/species groups
    (Wiederholm 1983)
    Ablabesmyia: Species (Roback 1985)
    Labrundinia: Species (Roback 1987)
    Tanypus: Species (Roback 1977)
    Corynoneura: Species (Simpson and Bode
      1980, Bolton In Prep.)
    Eukiefferiella and *Tvetenia*: Species groups
      (Bode 1983)
    Nanocladius: Species (Saether 1977,
      Simpson and Bode 1980, Bolton In Prep.)
    Parakiefferiella: Species (Bolton In Prep.)
    Rheocricotopus: Species (Saether 1985)
    Thienemanniella: Species (Simpson and
      Bode 1980, Bolton In Prep.)
     Chironomus: Species groups
      (Oliver and Roussel 1983)
    Dicrotendipes: Species (Epler 1987)
    Endochironomus and *Tribelos*: Species
      (Grodhaus 1987)
    Parachironomus: Species
      (Simpson and Bode 1980)
    Paracladopelma and *Saetheria*: Species
      (Jackson 1977)
    Polypedilum: Species groups/species
      (Maschwitz 1976, Bolton In Prep.)
    Tanytarsini: Genus/species groups/species
      (Simpson and Bode 1980, Bolton In Prep.)
  Muscidae: Species (Johannsen 1935)
Mollusca
  Gastropoda: Genus/species (Burch 1982)
  Pelecypoda
    Sphaeriidae: Genus (Burch 1982)
    Unionidae: Species (Waters 1993)

important are the actual physical conditions under which the units are placed in the aquatic environment. It is imperative that the artificial substrates be located in a consistent fashion with particular emphasis on sustained current velocity over the set. With the exception of water quality, the amount of current tends to have the most profound effect on the types and numbers of organisms collected using artificial substrates in Ohio. For an accurate interpretation of the ICI, current speeds should be no less than 0.3 ft/s (10 cm/s) under normal summer–fall flow regimes. These conditions can usually be adequately met in

Table 2.   Metrics Used to Calculate the Ohio EPA Invertebrate Community Index (ICI)

1. Total Number of Taxa	6. Percent Caddisfly Composition
2. Number of Mayfly Taxa	7. Percent Tribe Tanytarsini Midge Composition
3. Number of Caddisfly Taxa	8. Percent Other Dipteran and Non-Insect Composition
4. Number of Dipteran Taxa	9. Percent Tolerant Organisms (from Table 3)
5. Percent Mayfly Composition	10. Number of Qualitative EPT Taxa

Note: Scoring (6, 4, 2, or 0 points) for all metrics determined by basin drainage area (mi^2) at the sampling location. See Figures 1 to 10.

all sizes of perennial Ohio streams but can be a problem in small headwater streams or those streams so highly modified for drainage that dry weather flows maintain intermittent, pooled habitats only. In these situations, sampling can be conducted, but an alternative interpretation of the ICI value and/or the use of other assessment tools may be necessary.

An additional area of importance concerns the accuracy of identification of the sample organisms. The ICI has been calibrated to specified levels of taxonomy currently being used by the Ohio EPA. It is imperative that accurate identifications to those levels be accomplished. Otherwise, problems may arise in the ICI metrics dealing with the identity and/or number of taxa of a particular organism group. Inaccurate identifications can also be a problem in the ICI metric dealing with the percent abundance of pollution tolerant organisms. Table 1 lists current taxonomic keys and the level of taxonomy routinely used by the Ohio EPA for various macroinvertebrate taxonomic classifications.

3.0 DEVELOPMENT OF THE INVERTEBRATE COMMUNITY INDEX

3.1 Metric Selection

The principal measure of overall macroinvertebrate community condition used by the Ohio EPA is the Invertebrate Community Index (ICI), a measurement derived from the wealth of macroinvertebrate community data collected over the years by aquatic biologists at the Ohio EPA. The ICI consists of ten compositional and structural community metrics (Table 2), each of which receives a score of 6, 4, 2, or 0 points based on a comparison with a set of ecoregional reference sites. Metrics 1 to 9 of the ICI are generated from the artificial substrate sample data while Metric 10 is based on the qualitative sample data.

The selection of the ten metrics ultimately chosen for the ICI was facilitated by analyzing the process by which Ohio EPA biologists had subjectively judged the quality of a macroinvertebrate sample. In essence, the index and its final set of metrics effectively quantified a more subjective, narrative approach that was previously used. This allowed for a more objective and efficient level of assessment and decision making. Structural and compositional rather than functional metrics were selected because of their accepted historical use, simpler derivation, and ease of interpretation. However, a functional component is inherent in the index since watershed size at the sampling location affects metric scoring. In effect, scoring of metrics is strongly influenced by functionally based differences in the macroinvertebrates that are inhabiting the wide range of stream sizes of the reference sites used to calibrate scoring of each metric (e.g., populations predominated by collector/gatherers, scrapers, or filter feeders). The strength of the ICI lies in its ability to directly compare the biological performance of the subject stream site against performance measured at reference sites of similar watershed size and from the same ecoregion of Ohio. The ICI value, the summation of the metric scores, is a single number that reflects general biological condition, which has incorporated into it ten measurements that, with various degrees of effectiveness, can and have often been used to accomplish this task individually. It was thought that, used in the aggregate, these metrics would minimize the weaknesses and drawbacks that each has alone.

Application of an ICI-type multimetric assessment tool outside of Ohio should not be restricted by or limited to the set of metrics derived for use in Ohio. Rather, the flexibility of the multimetric approach allows for the utilization of differing collection methodologies with selection of metrics most appropriate for the diverse geographic settings and ecoregions of the United States. However, the common denominator of all applications should be the regional reference site approach and its use to calibrate scoring of the selected metrics and, ultimately, to set performance expectations and establish biocriteria.

3.2 Scoring of Metrics

The 6, 4, 2, or 0 point system is structured to score sample metrics against expectations derived from a database of least impacted regional reference sites. These sites were selected from the Ohio EPA database using guidelines developed by Hughes et al. (1986). Scoring criteria for each metric were developed through a quantitative calibration process in which reference site metric values were plotted against a log transformation of drainage area (a reflection of stream size) and scoring ranges determined after a method modified from Fausch et al. (1984). For example, six points are scored if a given metric falls in the range exhibited by exceptional stream communities, 4 points for those metric values characteristic of more typical, but good communities, 2 points for metric values moderately deviating from the expected range of good and exceptional values, and 0 points for metric values strongly deviating from the expected range of good and exceptional values. The summation of the individual metric scores, as determined by the relevant attributes of an invertebrate sample with consideration given to stream drainage area at the sampling location, results in the ICI value that ranges from 0 to 60. Four scoring categories were chosen because of the historical use by the Ohio EPA of four levels of biological community condition (i.e., exceptional, good, fair, and poor), a situation that, as defined above, is reflected by the metric score of a sample.

The four scoring categories were calibrated using data from 246 least impacted reference sites distributed across Ohio's five ecoregions as delineated by Omernik (1987) and Omernik and Gallant (1988). To determine the 6, 4, 2, or 0 values for each ICI metric, the reference site database was plotted vs. drainage area. Similar to procedures used to calibrate the IBI (Fausch et al. 1984), the scatter plot of each metric was examined by eye to determine if any sloped relationship existed with drainage area. When it was decided if a direct, inverse, combination of both, or no relationship existed, the appropriate 95th percentile line was estimated and the area beneath partitioned into four equal parts as determined by the distribution of the reference points. One difference between this procedure and that used by Fausch et al. (1984) to calibrate the IBI was the use of four scoring categories rather than three. Another difference involved some percent abundance and taxa richness scoring categories that were not equally divided since the distribution of data points showed a tendency to clump at or near zero. In these situations, a modified method was used where the zero scoring category included zero values only and the 6, 4, and 2 point categories were delineated by sequential bisections of the remaining wedge of reference data points. One final difference involved the use of drainage area as a scaling factor rather than stream order. The decision to use drainage area as an indicator of stream size rather than stream order, or a factor such as stream width used by Lyons (1992a), was based on the availability and ease of drainage area calculation and its relevance to stream size. Stream order was viewed as being too coarse (Hughes and Omernik 1981b) and stream width is simply not representative of stream size given the widespread historical modification of streams in Ohio. In other regions of the United States, these and other parameters may be appropriate as scaling factors. The ultimate decision must be determined by experts familiar with regional patterns of stream morphology.

3.3 Description of Metrics

3.3.1 Metric 1. Total Number of Taxa

The species area plot of the total taxa metric vs. drainage area is depicted in Figure 1. Taxa richness is a key component of several new-generation indices currently used to evaluate macroinvertebrate community integrity (Lenat 1988; Plafkin et al. 1989; Kerans and Karr 1994). The underlying reason is the basic ecological principal that healthy, stable biological communities in warmwater streams have high species richness and diversity. As can be seen by the species area curve, the total number of taxa collected from artificial substrates in Ohio tends to decrease in the larger rivers. This is consistent with the River Continuum Concept (RCC; Vannote et al. 1980), which predicts maximum taxa richness in midsized streams and decreased diversity in larger streams and rivers due to changes in organic inputs and plant growth. A contributing factor to the decline in taxa richness with increased watershed size is the more monotonous nature of substrate types in larger rivers. An additional consideration, however, is that even the best, larger rivers have been subjected to some degree of cultural degradation in Ohio.

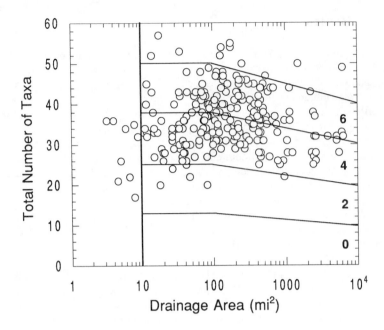

Figure 1. The relationship of ICI Metric 1, Total Number of Taxa, with the log transformation of drainage area at 246 Ohio reference sites. (An inverse relationship exists with drainage areas >100 mi².)

3.3.2 Metric 2. Number of Mayfly Taxa

Mayflies are an important component of an undisturbed stream macroinvertebrate fauna in Ohio. As a group, they are decidedly pollution sensitive and are often first to decline and eventually disappear from artificial substrate collections with the onset of environmental perturbation. Thus, they are a good indicator of high quality ambient conditions. Environmental requirements and pollution tolerances have been thoroughly documented for many species (Hubbard and Peters 1978). Taxa richness of mayflies is included as an individual metric of one current assessment protocol (Kerans and Karr 1994) but it most often appears as a component of an EPT (Ephemeroptera, Plecoptera, and Trichoptera) taxa richness metric (Lenat 1988; Plafkin et al. 1989). The species area plot of reference site mayfly taxa vs. drainage area is depicted in Figure 2. The general trend in mayfly diversity reflects highest variety of types in intermediate size streams with slight decreased diversity in the smaller and larger drainages. As predicted by the RCC, this is the result of the transitional nature of the intermediate streams and the corresponding increased variety of macrohabitat, microhabitat, and food sources. In effect, environmental conditions are highly diverse and support a mayfly fauna transitional between the smaller Ohio streams and the larger Ohio rivers.

3.3.3 Metric 3. Number of Caddisfly Taxa

Caddisflies are often a predominant component of the macroinvertebrate fauna collected from artificial substrates in larger, relatively unimpacted Ohio streams and rivers. Though generally thought to be slightly more pollution tolerant as a group than mayflies, they display a wide range of tolerance among genera and species (Harris and Lawrence 1978). Notwithstanding, few can tolerate heavy pollutional stress and, as such, can be good indicators of environmental conditions (Kerans and Karr 1994). The distribution of reference site caddisfly taxa vs. drainage area shows a clear, increasing trend with stream size (Figure 3). This can be explained by the predominance in Ohio of net spinning, filter feeding caddisflies of the families Hydropsychidae, Polycentropodidae, and Philopotamidae and case-making microcaddisflies of the family Hydroptilidae. Habitat preferences of the filter feeders include streams with abundant suspended organic matter while the micro-caddisflies feed mainly on periphytic diatoms and filamentous algae. These environmental conditions are best met in the larger streams and

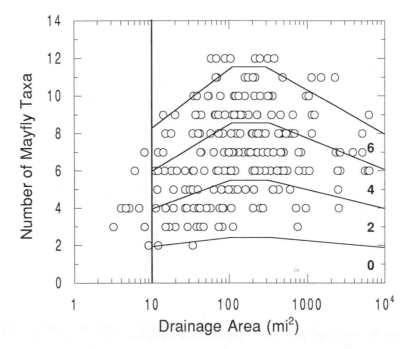

Figure 2. The relationship of ICI Metric 2, Number of Mayfly Taxa, with the log transformation of drainage area at 246 Ohio reference sites. (A direct relationship exists with drainage areas <100 mi²; an inverse relationship exists with drainage areas >300 mi².)

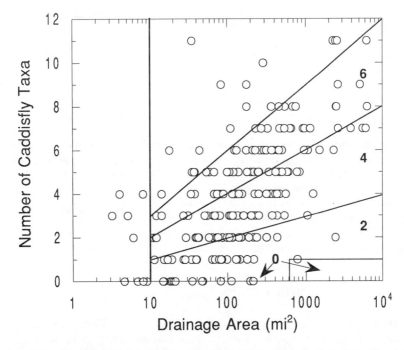

Figure 3. The relationship of ICI Metric 3, Number of Caddisfly Taxa, with the log transformation of drainage area at 246 Ohio reference sites. (A direct relationship exists with drainage area; a score of 0 for no taxa at drainage areas <600 mi²; a score of 0 for <2 taxa at drainage areas >600 mi².)

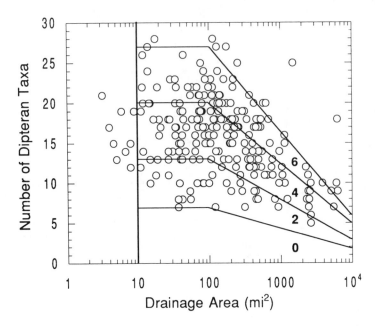

Figure 4. The relationship of ICI Metric 4, Number of Dipteran Taxa, with the log transformation of drainage area at 246 Ohio reference sites. (An inverse relationship exists with drainage areas >100 mi².)

rivers where the import of fine particulate organic matter is maximized and algal growth is optimal due to the availability of nutrients and more open canopies. As can be seen in Figure 3, for drainages less than 600 mi² (1550 km²), zero scores occur only when no caddisfly taxa are present. In artificial substrate collections from the smaller Ohio watersheds, it is normal to collect fewer kinds of the more common caddisflies even at unpolluted sites. For drainages greater than 600 mi² (1550 km²), habitat conditions at sampling sites are much more conducive to the proliferation of these caddisflies and, therefore, at least two taxa must be present to score higher than zero.

3.3.4 Metric 4. Number of Dipteran Taxa

Among the major aquatic macroinvertebrate groups, dipterans, especially midges of the family Chironomidae, rank high in faunal diversity and display a wide range of pollutional tolerances (Beck 1977; Berg and Hellenthal 1990; Lenat 1993). They are usually the major component of a macroinvertebrate collection using Ohio EPA methodologies and, under heavy pollutional stress, are often the only insects collected and, at the same time, are the predominant macroinvertebrate group. Larval taxonomy has improved greatly and, as a result, clear patterns of organism assemblages have become more distinctive under water quality conditions ranging from near pristine to heavily organic and toxic. The fact that they do not usually disappear under severe pollutional stress makes them especially valuable in evaluating water quality. The distribution of dipteran taxa vs. drainage area is shown in Figure 4. A clear, inverse relationship exists with watersheds greater than 100 mi² (260 km²). In the larger rivers, there is a tendency towards increased populations of proportionately fewer dipteran taxa. This is probably the result of abundant food supplies but fewer functional niches as habitat conditions and food types become increasingly monotonous.

3.3.5 Metric 5. Percent Mayfly Composition

As with number of mayfly taxa, the percent abundance of mayflies collected in an artificial substrate sample is often readily and rapidly affected by often minor environmental disturbances. Though much more reference site variability exists in this metric compared with the taxa metric, there is a strong relationship with water quality. The range of abundances in the relatively unimpacted reference site database varies from near zero to greater than 80% (Figure 5). However, data from slightly degraded (fair) and severely degraded (poor) stream communities in Ohio indicate that mayfly abundance is reduced

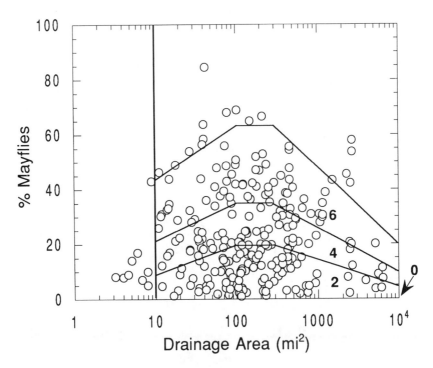

Figure 5. The relationship of ICI Metric 5, Percent Mayflies, with the log transformation of drainage area at 246 Ohio reference sites. (A direct relationship exists with drainage areas < 100 mi^2; an inverse relationship exists with drainage areas >300 mi^2.)

considerably under even slight impacts and is essentially nonexistent under more severe impacts. Thus, it was felt that even a few mayflies in low abundance should contribute a metric score of at least 2 points. Therefore, only those artificial substrate samples with no mayflies are scored zero for the metric. The distribution of reference site data points shows a trend similar to that observed for the mayfly taxa metric.

3.3.6 Metric 6. Percent Caddisfly Composition

As with the number of caddisfly taxa metric, the percent abundance of caddisflies collected from artificial substrates is strongly related to stream size (Figure 6). Again, optimal habitat and availability of appropriate food type seem to be the main considerations for stimulating large populations of the common Ohio caddisflies. As can be seen in Figure 6, caddisflies can make up a significant portion of the macroinvertebrate community, often exceeding 25% of all organisms collected. However, they are just as likely to be found in quite low numbers, at times comprising less than 1% of a sample. Because of their general disposition as an intermediate group (more tolerant than most mayflies and less tolerant than many dipterans) and because they disappear rapidly under environmental stress, zero scores are restricted to those sites with drainage areas less than 600 mi^2 (1550 km^2) where no caddisflies are collected. This scaling, similar to that used in Metric 3, is necessitated because low numbers of caddisflies are often collected from artificial substrates at the smaller stream sites even when resource disturbance is minimal. At sites with greater than 600 mi^2 (1550 km^2) of drainage, appropriate habitat conditions are much more likely to exist; therefore, individuals from the most prevalent Ohio caddisfly families should be well represented and must be present in at least minimal numbers to score greater than zero.

3.3.7 Metric 7. Percent Tribe Tanytarsini Midge Composition

The tanytarsini midges are a tribe of the chironomid subfamily Chironominae. The larvae are generally burrowers or clingers, and many species build cases out of sand, silt, and/or detritus. Many species feed on microorganisms and detritus through filtering and gathering, though a few are scrapers.

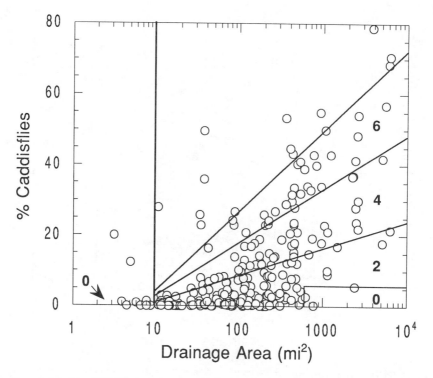

Figure 6. The relationship of ICI Metric 6, Percent Caddisflies, with the log transformation of drainage area at 246 Ohio reference sites. (A direct relationship exists with drainage area; a score of 0 for no individuals at drainage areas <600 mi^2; a score of 0 for minimal percent abundance at drainage areas >600 mi^2.)

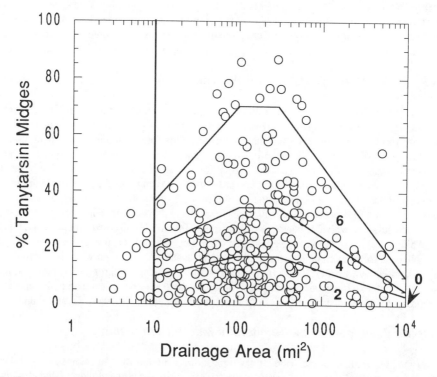

Figure 7. The relationship of ICI Metric 7, Percent Tanytarsini Midges, with the log transformation of drainage area at 246 Ohio reference sites. (A direct relationship exists with drainage areas <100 mi^2; an inverse relationship exists with drainage areas >300 mi^2).

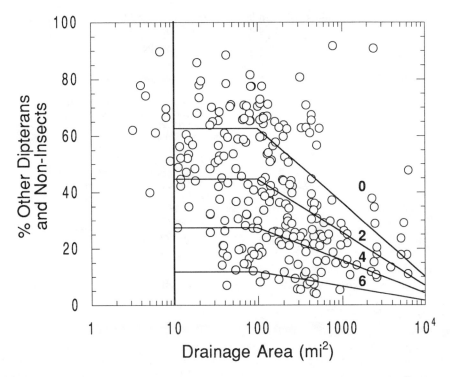

Figure 8. The relationship of ICI Metric 8, Percent Other Diptera and Non-Insects, with the log transformation of drainage area at 246 Ohio reference sites. (An inverse relationship exists with drainage areas >100 mi².).

Eleven genera and up to 140 species occur in North America, although only eight genera and 22 distinct taxa have been collected in Ohio streams and rivers. At the relatively unimpacted Ohio reference sites, they are often the predominant midge group and can exceed 50% of the total number of organisms collected from artificial substrates. As a group, they appear to be intermediate in pollution tolerance and often disappear or decline under moderate pollutional stress. However, some genera, species groups, and species have been determined to be quite sensitive to pollution (Anderson et al. 1980; Simpson and Bode 1980; Hilsenhoff 1987; Lenat 1993). As depicted in Figure 7, populations of tanytarsini midges tend to peak at reference sites in the 100 to 300 mi² (260 to 775 km²) range of watershed size. Because of their moderate intolerance to environmental disturbance, zero scores only occur when no tanytarsini midges are present.

3.3.8 Metric 8. Percent Other Dipteran and Non-insect Composition

This metric includes the community percentage of all dipterans (excluding the midge tribe Tanytarsini) and other non-insect invertebrates such as aquatic worms, flatworms, amphipods (scuds), isopods (aquatic sow bugs), freshwater hydras, and snails. This metric is one of two negative metrics of the ICI in that an increased abundance results in a lower metric score. Taxa in these groups of macroinvertebrates, though often present as part of a healthy stream community, are those that generally tend to predominate under adverse water quality conditions (Hynes 1966; Hart and Fuller 1974). Depending on the severity of the stress, these organisms will comprise over 90% of the individuals collected in an artificial substrate sample. Figure 8 depicts the distribution of reference site data for the metric. As indicated, reference site percentages are inversely related to stream size. However, this relationship does not seem to hold for seriously impaired situations; under these circumstances, the dipterans and non-insects defined by this metric usually predominate at a high percentage regardless of stream size. In cases where conditions are so severely degraded that no or only a few organisms (<50 individuals) are collected and the percentage of dipterans and non-insects is consequently at or near zero, the metric defaults to a zero score rather than

Table 3.   List of Pollution-Tolerant Macroinvertebrates Used to Determine Metric 9 of the Invertebrate Community Index (ICI)

Common name		Scientific name
Aquatic segmented worms	Annelida:	Oligochaeta
Midges	Diptera:	*Psectrotanypus dyari*
		Cricotopus (C.) bicinctus
		Cricotopus (Isocladius) sylvestris group
		Nanocladius (N.) distinctus
		Chironomus (C.) spp.
		Dicrotendipes simpsoni
		Glyptotendipes barbipes
		Parachironomus hirtalatus
		Polypedilum (Pentapedilum) tritum[a]
		Polypedilum (P.) fallax group
		Polypedilum (P.) illinoense
Limpets	Gastropoda:	*Ferrissia* spp.
Pond snails		*Physella* spp.

[a] New listing not included in original table from Ohio EPA (1987b).

a higher value. This adjustment is needed since a low number of organisms renders the proportional relationships between macroinvertebrate groups relatively meaningless.

3.3.9 Metric 9. Percent Tolerant Organisms

Values for this metric are generated using a predetermined list of organisms compiled by the Ohio EPA (1987b) and reproduced in Table 3. The list includes those organisms in Ohio that have consistently been observed to be extremely tolerant of a broad range of impacts and which tend to predominate artificial substrate collections from areas with severe perturbation. The list includes organisms tolerant to organic pollution as well as some taxa observed to withstand toxic impacts (Hynes 1966; Hart and Fuller 1974; Simpson and Bode 1980; Burch 1982). Thus, this should be a reasonable metric for the evaluation of community tolerance over a broad range of degradation types. This is a preferable difference over other measurements of community tolerance that have been developed primarily to reflect one type of pollution impact. Like Metric 8, this is a negative metric and, as such, a low number (<50 individuals) or an absence of organisms in a sample defaults to a zero score for the metric regardless of the presence or absence of the specified tolerant taxa. Figure 9 depicts the reference site tolerant organism percentages vs. drainage area. A strong inverse relationship with watershed size exists. For drainages greater than 1000 mi² (2600 km²), the percentage of tolerant organisms found at reference sites becomes so low that the scoring categories are quite restrictive. In fact, at a number of the reference sites, none or fewer than 1% of these organisms were present. However, as with Metric 8, watershed size tends to have little effect when pollutional disturbances are prevalent. Sites with minor to severe degradation can have large populations of these organisms regardless of stream size.

3.3.10 Metric 10. Number of Qualitative EPT Taxa

This is the sole ICI metric that is generated by the qualitative sample taken in conjunction with the artificial substrate sampling. Since the qualitative sampling utilizes a substrate and habitat dependent method, that is, a method affected by the variation in natural substrates and habitats available in the sampling area, the metric is a measurement of both habitat quality and diversity as well as water quality. The metric consists of the taxa richness of Ephemeroptera (mayflies), Plecoptera (stoneflies), and Trichoptera (caddisflies) and is commonly used as an index component (Lenat 1988; Plafkin et. al 1989; Resh, Chapter 12; Southerland and Stribling, Chapter 7). Since stoneflies are relatively uncommon in summer collections in Ohio, the metric is mostly dependent on the kinds of mayflies and caddisflies found. The depiction of qualitative EPT taxa vs. drainage area (Figure 10) reflects a trend similar to Metrics 2 and 5, total taxa richness and percentage of mayflies. As with Metrics 1 and 2, the higher numbers of EPT taxa occur in the midsized streams and rivers, a trend predicted by the RCC, and result from greater habitat and food type variety in the systems transitional between small streams and large rivers.

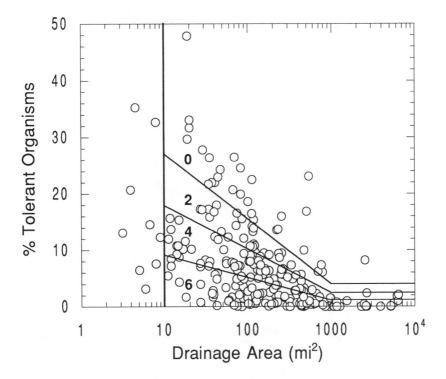

Figure 9. The relationship of ICI Metric 9, Percent Tolerant Organisms, with the log transformation of drainage area at 246 Ohio reference sites. (An inverse relationship exists with drainage areas >1000 mi^2.)

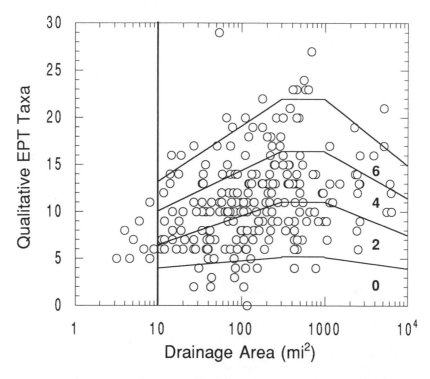

Figure 10. The relationship of ICI Metric 10, Qualitative EPT Taxa, with the log transformation of drainage area at 246 Ohio reference sites. (A direct relationship exists with drainage areas <300 mi^2; an inverse relationship exists with drainage areas >1000 mi^2.)

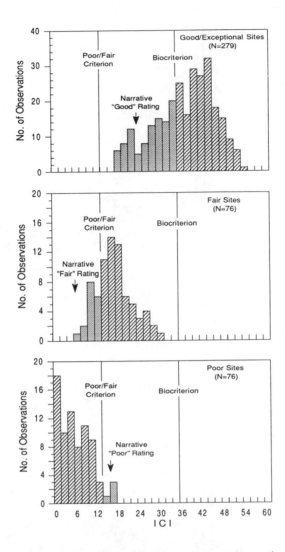

Figure 11. Frequency distribution of ICI scores for 431 sites rated as good/exceptional, fair, and poor using the original Ohio EPA narrative assessment protocol. The dotted bars are sites where narrative assessments rated differently than assessments using ICI scores and biocriteria derived from a regional reference site approach; slashed bars are sites where the two assessments were in agreement.

4.0 ICI VALIDATION AND TESTING

4.1 Comparison with Original Narrative Protocols

In an effort to determine the effect of using this more definitive, less subjective assessment methodology, evaluations using the original narrative protocols (i.e., exceptional, good, fair, or poor) from 431 sites sampled between 1981 and 1987 were compared to ICI-based biocriteria calibrated using regional reference sites. The results indicated that the original narrative approach rated a significant number of sites as being better than indicated by the calibrated ICI (Figure 11). The narrative approach rated as "good" (attaining the designated aquatic life use) 36% of sites classified by the ICI as impaired (less than the applicable ecoregional criterion and, thereby, not meeting the designated use). Additionally, 22% of the sites classified by the ICI as severely impaired (the numeric equivalent of poor) were placed in the "fair" category using the narrative approach. Conversely, only 5% of sites rated as "poor" by the narrative method were classified higher by the ICI and no ICI scores exceeded the ecoregional biocriteria at sites narratively rated as "fair." The primarily unidirectional error orientation of the narrative approach resulted in the rating of sites better than they should be when judged against the ICI score. In many cases, the tendency of the investigator was to give a stream location the benefit of the doubt and judge it acceptable when it may not have been. This is not a problem when using the more objective ICI assessment process. While it may seem premature to assume that the ICI is more accurate, the fact that it is a multimetric assessment mechanism designed to produce the essence of the narrative system, but with greater objectivity and precision, and that it extracts information directly from the regional reference sites, supports the contention that the ICI is the preferable assessment tool.

4.2 Variability Analyses

It is of critical importance in biological monitoring to collect a consistent and reproducible sample. Variation in data can be divided into sampling variation (i.e., errors in methods and techniques) and natural variation, both between sites and at a given site over time (i.e., temporal variability). Sampling error should be minimized to detect true spatial differences between sampling sites or at a site over time. Data from a special Ohio EPA methods study conducted in 1981 were used to estimate the ramifications of both natural variability at a sampling site and the inherent error of the macroinvertebrate sampling techniques and methodologies. Temporal variability of the ICI at a given site was assessed by analyzing sites in the Ohio EPA macroinvertebrate database having multiple year information. Within year variation (i.e., seasonality) of the ICI at a given location has not been thoroughly assessed; however, this should not be a significant consideration when evaluating community quality since macroinvertebrate sampling by the Ohio EPA is confined to a specified index period. Ohio EPA protocols restrict sampling to a mid-June through September index sampling period with all artificial substrate samples, in actuality, being retrieved over a six-week period from mid-August to the end of September. Thus, seasonal differences in macroinvertebrate communities at a given location over this relatively brief time frame should be minimal and not significantly affect ICI scoring as a consequence.

4.2.1 Site Variability Analysis

The 1981 study was conducted at a representative site in Big Darby Creek, a medium sized stream located in central Ohio. Big Darby Creek is a documented high quality aquatic resource and is populated by a very diverse macroinvertebrate fauna (Ohio EPA 1983). It was thought the potential for variation attributed to sampling error under these conditions would be significant and, thus, be a good test of the reliability of the ICI. Since external impacts are minimal, measured variability would be most likely due to sampling inconsistencies. Twenty-two artificial substrate sampling units were arranged in an X-shaped grid and colonized under similar conditions with regard to current velocity, water depth, and riparian canopy. Nineteen of the units were subsequently retrieved and analyzed; three units were not used in subsequent analyses because of differences in current velocity at the specific locations where these substrates were retrieved.

Initial examination of the data indicated that measured physical parameters (depth and current velocity) were relatively constant and should have had no significant effect on various biological parameters (e.g., total taxa, etc.). Similar results were found when the physical factors were compared to calculated ICI values. It seemed appropriate to assume that the same water quality conditions were affecting all the sampling units; thus, it was inferred that any variability in ICI scores was due to sampling error related to methodologies and/or natural biological processes such as predation, emigration, immigration, mortality, and natality. ICI summary statistics generated from the test data are presented in Table 4; the frequency distribution of ICI scores is depicted in Figure 12. ICI scores were reasonably consistent among the nineteen samples. The median ICI value was 36 and scores among the 19 samples ranged from 30 to 44. Though this appears to be a considerable range of ICI values, the 25th and 75th percentile scores were 36 and 38, respectively. Fourteen of nineteen (75%) and seventeen of nineteen (90%) scores were within plus or minus two and four points of the median, respectively. As such, it was determined that changes in ICI scores at test sites compared to an ecoregional biocriterion or to an upstream control station

Table 4.   Invertebrate Community Index (ICI)
            Summary Statistics Generated from
            Macroinvertebrate Data Collected at
            the 1981 Big Darby Creek Test Site

Sample size	19
Mean ICI	36.6
Standard error	0.7
Median ICI	36
Minimum ICI	30
Maximum ICI	44
ICI quartiles	
Lower (25%)	36
Upper (75%)	38

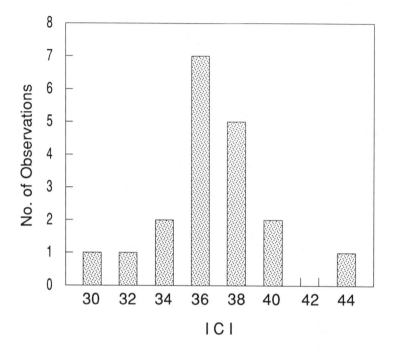

Figure 12. Frequency distribution of Invertebrate Community Index (ICI) scores derived from 19 replicate
samples collected at the 1981 Big Darby Creek test site.

should be considered in a zone of insignificant departure if the ICI difference is four points or less. Based
on the test data, this interval should adequately allow for the potential effects of natural variation and
sampling error on the ICI value.

4.2.2 Temporal Variability Analysis

Because of the inability to predict with absolute certainty that a given site has not changed in quality
over a period of years, it was difficult to assess the potential effect of natural, year-to-year variability on
ICI scoring. However, there are sampling locations in the Ohio EPA database where multiple-year data
are available and where little or no change in resource quality is believed to have occurred. ICI scores
from three such sites are depicted in Figure 13. Out of necessity, these sites needed to be high quality and
located on streams well removed from major anthropogenic pollution sources. In general, these sites
varied little from year to year. For the most part, ICI scores were consistent and differed by no more than
6 points over the sampling intervals which spanned 7 to 8 years and included 3 to 6 sampling events. Most
importantly, however, scores always exceeded the applicable biocriterion and the evaluation of aquatic

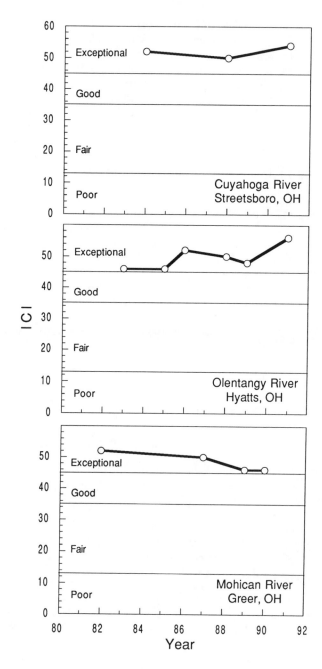

Figure 13. Invertebrate Community Index (ICI) scores from multiple-year sampling conducted at three high-quality Ohio stream sites.

life use attainment status did not change. A wide range of natural environmental conditions existed over the years at each location and included exceptionally high flow years as well as drought years with minimal stream flows and the potential for maximum biological stress. It was concluded that ICI scores are consistent at locations where little man-induced change has occurred. Thus, macroinvertebrate community condition, as reflected by the ICI, reacts minimally to natural biological processes that might otherwise be perceived as important factors which could potentially affect scoring and thereby the community assessment at a sampling site.

It is interesting to note that the Ohio EPA database also includes sites sampled over multiple years in areas of lesser quality and impacted by human activities. If degradation is not severe at these borderline quality sites, it is not unusual to see moderate fluctuations of ICI scores from year to year. However, far from indicating a "noise" problem with the ICI, this type of temporal variability can be of considerable importance in the diagnosis of pollution impacts. Oftentimes, these sites are located on stream reaches with significant nonpoint source problems or in urban/industrial influenced reaches where stream flows are nearly effluent dominated. In these cases, differing stream flow years and the corresponding effect on nonpoint source pollutant loadings and point source effluent dilution are apparently influencing the quality of the macroinvertebrate community that becomes established in a given year.

4.3 U.S. Environmental Protection Agency Evaluation

An independent evaluation of the ICI conducted by the U.S. Environmental Protection Agency (Davis and Lubin 1989) determined that the ICI, along with its associated metrics, seems to be a valid empirical indicator of macroinvertebrate community quality and is quite acceptable for its stated use. Various Statistical Analysis System (SAS Institute, Inc. 1985) procedures were use to test a number of aspects of the ICI including: (1) an evaluation of the reasonableness of the use and derivation of the invertebrate community measurements used to establish the ten metrics, (2) a determination if the drainage area relationships visually interpreted for the ten metrics were reasonable, (3) a determination if any of the ten ICI metrics are interrelated and, thereby, provide redundant information, (4) a determination if the assumption of equal weights for each metric was optimal, and (5) an evaluation of the overall accuracy of the ICI. The authors' analyses revealed no substantial faults or unnecessary redundancy in these various aspects of the ICI, and they concluded that there were no obvious changes which would significantly improve upon it.

5.0 APPLICATIONS OF THE ICI IN WATER RESOURCE ASSESSMENTS

The Ohio EPA has operated a program of intensive biological and water quality surveys of Ohio rivers and streams since 1979. In a 14-year period, over 550 different rivers and streams covering nearly 8000 miles have been assessed statewide. More than 90% of the principal pollution problem areas have been surveyed at least once. The Ohio EPA employs a multidisciplinary approach to the chemical, physical, and biological monitoring and assessment of surface waters. Biological evaluation methods include the use of the Index of Biotic Integrity (IBI) and Modified Index of Well-Being (MIwb) for fish, and the Invertebrate Community Index (ICI) for macroinvertebrates (Ohio EPA 1987b; Yoder 1989). Reference site-calibrated ecoregional biocriteria for all three indices were adopted into the Ohio Water Quality Standards and became effective in May, 1990. The rationale and procedures used in the development of Ohio's biocriteria have been reported elsewhere (Ohio EPA 1987a, 1987b, 1989a, 1989b). Besides biological monitoring, the Ohio EPA survey design also includes an assessment of physical habitat and the more traditional chemical/physical analyses of the water column and effluents. Additionally, assessments may include monitoring of toxic substances in the water column, effluents, fish tissue, and sediments as necessary. Together these data are used to support Ohio EPA program areas such as Water Quality Standards, NPDES permits, basinwide planning activities, natural resource damage assessments, and nonpoint source assessments. In 1990, the Ohio EPA initiated a five-year rotating basin approach to ambient biological and water quality monitoring and NPDES permitting. This cyclic and orderly approach to the assessment of Ohio's major watersheds not only makes the utilization of limited monitoring resources more cost effective, but assures that monitoring information will be available to support program areas when needed. Yoder (1991b) and Yoder and Rankin (Chapter 9) provide detailed discussions of the Ohio EPA's biological and water quality survey program and biocriteria applications.

Assessments of the ambient macroinvertebrate community using the ICI are primarily of two basic types: (1) intensive surveys of stream or river reaches using multiple sites in upstream to downstream longitudinal or synoptic subbasin configurations, and (2) multiple-year sampling at a specific fixed station on a stream or river. Intensive surveys are the basic design used in Ohio EPA's annual sampling program to address issues from a mainstem, subbasin, or basinwide perspective. Sampling sites are located based on the peculiarities of the stream or river and in accordance with the survey objectives. Assessments of

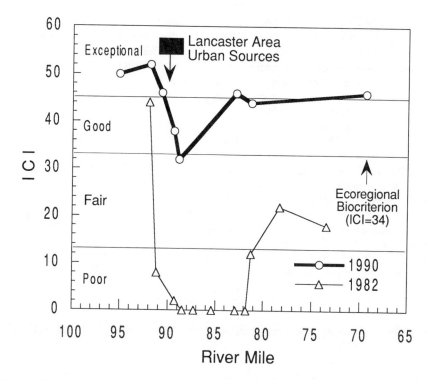

Figure 14. Longitudinal patterns and temporal trends in the Invertebrate Community Index (ICI) based on macroinvertebrate community data collected from the Hocking River within and downstream from the city of Lancaster, Ohio, 1982 and 1990.

point sources of pollution include upstream control(s) and downstream impact/recovery stations including mixing zone analyses to detect the potential for acutely toxic or rapidly lethal conditions. Additional sites are typically located within the study area to evaluate natural background conditions, to resample regional reference sites (once every ten years), or to monitor other issues such as pollution runoff from agriculture, mine drainage, urban, and construction site sources. With the advent of the five-year cycle to watershed monitoring, opportunities are arising to revisit previously sampled watersheds and to assess temporal changes in aquatic resource condition after major monetary expenditures for improvements in point source water pollution control. A similar opportunity exists over the long-term to evaluate the benefits of current efforts to implement best management practices and control nonpoint sources of pollution. In conjunction with the intensive biological surveys, macroinvertebrate sampling has also been conducted over multiple years since the early 1970s at over 30 key locations in major Ohio watersheds. Many fixed stations have had 10 or more annual samples taken over the intervening 18 years and significant long-term, mostly positive trends in ICI scores are being observed at some locations.

Results from these two distinctly different types of macroinvertebrate community monitoring are beginning to provide a uniquely comprehensive and standardized database from which changes in the quality of Ohio's water resources can be quantitatively evaluated over time. Of relevance to regulatory programs and others is the feedback provided about the success of efforts to control, reduce, and eliminate surface water pollution problems. The results of these types of analyses do more than simply answer the question of whether or not a waterbody is performing up to expectations with regards to biological integrity. Information is also derived that quantifies the degree of biological change, whether it be positive or negative, and, in the latter case, what amount of additional biological improvement will be necessary to achieve acceptable biological condition.

5.1 Use of the ICI in Intensive Surveys

An example of the use of macroinvertebrate community data in an intensive survey, including an assessment of temporal trends, is graphically presented in Figure 14. The subject waterbody is the

Hocking River, an Ohio River tributary originating in south-central Ohio, which was first surveyed and assessed in 1982 (Ohio EPA 1985a) and then again in 1990 (Ohio EPA 1991). Both surveys were focused on the river in the vicinity of the small city of Lancaster, Ohio, which for many years severely degraded the river. Inadequate wastewater treatment and raw sewage bypassing at the municipal treatment facility, combined wet and dry weather sewer overflows, and a heavy contribution of industrial effluents to the sewage collection system resulted in gross enrichment and heavy metals contamination, significant levels of instream toxicity, and periodic fish kills.

In 1982, the macroinvertebrate community was severely degraded in the vicinity of, and downstream from, Lancaster for over 20 miles (32 km). The impact of the various sources was dramatic and immediate. The ICI score decreased by 36 points between the very good quality upstream background sampling location and the first impacted site even though the sites were less than 1 mi (1.6 km) apart. The worst conditions were found in a 6 mi (10 km) stretch below the municipal treatment facility where ICI values of 0 were scored at five locations. Communities at these sites were almost exclusively composed of large numbers of tubificid worms populating extensive beds of sewage sludge throughout the river. Recovery in the macroinvertebrate community was incomplete at the farthest downstream site of the study area where only limited improvement was observed in the macroinvertebrate community.

Between the 1982 and 1990 surveys, numerous construction and operational improvements were instituted within the city's sewage system, which resulted in much better quality pretreatment of industrial effluents, higher-quality wastewater treatment plant effluent, and the virtual elimination of sewage bypass events. In 1990, macroinvertebrate communities, as assessed by ICI scores, reflected the vastly improved water quality conditions in the Hocking River as a result of these changes. Though there was still degradation in close proximity to the urban area and the municipal treatment facility, the differences between the 1982 and 1990 communities were visibly apparent. Beds of sewage sludge and tubificid worms were essentially absent and were replaced by much more diverse invertebrate populations including good numbers of mayfly and caddisfly taxa at sites where only worms had been found previously. Communities comparable to those found at upstream sites were collected only 6 mi (10 km) downstream and communities in between were not nearly as degraded as those in 1982.

Fish community assessments at comparable sites in the Hocking River reflected similar improvements between 1982 and 1990. However, changes were not nearly as dramatic as those observed in the macroinvertebrate community primarily due to a lagging and as yet incomplete recolonization process. However, complete community recovery to a level approaching that observed at reference sites appears further limited by the prevalence of severe macrohabitat degradation. Mainstem habitat in Lancaster remains in a state of recovery from prior channelization. Bank erosion is significant downstream from Lancaster where adjacent land use has encroached on the riparian zone. Although pockets of quality riffle, pool, and riparian habitat exists, much of the stream channel and riparian zone has suboptimal macrohabitat conditions, which will probably hamper the reestablishment of diverse populations of many fish species. The macroinvertebrate results, based on the artificial and natural substrate collections, indicate a strong positive response to the improved water quality and the availability of the few oases of quality macrohabitat. The net result is that continued fish community degradation precludes the full attainment of the designated beneficial aquatic life use of the Hocking River. The differential sensitivity and response shown by each organism group is advantageous when attempting to discern the effects of water quality problems in streams that also have serious habitat degradation. The use of this type of complementary biological data in assessments of streams and rivers in Ohio in conjunction with associated chemical and physical analyses has resulted in higher quality and more accurate information on which to base management decisions affecting the resource. Rankin (Chapter 13) provides a detailed discussion concerning the importance of macrohabitat integrity and its influence on the quality of Ohio rivers and streams.

5.2 Use of the ICI in Multiple-Year Fixed Station Sampling

Use of macroinvertebrate community data in multiple-year fixed station sampling are graphically presented in Figure 15. Two locations, the Black River at Elyria, Ohio, and the Cuyahoga River at Independence, Ohio, have been extensively monitored from the mid-1970s to the present. Over this interval, macroinvertebrate communities have been sampled nine times in the Black River and twelve times in the Cuyahoga River. In both cases, significant positive changes have occurred in the communities that are correlated with improvements in municipal wastewater treatment in each watershed. These rivers

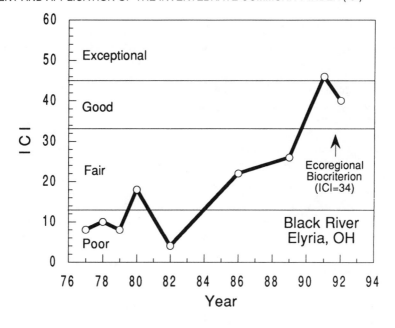

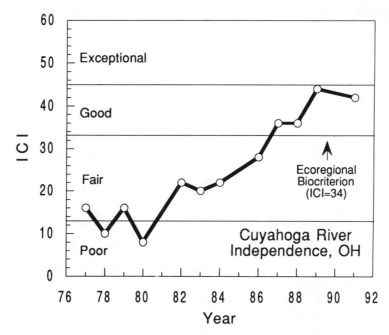

Figure 15. Multiple-year trends of the Invertebrate Community Index (ICI) based on macroinvertebrate community data collected from fixed stations located on the Black River at Elyria, Ohio, and the Cuyahoga River at Independence, Ohio, 1977–1992.

continue to be extensively investigated as both have been identified as Great Lakes Areas of Concern and have Remedial Action Plans in various stages of development.

5.2.1 Black River

The Black River is a major Lake Erie tributary originating in northeastern Ohio in the Erie–Ontario Lake Plain ecoregion. The long-term fixed station is located on the mainstem approximately 1 mi (1.6 km) downstream from the confluence of the East and West Branches and upstream from the Elyria wastewater treatment plant. This area of the river is located within the city of Elyria and, until recently, was impacted by a variety of sources including combined sewer overflows, industrial dischargers, on-site septic systems, and urban runoff. Sampling at the fixed station in the late 1970s and early 1980s resulted in low ICI scores indicating poor community performance. A biological and water quality survey of the Black River conducted in 1982 documented serious degradation throughout the mainstem (Ohio EPA 1985b). The predominant causes of the biological impairment were determined to be low instream dissolved oxygen levels and gross nutrient enrichment resulting from combined sewer releases and industrial discharges.

Since the 1982 survey and especially since 1986, significant improvements in the macroinvertebrate community at the fixed station have resulted in ICI scores achieving the biocriterion established for the Erie–Ontario Lake Plain ecoregion (Figure 15). Biological changes are correlated with improved combined sewer overflow controls and the elimination of industrial discharges due to plant closure or tie-in to the recently upgraded Elyria wastewater treatment plant. The 1988 plant expansion, coupled with the extension of a major interceptor sewer line, has helped to decrease the incidence of combined sewer discharges to the Black River within the Elyria wastewater collection system.

5.2.2 Cuyahoga River

The long-term fixed station on the Cuyahoga River is located at Independence, a small northeastern Ohio community located in the southern Cleveland metropolitan area and at the northern end of the Cuyahoga Valley National Recreation Area. The Cuyahoga River, like the Black River, is in the Erie–Ontario Lake Plain ecoregion of Ohio and is a major Lake Erie tributary. The basin is unique in that its headwaters begin near Lake Erie; the mainstem then flows south for about 50 mi (80 km) before flowing north near the City of Akron. Just north of Akron and at the southern edge of the National Recreation Area, the Akron wastewater treatment facility discharges to the river, essentially doubling river volume at baseflow conditions. The fixed station is located approximately 20 mi (32 km) downstream from the wastewater treatment plant; there are no other significant point sources in between.

Historically, the macroinvertebrate community was severely degraded at this site. From the late 1970s through the early 1980s, ICI scores were in the poor/fair performance category (Figure 15). An intensive survey in 1984 revealed biological impairment that extended from the Akron area all the way to Cleveland. The response patterns of both the fish and macroinvertebrate communities, a lack of observed exceedences of conventional chemical water quality criteria, and the fact that degradation extended for over 20 mi (32 km) downstream strongly suggested a persistent toxic impact. Bioassays of the Akron wastewater treatment plant effluent as late as 1985 confirmed the presence of acute toxicity. Additionally, combined and sanitary sewer releases were a problem upstream from the treatment facility at many overflow locations in the Akron urban area.

In 1986, acute toxicity in the treatment plant effluent essentially disappeared; chronic toxicity was not measured. Although the principal reason for this was not precisely known, a number of possibilities exist including the removal of several industrial sources to the collection system due to plant closures, additional pretreatment requirements of remaining industrial inputs to the plant, more stringent regulation of illegal "drop-in" dischargers, and major upgrades at the plant, both on-line and under construction. The latter upgrades resulted in decreased loadings of ammonia, suspended solids, biochemical oxygen demand, and, since 1988, heavy metals. Plant bypassing of raw or partially treated sewage has also decreased. As can be seen in the trend at the fixed station, a significant improvement has occurred in the quality of the macroinvertebrate community as reflected by the ICI (Figure 15). Beginning in 1987 and continuing to the present, annual sampling has documented attainment of the ICI ecoregional biocriterion at this site. Intensive surveys in the basin as recently as 1991 have documented substantial biological improvements in the entire reach of the river between Akron and Cleveland (Ohio EPA 1992d). Although the severity of degradation has decreased, biological improvement has been limited in closer proximity to the wastewater treatment plant and urban area. There remain serious problems with discharges of untreated combined and sanitary sewer overflows and frequent bypassing of partially treated secondary flows at the plant.

6.0 FUTURE CONSIDERATIONS FOR ICI DEVELOPMENT AND APPLICATION

6.1 Reference Site Resampling

Calibration of ICI metric scoring categories are based on the prevailing background conditions at 246 least impacted reference sites that were sampled across Ohio from 1981 to 1986. This follows the guidance of Hughes et al. (1986) and recognizes that attainable biological community structure and function in aquatic systems is influenced by widespread activities such as intensive land surface uses (e.g., row-crop agriculture, surface mining), natural stream channel alterations (e.g., channelization), human settlement, road and highway construction, and general land surface conversion (e.g., deforestation), all to suit socioeconomic desires. The use of least impacted reference sites is not intended to represent pristine, wilderness, or pre-Columbian conditions in Ohio but recognizes that the aforementioned factors have collectively or individually influenced the ability of watersheds to support a certain level of biological performance. This does not mean that the impacts from these land use activities are necessarily acceptable; however, metric scoring reflects what is reasonably attainable in Ohio rivers and streams. The calibration of metric scoring categories can and will change (i.e., become more stringent) if it becomes apparent that the impacts of these pervasive influences have lessened through improved nonpoint source control programs or other means.

To determine if the background reference condition has changed significantly, resampling of reference sites has become an Ohio EPA monitoring program priority. Between 1988 and 1993, over 100 macroinvertebrate reference sites were resampled. It is anticipated that enough sites will have been resampled by the late 1990s (based on a goal of resampling 10% of the reference sites each year) to reexamine the ten ICI metrics and determine if a scoring recalibration is in order. At the same time, it can be determined if refinements and advancements in macroinvertebrate taxonomy have been sufficient enough to warrant further adjustments to ICI scoring categories or the metrics themselves. Modifications of the metrics or scoring categories and any subsequent changes in ecoregional expectations will be subject to the requirements of the Ohio Water Quality Standards rule-making process and final approval by the U.S. Environmental Protection Agency.

6.2 Qualitative Community Tolerance Values

A new Ohio EPA assessment tool currently in use and a direct offshoot of the ICI is the Qualitative Community Tolerance Value (QCTV). The QCTV is envisioned as having application when a quick turnaround is needed to problem assessment or when a screening-level, less definitive technique is desired in lieu of the more complex ICI process. As such, the method utilizes the qualitative, natural substrate collection procedure, which necessitates one site visit and minimal laboratory analysis. The assessment of the qualitative data relies on taxon tolerance values that have been established for many macroinvertebrates collected in Ohio. The tolerance value of a given taxon was derived using numerical abundance data from sites around Ohio where that taxon had been collected with artificial substrates. To determine the tolerance value, ICI scores at all locations where the taxon was collected were weighted by the abundance data of that taxon. The mean of the weighted ICI scores for the taxon is its tolerance value. Thus, the tolerance value represents the relative level of tolerance of a particular macroinvertebrate taxon in terms of the 0 to 60 scale of the ICI. The most pollution-intolerant taxa, which tend to reach greatest abundance at undisturbed sites (i.e., sites with highest ICI scores), yielded high tolerance values. Conversely, the most pollution-tolerant taxa, attaining greatest abundance at highly disturbed sites with low ICI scores, resulted in lower tolerance values.

At a sampling location where only qualitative, natural substrate data has been collected, the QCTV score is determined using only those taxa for which QCTV values have been calculated. This provides a link to the ICI that cannot be calculated for natural substrate data. The QCTV is most commonly expressed as the median of the available tolerance values but can also be expressed at other frequency intervals such as the 25th and 75th percentiles. Thus examining the community performance at different percentiles of the QCTV increases the dimension of the analysis. This was recently demonstrated in an assessment of small urban and suburban streams in the greater Cincinnati, Ohio area (Ohio EPA 1992).

Currently under evaluation are interim performance criteria for the median QCTV determined by correlating QCTV scores with sites achieving and not achieving ecoregional expectations as determined by ICI biocriteria. Though still subject to revision, the QCTV procedure has shown promising potential as an additional Ohio EPA assessment tool that can be used under specified circumstances. It is currently being used to evaluate qualitative data in conjunction with other more traditional sample attributes such as total and EPT richness and overall community composition and balance. A list of tolerance values (i.e., average weighted ICIs) derived for commonly collected Ohio macroinvertebrate taxa is provided (Table 5).

ACKNOWLEDGMENTS

The author wishes to recognize and thank his fellow "bug pickers" at the Ohio EPA — Jack Freda, Mike Bolton, Chuck McKnight, Bernie Counts, and Marty Knapp. Their long years of dedication and effort contributed significantly to the success of the program. This work is dedicated to the memory of J. Pat Abrams whose years of service laid the foundation for our efforts.

Table 5. Tolerance Values for 314 Common Macroinvertebrate Taxa Derived Using the Invertebrate Community Index (ICI) Weighted by Abundance Data and Averaged Over All Sites (N ≥ 5) where Collected with Modified Hester-Dendy Multiple-Plate Artificial Substrate Samplers

Taxon	Tolerance value	N	Taxon	Tolerance value	N
Coelenterata			Orconectes (*Crokerinus*)	35.8	6
Cordylophora lacustris	22.5	24	propinquus		
Hydra sp.	33.5	707	Orconectes (*C.*) *sanbornii*	32.2	54
Platyhelminthes			sanbornii		
Turbellaria	23.0	805	Orconectes (*Procericambarus*)	32.3	114
Nematomorpha			rusticus		
Unidentified	22.0	6	Arachnoidea		
Bryozoa			Hydracarina	38.1	437
Fredericella sp.	45.2	5	Insecta		
Fredericella indica	40.9	6	Ephemeroptera		
Hyalinella punctata	43.3	20	Siphlonuridae		
Lophopodella carteri	42.6	11	*Isonychia* sp.	48.7	617
Paludicella articulata	41.7	29	Baetidae		
Pectinatella magnifica	25.2	6	*Acerpenna macdunnoughi*	53.1	9
Plumatella sp.	37.4	591	*Acerpenna pygmaeus*	46.3	66
Entoprocta			*Baetis* sp.	41.1	928
Urnatella gracilis	42.6	162	*Baetis armillatus*	50.9	5
Annelida			*Baetis flavistriga*	42.6	105
Oligochaeta			*Baetis intercalaris*	43.8	165
Unidentified	11.9	1289	*Callibaetis* sp.	26.1	27
Hirudinea			*Centroptilum* sp.	43.5	98
Dina sp.	28.6	8	*Cloeon* sp.	38.7	137
Erpobdella punctata punctata	8.8	28	*Diphetor hageni*	46.0	16
Helobdella stagnalis	10.8	44	*Pseudocloeon* sp.	49.5	55
Helobdella triserialis	12.6	60	Heptageniidae		
Mooreobdella fervida	8.0	11	*Heptagenia diabasia*	48.3	5
Mooreobdella microstoma	14.3	66	*Leucrocuta* sp.	45.8	246
Placobdella ornata	27.1	9	*Leucrocuta hebe*	43.8	18
Arthropoda			*Leucrocuta maculipennis*	47.5	15
Crustacea			*Nixe perfida*	29.2	9
Isopoda			*Stenacron* sp.	41.1	1019
Caecidotea sp.	18.8	231	*Stenonema exiguum*	48.4	204
Lirceus sp.	35.1	136	*Stenonema femoratum*	41.5	450
Amphipoda			*Stenonema mediopunctatum*	48.7	99
Crangonyx sp.	20.2	151	*Stenonema mexicanum integrum*	44.5	307
Gammarus sp.	15.8	10	*Stenonema pulchellum*	46.3	484
Gammarus fasciatus	18.7	45	*Stenonema terminatum*	46.4	455
Hyalella azteca	26.2	129	*Stenonema vicarium*	46.2	193
Decapoda			Leptophlebiidae		
Orconectes sp.	30.9	58	*Choroterpes* sp.	35.9	8

Table 5 (continued). Tolerance Values for 314 Common Macroinvertebrate Taxa Derived Using the Invertebrate Community Index (ICI) Weighted by Abundance Data and Averaged Over All Sites (N ≥ 5) where Collected with Modified Hester-Dendy Multiple-Plate Artificial Substrate Samplers

Taxon	Tolerance value	N	Taxon	Tolerance value	N
Paraleptophlebia sp.	43.4	225	*Polycentropus* sp.	39.7	144
Ephemerellidae			Hydropsychidae		
Dannella simplex	52.9	5	*Cheumatopsyche* sp.	42.5	1197
Ephemerella sp.	42.5	9	*Hydropsyche* (*Ceratopsyche*)	49.1	19
Eurylophella sp.	47.5	45	*morosa*		
Serratella sp.	56.3	5	*Hydropsyche* (*C.*) *morosa* group	44.6	747
Serratella deficiens	47.4	52	*Hydropsyche* (*C.*) *slossonae*	48.0	62
Tricorythidae			*Hydropsyche* (*C.*) *sparna*	40.6	58
Tricorythodes sp.	43.7	690	*Hydropsyche* (*Hydropsyche*) *aerata*	44.5	27
Caenidae			*Hydropsyche* (*H.*) *bidens*	45.2	179
Caenis sp.	41.6	976	*Hydropsyche* (*H.*) *depravata* group	39.4	404
Potamanthidae			*Hydropsyche* (*H.*) *dicantha*	39.2	293
Anthopotamus sp.	40.6	52	*Hydropsyche* (*H.*) *frisoni*	48.9	54
Ephemeridae			*Hydropsyche* (*H.*) *orris*	44.2	215
Ephemera sp.	41.4	82	*Hydropsyche* (*H.*) *simulans*	45.3	304
Hexagenia sp.	38.1	11	*Hydropsyche* (*H.*) *valanis*	41.8	153
Hexagenia limbata	44.2	16	*Hydropsyche* (*H.*) *venularis*	42.3	52
Odonata			*Macrostemum zebratum*	49.2	64
Calopterygidae			*Potamyia flava*	45.2	223
Calopteryx sp.	37.9	217	Hydroptilidae		
Hetaerina sp.	43.0	116	*Agraylea* sp.	25.6	6
Coenagrionidae			*Dibusa angata*	44.5	15
Coenagrionidae	23.2	297	*Hydroptila* sp.	40.7	488
Argia sp.	29.2	728	*Neotrichia* sp.	38.3	9
Aeshnidae			*Oxyethira* sp.	28.5	7
Boyeria vinosa	38.7	64	Limnephilidae		
Gomphidae			*Pycnopsyche* sp.	40.0	32
Gomphus sp.	44.8	10	Helicopsychidae		
Plathemis lydia	10.5	5	*Helicopsyche borealis*	49.1	16
Plecoptera			Leptoceridae		
Pteronarcyidae			*Ceraclea* sp.	44.4	41
Pteronarcys sp.	46.7	7	*Ceraclea maculata*	38.4	19
Perlidae			*Nectopsyche* sp.	45.8	20
Acroneuria carolinensis	45.9	6	*Nectopsyche diarina*	41.0	9
Acroneuria evoluta	42.9	73	*Nectopsyche pavida*	47.1	15
Acroneuria internata	47.0	10	*Oecetis* sp.	42.2	71
Acroneuria lycorias	45.3	8	Lepidoptera		
Agnetina sp.	49.0	50	*Petrophila* sp.	43.6	92
Neoperla sp.	41.4	11	Coleoptera		
Paragnetina media	47.4	27	Gyrinidae		
Perlesta sp.	32.9	16	*Dineutus* sp.	38.0	39
Megaloptera			*Gyrinus* sp.	38.1	12
Sialidae			Haliplidae		
Sialis sp.	30.6	201	*Peltodytes* sp.	17.7	5
Corydalidae			Dytiscidae		
Corydalus cornutus	45.7	312	*Hydroporus* sp.	35.6	8
Nigronia fasciatus	49.0	5	*Laccophilus* sp.	12.3	8
Nigronia serricornis	38.5	52	Hydrophilidae		
Trichoptera			*Berosus* sp.	20.0	223
Philopotamidae			*Laccobius* sp.	48.5	6
Chimarra aterrima	48.5	19	Psephenidae		
Chimarra obscura	39.3	127	*Ectopria nervosa*	36.2	16
Psychomyiidae			*Psephenus herricki*	38.9	41
Lype diversa	43.1	73	Dryopidae		
Psychomyia flavida	48.5	16	*Helichus* sp.	40.7	111
Polycentropodidae			*Lutrochus laticeps*	42.7	9
Cyrnellus fraternus	36.1	179	Elmidae		
Neureclipsis sp.	43.8	99	*Ancyronyx variegata*	35.7	414
Nyctiophylax sp.	45.1	72	*Dubiraphia* sp.	31.5	226

Table 5 (continued). Tolerance Values for 314 Common Macroinvertebrate Taxa Derived Using the Invertebrate Community Index (ICI) Weighted by Abundance Data and Averaged Over All Sites (N ≥ 5) where Collected with Modified Hester-Dendy Multiple-Plate Artificial Substrate Samplers

Taxon	Tolerance value	N	Taxon	Tolerance value	N
Dubiraphia bivittata	41.1	16	*Corynoneura "celeripes"*	45.9	120
Dubiraphia quadrinotata	40.9	8	(sensu Simpson and Bode 1980)		
Dubiraphia vittata group	37.8	316	*Corynoneura lobata*	40.0	536
Macronychus glabratus	42.8	886	*Cricotopus (C.)* sp.	34.4	488
Optioservus sp.	38.3	18	*Cricotopus (C.) bicinctus*	19.7	569
Optioservus fastiditus	48.4	10	*Cricotopus (C.) tremulus* group	32.4	477
Stenelmis sp.	41.7	870	*Cricotopus (C.) trifascia* group	19.0	33
Diptera			*Cricotopus (C.) vierriensis*	34.6	16
Tipulidae			*Cricotopus (Isocladius)*	26.7	11
Antocha sp.	46.9	75	*intersectus* group		
Hexatoma sp.	45.9	15	*Cricotopus (I.) sylvestris* group	11.9	26
Tipula sp.	17.0	54	*Doncricotopus* prob. *bicaudatus*	30.9	11
Tipula abdominalis	36.1	23	*Limnophyes* sp.	41.3	5
Psychodidae			*Nanocladius (N.)* sp.	28.2	97
Pericoma sp.	26.6	8	*Nanocladius (N.) crassicornus*	41.9	6
Psychoda sp.	37.6	14	*Nanocladius (N.) crassicornus* or		
Simuliidae			*N. (N.) rectinervus*	34.1	31
Simulium sp.	31.3	574	*Nanocladius (N.) distinctus*	23.1	425
Ceratopogonidae			*Nanocladius (N.) minimus*	33.2	52
Ceratopogonidae	25.8	307	*Nanocladius (N.) spiniplenus*	38.7	71
Atrichopogon sp.	32.9	11	*Nanocladius (Plecopteracoluthus)*	48.6	11
Atrichopogon websteri	33.7	9	n. sp.		
Chironomidae			*Orthocladius* sp.	36.5	6
Tanypodinae			*Orthocladius (O.)* sp.	43.5	9
Ablabesmyia sp.	28.6	32	*Orthocladius (O.) carlatus*	48.0	5
Ablabesmyia annulata	35.1	5	*Parakiefferiella* sp.	31.0	28
Ablabesmyia janta	30.7	7	*Parakiefferiella* n. sp 1	39.5	56
Ablabesmyia mallochi	30.1	388	*Parakiefferiella* n. sp 2	35.6	46
Ablabesmyia rhamphe group	21.4	90	*Parametriocnemus* sp.	43.0	119
Clinotanypus pinguis	15.0	13	*Paratrichocladius* sp.	48.1	7
Conchapelopia sp.	34.3	500	*Psectrocladius* sp.	25.9	8
Hayesomyia senata or			*Rheocricotopus* sp.	30.4	6
Thienemannimyia norena	32.2	706	*Rheocricotopus (Psilocricotopus)*	37.1	325
Helopelopia sp.	36.8	285	*robacki*		
Labrundinia sp.	36.9	112	*Synorthocladius semivirens*	44.8	5
Labrundinia pilosella	41.5	123	*Thienemanniella* sp.	38.0	7
Larsia sp.	34.2	69	*Thienemanniella* n. sp 1	46.9	82
Meropelopia sp.	43.5	73	*Thienemanniella* n. sp 3	41.6	125
Natarsia sp.	24.6	20	*Thienemanniella similis*	45.6	43
Natarsia species A	11.6	37	*Thienemanniella xena*	38.2	418
(sensu Roback 1978)			*Tvetenia bavarica* group	46.4	48
Natarsia baltimoreus	26.5	5	*Tvetenia discoloripes* group	46.4	103
Nilotanypus fimbriatus	43.2	453	Chironominae		
Paramerina fragilis	31.7	6	Chironomini		
Pentaneura inconspicua	30.8	60	*Chironomus (C.)* sp.	11.4	98
Procladius sp.	20.9	160	*Chironomus (C.) decorus* group	12.8	230
Psectrotanypus sp.	11.5	10	*Chironomus (C.) riparius* group	5.5	77
Psectrotanypus dyari	8.2	8	*Cryptochironomus* sp.	30.7	177
Rheopelopia paramaculipennis	42.8	66	*Cryptotendipes* sp.	34.8	8
Tanypus sp.	7.3	7	*Dicrotendipes* sp.	26.6	53
Tanypus neopunctipennis	14.8	8	*Dicrotendipes modestus*	24.8	12
Telopelopia okoboji	36.1	40	*Dicrotendipes fumidus*	25.4	5
Thienemannimyia group	29.6	526	*Dicrotendipes neomodestus*	32.7	598
Zavrelimyia sp.	35.6	28	*Dicrotendipes lucifer*	21.9	202
Orthocladiinae			*Dicrotendipes simpsoni*	15.8	199
Brillia flavifrons group	33.2	78	*Endochironomus* sp.	26.2	5
Cardiocladius obscurus	46.6	28	*Endochironomus nigricans*	28.3	24
Corynoneura sp.	38.9	114	*Glyptotendipes (Phytotendipes)*	22.5	568
Corynoneura n. sp 1	46.2	31	sp.		

Table 5 (continued). Tolerance Values for 314 Common Macroinvertebrate Taxa Derived Using the Invertebrate Community Index (ICI) Weighted by Abundance Data and Averaged Over All Sites (N ≥ 5) where Collected with Modified Hester-Dendy Multiple-Plate Artificial Substrate Samplers

Taxon	Tolerance value	N	Taxon	Tolerance value	N
Glyptotendipes (*G.*) *amplus*	40.8	21	group 5		
Harnischia curtilamellata	38.8	7	*Micropsectra* sp.	38.9	57
Kiefferulus dux	28.7	11	*Paratanytarsus* sp.	34.9	377
Microtendipes "caelum"	43.7	70	*Paratanytarsus* n. sp. 1	43.6	31
(sensu Simpson and Bode 1980)			*Rheotanytarsus* sp.	39.9	41
Microtendipes pedellus group	39.5	289	*Rheotanytarsus distinctissimus*		
Microtendipes rydalensis	48.0	6	group	47.0	179
Nilothauma sp.	41.1	16	*Rheotanytarsus exiguus* group	43.8	1034
Parachironomus sp.	35.2	10	*Stempellinella* sp.	44.1	46
Parachironomus abortivus	11.9	48	*Stempellinella* n. sp. nr. *flavidula*	40.1	24
Parachironomus carinatus	28.0	39	*Sublettea coffmani*	47.0	45
Parachironomus directus	12.5	5	*Tanytarsus* sp.	38.9	298
Parachironomus frequens	36.9	98	*Tanytarsus* Type 1	43.6	7
Parachironomus pectinatellae	37.9	23	*Tanytarsus* Type 3	37.1	32
Paralauterborniella nigrohalteralis	36.5	36	*Tanytarsus curticornis* group	44.4	126
Paratendipes sp.	25.7	12	*Tanytarsus glabrescens* group	40.4	818
Paratendipes albimanus	34.0	236	*Tanytarsus guerlus* group	40.6	506
Phaenopsectra obediens group	33.1	310	Tabanidae		
Phaenopsectra punctipes	39.1	11	*Chrysops* sp.	32.6	16
Phaenopsectra flavipes	25.8	133	Athericidae		
Polypedilum (*Pentapedilum*)	13.4	6	*Atherix lantha*	41.3	74
tritum			Empididae		
Polypedilum (*Polypedilum*)	37.0	14	Empididae	39.2	1002
albicorne			Mollusca		
Polypedilum (*P.*) *aviceps*	48.4	14	Gastropoda		
Polypedilum (*P.*) *convictum*	38.6	972	Hydrobiidae		
Polypedilum (*P.*) *fallax* group	30.6	945	Hydrobiidae	17.7	49
Polypedilum (*P.*) *illinoense*	18.4	449	Pleuroceridae		
Polypedilum (*P.*) *ophioides*	45.3	7	*Elimia* sp.	38.1	248
Polypedilum (*P.*) *n.sp.*	48.4	8	*Pleurocera* sp.	37.8	6
same as *tuberculum*			Lymnaeidae		
(*Maschwitz*, 1976)			Lymnaeidae	18.4	5
Polypedilum (*P.*) *ontario*	44.4	12	*Fossaria* sp.	44.6	11
Polypedilum (*Tripodura*)	37.2	31	Physidae		
halterale group			*Physella* sp.	14.3	536
Polypedilum (*Tripodura*)	26.1	871	Planorbidae		
scalaenum group			*Gyraulus* (*Torquis*) *parvus*	27.0	12
Stenochironomus sp.	38.2	75	*Helisoma anceps anceps*	24.0	16
Stictochironomus sp.	36.1	36	*Menetus* (*Micromenetus*) *dilatatus*	11.2	32
Tribelos fuscicorne	29.3	67	*Planorbella* (*Pierosoma*) *pilsbryi*	15.4	16
Tribelos jucundum	26.5	79	*Planorbella* (*Pierosoma*) *trivolvis*	3.1	5
Pseudochironomini			Ancylidae		
Pseudochironomus sp.	31.9	35	*Ferrissia* sp.	28.7	772
Tanytarsini			*Laevapex fuscus*	19.7	13
Cladotanytarsus sp.	33.5	76	Pelecypoda		
Cladotanytarsus species	39.5	14	Corbiculidae		
group A			*Corbicula fluminea*	40.8	89
Cladotanytarsus mancus group	37.1	20	Sphaeriidae		
Cladotanytarsus vanderwulpi	46.2	32	*Pisidium* sp.	32.5	63
group 1			*Sphaerium* sp.	31.5	155
Cladotanytarsus vanderwulpi	48.4	12			

Application of the Index of Biotic Integrity to Evaluate Water Resource Integrity in Freshwater Ecosystems

Thomas P. Simon and John Lyons

1.0 INTRODUCTION

The biotic and abiotic processes involved in the environmental degradation of freshwater ecosystems are often complex, and their combined effects are not easily measured. Many human activities across the landscape contribute to degradation, including discharge of domestic, agricultural, and industrial effluents; cultural eutrophication; acidification; erosion and sedimentation following human modification of watersheds; straightening, deepening, and clearing of stream channels; drainage of wetlands and impoundments of streams; flow alterations caused by dam operation and water diversions; overharvest of biota; and introduction of nonnative species. Estimating ecosystem health and integrity may be the best way to assess the total effects of these activities on aquatic environments (Karr 1991).

Fish communities are a highly visible and sensitive component of freshwater ecosystems, and have several attributes that make them useful indicators of biological integrity and ecosystem health (Table 1). Fish communities respond predictably to changes in both abiotic factors, such as habitat and water quality (e.g., Karr 1981; Hughes 1985; Karr et al. 1986; Leonard and Orth 1986; Scott et al. 1986; Berkman and Rabeni 1987; Steedman 1988; Simon 1990), and biotic factors, such as human exploitation and species additions (e.g., Hartman 1972; Colby et al. 1987; Rincon et al. 1990; Ross 1991). Many studies have identified the specific responses of fish communities to particular types of degradation (e.g., Forbes and Richardson 1913; Hubbs 1933; Millet et al. 1966; Gammon 1976; Karr et al. 1985a, b; Hughes et al. 1990; ORSANCO 1991; Hite et al. 1992).

Beginning around 1900 and accelerating greatly in the last 20 years, fish community characteristics have been used to measure relative ecosystem health (Fausch et al. 1990). Recent advances are attributed to the development of integrative ecological indices that directly relate fish communities to other biotic and abiotic components of the ecosystem (Karr 1981; Karr et al. 1986; Ohio EPA 1987a,b; Plafkin et al. 1989; Barbour et al., Chapter 6), the delineation of ecoregions that allow explicit consideration of natural differences among fish communities from different geographic areas (Hughes et al. 1986, 1987; Omernik 1987; Hughes and Larsen 1988; Omernik, Chapter 5), and the recognition of the importance of cumulative effects of degradation at the landscape scale (Hughes, Chapter 4; Larsen et al., Chapter 18).

In the United States, biological criteria or standards based on fish communities have been formulated by several state and federal agencies to assess and protect freshwater ecosystem health (Ohio EPA 1987a,b; Karr 1991; Southerland and Stribling Chapter 7). A variety of quantitative indices can define specific biocriteria including indicator species or guilds; species richness, diversity, and similarity indices; the Index of Well-Being; multivariate ordination and classification; and the Index of Biotic

Table 1.   Attributes of Fishes that Make Them Desirable Components of Biological Assessments and Monitoring Programs

Goal/quality	Attributes
Accurate assessment of environmental health	Fish populations and individuals generally remain in the same area during summer seasons
	Communities are persistent and recover rapidly from natural disturbances
	Comparable results can be expected from an unperturbed site at various times
	Fish have large ranges and are less affected by natural microhabitat differences than smaller organisms. This makes fish extremely useful for assessing regional and macrohabitat differences
	Most fish species have long life spans (2–10+ years) and can reflect both long-term and current water resource quality
	Fish continually inhabit the receiving water and integrate the chemical, physical, and biological histories of the waters
	Fish represent a broad spectrum of community tolerances from very sensitive to highly tolerant and respond to chemical, physical, and biological degradation in characteristic response patterns
Visibility	Fish are highly visible and valuable components of the aquatic community to the public
	Aquatic life uses and regulatory language are generally characterized in terms of fish (i.e., fishable and swimmable goal of the Clean Water Act)
Ease of Use and Interpretation	The sampling frequency for trend assessment is less than for short-lived organisms
	Taxonomy of fishes is well established, enabling professional biologists the ability to reduce laboratory time by identifying many specimens in the field
	Distribution, life histories, and tolerances to environmental stresses of many species of North American fish are documented in the literature

Modified from Plafkin et al. 1989; Simon 1991.

Integrity (reviewed by Fausch et al. 1990). Of these, the most commonly used and arguably the most effective has been the Index of Biotic Integrity or IBI.

In this paper, we review the evolution and use of the IBI, with particular emphasis on recent applications. Karr et al. (1986), Miller et al. (1987), and Fausch et al. (1990) have reviewed the early development of the IBI, but since these publications, several major "new" versions have been developed. Many new versions are not documented in the primary, peer-reviewed literature, and thus are unknown to most water resource professionals, although these versions are being used to make important management decisions in many states. Our goal is to describe the different ways in which the original IBI has been modified for use in different geographic regions and in different types of freshwater ecosystems.

With many different versions now in existence, the IBI is best thought of as a family of related indices rather than a single index. The IBI includes attributes of the biota that range from individual health to population, community, and ecosystem levels. We define the IBI broadly, as any index that is based on the sum or ratings for several different measures, termed metrics, of fish structure and/or function, with the rating for each metric based on quantitative expectations of what comprises high biotic integrity. For some metrics, expectations will vary depending on ecosystem size and location. The IBI is not a community analysis but it is an analysis of several hierarchical levels of biology that uses a sample of the assemblage.

2.0 A BRIEF HISTORY OF THE INDEX OF BIOTIC INTEGRITY

The Index of Biotic Integrity (IBI) was first developed by Dr. James Karr for use in small warmwater streams (i.e., too warm to support salmonids) in central Illinois and Indiana (Karr 1981). The original version had 12 metrics that reflected fish species richness and composition, number and abundance of indicator species, trophic organization and function, reproductive behavior, fish abundance, and condition of individual fish (Table 2). Each metric received a score of five points if it had a value similar to that expected for a fish community characteristic of a system with little human influence, a score of one point if it had a value similar to that expected for a fish community that departs significantly from the reference condition, and a score of three points if it had an intermediate value. Sites with high biotic integrity had

Table 2.   List of Original Index of Biotic Integrity Metrics (Capital Letters) Proposed by Karr (1981) for Streams in the Central United States, Followed by Modifications Proposed by Subsequent Authors for Streams in Other Regions or for Different Streams and River Types

Species Richness and Composition Metrics

1. Total Number of Fish Species (1–7, 11, 13–15, 18, 19, 22)
   A. Number of native fish species (8–10, 12, 16, 17, 20, 22)
   B. Number of fish species, excluding Salmonidae (13)
   C. Number of amphibian species (3, 13)
2. Number of Catostomidae Species (1, 4–6, 8, 9, 12, 15, 17, 19, 20)
   A. Percent of individuals that are Catostomidae (9)
   B. Percent of individuals that are round-bodied Catostomidae: *Cycleptus*, *Hypentelium*, *Minytrema*, and *Moxostoma* (9, 17, 19)
   C. Number of Catostomidae and Ictaluridae species (10)
   D. Number of Catostomidae and Cyprinidae species (17)
   E. Number of benthic insectivorous species (7, 11, 17, 22)
   F. Number of laterally compressed minnow species (21)
   G. Number of minnow species (4, 6, 9, 14, 15, 17, 22)
   H. This metric deleted from IBI (2, 3, 13, 14)
3. Number of Darter Species (Percidae genera: *Crystallaria*, *Etheostoma*, *Percina*, *Ammocrypta*) (1, 2, 4, 5, 8, 9, 12, 15–17, 19, 22)
   A. Number of darter and Cottidae species (9, 10)
   B. Number of darter, Cottidae, and *Noturus* (Ictaluridae) species (15, 16, 19)
   C. Number of darter, Cottidae, and round-bodied Catostomidae species (17)
   D. Number of Cottidae species (6, 13)
   E. Abundance of Cottidae individuals (3)
   F. Number of benthic species (11, 18)
   G. Percent of individuals that are native benthic species (11)(same as #2)
   H. Number of benthic insectivorous species (7)
   I. Number of darter species, excluding "tolerant darter species" (headwater sites) (21)
   J. Percent cyprinids with subterminal mouths (22)
   K. This metric deleted from IBI (14)
4. Number of Sunfish Species (Centrarchidae excluding *Micropterus*)(1, 4, 5, 14–17, 20)
   A. Number of native sunfish species (9, 12)
   B. Number of sunfish and Salmonidae species (10)
   C. Number of sunfish species and *Perca flavescens* (Percidae) (16)
   D. Number of headwater (restricted to small streams) species (9, 15, 22)
   E. Number of water column (non-benthic) species (7, 17, 18)
   F. Number of water column cyprinid species (17)
   G. Number of sunfish species including *Micropterus* (19, 22)
   H. This metric deleted from IBI (11)

Indicator Species Metrics

5. Number of Intolerant or Sensitive Species (1–4, 6–8, 10–12, 14, 15, 17–20, 22)
   A. Number of Salmonidae species (3, 15)
   B. Percent of individuals that are Salmonidae (11)
   C. Juvenile Salmonidae presence or abundance (3, 18)
   D. Large (>15–20 cm) or adult Salmonidae presence or abundance (3, 6, 18)
   E. Abundance or biomass of all sizes of Salmonidae (3, 13)
   F. Mean length or weight of Salmonidae (13)
   G. Percent of individuals that are anadromous *Oncorhynchus mykiss* (Salmonidae) older than age I (3)
   H. Presence of *Salvelinus fontinalis* (Salmonidae) (10)
   I. Presence of juvenile or large *Esox lucius* (Esocidae) (18)
   J. Number of Large River (restricted to great rivers) species (19, 22)
   K. Percent of species that are native species (3)
   L. Percent of individuals that are native species (3)
   M. This metric deleted from IBI (2, 14)
6. Percent of Individuals that Are *Lepomis cyanellus* (Centrarchidae) (1, 17)
   A. Percent of individuals that are *Lepomis megalotis* (Centrarchidae) (5)
   B. Percent of individuals that are *Cyprinus carpio* (Cyprinidae) (6)
   C. Percent of individuals that are *Semotilus atromaculatus* (Cyprinidae)(2)
   D. Percent of individuals that are *Rutilus rutilus* (Cyprinidae) (18)
   E. Percent of individuals that are *Rhinichthys* species (Cyprinidae) (10)
   F. Percent of individuals that are *Catostomus commersoni* (Catostomidae) (4, 7, 11)
   G. Percent of individuals that are tolerant species (8, 9, 12, 14–17, 19, 22)
   H. Percent of individuals that are "pioneering" species (9, 15, 22)

Table 2 (continued). List of Original Index of Biotic Integrity Metrics (Capital Letters) Proposed by Karr (1981) for Streams in the Central United States, Followed by Modifications Proposed by Subsequent Authors for Streams in Other Regions or for Different Streams and River Types

    I. Percent of individuals that are introduced species (4, 6, 12, 14)
    J. Number of introduced species (12, 13)
    K. Evenness (22)
    L. This metric deleted from IBI (3)

Trophic Function Metrics

7. Percent of Individuals that Are Omnivores (1–4, 6–8, 10, 12, 14–20, 22)
    A. Percent of individuals that are omnivorous Cyprinidae species (10)
    B. Percent of individuals that are *Luxilus cornutus* or *Cyprinella spiloptera* (Cyprinidae), facultative omnivores (9)
    C. Percent of individuals that are generalized feeders that eat a wide range of animal material but limited plant material (2, 9, 11)
    D. Percent biomass of omnivores (22)
    E. This metric deleted from IBI (3, 13)
8. Percent of Individuals that Are Insectivorous Cyprinidae (1, 17)
    A. Percent of individuals that are insectivores/invertivores (5–7, 9, 12, 14–19)
    B. Percent of individuals that are specialized insectivores (2, 4, 20)
    C. Percent of individuals that are specialized insectivorous minnows and darters (8)
    D. Percent biomass of insectivorous cyprinids (22)
    E. This metric deleted from IBI (3, 10, 13)
9. Percent of Individuals that Are Top Carnivores or Piscivores (1, 5, 7–9, 11, 12, 15–20)
    A. Percent of individuals that are large (> 20 cm) piscivores (10)
    B. Percent biomass of top carnivores (22)
    C. This metric deleted from IBI (2–4, 6, 13–15)

Reproductive Function Metrics

10. Percent of Individuals that Are Hybrids (1, 7, 8, 13)
    A. Percent of individuals that are simple lithophilous species: spawn on gravel, no nest, no parental care (9, 15–17, 19, 20, 22)
    B. Percent of individuals that are gravel spawners (18)
    C. Ratio of broadcast spawning to nest building cyprinids (22)
    D. This metric deleted from IBI (2–6, 10–12, 14)

Abundance and Condition Metrics

11. Abundance or Catch per Effort of Fish (1–8, 10–11, 14–15, 18, 19, 22)
    A. Catch per effort of fish, excluding tolerant species (9, 16, 20)
    B. Biomass of fish (6, 13, 22)
    C. Biomass of amphibians (13)
    D. Density of macroinvertebrates (13)
    E. This metric deleted from IBI (17)
12. Percent of Individuals that are Diseased, Deformed, or Have Eroded Fins, Lesions, or Tumors (1–7, 9, 11, 12, 14–16, 18, 19, 20, 22)
    A. Percent of individuals with heavy infestation of cysts of the parasite *Neascus* (10)
    B. This metric deleted from the IBI (13, 17)

Note: The numbers in parentheses correspond to the following references.

1. Karr (1981); Fausch et al. (1984); Karr et al. (1985a,b); Karr et al. (1986); Angermeier and Karr (1986); Berkman et al. (1986); Karr et al. (1987); Hite and Bertrand (1989); Angermeier and Schlosser (1988); Hite et al. (1992); Osborne et al. (1992).
2. Leonard and Orth (1986).
3. Moyle et al. (1986).
4. Schrader (1986).
5. Foster (1987).
6. Hughes and Gammon (1987).
7. Miller et al. (1988).
8. Saylor and Scott (1987); Saylor et al. (1988); Saylor and Ahlstedt (1990).
9. Ohio EPA (1987a,b).
10. Steedman (1988).
11. Langdon (1989).
12. Crumby et al. (1990).
13. Fisher (1990).
14. Bramblet and Fausch (1991).
15. Simon (1991).
16. Lyons (1992).
17. Hoefs and Boyle (1992).
18. Oberdorff and Hughes (1992).
19. Simon (1992).
20. Bailey et al. (1993).
21. Gatz and Harig (1993).
22. Goldstein et al. (1994).

relatively high numbers of total species, sucker (Catostomidae) species, darter (*Crystallaria, Ammocrypta, Etheostoma, and Percina*; Percidae), sunfish (Centrarchidae excluding *Micropterus*) species, and intolerant species; high relative abundance of top carnivores and insectivorous cyprinid species; high overall fish abundance; and low relative abundance of the tolerant green sunfish (*Lepomis cyanellus*), omnivores, hybrids, and fish with diseases or deformities. Expectations for species richness metrics increased with increasing stream order, and were derived from an empirical relationship between stream size and maximum number of species present, termed the maximum species richness (MSR) line (Fausch et al. 1984). The total IBI score was the sum of the 12 metric scores and ranged from 60 (best) to 12 (worst) (some authors have reduced the lowest score to zero, e.g., Simon 1991).

The original version of the IBI quickly became popular, and was used by many investigators to assess warmwater streams throughout the central United States (e.g., Berkman et al. 1986; Gorman 1987a, b; Bickers et al. 1988; Hite and Bertrand 1989; Simon 1990; Hite et al. 1992; Osborne et al. 1992). Karr and colleagues explored the sampling properties and effectiveness of the original version in several different regions and different types of streams (Fausch et al. 1984; Karr et al. 1986).

As the IBI became more widely used, different versions were developed for different regions and different ecosystems. These new versions had a multimetric structure, but differed from the original version in the number, identity, and scoring of metrics (Table 2). New versions developed for streams and rivers in the central United States generally retained most of the metrics used in the original IBI, modifying only those few that proved insensitive to environmental degradation in a particular geographic area or type of stream (e.g., Whittier et al. 1987; Saylor et al. 1988; Crumby et al. 1990; Rankin and Yoder 1990; Saylor and Ahlstedt 1990; Simon 1992; Hoefs and Boyle 1992; Lyons 1992). However, new versions developed for streams and rivers in France, Canada, and the eastern and western United States tended to have a very different set of metrics (e.g., Moyle et al. 1986; Schrader 1986; Langdon 1989; Steedman 1988; Bramblett and Fausch 1991; Oberdorff and Hughes 1992; Goldstein et al. 1994), reflecting the substantial differences in fish faunas between these regions and the central United States. Similarly, the metrics used in IBI versions developed for other types of ecosystems, such as estuaries, impoundments, and natural lakes, usually bore only a limited resemblance to those of the original version (Thompson and Fitzhugh 1986; Thoma 1990; Dionne and Karr 1992; Hughes et al. 1992; Dycus and Meinert 1993; Larsen and Christie 1993; but see Greenfield and Rogner 1984 for an exception) yet retain the ecological structure of the original IBI metrics.

3.0 CRITICAL FLOW VALUES AND BIOLOGICAL INTEGRITY

Water quality standards contain rules that define minimum stream flows above which chemical and narrative criteria must be met. This is most commonly the seven-day average flow that has a probability of recurring once every ten years (i.e., $Q_{7,10}$ flow). Other low-flow values can be used (95% duration flow, $Q_{30,10}$ flow) as well and these can approximate the $Q_{7,10}$ *relative to the annual hydrograph* for a given stream or river. Because the customary use of chemical and narrative criteria is essentially based on a steady-state, dilution-oriented process, a design "critical" flow is necessary. This has been widely accepted and essentially unquestioned practice in surface water quality regulation for many years. It is an inherently necessary component of the water quality based approach to limit and control the discharge of toxic substances. However, a direct ecological basis for such flow regulation is lacking and, furthermore, may not be relevant so that one flow duration determines ecological health and well-being. It is simplistic and ecologically unrealistic to expect that worst case biological community performance can only be measured under a $Q_{7,10}$ flow or some facsimile thereof.

There have been efforts to define ecologically critical flow thresholds using a toxicological rationale (USEPA 1986). This involved making judgements about the number of exceedences of acute and chronic chemical criteria that could occur *without causing harm to the aquatic community*. This effort attempts to establish a minimum flow at which chemical and/or toxic unit limits could be set and not have the aquatic communities in "a perpetual state of recovery" (Stephan et al. 1985). While there has been no direct experimental validation of the maximum exceedence frequency using complex ecological measures in the ambient environment, validation efforts were later directed at using experimental streams (USEPA 1991f). These efforts, while being experimentally valid, retain many basic limitations inherent to surrogate criteria, one of which remains that a single species serves as "surrogates" for community health.

Establishing a single critical flow (i.e., $Q_{7,10}$, $Q_{30,10}$, etc.) on an ecological basis, however, is not only improbable under current science, it is technically inappropriate. There are simply too many additional variables that simultaneously affect the response and resultant conditions of aquatic communities both spatially and temporally. Some can be estimated (e.g., duration of exposure, chemical fate dynamics, and additivity), but many cannot because of the intensive data collection and analysis requirements; other phenomena are simply not adequately understood, yet their influence is integrated in the biological result.

The ecological ramifications of low-flow conditions (particularly extreme drought) in small streams has probably contributed much of the attention given to critical low-flow. The results of low stream flow alone can be devastating in small watersheds (particularly those that have been modified via wetland destruction) during extended periods of severe drought (Larimore et al. 1957). The principal stressor in these cases is a loss of habitat via desiccation in which organisms either leave or die during these periods. Ironically enough, the sustaining flow provided by a point source discharge can mitigate the effects of desiccation if chemical conditions are minimally satisfactory for organism function and survival. While this may seem enigmatic in light of current strategies to regionalize wastewater flows, the presence of water with a seemingly marginal chemical quality can successfully mitigate what otherwise would be a total community loss. As was previously mentioned, this is dependent on the frequency, duration, and magnitude of any chemical stresses and local faunal tolerances. Small headwater streams (typically less than 10 to 20 mi^2 drainage areas) commonly experience near zero flows during extended dry weather periods, sometimes during several consecutive summers. Given the historical loss of wetlands that functioned to sustain flows during dry weather periods, strategies such as opting for small wastewater treatment plants instead of regionalization need to be considered if the aquatic communities in headwater streams are to be restored and maintained. In this situation, the discharge flow assumes the functional loss of the sustained dry weather flows formerly produced by wetlands. While this may seem contradictory the far worse consequences of repeated desiccation are far worse from a biological integrity standpoint.

Chemical-numerical applications necessarily have their basis in dilution scenarios. However, these types of simplified analyses are no match for the insights provided into the chemical, physical, and biological dynamics that are "included" in the condition of the resident biota. The resolution of steady-state chemical application techniques suffers when applied to extreme low-flow or high-flow conditions. Site-specific factors that outweigh the importance of flow alone include the availability and quality of permanent pools and other refugia, gradient, organism acclimatization, and riparian characteristics such as canopy cover. Together these and other factors determine the ability of a biological community to function and resist stress under worst case low-flow conditions and hence retain the essential elements of biological integrity. It would be a serious mistake to draw the conclusion that the only important function of stream flow is to dilute pollutant concentrations when in fact the influence on physical habitat, both flow and water volume, is a far more important factor. The misconceptions about the role of stream flow has not only hampered efforts to more accurately manage wastewater flows in small streams, but in some cases has actually led to policies that have resulted in far more devastating ecological impacts than that experienced under the original problems. The most frequently cited concept is that biological data collected during any time other than $Q_{7,10}$ critical flow does not represent the effect of worst case conditions and therefore has limited applications in water quality based issues. Sampling under worst case, low-flow conditions is simply not necessary when measuring the condition of communities that have relatively long life spans and carry out all of their life functions in the waterbody. It is inappropriate to expect biological community condition (which is the integrated result of physical, chemical, and biological factors) to be so dependent on a temporal extreme of a single physical variable.

4.0 CRITICISMS OF THE INDEX OF BIOTIC INTEGRITY

Suter (1993) critically evaluated ecological health and IBI. Although he states that his "paper does not attack the concept [IBI] but rather the much more limited belief that the best way to use...biosurvey data is to create an index of heterogenous variables [multimetric approach] and claim that it represents ecosystem health." The following is a list of his criticisms and a response to a potentially limited viewpoint of the IBI.

4.1 Ambiguity

Suter suggests that while using multimetric indices one cannot determine why values are high or low. The IBI utilizes multiple metrics to evaluate the water resource. One of the greatest advantages of IBI is that the site score can be dissected to reveal patterns exhibited at the specific reach compared to the reference community. Overall site quality can be determined from both the composite score and evaluation of each of the individual metrics. This reduces ambiguity compared to single metric indices such as the Shannon–Wiener Diversity Index.

4.2 Eclipsing

The eclipsing of low values of one metric can be dampened by the high values of another metric. Suter suggested that the density and disease linkage in epidemiology is an interrelated effect. He suggests that when toxic chemicals are involved, the disease factor may not be reflective of the density or quality of a otherwise unimpacted community. Studies by Ohio EPA (1987a,b) and other authors (e.g., Karr et al. 1985a,b; Karr et al. 1986) have shown that when the IBI is assessed properly each metric provides relevant information, which determines the position along a continuum of water resource quality. Thus, some sites may score well in some areas but poorly in other metrics depending on levels of degradation. Thus the reference condition is critical in determining the least impacted condition for the region.

4.3 Arbitrary Variance

Variance demonstrated in indices may be high due to the compounding of individual metric variances. Suter further suggests that other statistical properties of multimetric variables may be difficult to define. In studies conducted by Ohio EPA (Rankin and Yoder 1990; Yoder 1991b) they showed that IBI variability increased at highly degraded and disturbed sites but was low and stable at high-quality sites with increased biological integrity. The amount of variability within any of the component IBI metrics is irrelevant and does not necessarily have to be on the same scale assuming that proper metrics are selected and knowledge of how the metrics are applied are assessed by the field biologist. The high degree of resultant variability at sites that exhibit low biological integrity is an important indicator of site structure and function.

4.4 Unreality

Suter argues that using multimetric approaches results in values with "nonsense units." He suggests that the IBI does not use "real" properties to describe the status of the reach specific water resource. In contrast, he used an example of dose response curves or habitat suitability to better predict a real-world property such as the presence of trout in a stream following a defined perturbation. In Suter's simplistic approach to this complex problem he fails to recognize the multiple stresses that could potentially limit the possibility of aquatic organism uses of a stream. The water resource manager is not only interested in whether a species is present or absent from a stream reach, but also puts more weight on the species interactions in a web of dynamic interactions. The resultant hyperniche, defined by not only a single species but multiple species, becomes impossible with the limited amount of chemical specific information available. Likewise, the modeling of the synergistic and additive effects of multiple stressors suggests that the IBI is *only* sensitive to toxic influences. This has shown to be a poor assumption since the IBI can determine poor performance from point source, nonpoint source, and combinations of these effects. The IBI does use real-world measures, which individually are important attributes of a properly functioning and stable aquatic community. Additionally, the assessment of the aquatic community is enhanced by the acquisition of appropriate habitat information. It is highly recommended that all assessments include not only biological community information but habitat information (Davis and Simon 1988).

4.5 Post hoc Justification

Suter suggests that the reduction in IBI values is a tautology since the assessment of poor biological integrity is a result of the reduced score. He further suggests that the IBI will only work if all ecosystems in all cases become unhealthy in the same manner. The IBI metrics are *a priori* assumed to measure a specific attribute of the community. Each metric is not an answer unto itself and not all measure only attributes of a properly functioning community (e.g., percent disease). The metric must be sensitive to the environmental condition being monitored. The definition of degradation responses *a priori* is justified if clear patterns emerge from specific metrics. Although the probability of all ecosystems becoming unhealthy in the same manner is unrealistic it is important to note that response signatures are definable (Yoder and Rankin, Chapter 17) based on patterns of specific perturbations.

4.6 Unitary Response Scales

Suter suggests that combining multimetric measures into a single index value suggests only a single linear scale of response and, therefore, only one type of response by ecosystems to disturbance. Suter fails to recognize that the individual patterns exhibited by the various individual metrics usually reduces to single patterns in the community. For example, whether discussing siltation, reduced dissolved oxygen, or toxic chemical influences all reduce the sensitive species component of the community and reduces species diversity. Thus, although multiple measures of the individual metrics results in multiple vectors explaining those dynamic patterns, *a priori* predictions of the metric response will result in the biological integrity categories defined by Karr et al. (1986).

4.7 No Diagnostic Results

Suter suggests one of the most important uses of biological survey data is to determine the cause of changes in ecosystem properties. He further suggests that by combining the individual metrics into a single value causes a loss of resolution when attempting to diagnose the responsible entity. This is the same argument raised in the ambiguity discussion above. The greatest use of IBI is the ability to discern differences in individual metrics and determine cause and effect using additional information such as habitat, chemical water quality, and toxicity information. The inverse, however, is not apparent when attempting to reduce chemical water quality and toxicity test information into simple predictions of biological integrity based on complex interactions.

4.8 Disconnected from Testing and Modeling

Suter suggests that the field results determined from the IBI need to be verified in the laboratory using controlled studies such as toxicity tests. This is a narrow viewpoint of the complex nature of the multimetric approach. Seldom does the degradation observed at a site result from a single chemical contaminant. To suggest that a single-species or even multiple-species (usually run individually) toxicity test can predict an IBI is ridiculous given that the effects of siltation, habitat modification, guild and trophic responses cannot be adequately determined in a laboratory beaker. Those aspects of a community that can be tested in the laboratory has validated the individual metric approach, i.e., thermal responses. It is the compilation of the various attributes that gives IBI a robust measure of the community.

4.9 Nonsense Results

Suter indicates any index based on multiple metrics can produce nonsense results if the index has no interpretable real-world meaning. Suter suggests that green sunfish (considered a tolerant species in the IBI) may have a greater sensitivity to some chemicals than some "sensitive species," and that the reduction of these contaminants may enable increases in green sunfish populations which result in a reduction in biological integrity. However, Suter has mistakenly suggested that green sunfish have a greater position in the community than do sensitive species. Green sunfish and other tolerant species are defined by the species ability to increase under degraded conditions (Karr et al. 1986; Ohio EPA 1987a,b). Range extensions and the disruption of evenness in the community often occurs at the expense of other sensitive species. This suggests that scoring modifications and other mechanisms for factoring out problems when few individuals are collected are not real-world situations.

4.10 Improper Analogy to Other Indices

Since environmental health as a concept has been compared to an economic index several authors have argued that the environmental indices are not generally comprehensible and require an act of faith to make informed judgements or decisions. The IBI has greatly improved the decision-making process by removing the subjective nature of past biological assessments. By using quantitative criteria (biological criteria) to determine goals of the Clean Water Act (attainable goals and designated uses) the generally comprehensible goals of the IBI enable a linkage between water resource status and biological integrity. This does not require an act of faith; rather, it broadens the tools available to water resource managers for screening waterbody status and trends.

5.0 DEVELOPMENT OF METRIC EXPECTATIONS

The IBI requires quantitative expectations of what a fish community should look like under reference or least impacted conditions (Karr et al. 1986; USEPA 1990a). Each metric has its own set of expectations, and metric expectations often vary with ecosystem size or location (e.g., Fausch et al. 1984). Generating an acceptable set of expectations is perhaps the most difficult part of developing a new version of the IBI or effectively applying an existing version to a new geographic area. Usually, expectations have been developed on a watershed or regional basis, and have been derived from recent field data (Hughes et al. 1986, 1990; Plafkin et al. 1989).

Because fish communities may differ substantially between different geographic areas, accurate delineation of appropriate regions for development and application of expectations is critical. Early efforts defined regions based on watershed boundaries (Karr 1981; Fausch et al. 1984; Moyle et al. 1986; Karr et al. 1986), recognizing the major faunal differences that may exist among drainage basins (Hocutt and Wiley 1986). More recently, many IBI versions have used Omernik's (1987) ecoregions as their geographic framework for setting expectations (Hughes and Gammon 1987; Geise and Keith 1989; Langdon 1989; Ohio EPA 1987a,b; Simon 1991; Larsen and Christie 1993). Fish community composition in streams has been shown to differ among ecoregions (Larsen et al. 1986; Hughes et al. 1987; Rohm et al. 1987; Whittier et al. 1988; Lyons 1989; Hawkes et al. 1986), and several workers have argued that ecoregions may be more appropriate than drainage basins in developing regional expectations (Hughes et al. 1986; Hughes and Larsen 1988; Gallant et al. 1989; Omernik and Griffith 1991; Omernik, Chapter 5). Some IBI versions have used a combination of ecoregion and watershed boundaries to delineate regions (Fisher 1989; Simon 1991; Lyons 1992). Regardless of which regional framework is used the final tuning should utilize statistical multivariate approaches to determine patterns that may not follow any prescribed regional framework such as that observed among forested regions throughout the midwest (Fausch et al. 1984) and for large and great rivers in Indiana (Simon 1992).

Two general approaches have been used to generate quantitative metric expectations for a particular geographic area. The first approach requires identification and sampling of a limited number of representative sites in relatively undegraded or least impacted ecosystems (Hughes et al. 1986; Gallant et al. 1989; Warry and Hanau 1993). Hughes (1985) and Hughes et al. (1986) provided detailed guidelines for selecting appropriate least impacted sites. Data from these sites are then used to define expectations and establish metric scoring criteria. This approach has been used successfully with stream fish communities in Ohio (Larsen et al. 1986, 1988; Whittier et al. 1987; Ohio EPA 1987a,b), Arkansas (Rohm et al. 1987; Geise and Keith 1989), Vermont (Langdon 1989), and New York (Bode and Novak 1989).

The second approach does not involve delineation of specific high-quality or least impacted sites, but require more data. Under this approach, a large number or sites are surveyed in a systematic fashion to provide a representative view of the region. The best values observed for each metric, even if they do not come from the highest quality sites, are then used to define the expectations and set scoring criteria. This approach, which has been widely used (e.g., Karr 1981; Fausch et al. 1984; Karr et al. 1986; Moyle et al. 1986; Hite and Bertrand 1989; Maret 1989; Simon 1991; Lyons 1992; Osborne et al. 1992), has worked best either when it has been difficult or impractical to identify least impacted sites, or when a large database has already been present, but additional data collection has not been possible.

Within a particular geographic area, different metrics, expectations, and scoring criteria have been used for different types of ecosystems. Existing versions of the IBI have typically recognized several different types of lotic ecosystems, although the distinction between them has not always been clear and has varied among versions. Ohio EPA (1987a,b) distinguished among headwater streams, wadable streams and rivers, and boat-sampled rivers based on the expected fish fauna, the size of the drainage basin, and the fish sampling techniques that was most appropriate. Simon (1992) separated large and "great" rivers based on drainage basin area. Lyons (1992) discussed the differences in fish fauna, response to environmental degradation, and maximum summer water temperatures between Wisconsin warmwater and coldwater streams and concluded that a single version of the IBI would not be possible. For lentic ecosystems, the primary distinction has been between impoundments and natural lakes, although within each ecosystem type it seems that size- and temperature-based stratification are warranted (Hughes et al. 1992; Larsen and Christie 1993). Work on Tennessee Valley Authority impoundments has indicated that reservoirs with different hydrologic regimes and watershed position (mainstem, constant water level vs. tributary, fluctuating water levels) need to be treated separately and that different relative locations within large impoundments require different metric expectations (Dionne and Karr 1992).

Expectations for species richness metrics have often been an increasing function of ecosystem size (Fausch et al. 1984, 1990; Karr et al. 1986; Miller et al. 1988). For streams and rivers, ecosystem size has been expressed as stream order (Karr et al. 1986), drainage basin area (Ohio EPA 1987a,b), or mean channel width (Lyons 1992). Each of these measures has its pros and cons (Hughes and Omernik 1981, 1983; Lyons 1992). For impoundments and lakes, surface area has been used (Larsen and Christie 1993). Typically, the size vs. species richness function is assumed to be either log-linear or asymptotic or a combination of both, and is fit graphically by eye such that about 5% of the data points fall above the resulting maximum species richness (MSR) line. The by-eye method is imprecise, and often numerous but equally valid MSR lines could fit to the same data set. Some workers constrain the intercept to be at the graphical origin; however, since fish are found in even the smallest size streams problematic slopes result in the smallest headwater streams or the largest river. The MSR line should not be constrained to go through the origin (J. R. Karr, personal communication). Lyons (1992) developed an objective graphical technique for drawing MSR lines, but generally a more precise statistical procedure for generating MSR lines would be desirable. It is important to recognize what information is supported by the data and not to over extend the results beyond what can be supported.

Several other factors have been considered in the development of metric expectations. In some streams and rivers, gradient strongly influences fish species richness and composition, with high-gradient sites often having lower richness and different species than nearby low-gradient sites (Leonard and Orth 1986; Miller et al. 1988; Lyons 1989). Recent studies indicate that the relative position of ecosystems within the drainage network can also affect species richness. Fausch et al. (1984), Karr et al. (1986), and Osborne et al. (1992) demonstrated that small adventitious streams (Gorman 1986) that flowed directly into much larger rivers had higher species richness values and IBI scores than environmentally similar headwater streams that were distant from large rivers. They concluded that small adventitious and headwater streams needed different metric expectations. Lyons (1992) found that the number of sunfish species was higher in streams near lakes and large rivers than in those distant, and developed separate MSR lines for each case. Although drainage position has received only general, larger-scale consideration thus far in IBI versions for lentic habitats (Dionne and Karr 1992), the presence of tributaries and the relative "connectedness" of lakes with other bodies of water is known to have a major influence on lake fish community composition (summarized by Tonn et al. 1990).

6.0 MAJOR IBI MODIFICATIONS

6.1 Ichthyoplankton Index

As a rule, existing versions of the IBI have focused on juvenile and adult fish, and explicitly excluded larval and small young-of-the-year fish. For example, Karr et al. (1986) recommended against inclusion of fish under 20 mm total length, whereas Lyons (1992) excluded all fish below 25 mm. Angermeier and

Table 3. List of Original Ichthyoplankton Index Metrics Proposed by Simon (1989) for Streams in the United States Based on Early Life History and Reproductive Biology Characteristics

Taxonomic composition metrics
   1. Total Number of Families
   2. Number of Sensitive Families
   3. Equitability/Dominance
   4. Family Biotic Index
Reproductive guild metrics
   5. Percent Nonguarding Guild
   6. Percent Guarding Guild
   7. Percent Bearing Guild
   8. Percent Simple Lithophil Guild
Abundance, generation time, and deformity metrics
   9. Catch per Unit Effort
   10. Mean Generation Time
   11. Percent Deformity or Teratogenicity

Karr (1986) argued that inclusion of any size of young-of-the-year fish would reduce the accuracy of the IBI applications. The difficulties in sampling and identifying larvae coupled with the commonly high temporal variability in larval abundance have been the main arguments against the use of larvae and young-of-the-year fish in the IBI.

However, Simon (1989) developed a version of the IBI, termed the Ichthyoplankton Index, specifically for larval fish in lotic habitats. The early life history stage of fishes have been recognized as the most sensitive and vulnerable life stage (Blaxter 1974; Moser et al. 1984; Wallus et al. 1990), and, thus, would be particularly sensitive to certain types of ecosystem degradation.

The Ichthyoplankton Index is organized into 11 metrics covering taxonomic composition, reproductive guild, abundance, generation time, and deformity categories (Table 3). Since much of the North American fauna is incompletely described (Simon 1986), a family-level approach was designed to evaluate early life history stages. Reproductive guilds are based on Balon (1975, 1981). Simon (1989) assigned tolerance values based on sensitivity to siltation, sediment degradation, toxic substances, and flow modifications. Larvae of Clupeidae, Sciaenidae, and Osmeridae were excluded from metric calculations because they often reached such high abundances that their inclusion would obscure abundance patterns for other taxa.

The Ichthyoplankton Index has not been field tested, so little discussion of precision and accuracy is possible. Simon (1989) felt that its greatest value would be in assessing the integrity of nursery habitats, particularly backwaters of large rivers. The primary limits to its widespread and effective use will likely be the difficulty of collecting the necessary data to establish specific regional expectations and scoring criteria, and of establishing a standardized sampling period and technique.

6.2 Warmwater Streams

The majority of IBI applications have involved small- to medium-sized wadable warmwater streams, and numerous modifications have been made to the original version to reflect regional or ecosystem differences in fish communities. Some of the original metrics have been changed in nearly every subsequent version, whereas others have been largely retained (Table 2).

6.2.1 Species Richness and Composition

Changes in the original species richness and composition metrics have typically been limited, most commonly involving expansion of taxonomic groups to include other species. This has been done when the original taxonomic group had few species in the area or ecosystem type of interest. For example, the number of catostomid species metric has been expanded to include Ictaluridae in Ontario (Steedman 1988), all benthic insectivorous species in the northeastern United States (Miller et al. 1988), or all benthic species in France (Oberdorff and Hughes 1992). The number of darter species metric has been broadened to encompass Cottidae in Ohio headwaters (Karr et al. 1986; Ohio EPA 1987b) and in Ontario (Steedman 1988), or Cottidae and *Noturus* species (Ictaluridae) in northwestern Indiana and northern Wisconsin (Simon 1991; Lyons 1992), and has been replaced with the number of Cyprinidae species in Arkansas (Karr et al. 1986), Oregon (Hughes and Gammon 1987), Colorado (Schrader 1986; Bramblett and Fausch 1991), Ohio headwaters (Ohio EPA 1987a,b), Indiana headwaters (Simon 1991), and Minnesota headwaters (Bailey et al. 1993), or proportion of cyprinids with subterminal mouths in the Red River of the North basin (Goldstein et al. 1994). The number of sunfish species metric has been expanded to include all water column (nonbenthic) species in the northeastern United States (Miller et al. 1988) and in France (Oberdorff and Hughes 1992), and substituted by the number of headwater species (species generally restricted to good quality permanent headwaters) in Ohio and Indiana headwaters (Ohio EPA 1987a,b; Simon 1991) and the Red River of the North basin (Goldstein et al. 1994).

6.2.2 Indicator Species Metrics

Metrics proposed by Karr (1981) and Karr et al. (1986) to evaluate species sensitivity to human influence on watersheds have been the most frequently changed IBI metrics (Table 2). Usually the changes have resulted from differences of opinion in species sensitivity, differences in drainage area relationships, and use of other more regionally representative species to act as tolerant species surrogates. Oberdorff and Hughes (1992) substituted the number of intolerant species with the presence of northern pike.

The most frequently changed metric among IBI efforts has been the proportion of green sunfish. Green sunfish are usually abundant only in small creeks and streams and do not reflect deteriorating habitat in wadable rivers and larger rivers. Green sunfish are also restricted largely to the central United States, which makes them a poor choice for widespread application in other geographic areas. The proportion of green sunfish has been replaced with more regionally appropriate tolerant species, e.g., common carp (*Cyprinus carpio*), longear sunfish (*Lepomis megalotis*), creek chub (*Semotilus atromaculatus*), white sucker (*Catostomus commersoni*), blacknose dace (*Rhinichthys atratulus*), or roach (*Rutilus rutilus*)(Leonard and Orth 1986; Schrader 1986; Hughes and Gammon 1987; Miller et al. 1988; Oberdorff and Hughes 1992), or many versions have included all tolerant species to increase the sensitivity of this metric. Some have replaced the proportion of green sunfish with either the proportion or number of exotic species (Schrader 1986; Hughes and Gammon 1987; Crumby et al. 1990; Fisher 1990; Bramblet and Fausch 1991). Headwater streams typically do not provide sufficient habitat for many of the tolerant species, so a substitution of the metric includes the proportion of pioneer species in Ohio and Indiana (Ohio EPA 1987a,b; Simon 1991). Pioneer species are small tolerant species that are the first to recolonize headwaters following desiccation or fish kills (Smith 1971). Goldstein et al. (1994) suggests that tolerance is too subjective, so this metric may be substituted with evenness.

6.2.3 Trophic Function Metrics

The trophic composition metrics have been relatively unchanged across widespread geographic application of the IBI (Table 2). The omnivore and carnivore metrics have required the fewest modifications. Leonard and Orth (1986) substituted the proportion of generalized feeders because they found that few species in their streams fit the omnivore definition. Filter-feeding species were not considered omnivores for this metric nor were species that had a generally plastic response to diet when confronted with degraded habitat conditions, such as creek chub and blacknose dace. The proportion of individuals as insectivorous cyprinids has been modified in most versions to include all specialized invertebrate feeders or all insectivores. These broader groupings have proven generally more sensitive, particularly where the species richness of insectivorous cyprinids is low (Miller et al. 1988). The carnivore metric has been deleted from some analyses due to lack of occurrence (Leonard and Orth 1986; Fausch and Schrader 1987) or due to drainage area relationships (Ohio EPA 1987a,b; Simon 1991). Hughes and Gammon (1987) suggested that the primary carnivore in the Willamette River, Oregon, was an exotic and tolerant of degraded conditions indicating low integrity when abundant. Goldstein et al. (1994) substituted the percent of various trophic categories with the percent biomass of trophic guilds.

6.2.4 Reproductive Function Metrics

The proportion of hybrids metric has been difficult to apply in most geographic regions. The hybrid metric was designed to reflect tendencies for breakdown in reproductive isolation with increasing habitat degradation. Other researchers have not found hybridization to be correlated with habitat degradation (Pflieger 1975; Ohio EPA 1987a,b). Problems with field identification, lack of hybrids even in some very degraded habitats, and presence of hybrids in some high quality streams among some taxa have precluded this metric from being successfully applied. In western fish faunas, this metric has been modified with the number of introduced species (Fausch and Schrader 1987; Hughes and Gammon 1987). Courtney and Hensley (1980) consider the increase of nonnative species as a form of biological pollution which is similar to the original intent of the metric.

In some geographic areas the hybrid metric has been deleted from the index and replaced with the proportion of simple lithophils (Table 2), which are defined as those species spawning over gravel without preparing a nest or providing parental care. The proportion of simple lithophilous spawning species is believed to be inversely correlated with habitat degradation based on the destruction of high-quality spawning habitat (Berkman and Rabeni 1987; Ohio EPA 1987a,b; Simon 1991). Berkman and Rabeni (1987) found an inverse relationship between the number of lithophilous spawning species and siltation. Goldstein et al. (1994) utilized the ratio between broadcast spawning and nest-building cyprinids to determine reductions in substrate quality in the species depauperate Hudson River drainage.

6.2.5 Abundance Metrics

Catch per unit of effort (CPUE) has been retained in most versions of the IBI; however, some have excluded tolerant species (Ohio EPA 1987b; Lyons 1992). CPUE is influenced by stream size and sampling efficiency. Karr et al. (1986) suggested using relative CPUE to set scoring criteria for the total number of individuals metric; however, fish density and sampling efficiency tends to decrease as watershed and stream size increases (Thompson and Hunt 1930; Larimore and Smith 1963; Miller et al. 1988). Steedman (1988), Gammon (1990) and Lyons (1992) found abundance to be higher at moderate levels of degradation (i.e., nutrient enrichment) and lowest at severe levels. Lyons (1992) used the abundance metric as a correction factor for the overall IBI score only when very low CPUE was observed. Hughes and Gammon (1987) and Fisher (1990) included biomass and estimates for fish and amphibians, and the density of macroinvertebrates to increase sensitivity in their streams that had relatively low species diversity.

6.2.6 Fish Condition Metrics

The percent of diseased and deformed individuals has been retained in most versions of the IBI. However, infestation by parasites and protozoans has often been eliminated from this metric due to a lack of correlation between parasite burden and environmental quality (Whittier et al. 1987; Steedman 1991). Most researchers have included only obvious external anomalies such as deformations, eroded fins, lesions, or tumors. Lyons (1992) used this metric as a correction factor for the overall IBI score only when a relatively high percentage of diseased and deformed individuals were present at the sample site.

6.3 Large Rivers

Relatively little work has been directed towards modifying the IBI for use on rivers too large to sample by wading. Difficulties in accurately and easily sampling fish assemblages and the scarcity of appropriate least impacted reference sites for setting metric expectations have hampered IBI development for these types of systems. Only five published versions are currently available, covering the Willamette River in northwestern Oregon (Hughes and Gammon 1987), the large rivers of Ohio (Ohio EPA 1987a,b), the Seine River in north-central France (Oberdorff and Hughes 1992), the Current and Jacks Fork Rivers in southeastern Missouri (Hoefs and Boyle 1992), and the large rivers of Indiana (Simon 1992). The Indiana version distinguishes between "large" rivers, with drainage areas of 1000 to 2000 mi^2, and "great" rivers, with drainage areas greater than 2000 mi^2 (Figure 1).

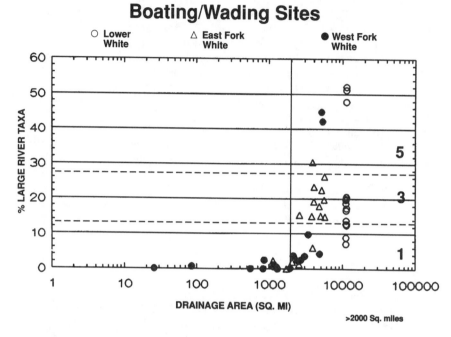

Figure 1. Maximum Species Richness (MSR) line plot of the proportion of large river taxa with drainage area. Drainage areas greater than 2000 mi² define the difference between large and great river metrics. (From Simon, T. P. 1992. *Development of Biological Criteria for Large Rivers with and Emphasis on an Assessment of the White River Drainage, Indiana.* USEPA, Chicago, Illinois.)

Generally, all four large river versions have metrics that are similar to those in IBI versions developed for smaller, wadable streams and rivers (Table 2). However, the Ohio, Missouri, and Indiana versions have incorporated two unique metrics especially tailored for larger rivers, the percent of individuals that are round-bodied catostomid species (*Cycleptus, Erimyzon, Hypentelium, Minytrema,* and *Moxostoma*), and the number of specialized large river species, which tend to be restricted to high-quality large rivers. Round-bodied catostomids are most common in relatively undegraded large rivers of eastern North America, and are sensitive to water pollution, thermal loadings, and habitat degradation (Gammon 1976; Karr et al. 1986; Ohio EPA 1987a,b; Simon 1992). Although round bodied in shape, the white sucker is not included in the round-bodied group because it is highly tolerant of poor water and habitat quality. Other common large-river catostomids, the carpsuckers (*Carpiodes*) and the buffaloes (*Ictiobus*), which are more laterally compressed, also are not included because they are omnivorous species that can survive in thermally stressed and degraded habitats.

The number of specialized large-river species metric has thus far been used only for great rivers in Indiana (Figure 1), although it seems a promising metric for large rivers in many different regions. This metric is based on studies in Missouri by Pflieger (1971), where a fish fauna characteristic of high-quality reaches of large rivers was identified. A similar fauna has also been documented for Indiana (Gerking 1945), Illinois (Smith et al. 1971), Arkansas (Matthews and Robison 1988), and Kentucky (Burr and Warren 1986).

6.4 Modification of the IBI for Coldwater and Coolwater Habitats

Coolwater streams have a mean maximum daily temperature between 22 and 24°C during a normal summer and coldwater streams normally have maximum daily means below 22°C, whereas warmwater streams exceed 24°C (Lyons 1992). High-quality coldwater streams are dominated by salmonid and cottid species. High-quality coolwater streams are often too warm to support large populations of salmonid and cottids, but too cold to support the full complement of species found in warmwater streams. Cool and coldwater streams are common in much of Canada and the northern and western United States.

Most versions of the IBI have been developed for warmwater streams, rather than cool- or coldwater streams. Most applications of the IBI for coolwater and coldwater streams have been for high-gradient areas of the eastern and western United States (Leonard and Orth 1986; Moyle et al. 1986; Hughes and Gammon 1987; Langdon 1989; Steedman 1988; Fisher 1990; Oberdorff and Hughes 1992), and versions developed by Leonard and Orth (1986) and Miller et al. (1988) for the eastern United States were for predominantly coolwater systems. Steedman (1988) and Lyons (1992) demonstrated that warmwater versions of the IBI were inappropriate for use in coldwater streams in the north-central United States and Canada. In this region, degraded cool- and coldwater streams often show increased species richness for many groups of fishes, the opposite of what normally occurs in warmwater streams.

The general trend for coolwater and coldwater versions of the IBI has been for a reduction in the number of metrics, e.g., 8 to 10 for coldwater (with the exception of Hughes and Gammon 1987, which has 12 metrics but is both a warm and coldwater version), vs. 10 to 12 metrics for warmwater. This reduction in metrics reflects the simplified structure and function of coldwater fish communities relative to warmwater communities. Moyle et al. (1986) and Fisher (1990) also included the number of amphibian species and the density of macroinvertebrates in their coldwater versions.

6.5 Estuaries

Very little work has been completed in the development of estuarine versions of the IBI. Only one ocean estuary version is currently available. Thompson and Fitzhugh (1986) made significant changes in the IBI to reflect estuarine/marine component of the Louisiana fauna that they studied, but they retained the general framework of the index. Apparently, little additional testing and application of this version has taken place and many questions remain about sampling, natural dominance of certain species, and high inherent temporal/seasonal variation in assemblage structure.

Preliminary investigations by Thoma (1990) on Lake Erie estuaries and Simon et al. (1989) for the Grand Calumet River and Indiana Harbor Canal are the only freshwater efforts. Thoma (1990) used Ohio EPA boat versions of the IBI and suggested replacement of the number of sunfish, number of sucker, proportion of round-bodied suckers, and proportion of simple lithophils with the proportion of exotics, number of vegetation associated species, and number of Lake Erie species. Simon et al. (1989) eliminated transient migratory species from their analysis in scoring the total number of species and proportional metrics.

6.6 Lakes

Little work has been published for lakes and reservoirs, but significant efforts are ongoing as part of the USEPA effort to develop consistent bioassessment protocols (Gerritsen et al. 1984). The USEPA's Environmental Monitoring Assessment Program (EMAP) is testing potential metrics on natural lakes in the northeastern United States (Hughes et al. 1992; Larsen and Christie 1993), and Karr and the Tennessee Valley Authority (TVA) (Dionne and Karr 1992; Dycus and Meinert 1993; Jennings, Karr and Fore, personal communication) are conducting research on IBI application in impoundments. Proposed metrics for lentic systems are listed in Table 4.

At present, EMAP has proposed eight potential metrics for evaluating the biotic integrity of natural lakes (Table 4; Hughes et al. 1993; Larsen and Christie 1993). Six of these metrics are the same as those used in some stream versions of the IBI, but two are unique, the overall age/size structure and the percentage of individuals above a certain size for selected species. These two metrics are designed to assess the reproductive success, survival, and vulnerability to increasing degradation for populations of key indicator species. Additional work is needed to more precisely define and set expectations for these two metrics, and to test their usefulness in field applications.

6.7 Impoundments

The original TVA version of the IBI for use in impoundments had 14 metrics (Table 4; Dionne and Karr 1992). Most of these metrics were taken from stream and river versions of the IBI, reflecting the

Table 4. List of Index of Biotic Integrity Metrics Proposed for TVA Reservoirs by Dionne and Karr (1992) (R1) and by Jennings, Karr, and Fore (personal communication) (R2) and for Natural Lakes in the Northeastern U.S. by Hughes et al. (1992) (L).

Species Richness And Composition Metrics
  1. Total Number of Species (R1, R2, L)
  2. Number of Catostomidae Species (R1)
  3. Number of Small Cyprinidae and Darter Species (R1)
  4. Number of Sunfish Species (R2)
Indicator Species Metrics
  5. Number of Intolerant Species (R1, R2, L)
  6. Age and/or Size Structure for Populations of Selected Species (L)
  7. Percent of Individuals Larger than a Certain Size for Selected Species (L)
  8. Percent of Individuals that are Tolerant Species (R1, R2, L)
  9. Percent of Individuals that are Exotic Species (L)
Trophic Function Metrics
  10. Percent of Individuals that are Specialized Benthic Insectivores (R1)
  11. Percent of Individuals that are Invertivores (R2)
  12. Percent of Individuals that are Omnivores (R1, R2)
  13. Percent of Individuals that are Piscivores (R1)
  14. Percent of Individuals that are Young-of-Year Shad and Bluegill (R1)
  15. Percent of Individuals that are Adult Shad and Bluegills (R1)
Reproductive Function Metrics
  16. Percent of Individuals that are Plant and Rock Substrate Spawners (R1)
  17. Percent of Individuals that are Simple Lithophilous Species (R2)
  18. Percent of Individuals that are Migratory Spawning Species (R1, R2)
Abundance and Condition Metrics
  19. Abundance or Catch per Effort of Fish (R1, R2, L)
  20. Percent of Individuals that are Diseased, Deformed, or Have Eroded Fins, Lesions, or Tumors (R2, L)
  21. Fish Health Score for Largemouth Bass (R1, R2)

intermediate nature of reservoirs between rivers and lakes. Six metrics were new: the number of small cyprinid and darter species, the percent of individuals as young-of-the-year shad (*Dorosoma cepedianum* and *D. petenense*) and bluegill (*Lepomis macrochirus*), the percent of individuals as adult shad and bluegill, the percent of species as plant and rock substrate spawners, the number of migratory spawning species, and the fish health score for largemouth bass (*Micropterus salmoides*). The number of small cyprinids and darters metric was based on the high species richness for these taxa that occurs in relatively undegraded waters of the region. The two shad/bluegill metrics indicated the key role that these two taxa play in structuring reservoir fish assemblages. The two metrics concerned with spawning species were included to assess the quality of littoral zone and tributary spawning areas, which can be degraded by excessive sedimentation, fluctuations in water level, and direct physical modification. The physical health score represented an effort to objectively evaluate the physical condition and physiological state of an important reservoir species through a thorough external and internal examination. The procedure followed was based on a scoring system developed by Goede (1988).

More recently, a second TVA reservoir version of the IBI has been developed, termed the Reservoir Fish Assemblage Index (RFAI; Jennings, Karr, and Fore, personal communication). The RFAI has a somewhat different set of 12 metrics (Table 4), with the changes in metrics designed to improve sensitivity to environmental degradation and to increase adaptability to different types of reservoirs. However, results from applications of both the original TVA version and the newer RFAI have often not accurately reflected what are believed to be the true patterns in environmental health within and among reservoirs, and additional modifications will probably be necessary to develop better versions of the IBI for impoundments (Jennings, personal communication). In the RFAI, the number of sunfish species has been substituted for the number of small cyprinid and darter species metric, the percent of individuals as invertivores has been substituted for the percent of individuals as specialized benthic insectivores metric, and the number of simple lithophilous spawning species has been substituted for the percent of individuals as plant and rock substrate spawners metrics. Three metrics have been dropped: the percent of individuals as piscivores, the percent of individuals as young-of-year, and percent adult shad and bluegill. Shad also are not included in the calculation of the four remaining percent abundance and catch-per-effort metrics. The physical health score metric has thus far been retained in the RFAI, but it has been recommended for deletion because of its low sensitivity to known environmental problems.

Metric expectations for the two TVA versions of the IBI differ between mainstem and tributary reservoirs and between the inflow, transition, and forebay (near dam) regions of individual reservoirs (Dionne and Karr 1992; Jennings, Karr, and Fore, personal communication). These differences in expectations reflect inherent differences in fish assemblages within and among reservoirs. Development of metric expectations for the TVA versions has been hampered by the high spatial and temporal variability typical of reservoir fish assemblages, the limited range of environmental quality and high interconnectedness among the reservoirs studied, and the lack of least impacted reference systems for comparison (Jennings, Karr, and Fore, personal communication). Generally, it will be difficult to define and identify appropriate least impacted reference waters for reservoirs, as reservoirs are, by nature, highly impacted and artificially modified rivers.

7.0 FUTURE DEVELOPMENTS

New versions of the IBI continue to be developed for wadable warmwater streams. To our knowledge, efforts are currently underway to generate IBI versions for streams in the coastal plain of Maryland and Delaware, the New River drainage in West Virginia, the Ridge and Valley Physiographic Province of the central Appalachian Mountains, Mississippi River tributaries in northwestern Mississippi, the Minnesota River drainage in central Minnesota, and the Red River of the North drainage in northwestern Minnesota and northeastern North Dakota. It is likely that we are unaware of additional ongoing efforts. Despite the large amount of current and past work on IBI versions for wadable warmwater streams, much remains to be done. In particular, versions are needed for most of Canada, the species-rich southeastern United States, and the species-poor western United States. Very small and intermittent streams have received insufficient attention everywhere but Ohio and Indiana. Low-gradient, wetland streams are another under represented habitat type. Finally, and perhaps most importantly, many existing versions have as yet not been properly validated with independent data. The development of an IBI version is usually an iterative, somewhat circular process, and each new version requires a thorough field test and critique on a new set of waters before it can be safely applied.

For those areas with validated versions of the IBI, the next challenge will be to incorporate the IBI into routine monitoring and assessment programs. Many versions still remain research-level tools that are not widely used in the region where they were developed. Although research has shown them to be useful and reliable approaches to assess resource condition, many states have not moved forward to implement them. However, this will change in the United States as the USEPA encourages and mandates development and application of biocriteria into water quality standards programs. Currently, Ohio EPA is the best example of how the IBI can be incorporated into a state water resource management and protection program.

Large rivers are also the subject of increasing efforts, although considerable work is still required. In the upper Mississippi River, the U.S. Fish and Wildlife Service's Long Term Resource Monitoring Program is attempting to develop an IBI version using existing large standardized data sets from multiple information sources. However, undesirable statistical properties of this version have stymied further development and application (Gutreuter, Lubinski, and Callen, personal communication). Significant unresolved issues include the determination of appropriate sampling techniques, temporal and spatial scales, and scope of sampling effort. Clearly, multiple sampling techniques will be needed to get a representative snapshot of the large-river fish community, but there are problems with aggregating results from different gears/techniques and amounts of effort. These are long-identified problems, and they await resolution before large-river IBI development efforts can move forward. Ohio River Valley Sanitation Commission's (ORSANCO) Biological Water Quality Subcommittee and Ohio EPA are developing standard operating procedures on the Ohio River that will enable representative sample collection and data interpretation.

Authors Lyons and Simon are independently working with colleagues on IBI versions for low- to moderate-gradient coldwater systems in the north-central United States. At present, the Lyons version is a greatly simplified warmwater IBI, focusing on the presence and relative abundance of obligate coldwater species, intolerant species, and tolerant species. The Simon version has more metrics, but also utilizes aspects of warmwater versions. Based on findings of Lyons (1992), Simon proposes a "reverse" scoring system, where greater numbers of species result in a lower metric score, the opposite of scoring procedures for warmwater versions. IBI streams versions are also being developed for areas outside North America. Lyons and colleagues are working on versions for streams in west-central Mexico, and Ganasan and Hughes (personal communication) are working on a version for central India.

Few new developments are occurring with estuarine IBI's. It appears that Thompson (Thompson and Fitzhugh 1986) never developed his Louisiana version any further. However, Thoma (1990) is in the process of validating his version for Lake Erie. Further efforts are needed to enable use of IBI's over broad geographic areas in other Great Lake and coastal estuaries.

Lake and impoundment IBIs are being improved, but have really only just begun. Natural lake IBI work is still concentrated on the preliminary stages of determining appropriate sampling methodologies, identifying least impacted sites, developing databases, and exploring the properties of potential metrics. The work is largely being done through EMAP, but up until now has focused exclusively on the northeastern United States.

Biocriteria for impoundments are being developed by TVA incorporating the RFAI into broader monitoring program. Dycus and Meinert (1993) have utilized dissolved oxygen, sediment, algae, benthic macroinvertebrates, and bacteria indices along with the RFAI to generate an overall picture of ecosystem health for reservoirs. However, Jennings, Karr, and Fore (personal communication) feel that the RFAI still requires significant work before it can be broadly applied. In particular, more data from additional interconnected reservoirs that range from poor to excellent in environmental quality are needed to improve the RFAI and ensure sensitivity to a wide range of environmental conditions.

8.0 SUMMARY

The Index of Biotic Integrity has been a widely applied and effective tool for using fish assemblage data to assess the environmental quality of aquatic habitats. The original version of the IBI has been modified in numerous ways for application in many different regions and habitat types, and the IBI is now best thought of as a family of related indices rather than a single index. The commonalities linking all IBI versions are a multimetric approach that rates different aspects of fish community structure and function based on quantitative expectations of what constitutes a fish community with high biotic integrity in a particular region and habitat type. All versions include metrics that address species richness and composition, indicator species, trophic function, reproductive function, and/or overall abundance and individual condition. Different metrics and metric expectations within each of these metric categories are what distinguish different IBI versions. A variety of approaches have been used to generate metric expectations, and the process of establishing appropriate, sensitive expectations is probably the most difficult step in preparing a new version of the IBI. At present, most existing IBI versions are for wadable warmwater streams in the central United States. However, versions have also been developed, or are in the process of being developed, for coldwater streams, large unwadable rivers, lakes, impoundments, and marine and Great Lakes estuaries in many different regions of the United States, and for streams and rivers in Canada, Mexico, France, and India. Despite the large amount of effort that has been directed towards IBI development, much remains to be done, both in terms of generating new versions for different regions and habitat types, and in terms of validating existing versions.

ACKNOWLEDGMENTS

The authors wish to thank the following individuals for providing information on research in progress: J.R. Karr, R.M. Hughes, D.L. Dycus, R.M. Goldstein, S.G. Paulsen, C. Yoder, R. Thoma, E.T. Rankin, E. Emory, J.R. Shulte, and G. Gibson. We appreciated critical review comments from three anonymous reviewers, which greatly improved an earlier draft of this manuscript. Support for John Lyons on this chapter was provided by the Wisconsin Department of Natural Resources, Bureau of Research and Bureau of Water Resources Management, Study RS60 FIS RS634.

## CHAPTER 17

BIOLOGICAL RESPONSE SIGNATURES AND THE AREA OF DEGRADATION VALUE: NEW TOOLS FOR INTERPRETING MULTIMETRIC DATA

Chris O. Yoder and Edward T. Rankin

1.0 INTRODUCTION

Biological criteria are used primarily as an environmental assessment tool and are intended to reflect the relative health and well-being of the resident aquatic community. The most appropriate use of biological criteria, and the attendant ambient bioassessment techniques, is in assessing the relative condition of the aquatic resource with regard to temporal and spatial variables and influences. Many states use biological assessment as the principal approach for determining the extent of impairments of the aquatic resource, and water quality in general. In Ohio, the appropriate role of biological criteria is defined by the Ohio Water Quality Standards as the principal arbiter of aquatic life use goal attainment/non-attainment (Yoder and Rankin, Chapter 9).

Interpretation of the biological results, and resulting use attainment status, is best done on a longitudinal basis (e.g., river reach). For example, a site exhibiting partial attainment in the midst of nonattaining sites that extend for some distance both upstream and downstream would be viewed as an exception, with the assessment of widespread nonattainment prevailing. Conversely, the interpretation of this marginal result would have been quite different if the attainment status of the other sites was full. Thus, the significance, severity, and spatial extent of any observed impairment is as important as determining the attainment status. Typically, the results are portrayed as a two-dimensional graph in an upstream to downstream format and interpreted visually. In the case of Ohio, longitudinal changes are compared to the ecoregional biocriteria (Yoder and Rankin, Chapter 9). Recently, a quantitative estimate of the degree of departure from a biocriterion along a longitudinal continuum (Area Of Degradation Value), was developed to enhance the use and interpretation of biological community assessments.

Few dispute the value of biological communities to demonstrate impairment due to any number of stressors in the environment. However, the capability to use the resultant community data and information to discriminate between different stressors is frequently questioned (USEPA 1985; Suter 1993). This diagnostic capability has not been demonstrated on a widespread basis although specific examples do exist (Eagleson et al. 1990). Discernable patterns in the response of aquatic community attributes were first described using Ohio EPA data from more than 250 sites from 25 different streams and rivers and were termed *Biological Response Signatures* (Yoder 1991). This chapter presents the technical basis, with examples, for using the Area of Degradation Value and Biological Response Signatures to further interpret multimetric data for water resource management programs.

2.0 LONGITUDINAL ASSESSMENT OF AQUATIC COMMUNITY PERFORMANCE: THE AREA OF DEGRADATION VALUE

The Area of Degradation Value (ADV; Rankin and Yoder 1992) was developed to measure and quantify the longitudinal extent and magnitude of aquatic life impairment based on biological monitoring results. Thus far the ADV has been used by Ohio EPA to (1) establish priorities for wastewater treatment plant construction funding, (2) assess trends over time for contiguous stream and river segments to document the effectiveness of pollution controls, and (3) to establish priorities for point source management (e.g., development of 303[d] list) as part of an overall risk assessment to prioritize limited administrative resources. Potential uses include the assessment of penalties to polluters on the basis of actual instream damage to aquatic communities and as a component of Natural Resource Damage Assessments.

ADVs are based on measures generated from longitudinal plots of the biological indices and biological criteria as described by Yoder and Rankin (Chapter 9). The length or extent of degradation is defined simply as the distance over which the applicable index is less than the biocriterion or other benchmark value applied to that stream (Figure 1). Magnitude refers to the vertical departure from the applicable biocriterion or other benchmark value. The total ADV is the area between the criteria value and the actual data values (shaded areas in Figure 1). The computational formula for the ADV value is (using the Ohio modified version of the Index of Biotic Integrity, as an example):

$$ADV = \Sigma \, [(pIBIa + pIBIb) - (aIBIa + aIBIb)] \, (RMa - RMb), \text{ for } a = 1 \text{ to } n$$

where: pIBIa = potential IBI at river mile a
        pIBIb = potential IBI at river mile b
        aIBIa = actual IBI at river mile a
        aIBIb = actual IBI at river mile b
        RMa = upstream river mile
        RMb = downstream river mile
        n   = number of sampling sites

The ADV equation assumes that the average of two contiguous sampling sites accurately integrates the aquatic community status for the distance between the points. This is supported by numerous examples in rivers and streams throughout Ohio. Sampling sites are also close enough to allow meaningful changes to be tracked along a longitudinal continuum. Aquatic communities gradually recover with increased distance downstream from impacts, although the pattern may vary according to the type of impact and discharge. The connection of the sampling results produces the lines drawn in Figure 1 and should yield a sufficiently accurate representation of real changes in community performance and quantifiable departures from ecoregional biological criteria. Sampling sites are spaced more closely as the complexity of the setting increases in order to enhance this longitudinal resolution.

Figure 1 also demonstrates how different types of impacts can be layered together in a segment. The darker shading represents the impact from a point source discharge, the lighter from a nonpoint source. The effect of the point source is immediate and severe with recovery occurring with increased distance downstream. Nonpoint source impacts are more diffuse and are spread more evenly throughout a segment. In this case the true magnitude and extent of the nonpoint source impact will not be known until it is "unmasked" by the reductions of the point source impact. If the two impact types occur together in the same stream segment as portrayed in Figure 1, then the relative difference between each type can be visualized. If the point source impact(s) are fully abated with little attention given to the nonpoint sources then the recovery will be limited. Conversely, there will be little benefit to abating the nonpoint source impacts in the absence of point source controls.

2.1 Case Examples

One of the best proven uses for biosurvey data is for spatial and temporal trend analysis. Unlike chemical parameters, multimetric biological indices integrate chemical, biological, and physical impacts to aquatic systems and portray both condition and status in terms of designated use attainment/nonattainment

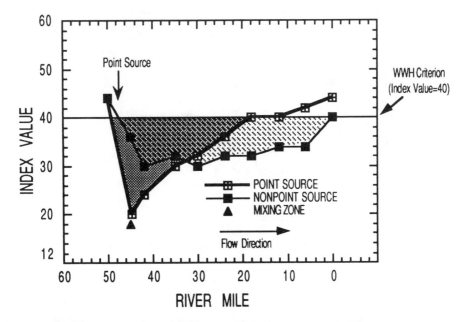

Figure 1. Portrayal of the Area of Degradation Value (ADV) concept used to estimate and quantify the severity of aquatic life impairment based on biosurvey data and ecoregional biocriteria. Examples include pre- and post upgrade of a point source (open squares) and the pattern expected for a nonpoint source impact (solid squares).

in direct terms. Frequency and duration considerations, which for chemical monitoring require large numbers of samples to adequately account for variation, are integrated by the resident aquatic life in the receiving water. Figures 2 through 6 portray biosurvey results from two Ohio rivers, each with a history of serious, but very different water pollution problems — the Scioto River and the Ottawa River. Each have been sampled over multiple years which is a prerequisite for trend analysis. The ADV statistics for these two rivers are summarized in Table 1. In both of these areas data is available before and after the July 1988 compliance deadline for municipal wastewater treatment plants (WWTPs) to meet water quality-based effluent limitations under EPA's National Municipal Policy. For most WWTPs throughout Ohio significant reductions in the loadings of pollutants such as oxygen demanding wastes, suspended solids, ammonia, nutrients, and in some instances heavy metals and other toxics has occurred. While overall loadings of conventional pollutants have declined in both rivers, differing biological results were evident in each. The Ottawa River has some significant and residual toxic chemical problems remaining, which have precluded the magnitude of improvements observed in the Scioto River. With the large number and varied types of chemicals (some of which go undetected) discharged into Ohio rivers like the Ottawa, measures that integrate the effects of all important physical and chemical perturbations are needed to accurately and realistically examine trends in water resource quality. The case examples will also be more easily understood after the reader has become familiar with the concepts and terms from Ohio EPA's biocriteria framework described in Chapter 9.

2.1.1 Scioto River

The Scioto River mainstem downstream from Columbus has been monitored frequently since 1979 over a distance of approximately 40 miles. The purpose of this monitoring has been to document changes as the result of the upgrading of the two major WWTPs, Jackson Pike (RM 127.2) and Columbus Southerly (RM 118.4). A major combined sewer overflow (CSO) discharges 2.6 miles upstream from the Jackson Pike WWTP (RM 129.8). Columbus also withdraws water for drinking purposes at RM 134. This effectively leaves the mainstem between this intake (RM 134.0) and the Jackson Pike WWTP (RM 127.1) with very little flow during the summer and early fall months. Two low-head dams form sizable impoundments in this area as well. CSOs discharge mostly during wet weather (dry weather overflows

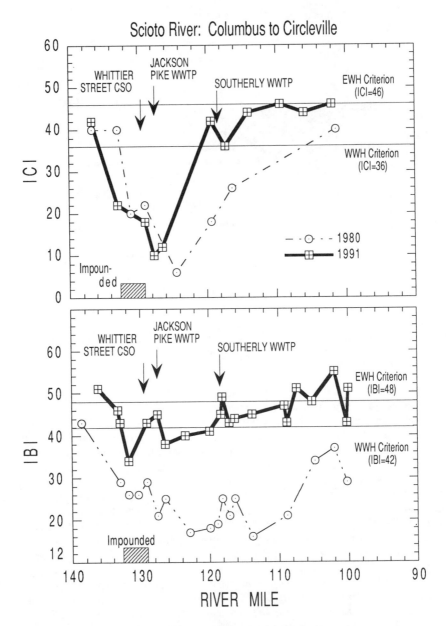

Figure 2. Longitudinal profile of the ICI (upper) and IBI (lower) in the Scioto River between Columbus and Circleville, Ohio during 1980 (circles) and 1991 (squares).

do exist, however) in the lower Olentangy River (confluence at RM 132.3) and to the mainstem between this point and the Jackson Pike WWTP. This arrangement of CSOs, impoundments, and water withdrawals is commonplace in Ohio cities and towns throughout the Eastern Corn Belt Plains, Huron/Erie Lake Plain, and portions of the Erie/Ontario Lake Plain ecoregions. Despite the aforementioned impoundments and flow alterations, overall habitat conditions in the Scioto River are good to excellent.

The Scioto River perhaps represents one of the best success stories of any river or stream in Ohio. Historically degraded since the early 1900s (Trautman 1981), a significant improvement in the fish community was evident between 1980 and 1991 (Figures 2 and 3). However, the macroinvertebrate community did not respond as quickly, although significant improvement was finally evident in 1991.

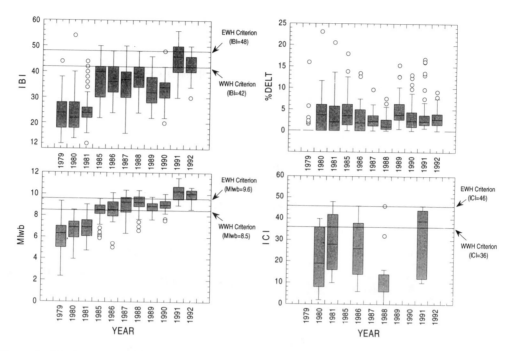

Figure 3. Boxplots of percent DELT anomalies on fish (upper left), MIwb (lower left), IBI (upper right), and ICI (lower right) for the Scioto River between Columbus and Circleville, Ohio for all years sampled between 1979 and 1992. Boxplots include the median (50th percentile), upper quartile (75th percentile), lower quartile (25th percentile), maximum, minimum, and outliers (values more than two interquartile ranges from the median).

Figure 3 portrays, in box-and-whisker plots by year, all 11 years of monitoring and reveals the broad improvement in the IBI and modified Index of Well-Being (MIwb) throughout the study area. The ICI trend has been less consistent and demonstrates the uncharacteristic lagging performance of this group in the segment most directly impacted by CSOs. Another important pattern with this example is the reversal in the response of the fish and macroinvertebrate communities (Figure 3). The years 1988 and 1991 were abnormally dry and both had extremely low flows. Conversely, rainfall and river flows during 1981 and 1986 were closer to normal. Although the ICI did not meet the Warmwater Habitat (WWH) biocriterion upstream from RM 109.4 in any year prior to 1991, the severity of the degradation was less in the normal flow years (Table 1). This was especially evident in the segment directly affected by the CSO discharges (Figure 2).

The combined effect of the upstream water withdrawals, impoundments, and remnants of past channel deepening have resulted in a physical environment that resembles a series of ponds during dry weather periods. The pools downstream from the Whittier Street CSO are unusually deep, which, combined with the artificially manipulated flow conditions, can result in an extremely long turnover rate. The duration of water turnover in each pool is correlated with the duration of the dry weather, low-flow periods. It is during these times that the products of the CSO discharges, urban runoff, and enrichment from upstream agricultural sources become concentrated in this area and the result is an extremely enriched aquatic environment. Attached algal growths completely smother and embed the coarse cobble/gravel substrates in the shallower run and riffle areas between pools and immediately downstream. The macroinvertebrate community in particular is composed of pollution tolerant taxa typical of this type of organic enrichment and, often, very high organism densities are observed. This is a key response signature of this type of impairment. The fish community has been less impacted than the macroinvertebrates as both the MIwb and IBI perform at least in the fair range and marginally met the WWH criteria in 1991.

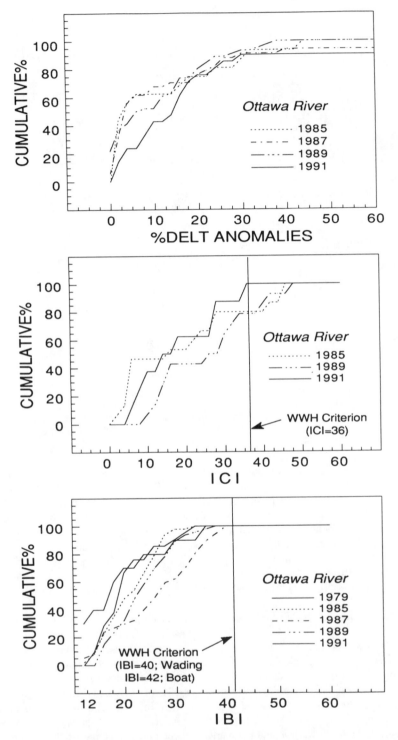

Figure 4. Cumulative frequency distribution of percent DELT anomalies in fish (upper), ICI (middle), and IBI (lower) in the Ottawa River between Lima, Ohio and the mouth based on data collected during five years between 1979 and 1991.

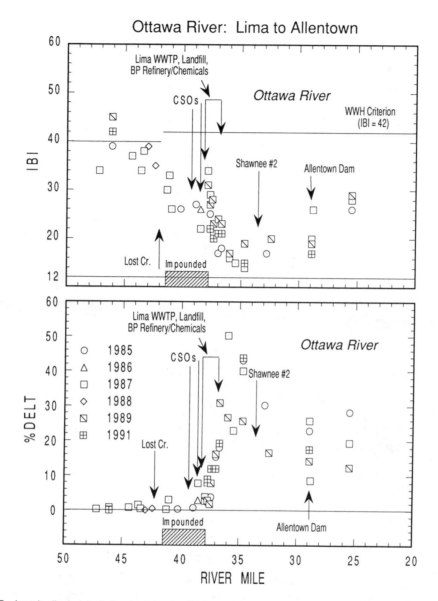

Figure 5. Longitudinal plot of all values for the IBI (upper) and percent anomalies on fish (lower) from Lima to Allentown, Ohio for six years between 1985 and 1991.

However, the incidence of external anomalies (mostly eroded fins and lesions) has been elevated (>5 to 13%) downstream from the CSOs and both WWTPs, and fish community indices (IBI and MIwb) tended to be lower during the high-flow years of 1989, 1990, and 1992.

The recovery exhibited by both organism groups since the late 1970s is best correlated with reduced loadings of sewage at both WWTPs and, to a lesser extent, the CSOs. The fish community showed a substantial response to the near elimination of raw sewage bypassing at the Columbus Southerly WWTP after 1982. The improvements observed in 1988 include partial attainment of the Exceptional Warmwater Habitat (EWH) criteria in the lower section of the study area. In 1991, full attainment of EWH was observed upstream as far as RM 109.4, and partial attainment at RM 114.0. This would likely increase if siltation impacts from nonpoint sources could be reduced. Historically, the Scioto River has been severely degraded with the entire mainstem from Columbus to the Ohio River having a dissolved oxygen

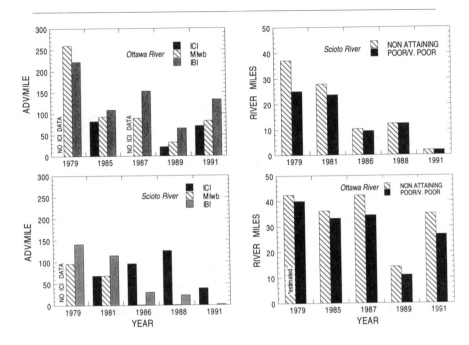

Figure 6. Comparison of ADV-per-mile statistics for the ICI, MIwb, and IBI from the Ottawa River (upper left) and Scioto River (lower left), and the miles of WWH nonattainment and miles of poor and very poor performance for the Ottawa River (upper right) and Scioto River (lower right).

content of near zero (and presumably very degraded aquatic communities) as recently as the late 1950s. In 1991, the total non-attaining miles shrank from 37.2 in 1979 to only 1.9 in 1991 (Table 1). The miles in full attainment of WWH increased markedly from 2.1 miles in 1979 to 24.1 miles in 1991, whereas the miles in partial attainment held steady after 1981. Thus, the increase in full attainment miles has resulted largely from a reduction in non- and partially attaining miles. While these observations demonstrate the beneficial effect of reduced loadings from the traditionally regulated point source discharges, underlying impacts are beginning to emerge. These include organic enrichment due to CSO discharges and urban runoff (which can be exacerbated by altered hydrology and habitat), and siltation from agricultural and urban sources.

2.1.2 Ottawa River

The Ottawa River in and downstream from Lima has been monitored frequently by Ohio EPA since the early 1970s, and infrequently since the turn of the century. The principal point sources include the Lima WWTP and CSOs, and two industrial facilities, a petroleum refinery, and a chemical manufacturing plant. There are similarities between this area and the Scioto River in that each river flows through an extensively urbanized area, each is impacted by CSOs, each has little upstream flow for dilution of point sources during dry weather, low-flow periods, and both are dominated by effluent flow downstream from the major point sources. The key difference is with the contributions from the two major industrial sources to the Ottawa River, which are essentially lacking in the Scioto River. Significant improvement in the biological indices has been observed during the past twenty years (Figure 4; Table 1), which has been due in large part to significant reductions in point source loadings. The improvement is even more remarkable when conditions since the turn of the century, when the river lacked any fish life for a distance of nearly 40 miles (Ohio EPA 1992a), are considered. However, the results of biological sampling conducted between 1985 and 1991 show that the Ottawa River has not attained full recovery in terms of the WWH biocriteria (Figure 4; Table 1). The most severely impaired segment remains downstream from the Lima WWTP and the British Petroleum (BP) Oil refinery and BP Chemicals facility where aquatic community performance remains poor. Compared to the Scioto River, the Ottawa River exhibits evidence that, despite some improvements, makes it one of the most severely impaired rivers in the state.

Table 1. Area of Degradation Value (ADV) Statistics for Selected Sampling Years in the Scioto River and Ottawa River Study Areas

Stream Index	Biological index scores				ADV statistics			Attainment status (miles[1])			
	Upper RM	Lower RM	Minimum	Maximum	ADV[2]	ADV/ Mile[3]	Poor/VP ADV[4]	Full	Partial	Non	Poor/VP[5]
Scioto River (1979)											
ICI (no data available)											
MIwb	140.0	100.2	4.0	8.7	3810	96	227	2.1	0.9	37.2	25.1
IBI			12	42	5618	141	1403				
Scioto River (1981)											
ICI			10	48	2691	68	151				
MIwb	140.0	100.2	5.2	8.4	2695	68	68	1.7	10.6	27.9	23.7
IBI			18	43	4582	115	541				
Scioto River (1986)											
ICI			6	46	3155	96	272				
MIwb	133.0	100.2	7.4	9.8	35	1.1	0	10.1	13.3	10.3	9.5
IBI			27	45	980	30	0				
Scioto River (1988)											
ICI			0	46	4147	126	689				
MIwb	133.0	100.0	7.4	10.0	35	1.1	0	9.7	11.6	12.4	12.4
IBI			28	47	750	23	0				
Scioto River (1991)											
ICI			10	46	1432	39	18				
MIwb	136.3	100.0	8.3	11.2	0	0	0	24.1	11.0	1.9	1.9
IBI			30	55	84	2.3	0				
Ottawa River (1979)											
ICI (no data available)											
MIwb	46.0	22.9	0.7	8.3	6015	260	806	0.6	0.0	23.6	22.2
IBI			12	36	5117	222	2255				
Ottawa River (1985)											
ICI			2	46	3734	82.4	532				
MIwb	46.1	0.8	2.3	8.8	4200	92.7	259	3.9	6.3	36.2	33.2
IBI			15	39	4959	109	1610				
Ottawa River (1987)											
ICI (no data available)											
MIwb	46.1	1.2	3.2	8.9	4025	89.6	296	3.1	0.6	42.3	34.4
IBI			14	42	6868	153	3757				
Ottawa River (1989)											
ICI			10	48	1001	22.1	7				
MIwb	46.1	0.8	3.9	9.0	1470	32.5	103	17.4	14.7	14.3	11.0
IBI			17	45	2978	65.7	517				
Ottawa River (1991)											
ICI			6	36	3209	70.8	85				
MIwb	46.1	0.8	3.4	9.2	3755	82.9	287	2.0	9.3	35.1	26.8
IBI			15	42	6019	133	1345				

Note: ADV is used here to demonstrate changes in the extent (miles) and severity (departure from the criterion for each index) of biological community impairment during representative survey years both before and after pollution abatement efforts at major point source discharges.

[1] Includes extrapolation of the results for a limited distance upstream and downstream from the upper and lower extremes of the study area.
[2] ADV: the Area of Degradation Value calculated for the study area.
[3] The average ADV per mile, which normalizes comparison between different years and study areas of differing length.
[4] The portion of the total ADV that is contributed by poor and very poor (VP) index values.
[5] The nonattaining miles within which biological index values are in the poor or very poor (VP) range of performance.

The combination of low fish species and macroinvertebrate taxa richness, a predominance of toxic tolerant species and taxa, and an unusually high incidence of external anomalies on fish point to a residual toxic impact in concert with low dissolved oxygen and organic enrichment. The incidence of deformities, eroded fins, lesions, tumors (DELT) anomalies was highest immediately downstream from the three major point sources exhibiting a precipitous increase (Figure 5). This corresponded to an almost equally precipitous decline in the IBI and MIwb. The ICI also exhibited the lowest scores in this area. A cumulative frequency distribution (CFD) analysis of different data aggregations (Figure 4), shows that the highest IBI scores occurred in 1987 and 1989, the lowest percent DELT in the same years, and the highest ICI in 1989 (no ICI results in 1987). Thus, the results from the most recent year (1991) have actually declined unlike what has been observed in the majority of streams and rivers throughout the state. In addition to the permitted point source discharges, spills and unpermitted discharges have been a prominent factor with 31 **reported** incidents since 1985. In addition, the cumulative pattern of noncompliance with NPDES permit limits by the three major point sources between 1989 and 1991 shows that there is a high probability that one or more of the sources will record at least one violation in any given month. Although none of the three sources is individually considered to be in significant noncompliance, the cumulative frequency of violations would constitute a significant violation. While the chemical sampling results showed exceedences of chemical water quality criteria mostly for dissolved oxygen and ammonia-N (chronic thresholds only) this alone did not indicate the full magnitude of the biological problems. Sediment chemistry results showed highly to extremely elevated heavy metals in the nine-mile long segment with the highest biological stress indications and biochemical markers from individual fish showed stress responses to organic chemicals (Ohio EPA 1992a). Thus, the cumulative frequency of NPDES violations, spills and other episodic discharges, elevated sediment metals, and low dissolved oxygen combine to create frequently occurring episodes of stress, which are much more apparent in the biological results than the chemical water column results. These results also point to the difficulties in relying on NPDES permit compliance **alone**, especially in complex areas such as the Ottawa River.

2.2 Trend Assessment

The statistics generated by the ADV and the longitudinal analysis of the community indices for different years provides information to evaluate the magnitude of change over time using several different dimensions of the change. Not only can the relative comparisons of magnitude (i.e., ADV statistics) be made, but the extent of various levels of condition (miles of nonattainment and miles of poor and very poor performance) can also be portrayed (Table 1). This is graphically illustrated in Figure 6 where the ADV per mile, the miles of nonattainment of the applicable aquatic life use designation, and the miles of poor or very poor performance for the Scioto River and the Ottawa River for selected sampling years during 1979 through 1991 are compared. Although the rivers are of different sizes, the relative magnitude of the impacts from point sources, flow diversions, and urbanization are similar. From the information portrayed in Figure 6 the Ottawa River is and has been more severely impacted than the Scioto River as indicated by higher ADV per mile values for each of the three indices. Also, whereas the decline in the ADV per mile has been consistent for the Scioto River, it has been much less so for the Ottawa River. This is even more evident in the comparison of the nonattaining and poor/very poor miles where the number of these miles was nearly eliminated in the Scioto River by 1991. The Ottawa River showed improvement in this category through 1989, but both categories increased in 1991. This information, combined with the biological community responses and knowledge of the sources and environmental setting, shows that the Ottawa River continues to be impaired by toxic impacts. The Scioto River has never had any significant problem with toxics, which is certainly evident in the biological responses. Problems do remain, but all of the indicators point to conventional problems associated with the municipal sewer system, particularly CSOs.

The preceding examples show how biosurvey information and biocriteria function as a direct measure of biological integrity, provide an unambiguous goal for a waterbody, and provides a way for water quality managers to visualize ambient conditions and evaluate the effectiveness of administrative programs and decisions. It also allows a quantification of observed departures from clean water goals beyond the traditional pass/fail framework. Integration of the trend assessment and ADV-based approaches outlined above into existing water quality management programs should increase accuracy, sensitivity, and discrimination in assessments of water resource integrity. This, in turn, should result in a more defensible allocation of funds and more efficient use of resources for accomplishing realized water quality improvements.

3.0 ABILITIES AND LIMITATIONS OF BIOSURVEY RESULTS TO DISCRIMINATE IMPACTS: BIOLOGICAL RESPONSE SIGNATURES

The availability of a comprehensive, standardized ambient biological database from a variety of environmental settings in Ohio has permitted certain patterns and characteristics of biological community response to perturbations to be identified. A common criticism of ambient biological survey data has been of its inability to determine the cause or source of an impaired condition (USEPA 1985; Suter 1993). While this is probably a valid concern for single-dimension indices (e.g., $\overline{d}$, $\overline{H}$, number of species, biomass, etc.) it does not apply equally to the new-generation multimetric indices such as the Index of Biotic Integrity (IBI) and Invertebrate Community Index (ICI).

When the response patterns of the various metrics and components of these indices were examined from areas where the predominant impairment causes and sources are well known, some consistent patterns emerged. Unique combinations of biological community characteristics that aid in distinguishing one impact type over another are referred to as *Biological Response Signatures*. These proved valuable in delineating predominant causes and sources of the aquatic life use impairments identified in the 1990 and 1992 305(b) reports (Ohio EPA 1990b, 1992b) and the many basin/subbasin assessments accomplished each year.

The original effort included a database made up of 25 similarly sized streams and rivers (drainage area range 90 to 450 sq. mi.) from the Eastern Corn Belt Plains (ECBP) and Huron/Erie Lake Plain (HELP) ecoregions (Yoder 1991). The data were from surveys conducted between 1982 and 1989 and which followed Ohio EPA procedures (Ohio EPA 1987a,b, 1989a,b). One of six general impact types were assigned to each sampling site. A parallel effort evaluated techniques by which even more subtle differences might be further defined within the concept of Biological Response Signatures (Anderson et al. 1990). This involved the use of genetic algorithms employing artificial intelligence and machine learning techniques. One initial finding of this effort was the distinctive response of the sensitive species metric. This metric of the headwaters IBI combines the intolerant metric of the wading and boat site type IBIs with moderately intolerant species (Ohio EPA 1987b). This aggregation of the community data was by itself found to consistently indicate the Complex Toxic impact type with a reliability of 82% in the stream and river sizes outside of its assigned use in the headwaters IBI.

This original analysis is updated here to include a database of more than 1200 samples from more than 70 streams and rivers spanning the period from 1981 through 1992. Although 20 different impact types have been classified thus far, eight of the impact types were assessed in this analysis. These represent the most common types of impacts that occur in Ohio rivers and streams. In addition, this analysis was limited largely to the Eastern Corn Belt Plains and Huron/Erie Lake Plain ecoregions in order to minimize the influence of this important factor. The nine impact types used in the analysis are described as follows:

1. **Complex Toxic** — Impacts from the complex combination and interactions of major municipal WWTP and industrial point sources that comprise a significant fraction of the summer base flow of the receiving stream *and* where one or more of the following have occurred: serious instream chemical water quality impairments involving toxics, recurrent whole effluent toxicity, fish kills, or severe sediment contamination involving toxics has occurred. This may include areas that have combined sewer overflows (CSOs) and/or urban areas located upstream from the point sources.

2. **Conventional Municipal/Industrial** — Impacts from municipal WWTPs that predominantly discharge conventional substances (these may or may not dominate stream flows) **and** where no serious or recurrent whole effluent toxicity is evident **or** small industrial discharges that may be toxic, but which do not comprise a significant fraction of the summer base flow; other influences such as CSOs and urban runoff may be present upstream from the point sources.

3. **Combined Sewer Overflows/Urban** — Impacts from CSOs and urban runoff within cities and metropolitan areas that are in direct proximity to sampling sites. This includes both free-flowing and impounded areas **upstream** from the major WWTP discharges. Minor point sources may also be present in some areas.

4. **Channelization** — Areas impacted by extensive, large-scale channel modification projects and where little or no habitat recovery has taken place. Some minor point source influences may be present.

5. **Agricultural Nonpoint** — Areas that are principally impacted by runoff from row-crop agriculture, which is the predominant agricultural land use in the ECBP and HELP ecoregions. Some minor point source and localized habitat influences may be present.

6. **Flow Alteration** — Sites affected by flow alterations including controlled flow releases downstream from major reservoirs or areas affected by water withdrawals as the predominant impact.

7. **Impoundment** — River segments that have been artificially impounded by low-head dams or by flood control and water supply reservoirs and where this is the predominant impact type present.

8. **Combined Sewer Overflows/Urban with Toxics** — The same as impact type 3 (CSO/Urban Conventional) except that there is a significant presence of toxics. Included are municipal CSO systems with significant pretreatment programs and sources of industrial contributions to the sewer system.

9. **Livestock Access** — Sites directly impacted by livestock operations where the animals (mostly cows) have unrestricted access to the adjacent stream.

These impact types were assigned based on our knowledge of the types of pollution and disturbance sources and other chemical (water column, sediments), physical (habitat), and toxicological (bioassays) information from Ohio EPA's various water and effluent databases. This information is contained in basin/subbasin specific reports, the biennial water resource inventory (305b report), and the Ohio EPA Water Body System. One of these was assigned as the primary impact type to each of the more than 1200 sites sampled for fish and macroinvertebrates. Assignments were based on the predominant impact that was directly influencing the site at the time the sampling took place. The assignments were also based on the site-specific knowledge of the study area gained by Ohio EPA while conducting biological surveys of the 70-plus streams and rivers. The extent of spatial overlap between different impact types throughout the database is somewhat variable ranging from a clear predominance of a single impact type to the overlapping influence of two or three impact types. Secondary and ancillary impact types were simultaneously assigned in these instances. The key objective of this analysis is to determine whether or not the feedback gained from the biological community can communicate about and characterize these differences. The results are provided in Figures 7 through 12. This project, thus far, has concentrated on two- and three-dimensional analyses of IBI, MIwb, and ICI metrics and selected subcomponents.

3.1 Fish Community Responses

Three components of the fish community data were included in three-dimensional plots that illustrate the concept behind examining the combined biological response characteristics of each impact type. The IBI, MIwb, and the frequency (%) of DELT anomalies on individual fish are different expressions of the relative health of the fish community at a given location. The Complex Toxic impact type was compared on a three-dimensional basis to each of 7 other impact types (Figure 7). In each comparison the Complex Toxic impact type exhibited a fairly distinct clustering pattern as compared to the other impact types. The amount of overlap was least with the Flow Regulation and Agricultural Nonpoint Source types and greatest with the CSO Toxic impact type. The response characteristics that most typified the Complex Toxic impact type includes IBI <18 to 22 (median and 75th percentile values), MIwb <4.5 to 5.9, and DELT anomalies >15.1% (median value), the combined occurrence of which should uniquely characterize a site impacted by sources which were characterized as Complex Toxic. The CSO Toxic is the only impact type to have IBI and MIwb values that exhibited any substantial overlap with the Complex Toxic. The key distinguishing characteristic was that the CSO Toxic impact type had lower percent DELT values (Figures 7 and 8). This overlap is not surprising given that many of the CSO Toxic impacted segments were in the same streams and rivers as some of the Complex Toxic impacted segments. Toxic substances discharged by industrial and other sources to the local WWTP also can enter the receiving streams and rivers via the CSOs. The other impact types also had some metric and biological index values that overlapped with the Complex Toxic impact type, but this was usually at the extreme ends of the ranges for each impact type and generally represented the transition from the Complex Toxic impact to another type of impact as distance from the source increased. Many of the Complex Toxic impacted sites with metric and index values that overlapped with other impact types were generally in the reaches of the various study areas where recovery was taking place along the longitudinal continuum. Thus, the differences between the Complex Toxic impact type and some of the other impact types are transitional, being more so for some types than others. In addition, even though one or more metrics or indices had overlapping values between the Complex Toxic impact and other types, very seldom do they have **all** characteristics in common. In the three-dimensional analysis (Figure 7) it was the percent DELT IBI metric that seemed to be the key discriminatory attribute in separating the Complex Toxic impact from the other impact types.

The classification of sites according to the previously described impact types was done strictly on a cultural basis, i.e., the selections were made based on the types of sources present in a study area. A few

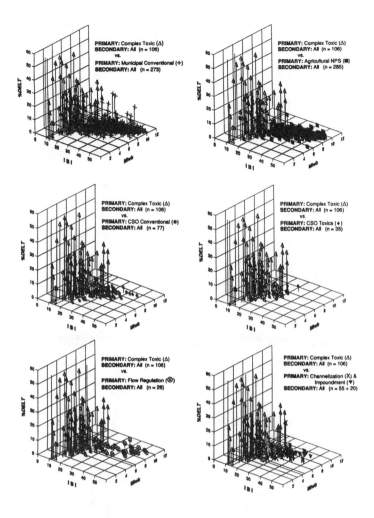

Figure 7. Three-dimensional comparison of the IBI, MIwb, and percent DELT anomalies among seven different impact types for 1214 electrofishing samples primarily from streams and small rivers in the ECBP and HELP ecoregions. The Complex Toxic impact type is compared to six other impact types described in the text. The sample size (n) is for primary and secondary impact types combined.

sites classified as the Agricultural Nonpoint Source impact type exhibited metric and index value combinations more consistent with the Complex Toxic impact type. Upon further investigation it was learned that one site was downstream from an experimental agricultural conservation tillage demonstration plot where pesticide usage was atypically intensive. The resultant biological response confirmed that the community response at this site better fit the Complex Toxic impact type in terms of the community response. Thus, the instream biota provided a more reliable indication of the type of impact than did the cultural characterization. Thus, the availability of the biological response signatures should fulfill the role as a guide for discovering and characterizing previously unknown problems.

We also examined all nine of the impact types in a two-dimensional framework. Some highlights of this analysis (Figures 8 and 9) of the IBI metrics and fish community indices are

1. Metric values and fish community index scores for the Complex Toxic and CSO Toxic impact types consistently indicated the lowest quality for the IBI, MIwb, darter species, percent round-bodied suckers, sensitive species, percent DELT anomalies, intolerant species, percent tolerant species, and density (less tolerant species).

2. Channelization had similarly low or even lower metric values for percent round-bodied suckers, intolerant species, sunfish species (lowest), percent top carnivores, percent simple lithophils, and

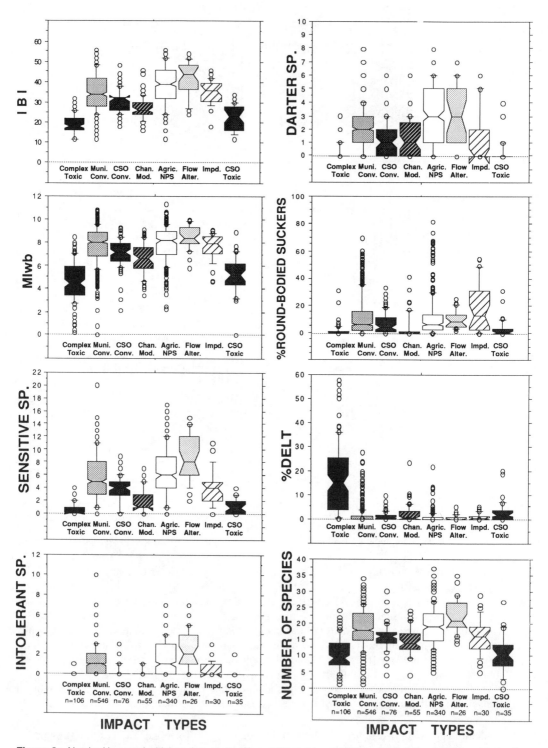

Figure 8. Notched box-and-whisker plots of six different IBI metrics, tne IBI, and MIwb by eight impact types for 1214 electrofishing samples primarily from streams and small rivers in the ECBP and HELP ecoregions.

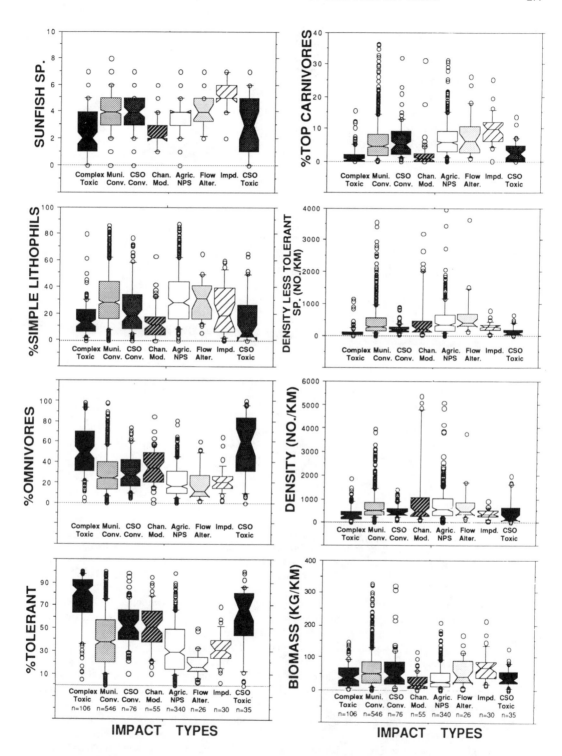

Figure 9. Notched box-and-whisker plots of six different IBI metrics, density (relative numbers), and biomass by eight impact types for 1214 electrofishing samples primarily from streams and small rivers in the ECBP and HELP ecoregions.

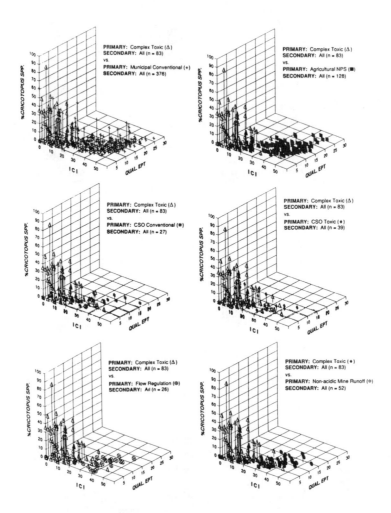

Figure 10. Three-dimensional comparison of the ICI, qualitative EPT taxa, and percent *Cricotopus* spp. among seven different impact types for 520 artificial substrate sampling sites primarily from streams and small rivers in the ECBP and HELP ecoregions. The Complex Toxic impact type is compared to six of the other impact types described in the text. The sample size (n) is for primary and secondary impact types combined.

biomass. In addition, Channelization exhibited some of the highest maximums and outliers for density (including tolerants). However, other metric values indicative of any toxic impacts were not evident for this impact type (e.g., percent DELT anomalies were very low).

3. Metric and index values indicative of good and exceptional performance were most frequently observed for the Flow Alteration and Agricultural NPS impact types followed by Conventional Municipal. Fair and poor performance was also observed under these types especially under extended low flows due to water withdrawals, higher effluent loadings, and/or more intensive land use and riparian impacts.

4. The incidence of extreme outliers was the most evident for the Municipal Conventional and Agricultural NPS impact types for darter species, percent round-bodied suckers, percent DELT anomalies, intolerant species, number of species, percent top carnivores, percent simple lithophils, density (less tolerants), density (including tolerants), and biomass. This indicates a wide range of biological performance within these two impact types from exceptional to poor due to the variable character of the local and segment-specific impacts. The extreme range of outliers in the Municipal Conventional impact type for percent DELT anomalies greater than 10% was observed mostly within or in close proximity to WWTP mixing zones.

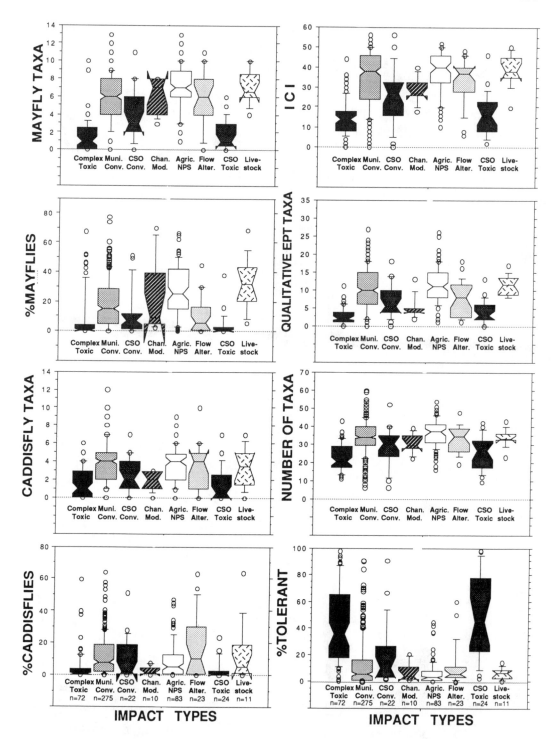

Figure 11. Notched box-and-whisker plots of seven different ICI metrics and the ICI by eight impact types for 520 artificial substrate sampling sites primarily from streams and small rivers in the ECBP and HELP ecoregions.

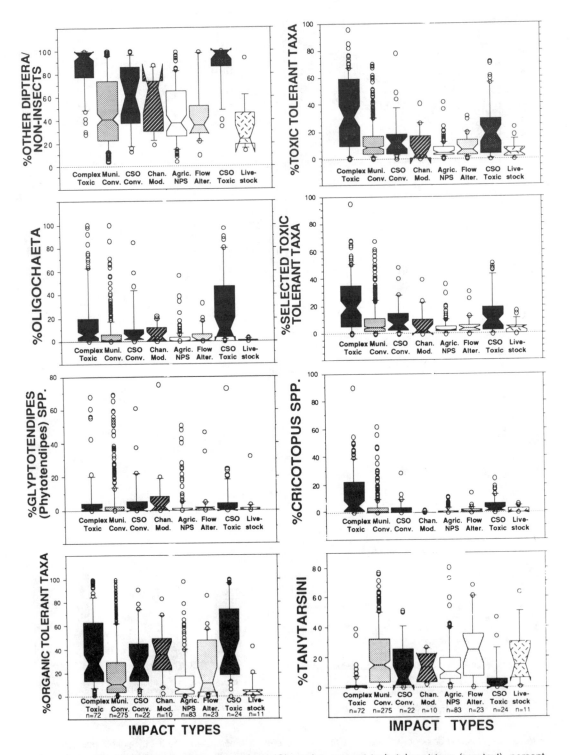

Figure 12. Notched box-and-whisker plots of two ICI metrics, percent toxic tolerant taxa (see text), percent selected toxic tolerant taxa (see text), percent *Glyptotendipes* (*Phytotendipes*) spp., percent organic tolerant taxa (see text), and percent *Cricotopus* spp. by eight impact types for 520 artificial substrate sampling sites primarily from streams and small rivers in the ECBP and HELP ecoregions.

3.2 Macroinvertebrate Community Responses

For macroinvertebrates we examined the relationship between the ICI, qualitative EPT taxa, and several other ICI metrics and other subcomponents on a three-dimensional basis for seven of the impact types. These included two of the ICI metrics (percent tolerant taxa and percent Tanytarsini midges) and three other macroinvertebrate community aggregations including percent Oligochaetes, percent toxic tolerant taxa*, and percent *Cricotopus* spp. Of the macroinvertebrate community aggregations the percent *Cricotopus* spp. seemed to differentiate the Complex Toxic impact type better than any other ICI metric or aggregation tested, much like the percent DELT anomalies did for the fish community results. The three-dimensional plots revealed a fairly reasonable separation of the Complex Toxic impact type from the six other impact types tested (Figure 10). For the Complex Toxic impact type the response charac-teristics include the combination of an ICI <14 to 18 (median and 75th percentile values), Qualitative EPT <2 to 4 (median and 75th percentile values), and percent *Cricotopus* spp. >5% (median value), which should uniquely characterize a site impacted by sources that were characterized as Complex Toxic. As with the fish community, the closest overlapping impact type was CSO/Toxic, with this impact type having a lower percent *Cricotopus* spp. Members of this latter group were found to comprise a toxic assemblage indicative of toxicity in North Carolina streams (Eagleson et al. 1990). However, Barbour et al. (1991) point out that *Cricotopus* (*Nostocladius*) spp. is an important component of pristine, western montane streams and is atypical of other members of this genus with regard to pollution tolerance. Simply eliminating this taxon would sufficiently protect the analysis where it is present.

This analysis demonstrates the need to access and interpret community information beyond the ICI and the individual metrics. Yoder (1991), using a smaller data set, showed the relationship between the ICI and organism density (number per sq. ft.) in a two-dimensional framework to demonstrate this point. In the comparison of the Complex Toxic and CSO/Urban impact types the ICI alone yielded similarly low results for each. Thus, the ICI alone was unable to discriminate between these impacts. However, organism density, which is not a direct component of the ICI, yielded a separation of the two impact types.

We also examined all nine of the impact types in a two-dimensional framework. This included the ICI, nine of the ten ICI metrics, and six additional aggregations of the macroinvertebrate data (Figures 11 and 12). The analysis revealed the following:

- The ICI, qualitative EPT taxa, percent tolerant, mayfly taxa, percent mayflies, percent caddisflies, percent other dipteran/noninsects, percent Tanytarsini midges, percent *Cricotopus spp.*, percent selected toxic tolerant taxa, percent toxic tolerant taxa, and percent Oligochaeta consistently indi-cated the lowest quality for the Complex Toxic and CSO/Toxic impact types. The inclusion of Oligochaeta as a toxic tolerant group contrasts with the original findings of Brinkhurst (1965), but agrees with the findings of Eagleson et al. (1990) and others (Chapman et al. 1980; LaPoint et al. 1984).
- The ICI metrics percent caddisflies and qualitative EPT taxa were as low for the Channelization impact type as for the Complex Toxic and CSO/Toxic types. However, for many of the other ICI metrics and other aggregations the results were much different. This shows the need to examine the response patterns of a number of metrics and other aggregations of the community data.
- The percent *Glyptotendipes* (*Phytotendipes*) spp. showed a tendency toward higher values for the Conventional Municipal impact type, particularly as outliers, which would indicate a high degree of organic enrichment. This group also exhibited high values for CSO/Urban, Channelization, and in the form of outliers, the Agricultural NPS impact type. None of these appeared to be entirely distinctive and overlapped with the Complex Toxic and CSO/Urban impact types. The other aggregations used to indicate organic enrichment (percent Oligochaeta and percent organic tolerant taxa*) did not discriminate as well and were more indicative of the Complex Toxic and CSO/Toxic impact types.

---

* Two aggregations of toxic tolerant taxa were used: (1) percent toxic tolerant taxa includes all *Cricotopus* spp., *Dicrotendipes simpsoni*, *Glyptotendipes barbipes*, *Polypedilum* (P) *fallax* group, *Polypedilum* (P) *illinoense*, *Nanocladius distinctus*, *Hesomyia senata* or *Thienemannimyia norena*, Conchepelopia, *Natarsia* sp. A, and *Ferrissia* sp. and (2) percent selected toxic tolerant taxa includes *Cricotopus* spp., *Dicrotendipes simpsoni*, *Glyptotendipes barbipes*, *Polypedilum* (P) *fallax* group, *Polypedilum* (P) *illinoense*, and *Nanocladius distinctus*.

3.3 Implications for Minimum Dataset Requirements

Minimum data requirements for using the Biological Response Signatures includes having sufficient information to employ the use of multimetric evaluation mechanisms, a standardized approach to data collection, and consistent and responsive management of the database. Other factors that further increase the analytical power of the biological data includes the use of multiple organism groups, an integrated approach to conducting assessments, and the inclusion of ancillary data such as biomass for fish. It is important to make these and other data collection decisions early in the process. An example of the importance of these early decisions was the recording of external anomalies on fish, which several years later allowed the development of the percent DELT metric of the IBI. This turned out to be a key metric in being able to discern the Complex Toxic impact type. At the time we did not realize that this use would exist; however, our failure to include it as a quantitative measurement early in the process would have constituted an unfortunate and irreplaceable loss of critical information. Another key decision was to identify midges to the genus/species level rather than limiting this to the family level. A failure to do this early on would have negated the use of the genus *Cricotopus* in the Biological Response Signatures. Thus, the ability to utilize biological data for diagnosis in Ohio was strengthened as the result of decisions made more than ten years ago not only about which organism groups to sample, but the types of information that were recorded. Frequently, biological monitoring programs are pressured to sacrifice data quantity and quality to meet regulatory and perceived financial constraints. As seen here, decisions made early in the process can have some far-reaching consequences in the long term.

4.0 CONCEPTUAL MODEL OF COMMUNITY RESPONSE

Another way to describe the attributes of ambient biological data for characterizing different types of environmental impacts is with a conceptual model. Figure 13 shows a model of the response of a fish community to increasing stress from a least impacted to severely degraded condition. The comparison of numbers and/or biomass with the IBI is used to demonstrate this conceptual relationship. The same general relationships would also apply to key macroinvertebrate community characteristics and the ICI. Beneath the graphic are narrative descriptions of biological community characteristics, chemical conditions, physical conditions, and examples of environmental perturbations that are typical of biological community responses across the five narrative performance classes. These are necessarily general and are not invariable. However, this model was developed from Ohio EPA's experience in analyzing biological, chemical, and physical data over a 15-year period and on a statewide basis and was further corroborated by the preceding analyses associated with the impact types and Biological Response Signatures. Thus the model has a good foundation in the observation of actual environmental conditions and associated biological community responses.

Results of the IBI from four similarly sized streams and rivers, each with different impact types and varying biological community performance, were plotted together (Figure 14) in an attempt further visualize these concepts. The results demonstrate the utility of using the theoretical range of the IBI (or ICI, MIwb, etc.) to differentiate between and interpret different types of impacts. The left column along the y axis in Figure 14 lists a gradient of impact types associated with the vertical scale of the IBI and the actual impacts present in each of the four areas listed in the right column. Lotic biological communities may experience spatially different impacts on a longitudinal basis with the degree of departure from the applicable biocriterion and recovery dependent on the severity and type(s) of impacts present; the Hocking River is one such example (Figure 14). Relatively unimpacted systems (e.g., Big Darby Creek) or those with moderate departures (e.g., Alum Creek) frequently reveal impairment within the fair range of performance while others may be uniformly devastated (e.g., Rush Creek), and reflect poor or very poor performance. These examples correspond to the narrative descriptions of community response and the attributes of the various impact types listed in the left column of Figure 14, and the range of

---

* Organic tolerant taxa includes the following: Oligochaeta, *Glyptotendipes* (*Phytotendipes*) spp., *Chironomus decorus* group, *Chironomus riparius* group, Turbellaria, *Physella* sp., *Simulium* sp., *Dicrotendipes lucifer, Dicrotendipes neomodestus,* and *Polypedilum* (Tripoda) *scalaenum* group.)

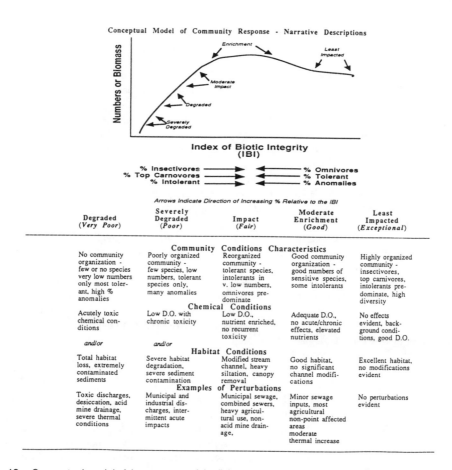

Figure 13. Conceptual model of the response of the fish community to a gradient of impacts in warmwater rivers and streams throughout Ohio.

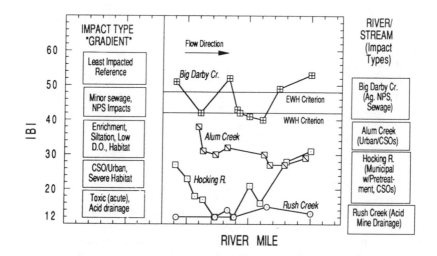

Figure 14. Biological response portrayed by the longitudinal trend in the IBI in four similarly sized Ohio rivers and streams with different types of point and nonpoint source impacts. Conceptual models of the impacts are also included on the left side under the heading Impact Type "Gradient."

performance described by the conceptual model (Figure 13). The important point here is that the biota integrates differing types and degrees of environmental impact along spatial and temporal scales and provides feedback that is inherently more accurate than we can predict or describe using cultural, surrogate, or process characterizations alone. Also, insight is gained on what can be expected as the predominant impact(s) in a particular segment change over time and/or space as a result of decreasing or increasing pollution levels (e.g., Hocking River). For example, we can expect a predominantly toxic impact to change temporally and spatially to a conventional impact when the sources of toxicity are controlled or eliminated. Different impact types are frequently "layered" in rivers and streams with the less severe impact types being masked by those that presently result in a more severe degradation (see Figure 1). As the more severe problems are reduced or eliminated the lesser problems may become evident in the results. Examples of this are presently being observed in Ohio streams and rivers where the abatement of historically severe municipal and industrial point source problems is leading to the unmasking of nonpoint source and other impacts (e.g., CSOs) that were always present, but not directly evident from the biological results.

Definite patterns in biological community data exist and can be used in the determination of whether or not a waterbody is attaining its designated use and in identifying the predominant associated causes of impairment. We have used this approach in producing the biennial 305b report, specifically the assignment of associated causes and sources of aquatic life use impairment (Ohio EPA 1990b, 1992b). While this does not represent a strict cause-and-effect analysis, it is intended to demonstrate causal associations based on a compilation of "lines of evidence." This involves utilizing all chemical, physical, biological, and source information as stressor, exposure, and response indicators to establish the most probable causes of observed biological impairments. Other uses of this approach include supporting enforcement and litigation proceedings and damage assessments. This information has been used to evaluate and verify claims made by NPDES permit holders that the degradation measured was due to factors (e.g., poor habitat or other sources) unrelated to their discharge. The Biological Response Signatures can be particularly useful in demonstrating that the observed degradation is likely related to specific discharges, especially those involving the Complex Toxic impact type. While the legal requirements of the Clean Water Act may be viewed as being sufficient to require entities to reduce pollutant loadings, the system of challenging these mandates requires the regulatory agency to defend the **reasonableness** of regulatory actions. This type of ambient biological response data can be particularly valuable in fulfilling that need.

The preceding discussion of the analysis of the capabilities of biological survey data to discriminate different types of impacts is contrary to several of the assertions of Suter (1993) in a critique of community indices. Although Suter (1993) made a number of thoughtful criticisms, the ones most relevant to the concept of Biological Response Signatures and the utility of the IBI and ICI include: (1) the IBI is composed of heterogeneous variables; (2) indices such as the IBI are too ambiguous to determine why an index value is high or low; (3) the inherent likelihood of eclipsing between different metrics (i.e., metrics cancel each other); and, (4) the IBI has no diagnostic power. The IBI comprises homogeneous variables as long as characteristics from a common organizational level such as fish, birds, chemicals, etc. are used. Many of Suter's concerns are based on the premise that they are heterogenous. There is merit to Suter's concern that an arbitrary mixing of variables, without any thought given to unintentional introductions of bias, compounding, and variance, is to be avoided. However, the charge that the IBI (and similar multimetric indices) are too ambiguous to determine why the index is high or low is refuted by the discussion of the Biological Response Signatures and the conceptual model of community response (Figures 13 and 14).

While we agree that an index value alone is almost meaningless, the information that results in an index value is obtained within the workings of a multidisciplinary framework where source, land use, and other chemical/physical water-related information are gathered and interpreted at the same time (e.g., lines of evidence within a causal associations framework). The chance that significant eclipsing will occur between different metrics is minimized and virtually eliminated by the careful construction of the IBI as modified by Ohio EPA (Ohio EPA 1987b). The example cited by Suter (1993) regarding the potentially eclipsing effect of organism density (numbers) and the frequency of diseases simply does not happen because of the use of only certain types of anomalies (e.g., the DELT anomalies) and the exclusion of highly tolerant species from the number of individuals metric. This latter modification prevents the mere increase in organism density from automatically producing a higher IBI value. The three-dimensional plots of the IBI, MIwb, and percent DELT anomalies also lend insight into and challenge the validity of Suter's concerns. The MIwb, comprising density and biomass components is **lowest** when the percent DELT is the **highest**, the antithesis of the eclipsing concern (Figure 7). This was also demonstrated in Figures 8 and 9 where organism density (including tolerants) was highest for the impact types (i.e., Municipal Conventional, Channelization, and Agricultural NPS) that exhibited much lower percent DELT anomalies than the Complex Toxic impact type, which generally exhibited the lowest density. A problem with the IBI occurs when organism densities are extremely low — the traditional scoring presumptions break down, particularly for the proportional metrics. To counter this problem we have devised a low-end scoring procedure (Ohio EPA 1987b). Thus, if the proper construction of the IBI (or similar evaluations) takes place up front, the eclipsing and other problems can be abated.

The most troubling of Suter's assertions is that the IBI has no diagnostic power. Over the linear scale of the IBI even gross changes have meaning (see Figure 13) with regard to the relative condition of the community and the types of impacts that are present. However, an IBI is not produced merely for the sake of generating a single-number-based evaluation. It is inherent in the IBI/regional reference site framework to simultaneously examine the aggregated and even the individual components in accomplishing an assessment of a particular waterbody. The use of key community attributes and metrics to examine if discernable patterns and combinations of responses exist for different types of cultural impacts (Figures 7 to 11) demonstrates that **some** level of diagnosis is possible. The Biological Response Signatures in combination with water quality, effluent, and habitat data was useful in demonstrating that the continuing impairment in the Ottawa River (Figures 4 and 5) was due to toxic impacts as opposed to habitat, the latter position taken by the regulated entities in an appeal of the use designation revision to WWH (see Chapter 9). While the application of this technology may not be entirely conclusive in every situation, it is informative and can lead to the cost-effective application of more powerful and intensive diagnostic tools (e.g., toxicity tests, chemical evaluation, and biomarkers). If the IBI is not used in this context then many of Suter's concerns will most likely retain their original validity.

5.0 SUMMARY

The preceding results demonstrate that discernable patterns in biological community information do exist and can be used to determine the probable cause of certain types of impairments, especially in combination with other source and instream chemical/physical data. This was best demonstrated for the Complex Toxic impact type, particularly for specific community indicators such as the percent DELT anomalies for fish and percent *Cricotopus* spp. for macroinvertebrates. The ability of these attributes **alone** was further improved when additional community indices, metrics, and other aggregations were included. The ability to discriminate between the other impact types varied widely with the ability to identify the CSO/Toxic impact type second only to the Complex Toxic type. Some attributes (number of sunfish species, percent caddisflies, and qualitative EPT) indicated the Channelization impact type reasonably well. However, there was a much broader overlap between the other impact types. Some of this is to be expected and is a result of the similarity of the environmental processes involved in each. For example, the Municipal Conventional, CSO/Urban, and Agricultural NPS impact types overlapped to a greater extent, which is not surprising since each involves the effects of nutrient enrichment. Examining finer aggregations of the community data than were done in this analysis could potentially increase the discriminatory power of the information to better separate these impact types.

These results point to the need to carry out biological monitoring to a prescribed level of effort, particularly in two areas: (1) taxonomic detail, and (2) the inclusion of multiple organism groups. With regard to taxonomy it was essential in our analysis to have the macroinvertebrates identified to the lowest possible level, particularly the midges. Some protocols do not require genus/species-level identifications for certain groups, which would have rendered the macroinvertebrate data much less useful in the analyses. For example, we would not have identified the utility of the genus *Cricotopus* in discerning the Complex Toxic impact type. National biological monitoring frameworks are currently debating this issue and should consider the future consequences of such choices. One possible conclusion from our analyses and the biocriteria framework (see Chapter 9) in general is that the increased "data dimensions" afforded by the more detailed taxonomy translates into more powerful and sensitive analytical tools such as the biological response signatures.

The inclusion of multiple organism groups also has significant benefits as well. The fish and macroinvertebrate assemblages each seem particularly well suited to defining different impact types and in combination further enhances this discrimination. For example, the fish community seems particularly well suited to identifying impairments due to macroscale influences such as habitat modification (i.e., Channelization and Impoundment impact types). In these situations the macroinvertebrates provide more specific information about water column effects as additional potential causes of impairment. Macroinvertebrates have been particularly responsive to indicating the severity of the CSO/Urban impact type. In addition, each group has different rates of recovery, the macroinvertebrates generally being quicker to respond over time than fish, but this has not been invariable. Eagleson et al. (1990), using a single indicator group (macroinvertebrates), cautioned about interpretations that could be confounded by seasonal variations and periodic absences of key indicator taxa. While we compensate for this potential problem via a standardized index sampling period, the inclusion of fish, which inhabit the receiving water year-round, can help to minimize this potential problem.

The Area of Degradation Value (ADV) provides a new tool to enhance the analysis of multimetric information beyond the traditional pass/fail assessments. By providing a quantified analysis of departures from a target index value (usually the ecoregional biocriterion) along a longitudinal continuum, both the severity and extent of an impairment can be evaluated. This allows for the analysis of incremental changes in aquatic community performance over space and time. As such, this has been a useful tool for demonstrating the effectiveness of pollution control efforts over time, and in comparing the severity of impairments between different stream and river segments. Future work is needed, however, to better utilize the ADV as part of resource value assessment such as NRDAs and in assessing other types of environmental damage claims.

ADV and Biological Response Signatures represent new tools that can be used as a key part of larger assessment efforts within which multiple goals are being pursued. For states, this would most commonly entail trend assessments and monitoring in support of various water quality management and regulatory programs such as water quality standards, NPDES permitting, and nonpoint source management and assessment. Clearly, other chemical, toxicological, physical, and source information (e.g., toxicity test results, pollutant loadings, permit violations, spills, etc.) must be used as part of the overall assessment process. The overall effort should focus on building improved "lines of evidence" in which the aquatic community response information is integrated with stressor and exposure indicator information to produce more accurate conclusions about likely causes and sources of observed biological impairments than would otherwise be accomplished relying on any single indicator class alone.

ACKNOWLEDGMENTS

This chapter would not have been possible without the many years put into field work, laboratory analysis, and data assessment and interpretation by members (past and present) of the Ohio EPA, Ecological Assessment Section. This includes the following: Dave Altfater, Mike Bolton, Chuck Boucher, Bernie Counts, Jeff DeShon, Jack Freda, Marty Knapp, Chuck McKnight, Dennis Mishne, Randy Sanders, Marc Smith, and Roger Thoma. We also acknowledge the helpful comments provided by Marc Smith and three anonymous reviewers.

## CHAPTER 18

The Role Of Ecological Sample Surveys In The Implementation Of Biocriteria

David P. Larsen

1.0 INTRODUCTION

Sample surveys have been used with great effectiveness in a variety of fields to describe the characteristics of populations that are too numerous to census efficiently. Best known are surveys associated with the national census, or with political polls (e.g., Gallup presidential election polls). Although sample surveys have been used for many years in a variety of arenas, their acceptance and use for characterizing natural resources has been limited. The National Agricultural Statistical Survey, not strictly a survey of natural resources, focuses on various aspects of crop production and produces annual summaries available to the public (Cotter and Nealon 1987). The Forest Inventory Assessment (Bickford et al. 1963; Hazard and Law 1989) also uses a survey approach and focuses on characteristics related to timber production. The National Wetlands Inventory uses a sample survey to estimate trends in the extent of wetlands in the conterminous United States (Dahl and Johnson 1991).

There are isolated cases in which sample surveys have been used to characterize aquatic ecosystems. The most notable of these are the national lake and stream surveys conducted by the U.S. Environmental Protection Agency (USEPA) under the auspices of National Acidic Precipitation Assessment Program (NAPAP) to characterize the status of lakes and streams relative to sensitivity to acid deposition (Linthurst et al. 1986; Messer et al. 1986; Landers et al. 1987; Kaufmann et al. 1988). These snapshots of the condition of lakes and streams covered one or two years. Some continuing surveys initiated with these lake and stream surveys extend over time to give a temporal dimension to the single-year snapshots. The USEPA recently extended these concepts in the creation of the Environmental Monitoring and Assessment Program (EMAP), which is designed to answer questions about the current status, changes, and trends in indicators of the ecological condition of the nation's ecosystems (Messer et al. 1991).

Some of the roots of the sample survey approach can be traced to the early part of this century when investigators were interested in describing the social and economic condition of urban and rural populations in this country and in Europe (Converse 1987). A variety of sampling approaches emerged, and there were numerous discussions and debates about which approach was best. Some were convinced that it was necessary to census the populations of interest, and there are some examples of incredibly time-consuming census surveys that still stand as remarkable achievements. Others recognized resource limitations and selected samples on which to make their observations. In some cases, samples of convenience were used: for example, interviewing the person on the street corner or sending out questionnaires and using whatever responses were returned. Others attempted to define a representative sample, such as a typical rural community or a typical city. The use of randomization in the selection of samples in these surveys was slow to emerge. One of the early statistical debates, between use of a representative sample and use of a random sample, eventually was resolved in favor of randomization (Converse 1987).

When the national population census was first established, the approach was to gather information about each person in the population. As survey techniques evolved, the national census changed from an extensive inventory to primarily a sample survey for most characteristics. However, certain characteristics are still censused. The evolution was slow, but the efficiency and utility of sample surveys indicated the effectiveness of this approach. Political polls are basically sample surveys to identify voter preferences and their changes over time. However, the evolution of survey techniques for polling followed much of the same track as the history of survey sampling oriented toward other social questions. It was not until the 1950s that Gallup achieved a refinement of his techniques that, with some modifications, are still used today, even though he began in the mid-1930s (Freedman et al. 1991). Marketing departments of many businesses use sample surveys to characterize populations and identify specific subpopulations that possess characteristics toward which particular products should be targeted.

This discussion of sample surveys is pertinent because it has been estimated that there are at least 460,000 lakes and 1.2 million perennial stream miles in the conterminous United States (USEPA 1992e). Clearly, to describe a national picture of the status of lakes and streams via an inventory of the resources is not feasible. Indeed, it is not feasible even at a regional or state level, for most states, to conduct an inventory of the condition of lakes and streams, particularly if regional or statewide trends in condition are of interest. Thus, there is clear benefit to adopting sample survey techniques for assessing the condition of lakes and streams.

This chapter introduces the elements of a sound survey design and illustrates the techniques for surface waters (lakes and streams) by describing several examples highlighting various components of sample surveys. In some cases, the focus of the surveys is on the chemical condition of lakes or streams; in others, the focus is biological. The role of sample surveys as part of a program assessing the biological condition of lakes or streams and incorporating biological criteria into the assessment process will be developed by hypothetical examples, drawn from surveys conducted for other purposes.

2.0 ELEMENTS OF A SAMPLE SURVEY

The fundamental feature of a sample survey design is that a portion of a population of interest will be surveyed and that measurements taken on the sample will be used to infer the character of the larger population. Effective sample surveys use randomization in the sample-selection process to assure that the selection of sample sites will be unbiased. The importance of randomization and the selection of a probability sample will be described somewhat later when rules for sample selection are discussed. Repeated application of the same sample survey over time allows the detection of trends in the condition of the target population.

Any field contains concepts and a vocabulary peculiar to that field. Survey sampling is no different. The concepts and vocabulary pertinent to survey sampling are described next, using examples illustrative of lakes and streams. Since this is a brief overview of a field that occupies numerous statistical texts, these concepts and illustrations can only be cursory. Four texts have been used as source material for the following overview of survey designs; they should be consulted for greater detail (Kish 1965; Cochran 1977; Williams 1978; Kish 1987). The primary intent is to alert the reader to the capabilities and utility of sample surveys for implementation of biocriteria, and to guide the reader to more extensive survey resources, even though those resources are not tailored to use of sample surveys of lakes and streams. Many of the finer points and details cannot be covered here, so the usual caveat holds: when you anticipate designing a sample survey, consult with a competent statistician familiar with sample survey techniques (a reasonably well-defined specialty within statistics). The illustrations are meant to ground the survey concepts and vocabulary to a resource type familiar to the readers of this book. The primary concepts to be described are populations, frames, sample units, and sample selection rules.

2.1 Populations

A *population* (also called a universe) consists of the collection of all the elements to be described by the survey. Sometimes these elements are called *sample units* or *members of the population*. The sample consists of a selected set of these population elements on which measurements are to be made. The population is the aggregate of all possible sample units. One of the first steps in developing any sample survey is to identify what is to be described as explicitly and unambiguously as possible: To what will the description of the sample apply? Although the concept appears simple on the surface, it can be deceptively complex in application because most natural resources are not unambiguously defined. It is important to refine the population definition as best as possible, and to debate the fuzzy edges until agreement is reached about the specifics of the definition. The inferred results are only as good as the specified population.

Defining a lake is a good place to begin discussing the dimensions of explicitly defining a population. A particular state agency may be interested in describing the condition of lakes within its borders; most people have a general perception of what is meant by a lake; however, the following types of questions are pertinent. Does the population of concern include reservoirs? Does the population consist of only publicly owned lakes? Are there size constraints on lakes to be considered in the population, e.g., lakes larger than 1 ha, but no more than 2000 ha; lakes larger than 10 acres? Is there a depth constraint on lakes, e.g., only lakes >2 m maximum depth are to be considered? How should lakes on state boundaries be considered — as part of the population or not? Should drinking water reservoirs be included? Should farm ponds meeting the specified size criteria be included in the lake population? Are there chemical conditions that exclude a lake from the population of interest, e.g., should saline lakes be included in the surveyed population? Whether a lake or stream is a member of the population of interest often cannot be determined until after reconnaissance or field measurements. Are there further geographic limits to the population of interest, such as specifying the population of lakes within wilderness areas, or within federal lands? The point is that careful thought is required in the specification of the target population.

From time to time, illustrations will be drawn from the surface waters component of EMAP designed to answer the status and trends questions about lakes and streams (Larsen and Christie 1993). For example, EMAP-Surface Waters adopted a very inclusive definition of a lake: a standing waterbody of >1 ha in surface area and >1 m depth at point of maximum depth, with >1000 m^2 open water at the time of field sampling, excluding the Great Lakes. Some aspects of this definition require further refinement. Under discussion is whether to include large lakes (>5000 ha) as part of the survey population, or to treat them differently. These lakes are rare, compared with other lakes, and from the perspective of character-izing their condition, could be considered separately. The interest is usually in the condition of individual large lakes and not in the collection of large lakes.

A set of decisions similar to those for lakes are also required for streams, in that the population of interest must be defined as explicitly as possible. Here again, it is important to state the definition in practical terms, so that a field sampler can recognize whether the site is a member of the population or not. For example, should the survey target wadable streams? Large rivers? Should intermittent streams be included? Irrigation canals? Tidally influenced segments? One consideration is whether to identify target subpopulations based on the primary type of sampling protocol. For example, backpack electrofishers or hand-held seines are generally used to sample wadable streams for fish. Nonwadable streams require small boats or rafts, and large rivers require larger boats. Thus, target subpopulations could be wadable streams (permanent flow), small rivers, and large rivers.

An initial decision faced in identifying stream populations is whether to describe the condition of streams in terms of stream segments or the total length of streams. More explicitly, is the resource agency interested in making statements such as, "25% of the stream segments do not attain specified biocriteria" or "25% of the stream miles do not attain specified biocriteria"? The decision about how the condition of the resource will be reported dictates how the sample selection rules will evolve. Most resource agencies seem to prefer to describe streams by the total length, e.g., miles or kilometers, in one condition or another, rather than describing a number of stream segments (USEPA 1992e).

2.2 Frame

Whereas the identification of the population or target population specifies what will be surveyed, the *frame* is the representation of that population. The frame sets up the mechanism by which the lakes or streams will be selected. The sample selection rules are applied to the frame to identify the actual set of lakes or streams to be visited. A simple example of a frame is a list of lakes in the particular area of interest. Often, it is not feasible, or necessary, to create an explicit list frame. In these cases, other devices serve the purposes of a frame well, e.g., maps such as the U.S. Geological Survey (USGS) topographic series, digital versions of these maps (the USGS digital line graphs), or the version contained in USEPA's

River Reach File (version RF3). Aerial photographs can also serve as a frame, as long as the elements of the population can be identified and selected from the aerial photos. A key feature of using maps or aerial photos as a frame is that a list could be created from them, that is, all elements of the population could be identified from the map or photo.

USEPA's Eastern Lake Survey (ELS), Western Lake Survey (WLS), and National Stream Survey (NSS), and U.S. Fish and Wildlife Service's National Wetlands Inventory provide some examples of lake and stream frames used in sample surveys. The ELS and the NSS used USGS 1:250,000-scale topographic series maps as the frame (Linthurst et al. 1986; Kaufmann et al. 1988; Landers et al. 1988; Kaufmann et al. 1991). For the ELS, all lakes appearing on these maps in the areas of interest were digitized and coded to create a list frame derived from the maps. For the NSS, the maps were used to select sites directly, so no lists were developed. For the WLS, 1:100,000-scale USGS topographic maps were used to derive a list frame (Landers et al. 1987). The National Wetlands Inventory used aerial photographs to identify and classify wetlands in a survey comparing the extent of wetlands in 1970 and in 1980. EMAP-SW is exploring the use of the USEPA's River Reach File (version RF3), derived from the digital version of the USGS 1:100,000-scale topographic series maps as a common frame for the selection of both lakes and streams.

Since a frame is only a representation of a population of interest, investigators must be on the lookout for errors in the correspondence between the frame and the true population of interest. Maps, as frames, can contain errors of two types: inclusion of erroneous population elements and omission of desired elements. Either of these types may be related to errors in coding or in identifying elements of the desired population. A mapped object might have been misclassified or the cartographer might have used a definition different from that specified for the survey. The cartographer might not have mapped an item that qualifies as a member of the population of interest. Similarly, when using aerial photographs, photointerpreters can misclassify or misidentify elements, and thereby include items that should not be included or exclude items that should be included. Well-designed surveys allow the calculation of both kinds of error rates — the frequency with which items are included that should not be included and the frequency with which omissions are apparent.

2.3 Sample Units and Samples

The term *sample unit*, although somewhat clumsy, refers to an element of the population — an individual lake, a stream segment, a point on a stream, or a lacustrine wetland. A survey sample consists of the collection of all the sample units selected for measurement; thus, a sample might consist of 50 lakes or 100 stream sites. As noted earlier, unambiguous description of the sample unit is critical, so that reconnaissance teams or field crews can determine whether the site visited is part of the surveyed population and ought to be sampled. Developing consistent definitions of the population, target population, frame, and sample units is critical to developing an effective survey. Any discrepancies should be resolved before actually drawing a sample to prevent as much confusion as possible.

A *sample selection rule* is the device by which the sample is drawn from the frame. Criteria that govern effective sample rules include:

- The sample is drawn by some method of random selection, or systematic selection with a random start.
- Every element of the frame population has a known probability (>0) of being included in the sample; otherwise, the sample is biased against that type of element.

Randomization is an important feature in sample selection because it ensures objectivity in the selection process. As a result, population inferences can be made with quantifiable uncertainty, i.e., objective statistical inference. Randomization also allows an objective evaluation of sources of error or sampling variability and fosters cumulative improvement by targeting troublesome sources of variability and refining designs or methods. Samples meeting these criteria are sometimes called statistical or probability samples.

In contrast, samples that do not meet these criteria (nonprobability samples) are limited to subjective inference; investigators are constantly challenged to identify possible biases, or to define how good the sample is. It is not possible to quantify the uncertainty with which population inferences are made. This type of sample is sometimes called an *informal* or *judgment* sample. Nonprobability sample selection includes:

- Samples of convenience, e.g, sampling accessible lakes or sampling streams from bridges
- Haphazard selection
- Typical or representative selection, which might be based on informed judgment

Inclusion probabilities quantify the chance that a particular sample unit will be selected. A simple example illustrates this concept. If a resource agency had sufficient resources to sample 50 lakes out of a specified target population of 1000 lakes, each lake would have a 50/1000 = 0.05 chance of being included in the sample, if a simple random sample were drawn from the population. The reciprocal of the inclusion probability is often called an *expansion factor*, because it describes how much of the population each sample element represents. If we were to make a measurement on a lake, such as its surface area, its mean depth, or any other property, we would infer that 20 lakes in the population would be similarly characterized. If half the samples contained a mean depth <5 m, we would infer that 500 lakes in the population had mean depths <5 m. Randomization and specification of inclusion probabilities permit objective, quantifiable population inferences. Without randomization and specification of inclusion probabilities, we are unable to make objective population inferences; we do not know what part of the population is represented by the sample.

3.0 EXAMPLES OF SAMPLE SURVEYS: LAKES AND STREAMS

This section illustrates the components of sound sample surveys by drawing from several lake and stream studies. The original citations should be consulted for a full description, as they contain much more detail than can be covered here. These examples are used to show how the important elements of a sample survey, described in Section 2, have been defined and to describe the kinds of conclusions that can be drawn from sample surveys.

3.1 Finnish Lakes (from Kämäri et al. 1990)

Finland contains about 56,000 lakes >0.01 km^2. In recent years, many researchers have been interested in the water quality of these lakes with regard to both acidification and nutrient enrichment. Since it was not feasible to monitor all the lakes, random subsamples of the lake population were selected. The target population consisted of all lakes meeting the following definition: a water pool with a surface area >0.01 km^2. For lakes in northern Finland, however, a definition of >0.1 km^2 was set. Lakes >10 km^2 and lakes heavily polluted by industrial and domestic wastewaters were excluded. These lakes are monitored individually, thus this subpopulation is censused.

The frame consisted of 1:50,000-scale topographic maps on which the lakes were depicted. The sample was selected in two stages. The first stage consisted of selecting target areas, proportional to lake density; the second stage consisted of selecting equal numbers of lakes from each of the areas selected during the first stage. This process delivered about 1000 lakes, or about 2% of the target lake population, on which chemical measurements were made. Because the lake sample was drawn using randomization for lake selection, results can be inferred to the entire target population of about 56,000 lakes. For example, this survey estimated a median total phosphorus concentration of 15 µg/l, with 79% of the lakes estimated to have total phosphorus concentrations <30 µg/l.

3.2 Eastern Lake Survey (from Linthurst et al. 1986; Landers et al. 1988)

As part of its National Surface Water Survey (NSWS), the USEPA focused attention on regions of the United States where acidic or acid-sensitive lakes and streams were expected to be found. One component of the NSWS, the ELS, targeted three regions in which to conduct lake surveys — the Northeast, the Upper Midwest, and the Southeast. A second lake survey, the WLS, addressed sensitive lakes in the west. Because the proportion of lakes that were acidic or sensitive to acid inputs was expected to be low, the ELS defined several strata before sample selection. Stratification permits a sampling effort to be allocated more efficiently than an unstratified sampling design, and if properly designed, stratification preserves the basic criteria of selecting a statistical sample. The three strata included (1) region (as just described), (2) ecoregion (areas of expected relative ecological homogeneity), and (3) within ecoregions, alkalinity class. In all, 33 strata were identified; 50 or more sample lakes were selected from each stratum. The initial population of interest included all lakes appearing on USGS 1:250,000-scale topographic series maps; all lakes appearing in each stratum were listed, creating a list frame from which to draw the sample lakes. Within each stratum, lakes were drawn using randomization to create the statistical sample.

A refinement to produce the target population excluded the following:

- Areas indicated as lakes on the maps but with no lakes present (map errors)
- Flowing water
- Bays or estuaries (high conductance)
- Lakes affected by intense urban, industrial, or agricultural activities
- Marsh/swamp areas
- Lakes too small to be consistently identified on the frame maps (<4 ha)

This refinement produced a clear definition of the target population to which inferences would be made subsequent to the field surveys. Not all the information for refining the target population was available before field visits; further refinement took place only after reconnaissance and field measurements. Because of the statistical nature of the lake selection, the proportion of the nontarget lakes could be tracked and inferences to the estimated target population could be made with identifiable uncertainty.

This design allowed the following kinds of inferences about the target populations of lakes:

- 12.9% of the lakes in the Northeast had pH <6.0, but in Florida, 32.7% of the lakes exhibited pH <6.0.
- Sulfate concentrations were high in the Northeast (>50 µeq/l in 97.3% of the lakes), but in subregions in the Southeast, sulfate concentration was estimated to exceed 50 µeq/l in <25% of the lakes.

3.3 Fish Communities in Lakes of the Upper Peninsula, Michigan (from Cusimano et al. 1989)

As part of the ELS (described in Section 3.2.), several satellite projects were developed. One of these described the status of fish populations in lakes in a region of the Upper Midwest, identified as Subregion 2B, with respect to lake acidity. This region covers much of the Upper Peninsula in Michigan. The first phase of ELS (ELS-I) characterized the chemical condition of lakes in the regions of interest. A subset of regions and lakes were chosen for the more detailed and targeted studies.

In Subregion 2B, the ELS-I lake population represented in the frame (1:250,000-scale USGS topographic maps) consisted of 1698 lakes, of which 1050 met the criteria for the target population. Of these, a sample of 254 was selected with probability methods; further screening and field visits reduced the number to 146 lakes, on which field measurements were made as part of ELS-I.

For the special study of fish communities, the target subpopulation was further refined by excluding another 41 of the 146 lakes; Cusimano et al. (1989) describes the specific exclusion criteria. Of the 105 remaining lakes, a subset of 50 (selected with probability methods) was chosen for the fish population surveys; one of these lakes was eventually dropped because of other ongoing fish survey activities. By virtue of the probability sampling (each lake selected had an associated expansion factor expressing the proportion of the population it represented), a revised target population size could be estimated as 597 lakes.

This is a fairly elaborate example of the application of probability sample selection through several stages. Important points to consider are (1) a clear definition of target populations was maintained throughout the selection process, and (2) explicit criteria were used to determine which lakes fell within the population of interest and which were to be excluded. Randomization was used whenever samples were selected from target subpopulations; individual lakes selected maintained their associated expansion factors throughout the selection process. This allowed inference to the full target population. In Subregion 2B in this example, the initial lake population was 1698, of which 1050 were targeted during ELS-I; this target population was further refined for purposes of the fish community assessment to 597 lakes. The information gathered from the lakes visited during the fish surveys was applicable to this refined target population of 597 lakes, not to 1050 or 1698 lakes.

The following are exemplary inferences of the survey, applicable to the target subpopulation of 597 lakes:

- Of these lakes 99.4% support fish.
- Game fish occur in 83.7% of this subpopulation of lakes.
- The most common species of fish in the subregion, yellow perch, occurs in 69.8% of the lakes.

3.4 National Stream Survey (from Herlihy et al. 1991; Kaufmann et al. 1991)

As part of the NSS, USEPA designed a project to estimate the regional extent, location, and chemical characteristics of acidic and low acid neutralizing capacity (ANC) streams in areas of the mid-Atlantic and southeastern United States sensitive to acidic deposition. A target population of 64,300 stream reaches defined the population of interest, from which a randomized systematic sample of 500 stream reaches was drawn. Chemical measurements were made on these 500 stream reaches during spring baseflow. The sample unit is a stream reach, defined as a blue-line headwater segment, or a segment between confluences, on USGS 1:250,000-scale topographic maps. The target population was limited to streams with watersheds <155 km^2 and not located within mapped urban areas. Also excluded were tidal reaches, small intermittent streams, and large streams and rivers. Each sampled reach was assigned a sample expansion factor (inversely proportional to its selection probability) equivalent to the number of reaches it represented in the target population.

Typical conclusions from this stream sample survey, conducted in 1986, included:

- Approximately 50% of the 64,300 stream reaches exhibited ANC < 200 μeq/l.
- 2.7% were acidic.

3.5 The EMAP Survey Design

One way to select a statistical sample of sites representing an ecological resource is to use a regular array of grid points randomly placed over a map of the resource of interest. This is similar to a systematic selection of sites along a linear feature, but in two dimensions. Systematic grids are useful for characterizing extensive resources. For example, the proportion (and area) of land in various land uses can be estimated by randomly placing a systematic grid over a representation of the landscape (such as an aerial photograph) and counting the number of grid points associated with each land type. Adding up the grid points of a particular land type and dividing by total number of grid points gives the proportion or extent of that land type. For resources that are discontinuous, such as lakes or streams, grid points rarely fall on the resource type of interest. In these cases, a search area defined around the grid point increases the probability of including the resource of interest. This is the basic philosophy used in EMAP in setting up a grid network that is usable for both extensive and discrete resource types. The size of the search area dictates the proportion of the population sampled.

EMAP's current design uses a triangular grid as the basic framework for sample selection; around each gridpoint is a hexagonal search area whose size (40 km^2) results in a $^1/_{16}$ (approximately 6%) area sample. Thus, intersection of this grid with a stream network captures about 6% of the stream length in the area of interest. If each lake is identified by a unique node point on a map, such as a point at the center of the lake, then the area sample will capture 6% of the lakes in the area of interest. Using this kind of framework on paper maps as frame materials is a slow process; thus, EMAP-Surface Waters takes advantage of digital versions of maps of lakes and streams. The sample selection can be performed electronically (with version RF3 of the USGS digital line graphs [1:100,000 scale]), facilitating the sample selection process.

The size of a 6% sample is too great for field visitation of all sites selected; however, some characteristics can be obtained relatively easily for this number of samples. For example, RF3 contains the surface area of lakes, and the Strahler stream order (as an approximation of stream size) can be calculated from the stream network. This information is useful for focusing the actual field sampling more efficiently than might be done otherwise. The proportion of small lakes and streams far outweighs the proportion of large lakes and streams; a straight random sample will draw sites in proportion to their distribution in the parent population. Thus the first-stage sample, drawn from the hexagon search areas, can be classified by size classes. Then a second-stage sample can be drawn, using randomization methods, to meet the sample size requirements for each size class. This is the basic process (excluding some technical detail) used to select lake and stream statistical samples for EMAP-Surface Waters. More information on the design and applications for lakes and streams can be obtained in Overton et al. (1991), White et al. (1992), and Larsen and Christie (1993).

4.0 ROLE OF SAMPLE SURVEYS IN DEVELOPING AND ESTABLISHING BIOCRITERIA

4.1 Development of Background Data for Establishing Biocriteria

The preceding sections describe how sample surveys can produce information that reflects the population from which the samples were drawn. Effectively designed sample surveys can be used to describe the range of conditions likely to be encountered in lakes and streams. The proportion of lakes or streams of particular types or condition occurring in the sample should reflect their proportional occurrence in the population: situations common in the population will be common in the sample; situations rare in the population will be rare in the sample. How might a sample survey be used in the development of biocriteria? Suppose a resource agency is in a position to begin a monitoring program that will gather information leading to the establishment of numeric biocriteria, and the agency has taken a number of preliminary steps that include the following:

- A decision is made to target fish and macroinvertebrates in streams as the organism groups on which to develop biocriteria, with ancillary information on physical habitat and water quality, and an identification of candidate metrics that might be used separately or in combination as an index of biological integrity of the system.
- Development of criteria to judge whether a stream meets a "least disturbed" condition and therefore could be used as a reference site; these criteria are largely based on land use, habitat information, and evidence of impact, including condition of the riparian zone and the stream channel (e.g., lack of channel modification); refer to Hughes, Chapter 4, which discusses approaches for characterizing least disturbed reference sites. Sites can also be characterized for "stressors" of various types.
- Decisions about site protocols (field collection methods, reach lengths), quality assurance/quality control procedures, logistics, and data management have all been made.
- Informal background monitoring on the biological condition of some streams has been conducted in conjunction with some site-specific studies, giving the resource agency some familiarity with the biota, but no basinwide, regional, or statewide surveys have been conducted.

In many cases, resource agencies can plan several years of background data collection as a stream biological survey is developing. Reasonable sample sizes of 50 to 100 streams per year can be anticipated (for the optimistic scenarios); over a four-year period, this produces a data set consisting of at least 200 sites if a rotational sampling scheme is used.

What kinds of information could a resource agency expect to obtain from a well-designed sample survey?

- The extent of the resource of interest can be characterized. How many kilometers of streams (more explicitly, target streams) are there in the region of interest? Statewide? Digital databases (e.g., RF3) are excellent resources for examining the magnitude of the resource and its spatial distribution. One advantage of using a database such as RF3 is that it enumerates the resource. The stream resource can be classified by size and flow regime. Selection of the sample and subsequent reconnaissance can check the accuracy of the digital database, and in so doing, modify the estimate of the extent of the resource based on the error rates for the sample.

- Land use and habitat characterization of the sample allows identification of the major stressor types and the extent of the resource affected by the various stressor types. More specifically, questions such as, "What proportion of the stream miles (and how many stream miles) are affected by point source discharges? Channel modification? Riparian modification? Stream bank instability?" can be addressed. As long as the stressor can be characterized at sites (by site visits, remote sensing platforms, or current maps such as land use/land cover), an estimate of the extent of the impact can be made. Here, *impact* is used to describe the presence and magnitude of "pollution" and *impairment* is used to describe the deleterious response of biota to the impact; impacts might be evident (e.g., modification to the riparian zone), but impairment might be negligible (the biota achieve index scores comparable to those of minimally impacted systems).

- Characterization of the sample sites relative to least impact criteria allows estimation of the extent or proportion of the stream resource that meets the "least impact" criterion and therefore qualifies as reference sites for biocriteria formulation. Is the stream resource extensively degraded in the region or are impacted sites infrequent? Thus, for any attribute that can be reasonably quantified for a stream site, the extent to which the population meets certain characteristics related to that attribute can be described from the sample. If the number of sites meeting a least impact criterion is minimal, sites can be supplemented by judgmentally identifying sites meeting the least impact criterion. In addition, gradients of impact seen in the sample reflect the overall extent of degradation in the area of interest. Important gradients can be targeted for refined information.

- In a similar way, the results of the biological survey on the sample sites allow description of the range of condition in the population of streams or lakes from which the sample is drawn. The proportion of streams (or extent) meeting specified biological goals can be estimated. The proportion meeting the least impact criterion could be used to establish the reference condition and therefore to create numeric criteria. If reasonable relationships between biological response variables and impact magnitude can be derived, then the effect of impact can be subtracted from impact sites and least impacted site condition can be inferred.

A recently completed survey, conducted by the Delaware Department of Natural Resources and Environmental Control (DNREC 1992a, b) illustrates many of these points. DNREC conducted the sample survey to quantify the relationship between biological quality and habitat quality in the nontidal coastal plain streams of Kent and Sussex counties in southern Delaware. Nontidal streams in these counties form the target population; individual sites were chosen randomly "from all available stream road crossings in the region and were believed to adequately encompass the range in existing biological and habitat conditions" (DNREC 1992a,b). Strictly speaking, the target subpopulation is the set of streams with bridge crossings. The extent to which this subpopulation reflects the target population would elicit discussion. DNREC chose invertebrate assemblages as the target group of organisms on which to make the assessments, following protocols described in Plafkin et al. (1989) Similarly, guidelines in that document were used to characterize the physical habitat of the target streams. Because numeric criteria had not yet been established for streams in Delaware, DNREC relied on comparisons with reference sites that appeared as part of the sample survey. The approach was to compare physical and biological quality at each site against reference scores to produce a "percent of reference" score for each site. A total of 93 stream sites was sampled over a three-week period during the fall of 1991.

Reference sites were chosen from the 93 randomly selected sites, based on those that exhibited the "best habitat and best biological measurements." From these, reference scores were chosen, for the individual site comparisons. Three biological condition categories were identified: good (>66% of reference scores), fair (34 to 66% of reference scores), and poor (<34% of reference scores).

Results of the survey indicated that approximately 29% of the sites were in good condition — attaining their biological potential, 43% were in fair condition, and 28% in poor condition, when compared with the reference condition, as defined in this study. Furthermore, a high proportion (80%) of the sites in poor biological condition also exhibited poor physical habitat. These sites might also have been affected by degraded water quality, but water quality measurements were not made as part of the survey. For the sites exhibiting good physical habitat, but poor biological condition, other causes had to be sought.

Thus, this relatively rapid (93 sites in three weeks) sample survey provided DNREC with important information in a timely manner (within a year of the time the survey was conducted) about the range of biological and physical habitat conditions of a subpopulation of streams without visiting each stream. It allowed DNREC to identify the proportion of streams in relatively good condition and to characterize the probable cause of those in poor condition. Along with an estimate of the total number of stream miles in the areas of interest, this information could also be converted into the number of stream miles in each condition category. Although numerous refinements can be suggested, this survey is a sound base on which to make the modifications. The information could also serve as a basis for developing numeric biocriteria.

Since the good sites were drawn from the target subpopulation, these sites represent a realistic goal or standard for the restoration of "fair" and "poor" sites. Review of the history of the "good" sites indicated that they had been channelized, but have been allowed to revert to a "natural" condition over the past 50 to 80 years. Consequently, one might expect that the "fair" and "poor" sites could be restored to "good" condition if left to revegetate and go natural over a similar time frame (J. Maxted, DNREC, Dover, Delaware, personal communication).

4.2 Using Sample Surveys to Implement Biocriteria

The Ohio EPA has been sampling fish and macroinvertebrate assemblages in streams in a consistent way for more than a decade to assess the condition of streams (Yoder and Rankin, Chapter 9). Elements of physical habitat are also measured. Furthermore, Ohio EPA condenses the information on the biological assemblages into indices, such as the Index of Biological Integrity (IBI) for fish. The information is available as an electronic database containing data on >7000 site visits. This database is useful for illustrating how a sample survey approach can be used to assess the condition of streams in a consistent way statewide. For this illustration, we will consider this set of 7000-plus sample records as the population of interest; the database can be characterized as a list frame from which a random sample can be drawn. By analogy, we can use the same process for conducting an actual sample survey of streams statewide, but with maps as the frame rather than a list of stream sites. The purpose of this illustrative sample survey is to characterize the population of the >7000 site visits. The specific characterization will show:

- How closely a sample, drawn according to statistical sampling rules, reflects the population's characteristics
- How, with numeric biocriteria in place, a sample survey could be used to estimate the proportion of a population of streams that is in poor condition, e.g., the streams that do not meet the specified biocriteria.

Part of Ohio EPA's assessment program included establishing numeric biological criteria. For a particular indicator, such as the IBI, the biocriteria are specific to aquatic life class, ecoregion, and sampling technique, all of which are contained in the database. The IBI biocriteria that we used in this illustration are summarized in Table 1 (Ohio EPA 1987a,b).

A sample size of 200 sites, randomly drawn from the list frame, was used. Although this might seem like a high number for a routine yearly survey, it might be reasonable to consider the survey extended over a four-year period, with 50 sites per year sampled.

Two figures and one table summarize the results. Figure 1 compares the sample and population characteristics via a cumulative distribution function (CDF); a CDF describes the overall population structure, containing information about the shape of the distribution and the range of scores. Other statistics can be calculated and represented on the graphical summary, such as means, medians, and percentiles. It is relatively easy to read off the proportion of sites that are above or below a selected score. For example, the proportion of sites with IBI scores <30 is illustrated by following the vertical line originating at 30 on the x axis to its intersection with the CDF, then following the horizontal line until it intersects the y axis. The y score is the proportion of sites with IBI scores < 30 (about 53%). One might also ask what score corresponds to a particular percentile, and simply reverse the order just described.

Table 1. Ohio EPA's Numeric Biological Criteria for the Index of Biotic Integrity (IBI)

| | WWH | | EWH | |
Ecoregion	Boat	Wading	Boat	Wading
Huron/Erie Lake Plain	34	32	48	50
Interior Plateau	38	40	48	50
Erie/Ontario Lake Plain	40	38	48	50
Western Allegheny Plateau	40	44	48	50
Eastern Cornbelt Plain	42	40	48	50

Note: Biocriteria were established by habitat type (WWH = warmwater habitat; EWH = exceptional warmwater habitat), type of sampling gear used (boat = boat mounted electrofisher; wading = backpack electrofisher), and ecoregion. Index of Biotic Integrity scores can range from 12 to 60.

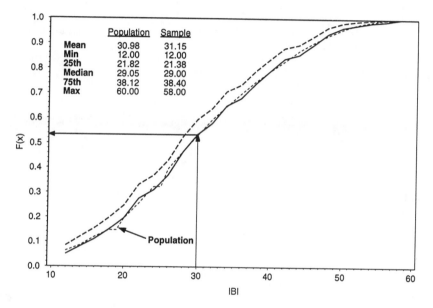

Figure 1. Cumulative distribution functions (CDF) compare the range of IBI scores and selected statistics between the population of >7000 site visits and the statistical sample (size 200) derived from the Ohio EPA's stream database. The sample CDF is delineated, by solid line, along with its upper 95% confidence limit (heavy dashed line) to show uncertainty of the sample estimate. F(x) is the proportion of sites.

A 95% confidence bound above the sample CDF is included to illustrate the uncertainty with which the sample reflects the population. Note that the sample CDF falls nearly identically with the population CDF, and that the sample and population summary statistics included in the figure are nearly identical, illustrating the capability of a statistically drawn sample to reflect the population's characteristics. This correspondence can also be seen in Table 2, which compares the proportion of the population and sample that falls into the different ecoregions or is assigned to the different aquatic life use classes. The 95% confidence interval is included to display the uncertainty of the sample estimates.

This characterization can be taken another step. Because Ohio EPA established numeric biocriteria, each IBI score can be compared with the relevant biocriterion score for the site as a ratio (Table 1). Then the distribution of these scores can be summarized as a CDF (Figure 2). The advantage of taking this additional step is that it is now relatively easy to see the proportion of sites that do not meet their respective biocriterion scores. This kind of display is particularly useful if sites are assigned different biocriteria. Whereas Figure 1 displays the range of IBI scores among all the sites sampled, it cannot be used directly to display the proportion of sites not meeting biocriteria, because different sites are assigned different biocriteria. Converting the information contained in Figure 1 by using a ratio of site scores to

Table 2. A Comparison of the Percent of Sites Classified into Different Ecoregions and into Different Aquatic Life Use Classes Between the Population of All Sites and the Statistical Sample

Ecoregion	Population percent	Sample percent	Standard error	95% Confidence limits Upper	95% Confidence limits Lower
Huron/Erie Lake Plain	11.37	7.00	1.80	10.54	3.46
Interior Plateau	5.01	4.50	1.47	7.37	1.63
Erie/Ontario Lake Plain	28.27	28.50	3.19	23.76	22.24
Western Allegheny Plateau	16.36	23.00	2.98	28.83	17.17
Eastern Cornbelt Plain	38.58	35.00	3.37	41.61	28.39
Aquatic life use[a]					
CWH	1.42	1.00	0.70	2.38	0.00
EWH	8.79	8.00	1.92	11.76	4.24
LRW	0.99	1.00	0.70	2.38	0.00
MWH	0.36	0.50	0.50	1.48	0.00
WWH	85.88	87.50	2.34	92.08	82.92

Note: Percentages do not add up to 100 because a small proportion of sites did not fall into any of the categories summarized here.

[a] CWH = coldwater habitat; EWH = exceptional warmwater habitat; LWH = limited warmwater habitat; MWH = modified warmwater habitat; WWH = warmwater habitat.

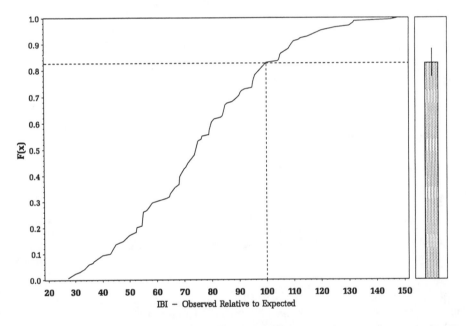

Figure 2. Site IBI scores can be compared with site numeric biocriteria to summarize the proportion of sites (or stream length) that does not achieve a biocriterion score, as illustrated by this CDF. This type of summary is useful if different sites have different numeric biocriteria, as do the sites in Ohio (Table 1). In this example, about 82% of the sites do not achieve their respective biocriterion scores, illustrated directly on the CDF, or as a histogram (with 95% confidence interval), illustrated to the right of the CDF.

biocriteria scores achieves the comparison across sites that are classified differently. If use impairment were described as failure to achieve the appropriate site biocriterion, then the proportion of impaired sites could be estimated along with the uncertainty of the estimate. In the case illustrated here, about 82% of the sites do not achieve their respective biocriterion scores.

Furthermore, if sample surveys were performed each year, trends in the proportion of impaired sites statewide (or over whatever area or subpopulation of interest) could be tracked. The expectation is that the aggregate of the regulations and policies intended to protect and restore the biological integrity of streams would produce improvement, which should show up as evident trends in the condition indicators.

4.3 Relationship of Sample Surveys to Clean Water Act Requirements

Several sections of the Clean Water Act require reports on the condition of the nation's aquatic resources. Section 106 requires states to report on the quality of navigable waters as a condition for receiving funds under Section 106(e) for water quality monitoring programs. Section 305(b) requires biennial reporting on the status of surface water quality; Section 314 focuses on the status and trends of significant publicly owned lakes. Section 319 is a requirement for a one-time assessment of the types and extent of nonpoint source pollution statewide; many states refine and update the original assessments.

As is probably evident at this point, the use of well-designed sample surveys is one way states can develop efficient monitoring programs for obtaining the information necessary to meet these reporting requirements. Statistical sample surveys allow inferences about the entire population of interest, not just the sites selected for the survey. In contrast, the use of nonstatistical surveys to describe populations always involves the question of the existence and extent of bias in the population inferences. Therefore, the extent to which the information can be used to characterize the resources statewide is open to question.

Routine use of sample surveys also allows states to refine the target subpopulations as information about lakes and streams statewide is gathered. Initial surveys might be designed to develop an overall picture of the condition of lakes or streams across the state. Subsequent surveys could then target particular, well-defined subpopulations about which there is greater interest. These subpopulations could be defined regionally, such as by ecoregion, by major watershed, or by lake or stream type, such as publicly owned lakes >4 ha in size or headwater streams. Then sample surveys could be conducted to characterize these subpopulations in greater detail than could be accomplished efficiently with a statewide survey. Set in the context of the statewide surveys, it would be possible to estimate the proportion of lakes or streams across the state accounted for by the subpopulation.

For example, there might be an interest in the biological condition of urban lakes. A statewide survey could show that urban lakes account for 10% of lakes in the state, and that the condition of urban lakes tends to be poorer than is generally seen across the statewide population of lakes. A refined survey (increased sample size; additional indicators of lake condition) could then be used specifically to characterize the urban lakes: Are those in poor condition associated with any particular cities? Are those in poor condition associated with major point source discharges? Thus, the flexibility offered by statistical sample surveys allows water resource managers to build efficiently on previous survey work, both to refine questions about lakes and streams of concern and to characterize those aquatic resources with increasing detail.

5.0 SUMMARY AND CONCLUSIONS

Statistical sample surveys can play an important role in characterizing the biological condition of lakes and streams. In this chapter, important concepts of sample surveys are described, including:

- Populations that identify the aggregate of the elements — lakes or streams — of concern
- Frames as representations of the populations
- Sample units, the individual elements of the population — individual lake or stream sites
- Samples, the collection of sample units drawn from the frame on which measurements will be made
- Selection rules that produce an objective collection of individual sample units from the frame with which to make unbiased inferences about population conditions

Relatively few statistical sample surveys have been carried out on lakes and streams; some of these are described to illustrate how the concepts of sample surveys have been implemented for a variety of purposes and the nature of the conclusions that can be drawn from sample surveys.

Sample surveys can form an important component of a state's program on the development and implementation of biocriteria. Specifically, sample surveys can be used to characterize the range of biological conditions found in lakes and streams statewide, or in specific parts of a state, without monitoring each lake or stream. These kinds of surveys can be used as a foundation for establishing numeric biological criteria. If a state has established numeric biological criteria, then sample surveys can be used to estimate the extent of impairment statewide by comparing the condition of a statistical sample of lakes or streams with the relevant numeric biocriterion for each lake or stream. The number of lakes, kilometers of streams, or proportions of either that do not achieve established numeric biocriteria can be estimated to describe the extent of impairment, and trends can be tracked if sample surveys are conducted over time. Sample surveys are an efficient way to meet many of the reporting requirements about the condition of lakes and streams set forth in the Clean Water Act.

ACKNOWLEDGMENTS

While the thoughts presented in this paper are my own, I am grateful for numerous discussions with Scott Urquhart, Don Stevens, and Tony Olsen, statisticians associated with EMAP, who have taught me about sample surveys. Danny Kugler ran the analyses of the Ohio EPA database. The Ohio EPA (represented by Ed Rankin and Chris Yoder) should be awarded some kind of commendation for the development of its extensive database on the biological condition of streams throughout the state and for its willingness to share that database with others. We all continue to learn a tremendous amount about the development of biological criteria from its monitoring efforts. John Maxted helped with the Delaware streams example. Susan Christie helped with editing and formatting, and George Gibson, John Maxted, Barry Rosen, and Don Stevens reviewed this chapter.

The research described in this report has been funded by the U.S. Environmental Protection Agency. This document has been prepared at the EPA Environmental Research Laboratory in Corvallis, Oregon. It has been subjected to the Agency's peer and administrative review and approved for publication. Mention of trade names or commercial products does not constitute endorsement or recommendation for use.

SECTION IV:
POLICY AND PERSPECTIVES

## CHAPTER 19

Monitoring Water Resource Quality Using Volunteers

Joyce E. Lathrop and Abby Markowitz

Increased interest in environmental issues in the mid-1960s, spurred by many factors including Rachel Carson's *Silent Spring* (1963), led to increased involvement of citizens with their government regarding environmental issues. This had an important impact on government legislation. The 1972 amendments to the Federal Water Pollution Control Act (known as the Clean Water Act) as well as the other major environmental legislation of the early 1970s (the Clean Air Act, the Federal Insecticide, Fungicide and Rodenticide Act [FIFRA], the Toxic Substances Control Act [TSCA], etc.) resulted from the influence of the public on Congress.

Another important outcome of increased citizen involvement in environmental issues during the 1960s was the establishment of the first organized citizen volunteer water quality monitoring programs in the United States. Since then, citizen volunteer monitoring programs have been initiated and coordinated by local watershed and community-based associations, by environmental organizations, and by state, federal and local agencies working with interested citizens and organizations (USEPA 1990e, 1992d, 1994a). By all accounts, volunteer programs are increasing throughout the country. About 65% of the 517 groups represented in the latest National Directory of Citizen Volunteer Environmental Monitoring Programs report being founded after 1988. Programs can be found in 45 states and the District of Columbia (USEPA 1994a). Citizen volunteer monitoring involves the gathering of water quality, water resource quality, watershed, or other environmental data by persons who are not paid to perform these assessments (Ellett 1988). Unpaid should not be confused with unskilled since many volunteer programs employ rigorous training protocols, which include training by water quality professionals. In fact, professionals often participate in citizen programs in a variety of ways, as both volunteers and paid staff (USEPA 1994a).

Volunteer monitoring programs have been developed to provide education and advocacy opportunities as well as basic data collection. They have also developed out of a desire to educate the members of the organization and the public. Some programs have been established to address specific local or watershed issues or projects. Data collection by volunteer monitoring programs includes casual visual observations for pollution sources; surveillance efforts ("watch dogging") to detect and deter ongoing pollution discharges; and traditional chemical physical and biological data collection for water resource quality assessment (USEPA 1990d, 1994a). Currently, there are monitoring programs for all types of aquatic and semiaquatic environments, including groundwater (USEPA 1994a). However, the only volunteer monitoring programs that are currently suitable for involvement in establishing or implementing biological criteria are those using quantitative or semiquantitative benthic macroinvertebrate sampling and analyses.

In this chapter, we offer an overview of volunteer monitoring program objectives and data collection. We specifically discuss benthic macroinvertebrate monitoring since it is most likely to be usable in

0-87371-894-1/95/$0.00+$.50
© 1995 by CRC Press, Inc.

biological criteria issues. We also discuss uses of biological data collected by volunteer monitors as illustrated by a case study of the Maryland Save Our Streams (SOS) biological monitoring program.

1.0 VOLUNTEER PROGRAM GOALS

Objectives of volunteer monitoring programs fit into three general categories. Volunteer groups monitor waterbodies to (1) educate themselves and others, (2) support action including advocacy and "watch dogging", and (3) to provide data for water resource quality assessment.

1.1 Education

Several sections of the Clean Water Act (USGPO 1989a,b) require public participation in water resource management decisions. Section 101(e) requires that the public be included "in the development, revision, and enforcement of any regulation, standard, effluent limitation, plan, or program established by the administrator or any state under this Act." Public participation in the form of public comment and public hearings is also an important part of Section 404 process, which addresses wetlands use and protection. Section 303, which addresses the establishment of water quality standards, includes requirements for public involvement in public hearings on proposed changes. If the public is to be effectively involved in activities related to these and other laws, the public must be able to form intelligent opinions. Volunteer monitoring programs are important in educating the public regarding the water environment and its resources (Newton 1992).

Although every program provides some form of educational benefits, the extent of educational opportunities varies among the programs, some of which exist almost entirely to provide education, some for which education is secondary to action or data collection, and others where the objectives of education and data collection are equally important. Not surprisingly, volunteer programs identify "education" as the most common use of their data, with 85% of programs reporting education as the primary application of collected data (USEPA 1994a). Benefits of educational programming include fostering individual and community responsibility and action to improve and protect water resources within watersheds. It also increases communication between citizens and government agencies and legislative bodies at all levels (Newton 1992).

1.2 Advocacy and Enforcement

Some volunteer programs focus on environmental advocacy or on enforcement of environmental permits and pollution deterrence (Herz et al. 1992). Advocates seek to persuade others of the importance of maintaining water resource quality. Motivating other citizens to become involved in water quality issues increases the influence the program will have on governmental policy and resource management decisions. Having data available improves the effectiveness of water quality advocates and provides a stronger base from which to speak (Iudicello 1992; Poten 1992). Observational data, water quality data collection, and surveillance activities by volunteer monitors can be important in supporting regulatory action by appropriate agencies (Herz 1992) and in supporting legal action against polluters by volunteer monitoring groups (Norris 1992).

Data collected by citizen volunteer monitors has influenced both the creation of new environmental legislation and the implementation of regulations under those laws. Citizens armed with accurate data are able to influence state and national legislators with greater effectiveness than when they lack data to support their statements. In New Hampshire, two state bills, one requiring septic tank inspections, the other limiting water skiing, were passed in part due to the influence of volunteers and their data (USEPA 1990e). Data collected by volunteers have also influenced the implementation of existing laws and regulations. Volunteers have used their data to fight water reclassifications and in the 404 permitting process. For example, data collected by River Watch Network and Ottauquechee Natural Resources Conservation District volunteers was useful in fighting reclassification of the Ottauquechee River in Vermont (Byrne 1992); and in North Carolina, volunteers also played a role in fighting reclassification of a waterbody (USEPA 1990e). As a tool for identifying problems, advocating for mitigation and enforcement of regulations, volunteer data have proven very useful. Recently, state and local governments and advocacy groups, were ranked the highest among users of volunteer data (USEPA 1994a).

1.3 Water Resource Quality Assessment

The collection of data by volunteer monitoring groups for use in water resource quality assessment is increasing (USEPA 1990d, 1994a). Volunteer-collected data are used for trend analysis, as screening tools prior to more intensive investigation by water quality professionals, as a basis for making local zoning and land management decisions, and in states' 305b water resource quality reports. The USEPA "Guidelines for the preparation of the 1992 State water quality assessments (305(b) reports)" allowed data collected by volunteers to be used in the "evaluation" of water quality but not in formal monitoring (USEPA 1991b). This is because most volunteer monitoring programs lack sufficient quality assurance to insure the collection of high quality data necessary to formal monitoring. The USEPA (1993b) "Guidelines for the preparation of the 1994 state water quality assessments (305(b) reports)" supports even greater use of volunteer data. Under these new guidelines "waters sampled under these [volunteer monitoring] programs could be considered monitored" rather than merely "evaluated." Only data from programs that implement rigorous quality assurance/quality control protocols would be acceptable in providing "monitored" waters.

While it is not important for volunteers collecting data for educational and advocacy purposes to adhere to strict quality assurance and control protocols, it is imperative that volunteers develop and maintain stringent quality assurance and control procedures if their data will be used in evaluation of water quality standards attainment, discharge permit regulation, support of legislation, establishment of databases, etc. (USEPA 1990e). The case study of the Maryland Save Our Streams Program in Section 5.0 highlights the use of volunteer data for water resource quality assessment.

2.0 DATA COLLECTION

Water quality data collected by citizen monitors are discussed below by the traditional categories of physical, chemical, and biological conditions. Many manuals are available that describe specific methodologies used by various monitoring groups. These may be obtained by contacting the group(s) at the addresses listed in the *National Directory of Citizen Volunteer Environmental Monitoring Programs* (USEPA 1990d, 1994a).

Increasingly, attempts are being made to integrate all three types (physical, chemical, and biological) of sampling data (Kopec 1992). Additionally, some programs are beginning to broaden the scope of sampling from a single type of waterbody such as a lake or stream to entire watersheds, including land-use monitoring. The advantages of these increased efforts are threefold. First, they help to maintain volunteer interest by providing additional challenges. Second, they increase the overall amount of data collected for the watershed. Third, they help volunteers understand the concept of nonpoint source pollution by emphasizing the idea that a watershed is a sum of its parts and that all actions within the watershed will influence water quality (Kopec 1992). When the program expands to all or a major part of a watershed, the effects on one area of adding pollution in another become more evident. The following discussion of the data collection is based on information provided by volunteer monitoring programs in the *National Directory of Citizen Volunteer Environmental Monitoring Programs* (USEPA 1990d, 1994a).

2.1 Chemical and Physical Data

Monitoring groups vary tremendously in what they observe or measure. However, most include observations and measurements of several parameters. The simplest programs monitor physical parameters such as turbidity and/or make qualitative visual assessments regarding such things as riparian habitat or watershed land use, including erodible soil conditions from construction or agriculture. Weather conditions, water color and odor, recreational suitability, and obvious sources of pollution such as drainage outfalls are also recorded by some groups. Physical parameters measured by volunteers directly related to the water resource include sediment composition, water temperature, turbidity (Secchi depth), flow, conductivity, and depth. Indirect physical parameters that may be measured include rainfall and air temperature.

Groups that have chosen to go beyond physical parameters or observations collect chemical and/or biological data. Chemical attributes such as dissolved oxygen, pH, alkalinity, phosphate, and nitrogen may be measured in the field using kits designed for water quality determination. Dissolved oxygen, nitrogen, and pH are also measured using field meters. Some groups collect water samples that are later analyzed by state, university/college, high school, or private laboratories. Total suspended solids, volatile suspended solids, phosphate, nitrite- and nitrate-nitrogen, iron and chloride concentrations, 5-day BOD, alkalinity, and hardness are also measured. Field measurements are somewhat less accurate than measurements made in a laboratory but they offer the important advantage of providing immediate results to the volunteers. Having results immediately available helps maintain volunteer interest. Laboratory analysis is generally more costly than field methods and does not provide on-the-spot information to the volunteers but in-lab analysis does allow a greater variety of constituents, such as chlorophyll content and heavy metals or pesticides, to be measured. Laboratory analysis also improves data quality through enhanced quality control and quality assurance.

2.2 Biological Data

Biological monitoring by citizens, like that by state biologists, gives a better indication of the effects of anthropogenic impacts including habitat alterations and chemical pollutants on the living organisms than does either chemical or physical sampling alone (see Karr, Chapter 2). In addition, it reveals cumulative effects of pollution from many sources on aquatic communities. Additional information is gained from population, bacteriological, and productivity studies. Biota monitored by volunteers include fecal coliform bacteria, fish, aquatic and semiaquatic vegetation, zoo- and phytoplankton, benthic macroinvertebrates, waterfowl populations, and other vertebrates (other birds, turtles, reptiles, and small mammals) associated with aquatic ecosystems. Volunteers monitor for the presence of nuisance organisms or to assess the health of individuals or the condition of assemblages or communities.

Nuisance organisms monitored by volunteers include coliform bacteria, algae, and unwanted higher plants. Algae are either measured directly through counting or indirectly by chlorophyll content of a water sample, which provides a measurement of algal biomass and production. In addition to coliform bacteria and algae, a number of programs measure other biological nuisances such as the presence of unwanted higher vegetation; for example, several programs monitor the growth and abundance of Eurasian milfoil or purple loosestrife.

Volunteers in some programs (e.g., New Hampshire Lakes Lay Monitoring Program; USEPA 1990d) monitor the health of individual organisms to assess chemical impacts on a waterbody. For example, fish are weighed and measured and comparisons are made between fish in contaminated and "clean" areas. In some cases scales are collected so that growth rates as well as general size can be calculated and compared. Abnormalities such as tumors or parasitic infections or abnormal behavior are recorded by a number of groups.

Volunteers also monitor the health of animal and plant populations including strictly aquatic organisms and semiaquatic organisms dependent upon the waterbody. Populations monitored include higher aquatic and semiaquatic plants, aquatic macroinvertebrates, and vertebrates such as fish, amphibians, reptiles, and breeding birds. In estuaries and other near-coastal waters shellfish, fish, zoo- and phytoplankton, and birds and small mammals are monitored. This kind of monitoring is especially import to evaluating and protecting rare and threatened species.

Community and assemblage condition monitoring ranges from simple qualitative surveys to more in-depth studies using a variety of collection and analytical methods. Community or assemblage conditions are most often monitored in streams; fish and macroinvertebrates are the most commonly monitored assemblages. Qualitative surveys enumerate the types of organisms present but not numbers of each type present. Some of these surveys incorporate pollution tolerance information in their analysis. Semiquantitative approaches such as those based on Rapid Bioassessment Protocols (Plafkin et al. 1989) are used by a few programs (e.g., Maryland Save Our Streams). The collection methods generally remain fairly simple allowing use by volunteers of a wide range of skill levels. Quantitative approaches (e.g., using Surber samplers or artificial substrates) are rare (see River Watch Network 1993).

2.2.1 Monitoring Macroinvertebrate Assemblages

Macroinvertebrate monitoring occurs primarily in streams and rivers because the sampling is relatively easy and protocols are well established. There are close to 400 volunteer programs nationally that monitor over 6,500 sampling stations in rivers and streams (USEPA 1994a). Benthic macroinvertebrates rank second (after water temperature) as the most common river and stream parameter sampled by volunteers (1994a). Most stream programs use the kick seine method to collect benthic macroinvertebrates from riffles. Surber samplers, artificial substrates, and dip nets are used by a few programs. Organisms are identified streamside and released or collected in alcohol for more precise identification in a laboratory. In 1991, the Maryland SOS and Delaware Inland Bay Citizen Monitoring programs began developing protocols for monitoring benthic macroinvertebrates in estuaries (Ely 1991).

The Izaak Walton League of America Save Our Streams (IWLA) programs, the Ohio Department of Natural Resources (ODNR) Scenic Rivers Program (Kopec and Lewis 1983) and others like them are qualitative approaches based on pollution tolerances of benthic macroinvertebrates. Organisms are collected by kick seine and identified streamside at the time of collection allowing the organisms to be returned to the stream alive. Since many taxa are difficult to distinguish even with a microscope, the levels of taxonomic identification in the field are limited. Worms, decapods, leeches, snails, and clams are often identified only to order or a higher taxonomic level. Insects are most often identified to order although some, especially the true flies (Diptera), are taken to the family level of taxonomy.

A more complete picture of the benthic macroinvertebrate community is attained by collection and preservation as is done by Idaho Streamwalk II volunteers (Rabe 1991), in the RBP Level 2 Analyses performed by Vermont River Watch Network (River Watch Network 1993), and Maryland Save Our Streams volunteers. These programs integrate procedures similar to the Rapid Bioassessment Protocol (RBP) II (Plafkin et al. 1992) used by some professional biologists and state agencies (e.g., Kurtenbach 1991; Primrose et al. 1991) into their volunteer methods. Modifications have been made to ease the quality assurance problems that are associated with volunteers. Volunteers identify preserved samples in a laboratory using microscopes and appropriate taxonomic keys; insects are identified to the family level. Taxonomic levels vary for other organisms but all organisms are taken to as low a taxon as volunteers can reasonably be expected to meet, in many cases family but in others order or class. Each of these programs also employs some form of physical habitat characterization or assessment as a component of their biological monitoring methods.

2.3 Analysis of Macroinvertebrate Data

The collection, identification, and analysis of macroinvertebrate data has the greatest potential for aiding development and implementation of biological criteria using volunteers. Simple qualitative analyses based on gross pollution tolerances can be done in the field by volunteers themselves. More detailed analyses can be done on samples that are collected and returned to the laboratory. More than one measure or metric (e.g., diversity, taxa richness, family biotic index, etc.) is examined in each program. These may be analyzed individually or composited into a single numeric value.

2.3.1 Simple Biotic Indices

Analysis of data by the IWLA, ODNR, and similar programs is based upon a modified three-group system of intolerant, facultative (moderately tolerant), and tolerant groups similar to Beck's original index (Beck 1954; Terrell and Perfetti 1989). Mayflies (Ephemeroptera), caddisflies (Trichoptera), dobsonflies and alderflies (Megaloptera), stoneflies (Plecoptera), water pennies (Coleoptera: Psephenidae), and snails (Gastropoda) other than the pouch snails (Gastropoda: Physellidae) are generally considered intolerant of organic enrichment pollution. Facultative organisms are considered to be less tolerant than those in the tolerant group and more tolerant than intolerant taxa. Facultative organisms include damselflies and dragonflies (Odonata), crayfish (Decapoda), sowbugs (Isopoda), scuds (Amphipoda), cranefly larvae (Diptera: Tipulidae), most beetles (Coleoptera), and all bivalves (Bivalvia). Tolerant groups include leeches (Hirudinea), midge larvae (Diptera: Chironomidae), blackfly larvae (Diptera: Simuliidae), pouch snails (Physidae), and aquatic worms (Oligochaeta).

Numbers of individuals of each taxon are estimated on a simple scale and recorded as letters, where A = 0 to 9, B = 10 to 99, and C = 100 or more. However, these values are not used in calculating the final index value for a site. Only the number of taxa in each of the three tolerance groups is used. The number of taxa in each group is multiplied by the group's tolerance value (intolerant = 3, facultative = 2, tolerant = 1). The sum of these group values, called the Cumulative (or Community) Index Value (CIV), is then put into a narrative scale of excellent, good, fair, and poor. The higher the value, the better the estimated water quality.

This three-tiered scheme works well for groups like stoneflies in which all species are nearly equal in tolerance to organic enrichment pollution. Unfortunately, tolerance varies considerably among the species and even within families of other insect orders like the caddisflies and mayflies (Hilsenhoff 1988; Lenat 1993; DeShon, Chapter 15), and with some of the other invertebrates (Klemm et al. 1990). Identification to the order level results in overestimating the quality of some waterbodies and underestimating others.

In addition, analysis is based solely on the presence or absence of a group and does not take into consideration the abundance of that group. This also lends bias to the evaluation. For example, the presence of a single aquatic worm lowers the overall evaluation. The presence of 100 worms has no greater effect. Likewise, the presence of a single mayfly would increase the cumulative score even if it is from one of the more tolerant families (e.g., Caenidae); the presence of several mayfly genera would have no greater effect.

2.3.2 Multimetric Approaches

Data analysis performed by several monitoring programs (Idaho Streamwalk II, Vermont River Watch Network, and Maryland Save Our Streams Project Heartbeat) incorporate a multimetric approach somewhat like that presented in the RBP II used by some state agencies. Replicate or composite (e.g., portions of a riffle) samples are taken and analysis is based on a 100-organism subsample, which may be taken in the field or in the laboratory. Macroinvertebrate identification and classification is performed primarily to the family level of taxonomy in a laboratory setting. Voucher collections are maintained allowing for more in-depth identification and analysis at a later date. Both structural and function attributes of the macroinvertebrate assemblage may be included in the analysis. Among the metrics used are diversity indices, a family biotic index, percent dominant taxa, Ephemeroptera/Plecoptera/Trichoptera (EPT) index, EPT/Chironomidae ratio, and scraper/filterer ratios. These metrics may be analyzed individually or composited into a single numeric value. Metric or composite values for each site are then compared to the appropriate reference condition values.

3.0 TESTING THE USE OF VOLUNTEER MONITORS

Few studies have been undertaken to compare data collected by volunteers with that of data collected by professionals. The few studies that have been done suggest that volunteers are capable of providing good-quality data for specific levels of use. Scientific support for biological monitoring in streams by nonprofessionals first came in 1971 and 1972. The Advisory Centre for Education in England sponsored stream water quality surveys by British school children using an analysis of benthic macroinvertebrates based on the presence or absence of indicator species in addition to chemical tests. These studies demonstrated that middle school-aged students could be used to sample water quality with reliability nearly equivalent to trained water quality specialists (Mellanby 1974). A similar demonstration project in Canada also provided support for this type of program (Reynoldson et al. 1986). High school-level students conducted benthic macroinvertebrate sampling with subsequent identification to the genus level in a laboratory. Students' samples were found to be very similar to those collected and analyzed at the same sites by professionals at the Alberta Environment, Water Quality Branch.

A more recent study in Ohio compared biological assessments made by volunteers in the ODNR Scenic Rivers Stream Quality Monitoring Program (SQM) with Index of Biotic Integrity (IBI) and Invertebrate Community Index (ICI) assessments made by the state's Environmental Protection Agency (Ohio EPA) water quality specialists (Dilley 1992). The study focused on data from 56 sites on streams that were part of Ohio's designated Scenic Rivers Programs and which were monitored by both Ohio DNR volunteers and by Ohio EPA biologists. These rivers are generally of moderate to high quality and few sites were found that were rated either fair or poor. Volunteers generally rated water quality higher at most sites than did state biologists. A similar study performed by Ohio EPA using data from a larger number of sites with a greater range of water resource quality found similar results (Yoder 1994). A consistently high percentage of agreement occurred only when both the volunteer and professional analyses rated the site in the fair/poor ranges. When the SQM analysis was used as a pass/fail screen to rate attainment or nonattainment of streams with Ohio's Warmwater Habitat (WWH) use designation, agreement improved. The study concluded that SQM methods have the "potential to be used as a screening tool for potential non-attainment of WWH use designation, particularly if operator error can be controlled and minimized."

The Georgia Adopt-A-Stream volunteer monitoring project is also evaluating the quality of volunteer biological monitoring data by working with the U.S. Geological Survey's National Water Quality Assessment (NAWQA) Program. This project is one of the first in the country to determine if volunteers can contribute quality data to the NAWQA Program. If successful, volunteers could provide additional data on benthic macroinvertebrate communities in streams, although at a more general level (USEPA 1994d).

To date no formal comparison has been made between data collected by volunteers in Maryland SOS Project Heartbeat and data collected by biologists with the Maryland Department of the Environment. However, preliminary examinations and comparisons show good agreement between the two at most common stations (see Section 5.3). Analyses are performed by both groups based on family level identifications converted to composite index scores. Comparisons were made by site on the percentage of reference condition score rather than on raw score.

4.0 USES OF VOLUNTEER MONITORING BIOLOGICAL DATA

Biological data collected by citizen volunteer monitors has the potential for use by both the monitoring programs themselves (e.g., as educational tools) and governmental agencies. The uses to which volunteer-collected biological data (or any other form of data) can be put depends largely on the quality of that data. The quality of data needed to meet each of the three volunteer monitoring program objectives varies but must suit the intended goals (Norris et al. 1992). Data that do not match the objectives of the program are not usable by that program. Data quality can be roughly divided into three categories based on program objectives: (1) data collected primarily for education and which lack sufficient quality assurance to be used in state and/or local water resource quality programs, (2) data for which the level of quality assurance is higher, allowing it to also be used by agency biologists in screening waters for beneficial use impairments, and (3) data with state-approved quality assurance protocols, allowing it to be used for determining whether waters meet their beneficial uses, and to support permit development and compliance studies (Davis et al. 1992). The data or information from any one category should not be viewed as better than the others in terms of increasing public involvement and education. However, only data from those programs with state-approved quality assurance and quality control protocols are appropriate for use in establishing and implementing biological criteria (Kellogg and Davis 1992, Davis et al. 1992). These high-quality data are comparable to family-level monitoring data collected by state and other professionals and can be one of the data sources used in establishment and implementation of biological criteria. In 1991, the USEPA published guidance for volunteers on lake monitoring methods (USEPA 1991a). Similar methods manuals on estuarine and stream and river monitoring are in various stages of development (USEPA 1992d, 1993e).

4.1 Watershed Management Decisions

Many local and state watershed management decisions have been based, at least in part, on data collected by volunteer monitors. Data can be used as a "blueprint" of where and what type of community and/or governmental action is needed. Examples from Maryland SOS experiences are found in Section 5.3 of this chapter.

In order to fully understand the potential impact of local resource management decisions, data should be available on the status of all the waterbodies within a watershed. Studies to provide these data may be prohibitively expensive, especially in a large watershed. In addition, professional personnel do not always have the time to perform these studies. Volunteers can be extremely useful in providing the data necessary to reasonable decision making.

Volunteer monitoring data can be used to screen for potential nonpoint source pollution impacts as has been done in Ohio. The Ohio Stream Quality Monitoring Program (SQM) is employed extensively throughout the state by the state's Scenic Rivers Program, schools, designated planning agencies, community associations, and nearly onehalf of Ohio's Soil and Water Conservation Districts (SWCDs). Some of the SWCDs and the designated planning agencies are attempting to use data analyzed with these methods to assess water quality although the quality of data are not high enough for this use due to the insufficient quality assurance and low level of analysis involved.

4.2 Expanding Biological Data Bases

As our water resource management philosophies have evolved from regulation of point sources to watershed management, already-stressed budgets are stretched even further. It is not reasonable to expect each state to monitor every reach of river or every lake on a timely basis. In fact, as of 1992, only about ⅓ of all river miles, less than ½ of lakes and only ¾ of estuary areas were monitored by water quality specialists (Stokes 1992). On the other hand, it is imperative to management decisions that data be available on each watershed. Volunteer monitoring can help provide that data. Volunteers have monitored waterbodies that would not be monitored at all and they have increased the frequency of monitoring compared to state schedules in many areas.

The value of using volunteers to monitor waterbodies is easily seen when data are desired from storm events or during unusual hours and weekends when professional personnel are not normally available. For example, storm event data has been collected by volunteers in Minnesota that would not have been otherwise collected (Tippie 1992). Remote, difficult to reach areas have also been monitored by citizen volunteers when they were too difficult for state personnel to reach on a timely basis. For example, data were collected by citizens in parts of the Pecos River (Texas) that are remote and difficult to reach; these data were used by the state in a study of fish kills in that river (Buzan 1992).

Volunteers can make significant contributions to state water resource data banks. They can add to the number of stations that are monitored without significantly increasing monitoring costs. Although they are cost-effective, they are not free. In 1991, volunteers sampled 85 stations not sampled by the West Virginia Division of Natural Resources, effectively doubling the number of stations with biological sampling at a cost of approximately $55,000 (Firehock 1992). In some states, volunteer data has been the only data available on many lakes (USEPA 1990c). For example, over half of the 400 lakes that are monitored in Illinois and included in the state's 305(b) reports are sampled by citizen volunteers rather than state water quality specialists (Sefton 1992).

5.0 CASE STUDY: MARYLAND SAVE OUR STREAMS PROJECT HEARTBEAT

Founded in 1970, Maryland Save Our Streams (SOS) is a statewide membership organization of volunteer stream advocates. SOS works at the community level to preserve, protect, and enhance Maryland's 17,000 miles of flowing waters, most of which lead to the Chesapeake Bay. Since its inception, SOS has encouraged a stewardship ethic by working with communities to "adopt" local waterways.

5.1 SOS Activities

SOS promotes healthy neighborhood streams by building citizen advocates for those waterways. We recognize that the protection and restoration of our streams, creeks, and rivers require the support of an active and educated citizen constituency working in partnership with government agencies and business communities. Constituencies are built through the promotion of a "watershed mentality". Responsible watershed management requires that citizens and governments alike regard a watershed as a sum of its parts and that all components of a watershed population work together to reduce water quality impacts. An essential element of citizen involvement is the development and support of volunteer leadership in SOS programs.

SOS monitoring activities include Watershed Surveys where volunteer teams drive sections of the watershed identifying and recording potential land-based pollution sources. In our Stream Survey, a companion project, volunteer teams walk sections of a waterway observing and recording potential problems in both the stream and its riparian zone. Many communities use the data from these monitoring activities as a blueprint for future community action (i.e., identifying areas in need of reforestation, bank stabilization, and waterway trash cleanups). These projects are funded, in part, through the *Adopt-A-Stream* program, a partnership between Maryland SOS and the Maryland Department of Natural Resources (DNR). All data is forwarded to the DNR and the relevant local agencies. For over twenty years, SOS has been training citizens to be "Mud Busters". In this program, volunteers are trained to monitor construction sites by observing whether sediment and erosion control measures are present and functioning. These volunteers are also trained to report potential violations to the appropriate enforcement agency and to follow-up with the agency and, in some cases, the developer.

In 1989, SOS joined forces with the Baltimore County Department of Environmental Protection and Resource Management (DEPRM) to initiate the countywide *Citizens for Stream Restoration Campaign*. This project was designed to involve the county's citizens in ongoing waterway education and restoration and to foster the relationship between citizens and government to ensure that both are more responsive to the needs of the environment. Project Heartbeat, a volunteer biological monitoring program, developed in conjunction with both DNR and DEPRM, was piloted in Baltimore County as part of the *Citizens for Stream Restoration Campaign*. The importance of these types of programs is emphasized in *President Clinton's Clean Water Initiative* (USEPA 1994b) which recommended "encouraging neighborhood or regional nonprofit watershed citizen groups and councils to develop consensus watershed restoration strategies, conduct volunteer monitoring programs, and build long-term commitments with communities to protect water resources."

5.2 Biological Monitoring: Project Heartbeat

Project Heartbeat, an outgrowth of the watershed management concept, involves volunteers in taking the "pulse" of the Chesapeake Bay by monitoring the Bay's freshwater tributaries, which are the veins and arteries of the Bay, providing habitat, freshwater, and unfortunately, the majority of the pollutants entering the estuary. The Project Heartbeat philosophy is to complement water quality initiatives undertaken by government agencies. The concept for Project Heartbeat was submitted to Maryland DNR for technical review in 1988. Through Project Heartbeat, volunteers perform cost-effective and technically credible assessments of freshwater streams using the collection and identification of benthic macroinvertebrates and physical habitat assessment to monitor water quality. Preliminary comparisons between evaluations made based on data collected by volunteers in Maryland SOS Project Heartbeat and evaluations made by the state's professional water quality specialists indicate that volunteer data are usable in designated use assessments.

The pilot Project Heartbeat program was implemented in Baltimore County in 1990, and was locally dubbed the "100 Points of Stream Monitoring." All sites were selected in conjunction with DEPRM's Waterway Improvement Program (WIP). Selection criteria included: sites in watersheds where mobilized SOS community constituencies already existed; sites that provided a representation of freshwater stream systems; sites that were accessible to volunteers; sites that corresponded with sites being monitored by the Maryland Department of the Environment (MDE) as part of its RBP II water quality monitoring program (approximately 20%); and sites in watersheds where WIP had current or scheduled capital stream restoration projects.

Volunteers conduct field assessments on 100 permanent sites each summer during July and August. From 1990 to 1992, spring and fall assessments were also conducted on all sites. Beginning in 1994, spring and fall assessments only involved approximately ¼ of the sites. This reduction in sites was made due to problems with seasonal weather conditions and to refocus limited resources on an expansion in training and leadership development, data management and presentation, and monitoring communications.

Project Heartbeat methodology (including insect collection, taxonomic identification, habitat assessment, and equipment) is modeled on USEPA's RBP II outlined in the guidance document, *Rapid Bioassessment Protocols for Use in Streams and Rivers: Benthic Macroinvertebrates and Fish* (Plafkin et al. 1989). The RBPs were originally intended for professional use and the intent, philosophy, and procedures outlined in the RBPs were left fundamentally intact in designing Project Heartbeat. Specific procedures have been modified to allow for limited resources and to compensate for lack of volunteer technical expertise. For example, language used, especially for the habitat assessment, was modified for the layperson's vocabulary. All data forms, training materials, and field supplies used by SOS volunteers have gone through technical review by state and county agencies.

Project Heartbeat volunteer field monitoring procedures consist of a streamside physical habitat assessment, collection of macroinvertebrates using a kick-seine with a 500-μm mesh, and a general insect identification to the taxonomic order level. Three areas of the riffle, at the upstream, middle, and downstream ends, are sampled and the organisms composited. A minimum of 100 macroinvertebrates are then removed from the gridded net and preserved in 70% ethanol. Subsequent to field collection, volunteers classify all macroinvertebrate samples to the taxonomic family level as described in Protocol II (Plafkin et al. 1989). This is accomplished in a university laboratory under the training and supervision of professional biologists and entomologists. (For a full description of Project Heartbeat methodologies, refer to the SOS methods manual available from Maryland SOS in summer 1994.)

The identification of the collected organisms in a laboratory is another example of a modification of the RBPs. Protocol II, which was written for professional biologists, calls for a family level taxonomic identification in the field. Although Project Heartbeat monitors are enthusiastic, dedicated and pay attention to detail, it is unlikely that this level of identification could be accurately carried out by volunteers in a field setting. All Project Heartbeat macroinvertebrate samples are archived as a voucher collection and are available upon request to corroborate volunteer findings. A reference collection of the types of organisms collected is also maintained.

Two committees oversee specific activities for Project Heartbeat: the Technical Advisory Committee (TAC) and the Steering Committee (SC). The TAC, made up of agency personnel and volunteer scientists, addresses quality assurance, quality control, data coordination, data management, and data analyses. The SC is comprised of educators, community leaders, and veteran volunteers who oversee volunteer recruitment, training and retention strategies, publicity and public relations, periodic site evaluation, and other aspects of volunteer organization.

In 1991, when resources became available to initiate a Baltimore City Stream Restoration project, SOS expanded Project Heartbeat into the city with 14 additional sites on the downstream reaches of streams and rivers monitored in Baltimore County. Since 1990, over 650 volunteers have been trained as Project Heartbeat monitors in Baltimore County and Baltimore City. In 1992, SOS approached private environmental consulting firms and asked what it would cost for them to replicate this program — without the educational and outreach components. The response confirmed what we had suspected: using volunteers, Project Heartbeat costs approximately 1/10 of what a consultant would charge for similar monitoring services. However, the benefit of an educated constituency is priceless and cannot be achieved through private consultants.

Due to the success of this program in Baltimore County and Baltimore City, SOS has been contracted by the Maryland State Highway Administration (SHA) to implement Project Heartbeat on proposed and current road projects in order to document stream conditions and to evaluate potential impact. This project will also include tests on selected physical and chemical parameters such as temperature, pH, turbidity, and dissolved oxygen. Project Heartbeat conducted in conjunction with SHA will involve volunteers in central Maryland in the monitoring of up to 90 additional stream and river sites in 1994.

Beginning in 1994, SOS expanded its training protocols for Project Heartbeat by offering annual or twice-yearly, day-long intensive classroom and training workshop for volunteers based on the RBP workshops for water quality professionals. Training will be conducted by members of the TAC as well as other professionals from MDE, DNR, USEPA, as well as private environmental consulting firms. Volunteers completing this training will be certified as monitoring team captains. Each monitoring team will be required to have at least one certified member. Shorter refresher training sessions will be offered in conjunction with each sampling session.

5.3 Data Use

Baltimore County DEPRM includes data from many volunteer activities, including Project Heartbeat, as part of its Resource Management Project in support of regulatory and capital improvement programs conducted by DEPRM's Bureau of Water Quality and Resource Management. Potential applications of Resource Management Project data include implementation of the state Forest Conservation Act; implementation of the NPDES Stormwater permit; prioritization of capital projects for the Waterway Improvement Program (including stream restoration, stormwater retrofitting, shoreline erosion control); and determination of overall trends in resource management for evaluation of the bureau's performance with respect to its goals and objectives. Where appropriate, resource management analysis will be used in the formulation of DEPRM's recommendations regarding countywide land use and growth management decision making.

In the upcoming months and years, SOS anticipates working with state and local agencies to further develop and implement Project Heartbeat in more watersheds as a monitoring component of demonstration and permanent characterization and impact studies, to incorporate Project Heartbeat data into 305(b) reports and designated-use attainment evaluations, and into the development of biocriteria. An informal review of the Project Heartbeat data collected from 1990–1993 revealed that 77 of the 100 monitoring locations had data suitable for determining aquatic life use attainment using MDE methods.

In the past, most volunteer biological monitoring programs have concentrated on field identification to the taxonomic order level. The primary goal of these programs has been to facilitate community education and stewardship while providing basic screening data to local, state and federal agencies. The Project Heartbeat experience demonstrates that it is not only possible, but also imperative in order to produce the most widely usable data possible, to combine scientific credibility with community education and leadership development in a volunteer monitoring program. Only in this way will volunteer monitoring reach its full potential by providing essential and cost-effective data on our waterways, promoting partnerships between the community and government agencies, fostering a stewardship ethic within the community, and developing citizen leadership.

6.0 CONCLUSIONS

In the 1960s, Americans awoke to the dangers of ignoring environmental concerns and problems in estimating economic costs and benefits. Since then, we have begun to realize the extent to which human activities can influence the environment. As we move into the 21st century, the need to create a "sustainable economy" based on a "sustainable environment" has become critical. Creating a sense of stewardship of the environment among citizens so that more ecologically sound choices can be made is essential to the success of sustainable development. Volunteer monitoring programs can be a great asset in creating this sense of stewardship.

Growing demands by citizens for improved water resource quality coupled with increased budgetary constraints put on governmental agencies have focused the need for citizen involvement beyond traditional nontechnical roles. Citizen volunteer monitoring programs are a cost-effective way for states to increase water quality monitoring without the high cost of additional professional personnel. Volunteer monitoring data can be used to determine which waterbodies are most in need of intensive surveys by water quality specialists and potential remediation. Sites that would otherwise be sampled only intermittently can be sampled more often to develop overall trends. Data can be added from sites that otherwise would not be sampled at all. Development of quality assurance protocols will allow data from volunteer monitoring programs to be used to a greater extent by state monitoring programs. The USEPA and an increasing number of states have begun to recognize the ability of citizen monitors to produce valuable, credible data (Abe et al. 1992; USEPA 1993b; 1993e).

The Environmental Monitoring and Assessment Program (EMAP) represents an attempt at assessing water quality on a national basis. Citizen volunteer monitoring lends itself to these types of watershed characterization and broad geographic studies because of the comparatively low cost (Hughes et al. 1992) and could be a potentially valuable resource for this program. Many individuals are willing to sample close to their homes (and gain the added benefit of becoming better stewards at the same time). It is important to stress the importance of maintaining proper quality control through training sessions and other educational opportunities.

Along with the change in focus from upstream–downstream studies to watershed surveys, there has also been an evolution in the way in which waterbodies are monitored (USEPA 1993c). The evolution of water quality assessment has progressed from chemical and microbiological assessment to the inclusion of biological surveys by state water quality personnel. The Clean Water Act demands that the nation's waters be restored to the extent that they are capable of supporting a variety of aquatic life. Biological measurements are most appropriate for assessing whether bodies of water are meeting this goal (see Karr, Chapter 2; Adler, Chapter 22). This task is almost impossible without the support and involvement of the nation's citizens.

Quality assured volunteer monitoring data could be used to supplement monitoring by state biologists. To be used in establishment or implementation of biological criteria, volunteer monitoring programs must meet certain requirements. We recommend the following: (1) a written quality assurance/quality control plan approved by their state; (2) a minimum of family-level identification for macroinvertebrates; and (3) the establishment and maintenance of a voucher collection. While volunteer data will not replace surveys by professional biologists, citizen volunteer monitoring has the potential to become an important supplement to state and other agency programs.

ACKNOWLEDGMENTS

We would like to thank the many thousands of volunteer monitors and volunteer monitoring program staff, especially those at Maryland Save Our Streams, for their tireless efforts on behalf of the nation's water resources. We would also like to thank Eleanor Ely (editor, *The Volunteer Monitor*) for her encouragement and for the information she provided on current volunteer monitoring programs. We would also like to thank the editors for the opportunity to participate in this effort on biocriteria.

## CHAPTER 20

MERGING THE SCIENCE OF BIOLOGICAL MONITORING WITH WATER RESOURCE MANAGEMENT POLICY: CRITERIA DEVELOPMENT

David L. Courtemanch

1.0 INTRODUCTION

Water resource management will require a change in management strategies and policies for effective environmental improvements to continue. The scope of problems has shifted from a predominance of discrete pollution sources to one of diffuse sources. Coupled with that is a legacy of water resource abuse still found in the sediments, in habitat alterations, and species introductions that continue to alter and suppress the value of our water resources and diminish the return on investments in water quality improvement. The perspective of water resource management needs a stronger foundation in ecological information. Restoration and maintenance of ecological integrity (variously defined in Frey 1977; Patrick 1977; Karr and Dudley 1981) is now being advanced as a primary goal of water resource management (SAB 1990; Karr 1991). This will require new approaches to management with an emphasis on the assessment of a wide expression of ecological impacts.

2.0 IMPLICATIONS FOR WATER RESOURCE POLICY

Contemporary water resource management has relied upon performance (technology-based) standards, which establish allowable pollutant loads, compliance monitoring requirements, and enforcement actions, but do not measure the effectiveness of achieving water resource goals. These performance standards were designed to alleviate distinct pollution problems, especially to reduce waterborne pathogens, maintain minimum dissolved oxygen levels, and control selected toxic substances. The traditional use of performance standards is inadequate because they are not designed to address broad goals such as ecological integrity. Biological assessment, focusing on population and community level response, addresses impact rather than only discharger performance. With clearly defined goals for the maintenance of different levels of integrity through impact standards, the overall success of a water resource management program can be evaluated.

2.1 Performance Standards

Performance standards have been used extensively to regulate specific activities. They are characterized by a focus on each pollutant (Table 1). An underlying assumption of this approach is that the patchwork of performance standards, when implemented as a whole, will be sufficient to restore and maintain the physical, chemical, and biological integrity of the water. Unquestionably, the use of

Table 1. Comparison of Performance and Impact Standards for Water Quality Management

	Performance standard	Impact standard
Definition	Regulation of select, project-specific output and/or activity	Regulation of multivariate and interdependent project outcomes
Characteristics		
Scope	Limited, discrete	Broad, interactive
Relationship to goal	Indirect	Direct
Application	Point sources, discrete effluents	Point and non-point sources, complex and compound effluents
Interpretation	Simple, objective	Often difficult, objective and subjective
Models	Many	Few, weak
Liability	Regulator	Pollutant source

performance standards has brought us a great distance toward the improvement of water quality. As the overwhelming effects of oxygen demand and acute toxicity were overcome, a large number of previously unrecognized impacts were found. This raised the question of whether biotic integrity had really been restored.

Performance standards offer several advantages that will perpetuate their use in water resource management. They are regarded as simple to apply and interpret; thus, regulation of an activity can be undertaken incrementally through permits. Mathematical models are readily available that can predict the treatment necessary to meet each performance standard. Performance standards are generally acceptable to the regulated community because of their relative ease of application, accustomed use, and lack of ambiguity.

There are a number of deficiencies, however, associated with the imposition of performance standards alone. The criteria currently available to assign performance standards represent a very small fraction of the number of potential pollutants for which standards might be developed and are based largely on laboratory studies that test only selected effects and may not reflect ambient impacts. Nor do the criteria account well for the interaction of pollutants in complex or compound effluents and their relationship to water use.

The burden of development and implementation of each performance criterion is on the regulatory agency. Because these criteria must be developed for application to wide-ranging circumstances, they are often conservative, with built-in factors of uncertain reliability. Unless the problem to be addressed by the criteria is uniform across all groups affected, the results of application will be inconsistent, thus leading to a search for accommodation. Because the burden of development and implementation rests with the regulatory agency, liability for protection from unacceptable impact also shifts from the pollutant discharger to the regulator.

2.2 Impact Standards

Impact standards, in contrast, require that a certain result must be achieved. However, the means to achieve the standard may be flexible, taking into consideration site-specific factors, and using a planning approach to develop these means. Twenty years ago, before widespread waste treatment had been achieved, use of impact standards would have been difficult. There was too little experience with different levels of waste treatment to know the consequences of these treatments in an aquatic ecosystem. Since that time, a much greater understanding has developed of how each level of treatment manifests itself in the aquatic environment. Impact standards overcome many of the deficiencies noted for performance standards. Therefore, they can be used to complement performance standards.

The imposition of impact standards may create certain tensions for the discharger. Impact standards usually do not lend themselves well to predictive models because control of variables is less certain. Impact standards provide interpretation of the impact of all stressors and the relationship to specific goals; however, this interpretation may be dependent on both objective and subjective decisions by the regulator (quantitative criteria combined with professional judgment) and may be influenced by factors that are beyond the control of the discharger. In a legal environment, this is an unacceptable situation. In a planning environment, this can be tolerable and useful by employing analytical techniques to interpret the multivariates.

A transition from performance standards to impact standards may provide the opportunity to employ a planning approach to the implementation of water resource policy. The regulatory-intensive approach, with which performance standards have traditionally been associated, has come under substantial criticism (Daneke 1982). The contemporary regulatory-intensive approach relies heavily on a legal style of rule setting. Policy options tend to be formulated in legal rather than policy or technical terms, and legal definitions tend to displace policy or scientific/technical considerations. This style of regulation stresses rules and standards regarding process instead of product. Regulatory policies tend to evolve on a case-by-case basis due to the *ad hoc* nature of permitting. Therefore, it has been argued, insufficient attention is directed toward assessing cumulative impact of these policies in terms of costs and outcomes (Benveniste 1981; Daneke 1982).

2.3 Planning

Planning techniques for identifying goals and objectives, and accommodating them in the policy implementation process, may offer distinct advantages over existing regulatory-intensive procedures. Planning is the elaboration of a set of related programs designed to achieve certain goals. The main distinction between planning and regulation is that planning addresses complex situations where government intervention has to take into account the many parts of a problem while regulation addresses parts of a problem independently. Planning requires simultaneous and continued government involvement intended to affect the behavior of many different organizations or individuals to achieve collective goals. In contrast, regulation is a one-to-one relationship between the regulator and the discharger and is employed in situations where cumulative control can be used to achieve collective goals. Therefore, planning is by definition more concerned with integration and with the specification of targets and goals. Planning integrates regulation into the process. It focuses on implementing output goals, instead of relying on process rules, and can establish goals from which meaningful regulatory standards can be deduced (Benveniste 1981). In water resource management, planning and the use of impact standards may be beneficial to the individual discharger. Because the planning process focuses on complex factors simultaneously, both point and nonpoint sources, there is an expectation that management will be more equitable.

3.0 DESIGN OF BIOLOGICAL CRITERIA

The vertebrate and invertebrate communities are excellent sources of information to set goals and provide the required feedback needed for a planning style of management, and, therefore, are areas where development of impact standards is appropriate. First, they provide a real rather than surrogate measure of goals expressed in the Clean Water Act. The organisms are long-lived, integrating the effects of pollutants from all sources over a period of time rather than producing instantaneous assessments of quality. They are also interactive, integrating the effects of species requirements, habitat, biotic interaction, and pollution interference. While the dynamics of aquatic communities are not completely understood, sufficient understanding of the basic controlling principles of these communities has been established to allow predictive evaluations to be made and to bring an array of quantitative measures to assess integrity of a community (e.g., Karr 1981). Specific ecological properties of the community should be identified in the standards. This ensures that the selection of quantitative criteria will be relevant to ecological integrity as defined in the standard.

Biological standards and criteria can be used most effectively for comprehensive impact evaluation. Therefore, the traditional regulatory role of standards and criteria should be relaxed. This may alleviate some of the legal tensions, thus allowing the use of a broad scope of measures, both objective and subjective, and the maximum use of the information available within the biological community for management decisions.

3.1 Managing Uncertainty

Water resource management has evolved by a regulatory-intensive design. While the argument is strong for a timely infusion of planning in decision making, there are certain reservations that apply when merging ecological information into the present water resource management programs. Uncertainty is a common element in ecological predictions, but is not well tolerated in a legalist/regulatory environment because the focus there is so strong on the process of control. Uncertainty is unavoidable but can be managed.

There are a number of sources of uncertainty (NAS 1986). Relationships among species are complex and changes in one can lead to unexpected changes in another. Indirect effects, those effects occurring at distant species linkages, are difficult to assess and predict. Response is nonlinear (Odum et al. 1979). Population or community response can be differential depending on the nature and magnitude of the perturbation and the various relationships existing within a particular community. Natural variability also contributes to uncertainty. Populations and communities vary spatially and temporally because of both intrinsic processes and changes in the physical and biological environment. Errors of estimation occur whenever sampling of a population takes place. The type of error can be of significant concern in a regulatory environment. Concluding that an effect has occurred when it has not (type I error) has significance to regulated interests, and costs associated with it that will be different from a type II error (conclusion that no effect has occurred when one has). Each of these problems should be considered in the design of criteria.

System complexity is most evident as one works with a large database. Biological samples are rich in information. The difficulty of dealing with a large mass of information is that of the many interactions occurring within the community, some may be related to water quality while others may not. This is part of the problem with biological data that has made criteria development look forbidding in the past. In developing the criteria, it is necessary to eliminate the irrelevant factors and reduce multicorrelation. Criteria should rely on an array of measures that test elements of integrity. Criteria designs using multiple measures of a community, such as Ohio's multimetric approach (Ohio EPA 1987a,b) or Maine's multivariate approach (Davies et al. 1993), provide a robustness to assess ecological integrity that single measures lack and diminishes the influence of error within any one measure on the overall assessment outcome. Sampling error is controlled by establishing a rigid sampling protocol designed to standardize the bias which does exist.

Because attainment of water quality criteria is essentially a one-tailed examination (does a sample attain the specified criteria, or better, vs. nonattainment of the specified criteria), the criteria can be built to reduce error on the one tail. This has importance to the balance of error types and the uncertainty that these present in management decisions. Because the emphasis of error control is on one tail, the amount of error expected in the criteria may be more easily adjusted.

3.2 Conflicts with Other Criteria

Presently, there are two classes of monitoring criteria used to assess water quality: (1) pollutant-specific numeric criteria, which include numeric concentration limits for a number of chemical and microbiological substances, and (2) whole effluent toxicity test requirements, which expose test organisms to effluent mixed to various concentrations with receiving water. Each of these types of criteria was developed to address specific regulatory needs of management as related to the permitting of point source discharges.

The use of biological criteria based on community response in the receiving water should have a substantially different focus. The biological criteria are used to supplement the other existing criteria by providing information, not only on permit performance, but on all water resource management factors (e.g., unpermitted sources of pollutants, flow and other habitat modifications, and biotic interactions) to determine the quality of the aquatic community (ecological integrity).

The potential for conflict between the different criteria has been identified (USEPA 1990a). Numeric chemical, whole effluent toxicity, and ambient biological criteria may not show agreement in the assessment of attainment of water quality standards (Ohio EPA 1990a). Biological criteria cannot be used interchangeably with other types of criteria. Each criterion is able to detect different types of water resource impairment and has different levels of sensitivity for detecting certain types of impairment. The principles of independent applicability must be applied. Each criteria should be independently applied and satisfied for standards to be achieved.

4.0 POTENTIAL APPLICATIONS

Biological information has traditionally been used to make site-specific assessments of the impact of discrete activities (e.g., comparison of a downstream community to an upstream community). The magnitude of difference between the communities is inferred to represent the impact of the activity. By putting biological information into the form of impact standards, it is possible to set standardized goals for the water resource management. The standards and criteria can be designed to assess goals on larger scales: watershed, ecoregion, statewide, and regional.

4.1 Planning

As previously stated, planning is the elaboration of a set of related programs designed to achieve certain goals. Biological standards provide the needed definition for the goal of restoration and protection of biological integrity and the biological criteria provide the mechanism for measurement of attainment of those goals. Biological standards can fulfill the needs of problem definition and prioritization needed in a systems analysis approach. By providing several standards (as Maine has in its use classification system; Courtemanch et al. 1989), it is possible to provide a choice of management goals, from maximum preservation of the aquatic ecosystem to varying levels of human intrusion, but with important provisions for protection of the essential elements of biological integrity.

4.2 Assessment

Establishment of biological standards can extend the use of water resource assessments. Presently, there is no standardized assessment method for measuring biological integrity. Assessments, such as required in Section 305(b) reporting, are based on water chemistry information with the assumption that these criteria are providing information about integrity. That assumption has been found to be only partially correct. Biological criteria will allow assessment of whole classes of water to determine if management programs are collectively achieving desired goals. This provides important system feedback that can, in turn, determine allocation of resources including research, permit revision, project designs, facilities funding, and enforcement.

4.3 Permitting

Ambient biological criteria provide limited value for establishing permit limits. This is due to the low predictability of the ambient biological response, especially at the community level, to effluent load estimates. Permit limits constitute a small number of potential variables that will determine community composition. In fact, the need for ambient biological criteria is due to the limited scope of present effluent permits, and the water quality models used to produce them, to account for ecosystem condition. Ambient biological criteria may provide useful feedback for the evaluation of permit performance in combination with other factors. A finding of impaired community quality may provide the necessary feedback to reevaluate permit limits or to prevent issuance of a new permit in the area of impairment.

4.4 Enforcement

The value of biological information for enforcement purposes is not lost by changing from a performance style to impact style standard. While the impact standard is designed to collectively assess the condition of a community resulting from all factors acting on it, monitoring can be designed to assess discrete activities for enforcement purposes. Like traditional assessment studies, this requires the need to establish reference samples where attainment of the standard can be demonstrated and test samples where there is a high degree of certainty that the effects of the specific target activity is significant. Having criteria that are universally accepted relaxes the subjective nature of an enforcement decision and the extended negotiation that might be required to address a water resource problem.

5.0 NEW POLICY FORMULATION: EXPERIENCE IN MAINE

As science and technology have evolved to hold a dominant position in the function of present-day society, there is an assumption that political thought has likewise evolved to utilize this technical information. However, this is not universally true and the means of incorporating scientific advancements into the policy-making process continues to develop. Implementation of scientific information into public policy can be found to follow certain predictable courses. It is important for the scientist and the policy planner to understand the elements of implementation. The evolution of biological standards and criteria in Maine's water resource management program provides a good demonstration of these necessary elements.

Policy making is more than a political compromise of competing interests, and certain organizational theories may be applied to describe the process. Kingdon (1984) presents a model of environmental policy innovation that appears to define the multiple factors involved in formulation and implementation of biological standards in Maine's program. The model proposes that three "streams" must converge before a policy forms. A "problem stream" must define some dilemma or crisis for which a new policy is warranted, a "politics stream" must include various groups receptive to policy change, and a "policy stream" must generate the ideas that lead to an acceptable proposal. Science plays a role in each of the "streams." Examples of each stream can be identified in Maine's implementation approach and similar examples can be identified in other states where implementation has occurred.

5.1 Problem Stream

Water resource quality has changed since the implementation of the Clean Water Act, and management policies should be reviewed. In Maine, there was common agreement that sizable progress had been made in abating the most obvious water pollution problems that existed 20 years ago. The state needed an expression of its future water resource goals and a means to assess those goals so it could plan the direction of its programs (Courtemanch et al. 1989). Having implemented the requirements of wastewater treatment throughout the state, there was a need to better assess water quality and decide what further courses of action should be taken for remaining problems, many of which had been revealed after implementation of waste treatment. Continued emphasis on the application of performance standards would have limited value. It was well understood that nonpoint source pollution and habitat alteration were contributing a great share of the remaining water quality problems. Best Management Practices for nonpoint source controls were being proposed to reduce these sources, but these practices are highly variable in effectiveness; thus, the predictability of results is low. A new means of assessing water resources toward a well-defined goal of biological integrity was needed.

Federally directed water resource management policy was also changing during this period. The U.S. Environmental Protection Agency was imposing greater performance monitoring requirements on facilities to assess their compliance but was not requiring a complementary assessment of the ambient state of the waters. Construction-grants money for public treatment works was in decline, federal cost-share percentages were being reduced, and a revolving loan system was being proposed for future construction. Future management had to be directed in a more structured planning environment that can direct increasing compliance requirements and diminishing state and federal resources toward those problems that are identified as the greatest and where improvement is most assured.

5.2 Politics Stream

Recognizing the need to manage the state's water resources more closely, the Maine Legislature was open to the use of new types of information if it could be shown to contribute to these decisions. Aquatic life standards were proposed to improve impact assessment and provide greater emphasis on the management of ecological resources. Politics formed around an "iron triangle" of interests (business, environmental advocates, and the environmental agency, Maine Department of Environmental Protection), each of which could identify advantages for implementing aquatic life standards.

The business community was interested in promoting the use of biological standards and criteria. Significant investment had been made to clean up pollution generated by municipalities and industry. It was in the interest of commerce to be able to demonstrate these benefits to the public; first, as important public relations information, but more importantly as a means of convincing lawmakers, environmental agencies, and the environment interest groups that their actions had been sufficient and that further pollution controls should only be imposed where problems could be demonstrated, a water quality-based approach. A set of commonly accepted goals and a standardized means of assessment was needed. Industry had made numerous studies to demonstrate the improvements they had made, but because there was no standard assessment method these studies were often suspected of biased reporting. By demonstrating achievement of a clear set of goals in the law, the business interests could avoid a continued defense of their actions.

While business interests could see the advantage of ambient biological criteria to assess the effectiveness of treatment, environmental advocates could also see the advantage of its use to force added improvements based on defined goals. Biological criteria were regarded as a means of measuring important uses to be protected. The agency wanted to get rid of some obsolete standards in the law and introduce some impact standards that could lead the agency toward a planning environment. The debate over the final design of the criteria would determine which advantage was given the greater weight, an important consideration in negotiations in the "policy stream."

5.3 Policy Stream

With the need for this planning approach to management established, the debate focused on development of the standards. The Department proposed that three levels of biological standards (integrity) be applied to the rivers and streams in the state's classification statutes, and one standard was established for lakes (Table 2). The debate focused on the stream standards. All three levels for the rivers and streams were considered to be acceptable interpretations of what was intended by the Clean Water Act for protection of biological integrity, but allowed important management options.

The "natural" community proposed for Classes AA and A waters was consistent with other existing performance criteria that strictly limited activities in those waters. The "without detrimental change" in the standard for Class B waters recognized that human activities, particularly the discharge of wastewater and significant land-use activities, would almost always induce a change in the community but that certain types of change could be regarded as benign or possibly beneficial. Maine waters are typically low in ionic strength and support relatively lower numbers of organisms and species than many waters elsewhere. Certain nondetrimental community responses, such as the recruitment of species and increased populations, can be expected and may be desirable. The proposal for Class C waters to maintain structure and function of the community was initially viewed by the environmental advocacy groups as a lowering of the previous standard for Class C, which had stated "no disposal...harmful...to aquatic life." Since Class C waters are the populated corridors of the state with industrial and municipal development, it was unlikely that meaningful criteria could be applied that would protect aquatic life from any harm. The Department argued that by focusing on structure and function of the community, the important elements of integrity could be preserved without a rigorous species by species test of harm. Business interests supported the Department's concept. The environmental advocacy groups agreed that the previous Class C standard was not consistent with existing management, and accepted the proposal pending the outcome of debate over the definitions.

That debate concerned the reference condition to which the criteria would apply. A reference condition is an essential element of biological evaluation and is the basis of any decision. Each party realized that the reference condition definition would be important to their interests. The Department was interested in having a broad definition for a reference condition, since this would allow greater freedom for development of the numerical criteria. By allowing a broad definition of reference condition, the environmental advocacy groups could bring attention to a greater breadth of potential impacts. By narrowing the definitions, business interests could bring some limit to the kind and magnitude of impact that would be regulated.

There are basically two approaches to establishing reference condition: (1) site-specific references that measure differences between a test site and a reference immediately upstream, and (2) geographic references that define the reference from a characterization of many waters in a region. The Department wanted to use both approaches and had already invested considerable resources toward developing a database with both paired analysis and statewide characterizations. While pairwise analysis is considered more definitive and important to assess discrete activities, suitable paired sites are often not available. The

Table 2.   Maine's Water Quality Classification System for Rivers, Streams and Lakes, and Associated Biological Standards

Class	Management perspective	Biological standard
AA	High-quality water for recreation and ecological interests. No discharges or impoundments permitted	Habitat natural and free flowing. Aquatic life as naturally occurs
A	High-quality water with limited human interference. Discharges restricted to noncontact process water or highly treated wastewater equal to or better than the receiving water. Impoundments allowed	Habitat natural. Aquatic life as naturally occurs
B	Good-quality water. Discharge of well-treated effluent with ample dilution permitted	Habitat unimpaired. Ambient water quality sufficient to support life stages of all indigenous aquatic species. Only nondetrimental changes in community composition allowed
C	Lowest water quality. Maintains the interim goals of the Federal Water Quality Act (fishable/swimmable)	Ambient water quality sufficient to support life stages of all indigenous fish species. Change in community composition may occur but structure and function of the community must be maintained
GPA (lakes)	Protection of present high quality for recreation and ecological interests. No discharges allowed. Change of land use may not cause trophic state increase	Habitat natural. Aquatic life as naturally occurs. Maintain stable or decreasing trophic state

business interests favored limitation to just pairwise analysis, limiting the definition for "resident biological community" to the community existing upstream of an assessment site. They argued that this was the only way to confidently assign specific responsibility if a site were found to be in nonattainment. From a regulatory perspective, this was not unreasonable, but the Department was interested more in the general assessment of its water resources and had less interest in specific regulatory outcomes. Therefore, the broad application of reference condition methodology was argued. The environmental advocacy groups wanted a powerful assessment tool, but also the ability to use these new standards in a regulatory mode for permitting and enforcement of specific activities.

The final statutory definitions provided the Department with the ability to pursue a broad reference condition methodology (Table 3). Criteria design is based on a statewide reference model for each class (Davies et al. 1993); however, paired analysis around a site might be used where enforcement is contemplated.

5.4 Policy Entrepreneur

In addition to the "problem, politics and policy streams" of policy formulation, Rabe (1986) proposes that another component of policy formulation is required. A policy entrepreneur (an individual with technical expertise and negotiating ability) is needed to get the three streams to converge. Benveniste (1977) found a similar need for this functional individual if science and technology were to have prominence in public policy development. Biological staff from the state agency provided this function. Similar individuals can be identified in each of the states where biological criteria have been developed, and are probably a necessary component. Without this function, it is unclear if the standards would have been enacted or if they would have contained the necessary scientific design. Nonagency biologists were also involved as entrepreneurs in the design and implementation process. Department biologists and corporate biologists were allowed direct involvement in the negotiations of the statutory standards before the legislature. The Department also convened a technical advisory committee of non-agency biologists to oversee development of the numerical criteria. This provided an objective forum of biological expertise to provide oversight of the criteria development process. The Department presented each step of the criteria development process for peer review to various scientific forums. In doing so, the Department hoped to be regarded as an objective designer and implementor of public policy, rather than as a specific interest group.

Table 3. Definitions of Key Terms Found in Maine's Water Classification Law (38 M.R.S.A., Section 466)

Aquatic life	Any plants or animals that live at least part of their life cycle in fresh water
As naturally occurs	Conditions with essentially the same physical, chemical, and biological characteristics as found in situations with similar habitats free of measurable effects of human activity
Indigenous	Supported in a reach of water or known to have been supported according to historical records compiled by state and federal agencies or published scientific literature
Natural	Living in, or as if in, a state of nature not measurably affected by human activity
Resident biological community	Aquatic life expected to exist in a habitat that is free from the influence of the discharge of any pollutant. This shall be established by accepted biomonitoring techniques
Unimpaired	Without a diminished capacity to support aquatic life
Without detrimental changes in the resident biological community	No significant loss of species or excessive dominance by any species or group of species attributable to human activity

The final product was a set of standards and criteria that have broad application to current program needs. By emphasizing their value for planning water resource programs, they can be used for an integrative assessment of conditions while still offering traditional site-specific evaluations. By establishing an array of management choices (goals), the criteria aid in decision making.

6.0 MAINE'S BIOLOGICAL CRITERIA MODEL

Maine has adopted numerical criteria based on a probability model for each of its four water quality classes for rivers and streams (see Table 2). The models are linear discriminant functions that collectively use 25 variables (Table 4) from samples of benthic invertebrates collected by a defined sampling protocol. The models were developed using data from 145 sites throughout the state representing the span of water quality conditions that exist. Since water classification is an artificial system for describing differences in water quality (placement of continuous states of condition into discrete categories), it was decided that an interpretation of the data relative to the law should be made and an algorithm designed to describe the best fit of the interpretation of each classification. The data from each site was interpreted by both agency and nonagency biologists to determine which sites fit into each classification, a consensus classification. Using this consensus classification of the sites, linear discriminant analysis was used to select the variables with the greatest predictive value and coefficients that weighted their value. By constructing the models in this fashion, certain advantages are gained. First, the cognitive process of interpreting data from each site involves the use of many potential variables. While each analyst is aware of the variables they have considered, there is no certain way of determining which variables are significant to the final determination. Linear discriminant analysis provides an objective means of identifying the significant variables. Secondly, since the variables have different information content, the weight of each variable used in classification is expected to be different. Discriminant analysis provides a weight for each variable in the form of a coefficient. Finally, because the linear discriminant function is expected to exhibit a multivariate normal distribution, probabilities of fit can be calculated that provide the analyst with information about the strength of a classification assignment when a new site is evaluated using the models.

The use of probabilities offers certain advantages for implementation. It is possible to use differing probabilities for regulatory and nonregulatory decisions. Assessment, which is used for status reporting, program management, and prioritization can tolerate the use of weaker probabilities since these activities are administered continuously and interactively. Enforcement, on the other hand, is a discrete function, where the action should not allow for much indecision in the model. Therefore, the use of differing decision probabilities from the model can satisfy these management needs.

An important final step of the numerical criteria design involves professional judgment. This is provided to assure that the scientist is not separated from the policy, a common fault noted by Benveniste

Table 4.   Variables (and Natural Log Transformations) Used in Maine's
           Aquatic Life Classification Models

Variable	Classification
Total abundance (nlog)	AA/A, B, C
Generic richness	AA/A, B, C
Plecoptera abundance (nlog)	AA/A, B, C
Ephemeroptera abundance (nlog)	AA/A, B, C
Shannon-Wiener diversity	AA/A, B, C
Hilsenhoff biotic index	AA/A, B, C
Relative Chironomidae abundance	AA/A, B, C
Relative Diptera richness	AA/A, B, C
Hydropsyche abundance (nlog)	AA/A, B, C
Relative Plecoptera richness	AA/A
Relative *Brachycentrus* abundance	AA/A
Plecoptera/Ephemeroptera richness	AA/A
Abundance of *Cheumatopsyche, Cricotopus,*	AA/A
Tanytarsus, and *Ablabesmyia* (nlog)	
Abundance of *Acroneuria,* and *Stenonema* (nlog)	AA/A
Dominant indicator taxa	AA/A
Presence/absence of indicator taxa	AA/A
Perlidae abundance (nlog)	B
Tanypodinae abundance (nlog)	B
Chironmini abundance (nlog)	B
Relative Ephemeroptera abundance	B
Ephemeroptera/Plecoptera/Trichoptera abundance	B
Abundance of *Dicrotendipes, Micropsectra,*	B
Parachironomus, and *Helobdella* (nlog)	
Cheumatopsyche abundance	C
Relative Oligochaeta abundance	C
Ephemeroptera/Plecoptera/Trichoptera and Diptera richness	C

Note: Algorithms and coefficients for each classification can be found in Davies
       et al. (1993).

(1977) in regulation design involving technical information. Professional judgment allows the biologist analyzing the sample to reject or modify the findings due to circumstances for which the criteria were not designed. While this creates some obvious apprehension from parties affected by a decision, an objective basis must be demonstrated to limit the use of professional judgment to appropriate situations.

6.1 Maine Regulations Using Biological Criteria

The structure for a state regulation using ambient biological information involves three steps of development:

Sampling step: development of standardized sampling protocols, data management and quality assurance

Analytical step: development of a decision criteria using quantitative measurements of the community relevant to standards in the water quality law

Decision step: development of rules for the application of information derived from the analytical step

The rules for application determine how the information will be used. These vary for each of the applications described: assessment, permitting, and enforcement. Development of quantitative biological criteria is important for consistent review and assessment of water resource quality. This has been lacking at both the federal and state level, notably for requirements of Section 305(b) of the Clean Water Act, but also for many other planning purposes including nonpoint source assessment, toxics assessment, and program prioritization (USEPA 1990a).

For permitting decisions, Maine requires that biological monitoring (ambient biological assessment and effluent toxicity testing) be conducted two years prior to permitting an existing facility as a means of determining performance. Where it can be demonstrated that an activity is responsible for nonattainment, that activity would not be given a permit renewal until attainment is achieved or modifications are made in the permit that would bring water quality into attainment. Where multiple sources contribute to nonattainment, it becomes the responsibility of the agency to implement a plan. The plan would direct the involvement of all appropriate programs to achieve attainment. Issuance of a permit would be contingent upon compliance with those aspects of the plan that are the applicant's responsibility. Biological criteria have been used in Maine for all major wastewater permit and water quality certification (e.g., hydropower relicensing) decisions since 1991.

Biological criteria are used for enforcement where it can be demonstrated that a specific activity is responsible for nonattainment. This may be determined using paired samples (above/below, before/after), where there is clear evidence of detrimental change and nonattainment of the designated class criteria. In Maine, the use of probability scores for attainment provides important information about the strength of a nonattainment decision. The criteria have been used in one enforcement case.

7.0 SUMMARY

Biological information can be expected to have a greater role in water resource decisions in the future for several reasons. There is an increasing demand from the public to improve our ecological resources and fulfill the goal to restore and maintain biological integrity that will force management to seek new methods of decision making. There is a realization that protection of ecological resources has not been effectively implemented because of the limitations of the performance standards presently used for management decisions. More effective and efficient water resource management will occur with a transition toward a planning approach that can provide a better balance of program initiatives to solve identified problems on a priority basis. Ambient biological monitoring can contribute the type of integrated information to direct management toward fulfilling the goal of restoration and protection of biological integrity through this planning approach.

## CHAPTER 21

Policy Issues and Management Applications of Biological Criteria

Chris O. Yoder

1.0 INTRODUCTION

1.1 Overview: Should Biocriteria Be Part of State Water Programs?

Biological criteria as a part of ambient chemical, physical, and biological assessments (i.e., biosurveys) have a potentially broad role in many surface water resource management and policy issues as has been demonstrated throughout this volume (Karr, Chapter 2; Bode and Novak, Chapter 8; Courtemanch, Chapter 20; DeShon, Chapter 15; Yoder and Rankin, Chapter 9, 17; Rankin, Chapter 13). Despite these proven uses the role of biological criteria in surface water resource management and policy has yet to be fully implemented by most states and USEPA. This chapter is an overview of the management and policy uses of biological criteria in Ohio and an examination of the challenges to broader usage throughout the United States.

Recently, the cumulative costs associated with environmental mandates, many of which are the result of prescriptive regulations, have come into question. Former USEPA Administrator William Reilly frequently cited the need for the increased use of good science in formulating regulatory requirements. Both the regulated community and the public desire evidence of real world results in return for the pollution control expenditures made necessary by federal and state mandated requirements. Biological criteria seem particularly well suited to meet some of these needs in that the underlying science and theory is robust (Karr 1991) and biocriteria certainly qualifies as real world.

The administrative dominated direction of the traditional surface water regulation strategies partially emanates from the belief that it is neither practical nor feasible to directly measure compliance with the CWA goal of biological integrity (Jorling 1977). Since concise and practical frameworks for using direct biological measures were not forthcoming in the early days of the CWA, surrogate approaches were relied upon. Two reasons for this include the perception that direct biological information is simply not obtainable from a technical and resource/cost standpoint, and natural biological communities are simply too complex to measure and too poorly understood to use. The alternative was to use surrogate indicators of *potential* impairment as the basis for regulation and to provide feedback about current and future conditions. However, a continued reliance on this approach *alone* is questionable when considering the growing body of information that demonstrates the usefulness of the presently available bioassessment approaches.

The tendency in water quality management has been to make biological measurements fit the perceptions and use of chemical criteria, rather than the reverse. This is a paradox because an aquatic community is the embodiment of the temporal and spatial chemical, physical, and biological dynamics (i.e., the "pieces") of the aquatic environment, not the reverse. Perhaps the inability of biologists to agree on a set of empirical measurements of biological integrity, or at least a common framework, has resulted in this tendency (Karr et al. 1986). One solution to this deficiency is to employ biological criteria that can directly indicate the degree to which biological integrity is or is not being achieved. This does not mean that biological criteria are a substitute for chemical criteria and bioassay techniques as these will *always* play an important role in water quality management. Their value, however, is greatly enhanced when used in combination with biological criteria. Such an approach will undoubtedly lead to more effective regulation of pollution sources, improved assessment of diffuse and nonchemical impacts, and a broadened capability to implement management strategies for protecting and restoring watersheds.

In both the 1988 and 1990 Ohio Water Resource Inventories (305[b] reports) a comparison of the indications of aquatic life use attainment/nonattainment with the Ohio EPA numeric biocriteria was made with ambient chemical indications of the same. The ambient database for this analysis was generated from biosurveys dating to the late 1970s (Ohio EPA 1988, 1990b). Out of 645 waterbody segments analyzed, biological impairment was evident in 49.8% of the cases where no impairments of chemical water quality criteria were observed (Ohio EPA 1990b). While this discrepancy may at first seem remarkable, the reasons for it are many and complex. Biological communities respond to and integrate a wide variety of chemical, physical, and biological factors in the environment of both natural and anthropogenic origin. Simply stated, controlling chemical water quality *alone* does not assure the ecological integrity of water resources (Karr et al. 1986). The results of this analysis indicate not only the broad ability of the aquatic biota to reflect and integrate multiple chemical, physical, and biotic influences, but also the more important issues of accuracy and comprehensiveness within a state water quality management program.

What role, then, can states play in improving the process? The most logical avenue is through state water quality standards (WQS). As part of the state WQS regulations the management and implementation tools associated with biological criteria become legitimized beyond their present "optional" status. An understanding of how the fundamental goals of the Clean Water Act (CWA) influence state WQS is additionally necessary to comprehending the broad role that biological criteria can play in the entire surface water quality management process.

1.2 Goals of State Water Quality Standards

A principal objective of the CWA is to restore and maintain the biological integrity of the nation's surface waters. Although this goal is fundamentally biological in nature, the specific methods by which regulatory agencies have attempted to reach this goal have been predominated by such nonbiological measures as chemical/physical water quality (Karr et al. 1986). The presumption is that improvements in chemical water quality will be followed by a restoration of biological integrity. This approach does not directly measure the ecological health and well-being of surface waters nor does it follow the definition of pollution in Section 502 of the CWA as "man-made or man-induced alteration of the chemical, physical, biological or radiological integrity of water" which clearly is broader than a singular concern for chemical pollutants. The notion that controlling point source discharges of chemicals as the cornerstone of regulatory efforts towards attaining the biological integrity goal of the CWA has become so ingrained into the system that some interesting misconceptions about water quality standards and CWA goals have arisen.

Water quality can easily become a confused and nebulous concept, especially when no demonstrable or tangible endproduct can be identified. Regulators assert that the attainment of administrative goals will logically be followed by actual environmental improvements. But how can this be verified? Do we simply continue to assume that improvement occurs without making an effort to confirm this with environmentally based measures? The presumptions of an administrative, surrogate indicators-dominated approach follow: (1) any chemical water quality criteria exceedence is bad; (2) the observation of no exceedences is good; and (3) an emphasis on the control of toxic chemicals will result in the attainment of CWA goals. In fact, well-intentioned, but simplistic quests for clear and/or chemically cleaner water have fostered management strategies that have actually resulted in *increased* damage to the environment (Ohio EPA 1992a) because of a reliance on these sometimes-flawed presumptions.

1.3 The Role and Applicability of Biological Criteria

The existing status of the biota resident in any surface waterbody is the integrated result of complex and interrelated chemical, physical, and biological processes over time and is the summation, or result,

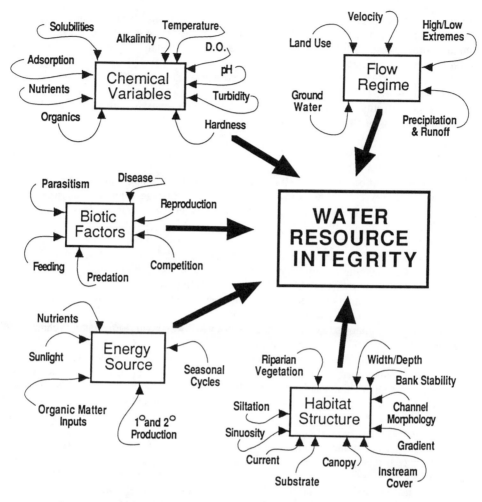

Figure 1. The five principal factors, with some of their important chemical, physical, and biological components that influence and determine the integrity of surface water resources. (Modified from Karr et al. 1986.)

of these processes in their dynamic sequences (Figure 1). Biological communities are precise indicators of actual conditions since they inhabit the receiving waters continuously and are subject to the array of chemical and physical influences that occur over time including both common and extreme events. This includes all of the chemical and physical variables that are commonly measured by most ambient monitoring programs plus additional important variables that are frequently not measured. If these chemical, physical, and biological variables are considered as "pieces" then the resultant biological condition is the integrated result of the assembly of these pieces in the proper dynamic sequence. In this sense biological criteria represent a top-down evaluation where the end product (biological community performance) is used to characterize the summed or integrated result of *all* the pieces (i.e., the chemical, physical, and biological processes that affect biological performance). By comparison the reductionist chemical/toxicity approach represents a bottom-up evaluation where *some* of the pieces are used in an attempt to simulate, predict, or explain complex processes using surrogate end points.

Adopting an increased reliance on direct measurements of biological community performance to establish regulatory direction and priorities may require a modification of some current regulatory attitudes and approaches. In addition to attempting to estimate a protection level for the end point of concern (i.e., biological integrity) via the chemical-specific and/or narrative (i.e., "free from") approaches, this process should ultimately involve the development of pollution abatement strategies to achieve or maintain biological community end points by having *prior* quantitative knowledge about that end point. This will involve linking treatment processes and performance, ambient water quality, habitat, toxicity units, best management practices, etc. with observed biological community response in a feedback-loop arrangement. Steedman (1988) provided a good example of how empirical data was used in a similar fashion to establish land use/riparian zone criteria for attaining prescribed levels of performance for Lake Ontario tributary fish communities.

USEPA regulations (40 CFR Parts 35 and 130) encourage the use of ambient biological data in water quality decision making. The USEPA technical guidance manual for performing wasteload allocations (USEPA 1984a) specifically states that it is preferable to coordinate the determination of Total Maximum Daily Loads (TMDL) with a biological survey because:

> As the numerical criteria of water quality standards are mostly derived from single species laboratory tests, an observation that a criterion is violated for a certain time period may provide no indication of how the integrity of the ecosystem is being affected. In addition to demonstrating the impairment of use, a biological survey, coordinated with a chemical survey, can help in identifying culprit pollutants and in substantiating the criteria values. The resulting database may also provide information transferable to other sites.

Any of these liabilities are further compounded when no violations of chemical water quality are detected especially considering our findings in comparing the relative abilities of chemical criteria and biocriteria to reveal impairments (Ohio EPA 1990; Yoder 1991a). As water quality management expands into watershed-level applications the need for biocriteria-based assessment becomes even more urgent. These shortcomings are serious enough for point sources, but are further compounded with nonpoint and intermittent sources of pollutants (e.g., combined sewer overflows, urban storm water discharges, spills/dumping, etc.). Nonpoint sources tend to be predominated by natural constituents (e.g., nutrients, sediments, etc.), overlap with physical impacts (e.g., habitat modification, riparian zone degradation), and are frequently more subtle than are point sources. Even with near-continuous chemical monitoring the need remains to interpret the biological meaning of the chemical results. Simply put, biological communities are broader indicators of environmental problems than is chemical sampling *alone* because they reflect the integrated dynamics of the chemical, physical, and biological processes that are constantly at work in aquatic ecosystems (Figure 1).

No single monitoring component can "do it all," particularly in the more complex situations (i.e., multiple discharges, habitat alterations, presence of toxic compounds, etc.). A lack of information from any one component or an overreliance on a single component can result in environmental regulation that is less accurate and potentially underprotective of the water resource. Accounting for cost is not only a matter of dollars spent, but is also a question of environmental accuracy and technical validity. In short, a credible and genuinely cost-effective approach to water quality management should include an appropriate mix of all monitoring components. Prioritizing the use of these components must be based on experience, existing information, and best professional judgement.

2.0 POLICY AND PROGRAM APPLICATIONS

2.1 The Role of Biological Criteria in The Management of Aquatic Resources

We define the management of aquatic resources here as being broader than the traditional purview of water quality management and that efforts to attain the goals espoused by the CWA and other initiatives (e.g., maintenance and recovery of aquatic biodiversity) ought to recognize the potentially broad role that biological criteria and assessment have in each area. It would be unfortunate to limit biological criteria to the traditional regulatory focus of water quality management programs (i.e., NPDES permits) as it has the demonstrated ability to be useful in virtually any issue involving water resources where a goal is to protect, enhance, or restore aquatic communities and ecosystems. We believe that biological criteria and the attendant concepts of regionalization and reference sites have a broad applicability beyond the CWA.

The Ohio EPA water programs have relied extensively on ambient bioassessments since the late 1970s. The program areas within which biological criteria have found the most widespread uses are the biennial water resource inventory (305b report), water quality standards (aquatic life use classifications),

Biocriteria: Ohio EPA Surface Water Program Applications

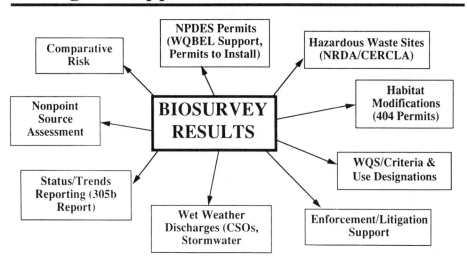

Figure 2. The various environmental management programs at Ohio EPA which are supported by information from biological surveys (biosurveys).

NPDES permits (includes enforcement and litigation support), the construction-grants program (now the State Revolving Loan Fund), the Ohio Nonpoint Source Assessment (CWA section 319), evaluation of wet weather flow impacts (stormwater, CSOs), the state certification of CWA section 404 permits (401 program) and petitioned ditches, ranking of CERCLA sites, and comparative risk (Figure 2). In addition the biological data has proved useful to other state agencies including the Ohio Department of Natural Resources (rare, threatened, and endangered species, scenic rivers, nonpoint source management, and fisheries management) and the Ohio Department of Transportation (environmental impact statements). Some of these applications were discussed in detail by Yoder and Rankin (Chapter 9).

2.2 Technical Concerns With Biocriteria

Some of the technical concerns that have been expressed by USEPA (1991a; Jackson 1992) and others (Schmidt 1992) about the application of biocriteria include:

1. Bioassessments are more costly than chemical-specific assessments or bioassays and data interpretation is more difficult.
2. Biocriteria and bioassessment have not been sufficiently developed for uniform application and specified levels of protection are lacking.
3. The scientific basis of biocriteria remains untested over a sufficiently broad range of conditions.
4. Bioassessments and biocriteria are not applicable to other water use concerns such as human health, wildlife health, fish consumption, etc.
5. Biocriteria, chemical criteria, and whole effluent toxicity each measure different end points.
6. Bioassessments only provide a short-term evaluation and may not encompass critical conditions (i.e., the $Q_{7,10}$ flow used to portray critical conditions for wasteload allocation purposes).
7. The ability of biocriteria to indicate cause-and-effect relationships and instream response has not been sufficiently established.
8. Criteria are lacking for what constitutes a sufficiently comprehensive bioassessment.
9. Bioassessments are limited in their ability to define unimpaired waters, i.e., no definitive statement pertaining to unimpaired status can be made with bioassessments.
10. Bioassessment lacks a predictive capability as compared to chemical-specific and whole effluent tests.

Based on our experiences in Ohio and familiarity with other state efforts we take some issue, to varying degrees, with all of the above stated reservations.

2.2.1 Cost and Resource Issues

USEPA (1985, 1991a) and others previously have pointed to the "high" costs associated with biological surveys. Chemical surveys are viewed by USEPA (1991a) as being cheaper because of market availability. Independently, Yoder and Rankin (Chapter 9) examined the relative cost of chemical-specific, bioassay, and biosurveys and found that this was not the case. In fact, chemical assessment was the most expensive of the three tools examined based on costs incurred by Ohio EPA. Costs for chemical analysis will continue to increase not only because basic analytical costs are rising, but also because the number of analyses and parameters required is likewise increasing. As long as a cost-effective bioassessment approach is used, cost should not be a concern.

2.2.2 Biocriteria Development and Scientific Basis

Criticisms 2 and 3 pertain to the apparent lack of a framework and procedures for developing biocriteria, establishing specified levels of protection, and the untested scientific basis for biocriteria. These criticisms tend to ignore the significant progress made in a number of important areas, not the least of which has been the operational definition of biological integrity (Karr and Dudley 1981). Add to this the development of robust and information-rich multimetric indices, regionalization, and tiered aquatic life uses that are embodied in the peer-review literature (Karr et al. 1986; Hughes et al. 1986), USEPA guidance documents (e.g., Plafkin et al. 1989; Gibson 1994), and state frameworks (Davies et al. 1993; Ohio EPA 1987, 1989a, 1989b; Bode and Novak, Chapter 8; Courtemanch, Chapter 20).

Specified levels of protection defined by Yoder and Rankin (Chapter 9) employ threshold criteria that are not unlike those that have routinely been used in chemical-specific and whole effluent toxicity applications. In addition, biocriteria are not subject to the assumptions necessary in the chemical-specific criteria derivation and application process thus the need for artificial safety factors is greatly diminished. The fundamental nature of the calibration of multimetric indices and the biocriteria setting process each contain sufficient analogs to the safety factors specified as being necessary for water quality criteria by the Section 303 of the CWA.

With regard to the untested scientific basis of biocriteria we point first to the long history of water pollution biology. While the early studies may not "look" like what is being proposed today, the fundamental tenets of using the indigenous biota as an indicator of aquatic ecosystem quality is well developed. The usefulness of biocriteria as an ambient assessment tool to show impairment (Ohio EPA 1990a; Yoder 1991b) and portray changes over time (Ohio EPA 1992b) demonstrates the ability to fulfill basic and practical needs. In addition, we have shown extensively throughout Ohio the different and consistent patterns of response to different types of impacts (Yoder and Rankin, Chapter 17). We firmly believe that many of these patterns, and more importantly the basic concepts, are transferable to other states in the Midwest and possibly elsewhere. This is also something that can be further refined as the process evolves much like it has with the chemical-specific and whole effluent toxicity tools.

2.2.3 Applicability of Biocriteria

We agree that bioassessments are not directly applicable to other uses including human health, wildlife health, and other nonaquatic life uses (criticism 4). However, biocriteria should not be held accountable for failing to perform tasks other than that for which they were originally designed. We would be remiss in failing to point out that healthy aquatic communities, though, occur in waters where other uses such as public water supply, recreation, and wildlife flourish. With respect to the latter, the health of the ecosystem as portrayed by the instream biota is more often than not relevant to the habitat requirements of most aquatic and semi-aquatic avian, reptilian, amphibian, and mammalian species. For example, the riparian zones associated with our high-quality streams and rivers correlates well with critical habitat for a number of Ohio breeding birds (Ohio EPA 1992b).

2.2.4 Measurement End Points and Inferences About WQS Goal Attainment

In the strictest technical sense we agree with criticism 5 that each of the three tools measure different end points. However, this is not entirely germane to the debate over biocriteria applicability. With regard to the very practical and widespread purpose of each tool to indicate aquatic life use impairment, each is attempting to measure the *same* end point. Is each tool equally capable of determining whether a particular waterbody is attaining a designated aquatic life use? It is unrealistic to expect three different approaches to each have the same power of assessment, simply because they measure different end points. Our comparisons of ambient chemical assessment, acute bioassays, and instream bioassessment (Ohio EPA 1990a; Yoder 1991a) showed extensive disagreement between the three tools with regard to reflecting impairment. Furthermore, the inherent error tendencies of the ambient chemical and bioassay methods are to miss or underrate impairments. While none of the three tools are perfect measures of actual conditions, ambient bioassessment and biocriteria are the nearest to being a direct measurement. Instead of pitting the strengths and exploiting the weaknesses of each tool we would be better served in the long run by defining under what conditions each of the three tools are the most powerful. The current debate seems to result more in diminishing the strengths of each tool by simply equating their status in policy matters.

A frequent criticism of bioassessments and biocriteria has been the relevance of results obtained at stream and river flows other than the critical design flows (criticism 6) used in the TMDL process, i.e., the $Q_{7,10}$ flow. One problem with the logic behind this assertion is that biological communities do not respond along a linear continuum with flow and chemical concentrations, but rather are more threshold oriented. This assertion does not make ecological sense in that communities that are functionally intact can withstand extreme conditions; thus, the history of coping with such rare events should be reflected in the community performance measured at other flows. We would also point to the characteristics of intact communities outlined by Karr et al. (1986), particularly the capacity for self-repair. This is not to say that certain elements of an aquatic ecosystem cannot become unacceptably stressed over the short term during critical conditions. We frequently observed such stress in our results, particularly in the functional metrics of the biocriteria indices. However, the observation of an intact community following such events indicates that environmental health has been maintained to the point that no discernable effects took place or that the capacity for self-repair was not exceeded. We agree with Stephan et al. (1985) that a community should not be kept in a perpetual state of recovery. However, a community undergoing recovery would yield indications of this state at times other than critical low flows. It would be irresponsible to categorically deny the importance of employing critical flows as NPDES permit design criteria, but this is much different than maintaining that bioassessments are valid only under these critical low flows.

2.2.5 Cause-and-Effect Relationships

Another criticism (number 7) is with the perceived inability of the biota to determine cause-and-effect relationships, i.e., bioassessments can detect impairment, but provide no insights as to the sources or causes (Suter 1993). While this hypothesis seems plausible there are consistent patterns that emerge from the rich information contained in biological data, as we have demonstrated with the biological response signatures (Yoder and Rankin, Chapter 17). Others (e.g., Eagleson et al. 1990) have also observed discernable patterns in biological community data. Indeed, bioassessment has some very distinct limitations in pinpointing specific causative chemicals and, in complex areas, specific sources. However, bioassessments, when properly planned, designed, and implemented are not performed in a vacuum. Proximity to sources, source loadings, chemical results, toxicity tests, sediment chemistry, and habitat quality are all important factors and we routinely collect this type of information during biosurveys. Hence the integrated use of chemical, physical, and biological data. Even with this information at hand, the resolution of all cause-and-effect relationships is not completely accomplished in one or even multiple years of monitoring. However, this should not be construed as a failure of the process, but rather a statement of the complexity of some situations and the inherent limitations of all tools. The strength of bioassessment in these situations is the feedback provided in terms of an ambient, instream reality check on the application of the chemical-specific and whole effluent tools. For many situations this will only become evident through an iterative process.

2.2.6 Bioassessment Levels

The question about what constitutes a sufficiently comprehensive bioassessment (criticism 8) is another key contemporary issue facing the implementation of bioassessments and biocriteria. Yoder (1994) discussed the relative abilities and biases of different levels of bioassessment to discriminate varying levels of aquatic community performance. A hierarchy of bioassessment types was identified and is related to the number and complexity of data dimensions generated by each (Table 1). The need to recognize the existence of different levels of bioassessment is illustrated by the experience of Ohio EPA in their reporting of the status of stream and river miles attaining and not attaining designated uses (i.e., Clean Water Act goals) between the 1986 and 1988 Ohio Water Resource Inventories (305[b] report). Because of a change in the type of bioassessment used between the two 305(b) report years from level 5 to level 8 (Table 1), the miles of streams and rivers *failing* to attain their designated uses changed from 9% in 1986 to 44% in 1988, an *increase of nearly five times*. This remarkable change illustrates the important influence that the differing capabilities of the various bioassessment types listed in Table 1 can exert on the relative level of accuracy of an assessment and the need to categorize and classify each according to their respective abilities to detect and discriminate impairments both spatially and temporally.

2.2.7 Basic Hypothesis Testing and Bioassessments

USEPA (1990a, 1991f) employs the null hypothesis that there is an effect on water resource integrity, and the alternative hypothesis that there is no effect. Rejection of the null hypothesis is not interpreted to mean that the alternative is accepted since the initial rejection is not entirely decisive, but leaves the possibility that there is an effect too small to detect. In fact, others (Schmidt 1992) maintain that there is always an effect no matter how difficult it is to measure, with the implication that it is always significant. This approach can lead to the presumption that only negative findings, i.e., those that indicate an adverse effect, are the only valid findings. Thus, any findings of no effect are potentially invalid and, if taken to extremes, could mean that showing *attainment* of a goal or standard is a statistical impossibility.

There are some practical problems with this position: (1) it is impractical in a regulatory and administrative framework to become incapable of affirming that a goal or standard has been attained; (2) this approach assumes that each assessment tool is a unique and perfect indicator of adverse effects in the sense of CWA goal attainment (Ruffier 1992); and (3) this approach tends to dismiss any positive findings from one tool when it is contradicted by the negative findings of another. Miner and Borton (1991) argued that some effects are indeed insignificant in practical terms. There is precedence for this type of reasoning, e.g., in the declaration by USEPA under the general permitting regulations that certain discharges are classified as *de minimis*, hence they have little risk of a *significant* impact and are subjected to comparatively less scrutiny in the regulatory process.

In the application of the three tools the relative strengths (not weaknesses) of each need to be emphasized in determining the extent to which we should trust a finding of no adverse effect. Based on comparisons of the performance of each tool (Ohio EPA 1990a,b; Yoder 1991a) this should be a very different approach for the chemical-specific and whole effluent results than for bioassessment results. Because of the inherent error biases of each tool, a finding of no adverse effect based on the chemical-specific and whole effluent tools should be regarded as inconclusive without a concurrent finding of full attainment from the bioassessment. It is difficult for a bioassessment, assuming an adequate bioassessment level and biocriteria framework is employed, to give a false indication of attainment since the basic method is contingent upon finding a sufficient number and the right types of organisms. The error propensity inherent to bioassessment is to fail to find enough of the right organisms because of poor sampling and ignorance of the admonishments concerning certain prohibitive field conditions, and, hence, the false determination of an adverse effect. This error propensity is the *opposite* of the chemical-specific and whole effluent tools, which are prone to missing effects, particularly if they are episodic and/or sampling is conducted at an insufficient frequency. Thus, the use of bioassessments and biocriteria is viewed as a safeguard against being "fooled" by a showing of no effect by chemical and physical measures. Simply stated, the three tools are not equal in their respective abilities to detect adverse effects

Table 1. Hierarchy of Ambient Bioassessment Approaches That Use Information about Indigenous Aquatic Biological Communities

Bioassessment Type	Skill Required[1]	Organism Groups[2]	Technical Components[3]	Ecological Complexity[4]	Environmental Accuracy[5]	Discriminatory Power[6]	Policy Restrictions[7]
1. Stream walk (visual observations)	Nonbiologist	None	Handbook[8]	Simple	Low	Low	Many
2. Volunteer monitoring	Nonbiologist to technician	Invertebrates	Handbook[9], simple equipment	Low	Low to moderate	Low	Many
3. Professional opinion (e.g., RBP Protocol V)	Biologist w/experience	None or fish/inverts.	Historical records	Low to moderate	Low to moderate	Low	Many
4. RBP Protocols I and II	Biologist w/training	Invertebrates	Tech. manual,[10] simple equip.	Low to moderate	Low to moderate	Low to moderate	Many
5. Narrative evaluations	Aquatic biologist w/training and experience	Fish and/or inverts.	Std. methods, detailed taxonomy specialized equip.	Moderate	Moderate	Moderate	Moderate
6. Single dimension indices	(Same)	(Same)	(Same)	Moderate	Moderate	Moderate	Moderate
7. Biotic indices (HBI, BCI, etc.)	(Same)	Invertebrates	(Same)	Moderate to high	Moderate to high	Moderate	Moderate to few
8. RBP Protocols III and V	(Same)	Fish and inverts.	Tech. manual,[10] detailed taxonomy, specialized equip., dual organism groups	High	Moderate to high	Moderate to high	Few
9. Regional reference site approach	(Same)	Fish and inverts.	Same plus baseline calibration of multimetric indices and dual organism groups	High	High	High	Few
10. Comprehensive Bioassessment	(Same)	All organism groups	Same except all organism groups are sampled	Highest	High	High	Few

Note: This applies to aquatic life use attainment only — it does not apply to bioaccumulation concerns, wildlife uses, human health, or recreation uses.

[1] Level of training and experience needed to accurately implement and use the bioassessment type.
[2] Organism groups that are directly used and/or sampled; fish and macroinvertebrates are most commonly employed in the midwestern states.
[3] Handbooks, technical manuals, taxonomic keys, and data requirements for each bioassessment type.
[4] Refers to ecological dimensions inherent in the basic data that is routinely generated by the bioassessment type.
[5] Refers to the ability of the ecological endpoints or indicators to differentiate conditions along a gradient of environmental conditions.
[6] The relative power of the data and information derived to discriminate between different and increasingly subtle impacts.
[7] Refers to the relationship of biosurveys to chemical-specific, toxicological (i.e., bioassays), physical, and other assessments and criteria that serve as surrogate indicators of aquatic life use attainment/non-attainment.
[8] Water Quality Indicators Guide: Surface Waters (Terrell and Perfetti 1989)
[9] Ohio Scenic River Stream Quality Monitoring (Kopec and Lewis 1983).
[10] U.S. EPA Rapid Bioassessment Protocol (Plafkin et al. 1989).

and each has varying degrees of uncertainty; thus, the foundation for relying too much on the null hypothesis concept can be flawed. If nothing else, the position that a significant effect is always present despite the showing of no impairment by a bioassessment or other test, will become increasingly difficult to defend.

2.2.8 Predictive Abilities of Bioassessment

Bioassessments and biocriteria are widely perceived as not being predictive in the sense that a specified level of protection can be anticipated in the same manner as using chemical-specific criteria via the TMDL process or whole effluent toxicity test results through the use of toxic unit limitations (Suter 1993). The concerns expressed by some is that while biocriteria are a valid and useful measure of instream impairment, by the time it is detected the environmental damage is already done. While this may be true, we argue here that a significantly larger number of environmental problems will go undetected if bioassessments and biocriteria are not widely used by states.

There are some problems with the perception of a lack of predictive ability. Rankin (1989; Chapter 13) developed a relationship between the IBI and the Qualitative Habitat Evaluation Index (QHEI) sufficient to forecast nonattainment due to habitat degradation with a reasonable degree of certainty. This is most applicable to the review of proposed activities that would alter riparian and instream habitat. By knowing the present habitat condition using the QHEI, changes can be projected to the various QHEI attributes based on the details of the proposed activity. It can then be determined if there is a reasonable potential to degrade the habitat so that the applicable biocriteria will be violated.

Another example of a predictive capability is with the development of a model relationship between riparian corridor condition and percentage of urban land use (Steedman 1988). The model was "calibrated" using data from a cross section of watersheds and developing a linear relationship with the two landscape attributes as covariates. The usefulness of this to watershed planning is obvious since quantifiable estimates of land use and riparian corridor compatible with IBI values that attain a particular use or goal can be made with a known degree of certainty. The predictive abilities of biocriteria lie primarily in establishing precedents in the ambient environment. Thus, if we are to use and improve biocriteria as a predictive tool, robust databases will need to be amassed in order to develop model relationships.

2.3 Policy Applications of Biological Criteria

This is perhaps the most controversial and certainly the least understood aspect of biological criteria, at least at the national level. When addressing the policy implications of biological criteria it is important to understand the applications of biocriteria and how this overlaps with the uses of the more traditional chemical/physical and toxicological tools and criteria. Biological criteria are largely limited to ambient assessment applications whereas chemical and toxicological criteria can be used in ambient assessments and as design criteria. Understanding the basis behind these differences is important. For example, biocriteria are not intended to function the same as a chemical criterion from which effluent limitations for specific chemical substances are derived, even though both employ the common term "criterion." Also, biocriteria are limited to aquatic life use issues; thus, they play no more than an ancillary role in human health risk assessment. Despite these intuitively obvious limitations, biocriteria are frequently criticized for not being able to function for purposes that they were not originally designed to address. Hopefully the following discussions will provide a firmer definition of the appropriate role of biocriteria in a state water quality management program.

There is a consensus that application of bioassessments and biocriteria is one of the best ways to determine and characterize aquatic life use impairment. Beyond that, however, there are varied opinions about the policy and regulatory role of biocriteria. USEPA and most environmental groups favor the policy termed "independent application" when considering how to apply the results of bioassessments, ambient chemical data, and whole effluent toxicity. Others, principally states and the regulated community, favor a "weight of evidence" approach. We attempt here to examine the technical issues underlying this debate. Also, we emphasize that the following deals entirely with aquatic life use issues and does not transcend the importance of criteria for persistent toxicants as they pertain to human health, wildlife, and other nonaquatic life uses. These uses are truly independent of the aquatic life concerns in terms of how the criteria are applied.

2.4 Policy of Independent Application

The USEPA policy of independent application was first outlined in the national biocriteria program guidance (USEPA 1990a) and later reaffirmed in the "eco policy" statement (USEPA 1991c) and the revised *Technical Support Document for Water Quality-Based Toxics Control* (USEPA 1991f). A panel discussion at the Third National Water Quality Standards for the 21st Century conference dealt with this issue and included perspectives from USEPA (Jackson 1992), the regulated community (Ruffier 1992), a state (Schregardus 1992), and an environmental group (Schmidt 1992).

Since biological criteria are applied as a direct measure of aquatic life use attainment/nonattainment, an obvious overlap with chemical/physical and toxicological surrogate criteria occurs. This can happen in at least two different ways: (1) where concurrent biological, chemical and/or toxicological ambient data are being used to assess aquatic life use attainment/nonattainment, and (2) in determining appropriate wasteload allocations for point sources or load allocations for nonpoint sources based on the reasonable expectation that one or more criteria (including whole effluent toxicity) might be exceeded based on worst-case assumptions about receiving water and watershed characteristics. In both cases conflicts may arise between the three major assessment tools (chemical-specific, whole effluent, and biocriteria). USEPA's definition of independent applicability means that the validity of the results of any one of the three approaches is independent of any confirmation by the others. Each assessment operates independently with none being viewed as superior or more powerful than another regardless of the situation. USEPA bases this policy on the "unique attributes, limitations, and program applications of each of the three tools." Jackson (1992) asserts that each method independently provides sufficient evidence of aquatic life use impairment irrespective of what the other tools show or fail to show. Thus, appropriate regulatory action should be taken when any one of the assessments determines that a standard is not attained.

USEPA (1991a) bases some of their equivalency of the biological, physical, and chemical components on a conceptual model of ecological integrity (Figure 3). However, we agree with Karr (1991) that this model is inadequate and not representative of the variable interaction of the three key components. We offer an alternative model, which shows that the overlapping influence of the three major components is not only disproportionate, but is dynamic (Figure 3). This depends on a given situation relative to the five major factors that influence water resource integrity (see Figure 1); thus, the influence of any one component will likely be disproportionate to the others in most situations. We also believe that for the purpose of aquatic life protection the biological component will dominate the other two because it is the product of the integrated interaction of the chemical and physical components.

2.5 The Case for Weight-of-Evidence

The alternative to independent application is to employ what has been termed a weight-of-evidence approach in which no one tool is assumed to be either equal or superior *a priori*, but an informed examination of the results may lead to giving one of the tools more "weight" in the decision-making process. In this process the respective power and site-specific applicability of each tool is considered and no prior decision about the independence of one tool from the others is made.

Based on the evidence we have examined and our own experiences in conducting bioassessments within a regulatory framework for more than 15 years, we believe there is a case to be made for employing a weight-of-evidence approach *with some important restrictions and limitations*. Unlike the policy of independent application, where all of the tools are considered to have an equal weight and ability, this approach acknowledges and accounts for the attributes of each tool. Weight-of-evidence takes advantage of the strengths of each, emphasizes the role of site-specific data, and promotes controlled flexibility in the process, an attribute that has the advantage of allowing new advances to be incorporated into the process (Ruffier 1992). This approach also seems to be consistent with the USEPA emphasis on ecological risk-based management and the well-known call for ensuring good science in the process. The policy of independent application is admittedly a regulatory approach (Jackson 1992) and has the appearance of being administered for the sake of regulatory expediency, allowing water pollution measures to keep pace with science without having to modify the permit program, and without the need to reconcile or justify discrepancies (Ruffier 1992). Independent application certainly simplifies the process and minimizes best professional judgement and site specificity, but at the potential loss of accuracy in the process. Weight-of-evidence does not make assumptions ahead of time, but, rather, relies

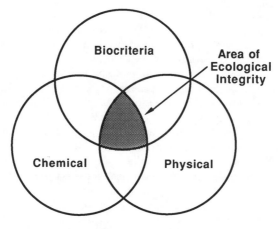

**U.S. EPA (1990) Model
of Ecological Integrity (Static)**

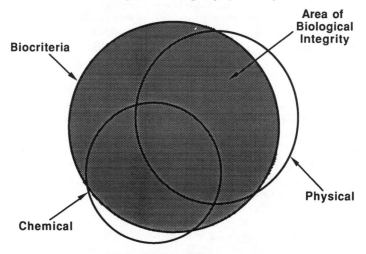

**Alternative Model of
Biological Integrity (Dynamic)**
For Aquatic Life Uses the Area of Biological
Integrity is Best Expressed by Biocriteria

Figure 3. The elements of ecological integrity as envisioned by USEPA (1990a, 1991f; upper) and our view of
the dynamic relationship between the biological, chemical, and physical elements of overall biological
integrity as it pertains to aquatic life uses in Ohio (lower).

on the strengths of each assessment and application tool in a complimentary manner. One of the most
serious concerns of independent application is that it limits the abilities of any one tool and reduces the
process to possibly relying on the least powerful. Even though it may appear that the most stringent result
is being applied, stringency does not necessarily imply accuracy or legality. Ruffier (1992) ranked each
tool in a hierarchical manner as chemical-specific < whole effluent < bioassessment. Although we
generally agree with this hierarchy, especially based on ours and others comparison tests, this is not
invariable. To make it so in all cases would violate an important precept of weight-of-evidence.

Another concern with independent application is that it does not promote incentives for addressing
sources or causes of water resource impairments other than those that are identifiable and controllable by
the chemical-specific or whole effluent approaches. Ruffier (1992) presented the situation in which a
relatively minor exceedence of a chemical-specific criterion (copper) was *predicted* alongside an *absence*

of whole effluent toxicity and instream biological impairment. Independent application would require that the copper exceedence be the driving force behind any regulatory solution, thus essentially omitting the results of the other two tools. Any contribution of information from these tools is excluded from the development of permit limits. Furthermore, the uncertainties (i.e., weaknesses) of the chemical-specific criteria application are amplified because of the inherent liabilities imposed by the *predicted* chemical-specific violation. No investigation into why the discrepancy occurred would be pursued, which could effectively rule out other important information. Under the weight-of-evidence approach the reasons why the discrepancy existed would be investigated prior to initiating regulatory action. It is the *predicted* (as opposed to actual) exceedences of chemical-specific criteria, and subsequent indications of the *potential* for non-attainment, that pose the most frequent conflict. A predicted exceedence would certainly carry more weight if it was accompanied by a measured instream exceedence and more so by an instream biological impairment and/or effluent toxicity. Oftentimes it is the wasteload allocation assumptions, which are inherently conservative, that result in the prediction of an exceedence under specific conditions. If the other tools fail to show any risk of impairment, the reasonable potential to violate may not be so reasonable after all. USEPA (1991f) acknowledges this scenario and recommends that a more detailed and sophisticated monitoring and modeling approach be undertaken in an attempt to bring closer agreement. We certainly endorse this approach as being consistent with weight-of-evidence.

The concern is not solely with the potential for overregulation, but the far greater risk of "hidden" degradation. As was previously pointed out, relegating ambient bioassessment information to an equal place with the chemical-specific and whole effluent tools will lead state program managers to the conclusion that biocriteria are a nonproductive addition and are hence unnecessary. Thus, an opportunity to expose permit writers, plant operators, administrators, and others involved in decision making to more direct information about the aquatic resource will be lost. The NPDES process will essentially remain detached from the ambient environment. Not only will this result in a loss of valuable information, it will likely result in the continued degradation of the aquatic resource because the superior ability of bioassessment and biocriteria to detect and characterize problems will be missing.

Much of the concern about weight-of-evidence is with the potential for misuse and abuse by attempts to justify increasing loadings of pollutants. However, given the ability of the higher level of bioassessment types (Table 1) to detect degradation we doubt that significant increases in pollutant loadings will be justified with biocriteria on a widespread basis. What is more likely is that the magnitude of required pollutant load *reduction* may in a few cases be modified by the application of weight-of-evidence. In our experience the following are the situations where conflicts have arisen in Ohio:

1. Oxygen-demanding substances, ammonia-N, and copper are the most frequently involved parameters. Although the latter is a priority pollutant the chemistry is complex and is likely the root cause of many of the conflicts. We have yet to encounter a situation where a truly bioaccumulative toxicant has been involved; where such toxicants are elevated the bioassessments usually show severe degradation.
2. Conflicts can arise with the non-WWH uses, particularly for the MWH use. Presently the Ohio WQS have chemical-specific criteria for the MWH use different from WWH for dissolved oxygen (D.O.) and ammonia-N only — all other chemical parameters are equivalent to the WWH use designation.
3. The strict adherence to the results of chemical-specific criteria applications without the site and regional specific information fostered by a biocriteria approach can lead to abatement measures which actually result in increased environmental degradation. We observed this in a major suburban area where there was a "need" to eliminate small wastewater treatment plants due to noncompliance with their NPDES permits. This resulted in the destruction of stream habitat by the instream construction of interceptor sewers intended to deliver the sewage flow to newly constructed regional facilities (Ohio EPA 1992a). While this eliminated the "paper compliance" problem of the small plants, biosurveys showed that the zones of degradation below each were small in comparison to the miles of habitat destruction wrought by the sewer line construction and subsequent maintenance activities.
4. A "reverse conflict" can also occur with the application of the EWH use designation. Here the chemical-specific criteria for most substances are the same as the WWH use (D.O. is an exception), but the communities are more sensitive. This has led to efforts to oppose redesignation of stream segments to EWH from WWH based on the argument that the chemical-specific criteria cannot be attained, even though the more difficult-to-meet EWH biocriteria are attained, this being a prerequisite to assigning the EWH use.

With regard to the latter example, the projected inability to meet the EWH D.O. criterion has been presented as an argument for not adopting the EWH use. In order to redesignate a segment as EWH there must be a demonstration of the ability of the segment to attain the EWH biocriteria. Thus, we are faced here with the classic independent application scenario — the biocriteria are attaining, but the chemical-specific criteria are predicted to be exceeded under the critical, worst-case design conditions. However, the stakes in this situation are different than with the other use designations. Do we want to leave a stream segment that is performing at the EWH biocriteria levels vulnerable to future degradation by nonchemical impacts or impacts that are not regulated via the NPDES process? This is the risk if the redesignation to EWH cannot take place because of the *predicted* inability to meet a chemical-specific criterion. Thus, a strict adherence to independent application in this case can leave stream and river segments underprotected and vulnerable to future degradation.

The flaws inherent to the policy of independent application lie primarily in the equating of the three major tools, chemical-specific, whole effluent toxicity, and bioassessment/biocriteria. In many respects this policy is only a more complex facsimile of the long-abandoned USEPA policy of presumptive applicability of the 1976 water quality criteria. The site-specific circumstances relevant to the five factors that determine water resource integrity (see Figure 1) should determine how much weight should be given to each tool. This type of decision can be made only when comprehensive and adequate data from all three tools are available. Under independent application a decision may only be as good as the least powerful assessment tool, whereas under weight-of-evidence the strongest data play a more appropriate role. In many situations the bioassessment and biocriteria will provide the most powerful information, but may not be entirely conclusive. Thus, the integrated application of the chemical-specific, whole toxicity, and possibly other tools (e.g., habitat, biomarkers) will be required to successfully employ this policy.

2.6 Recommendations for Resolving the Independent Application/Weight-of-Evidence Conflict

Based on a review of the existing policy debate we believe the conflict can be resolved by establishing program criteria that states must follow in order to gain the desired policy flexibility. There are several advantages to codifying biocriteria in the WQS, not the least of which is the legal standing relative to other criteria. For aquatic life uses the basic narrative descriptions are generally written in biological terms and in the more sophisticated frameworks these qualify as narrative biocriteria. The addition of numeric biocriteria in this scheme then provides the quantitative benchmarks of aquatic life use attainment and nonattainment. For applications to habitat-modifying activities the biocriteria provide a powerful tool for minimizing degradation or in some cases preventing it altogether. This is of critical importance in states like Ohio where habitat and related sedimentation impacts are among the leading causes of impairment (Ohio EPA 1992b). Thus, one of our strongest recommendations is for numerical biocriteria to be part of the states' WQS.

Policy restrictions on the use of biocriteria in the overall water quality management program should be based on the level of bioassessment employed (Yoder 1994; Table 1). This is dependent on the relative sophistication of the bioassessment framework, which consists of methods, level of taxonomy, number of organism groups, number of data dimensions involved, and the biocriteria derivation framework. In short, it is not just the use of biological information, but the level of bioassessment and the framework within which biocriteria are developed that is most important. As the complexity and sophistication of the bioassessment and biocriteria framework increase, policy restrictions should decline accordingly.

This in no way implies an unrestricted use of bioassessments over chemical-specific and whole toxicity approaches. Rather, this approach advocates an informed and integrated use of each tool with the recognition that bioassessment is more likely to detect impairment than the other tools. We also recommend that special consideration be given to reserving policy flexibility for specific classes of pollutants such as conventional parameters (e.g., D.O. and suspended solids), nutrients and ammonia-N, and other "troublesome" parameters such as copper. Nearly all CWA Section 307(a) pollutants would be excluded from such a policy since most are persistent and bioaccumulative, most result in severe impairments, and ambient information about all except the most common heavy metals is generally lacking. While we do not advocate the unrestricted use of a weight-of-evidence approach, we believe a controlled application will be necessary to better deal with some of the emerging problems, particularly with nonpoint source constituents. Another safeguard could include making the use of a weight-of-evidence approach subject to a case-by-case justification similar to a use attainability analyses (USEPA 1984b), which is required prior to the designation of segments for less than a CWA goal use. Requiring a showing of biocriteria attainment for three consecutive years might be another safeguard to ensure that any showing of attainment was not spurious.

The allowance of policy flexibility, which is contingent on using a higher level of bioassessment (Table 1), might serve as an incentive for states to invest the resources necessary to develop a reference site database, numerical biocriteria, and sufficient case histories to fully implement bioassessments and biocriteria. Since the resultant accuracy of each bioassessment type is different, placing policy restrictions on the use of a particular level of bioassessment becomes an important issue. This is an unprecedented area of opportunity for USEPA to resolve some of the state and regulated entity objections to the policy of independent application in that policy flexibility would be influenced by the level of bioassessment. Under this framework policy restrictions decline as the level of bioassessment increases in power and accuracy. This would not only provide an incentive for states to develop adequate biocriteria frameworks, but would also benefit USEPA and the public in general in that the improved bioassessment capabilities would provide a more accurate and comprehensive assessment of the states' and nation's waters. We believe that a rigid adherence to the policy of independent application will discourage already reluctant states to develop adequate bioassessment programs since biocriteria will simply become another layer in the water quality management process (Pifher 1991). This would not only serve the needs of individual states, but would in turn provide a better statewide assessment of water resource conditions, which would lead to a much improved national assessment, something that has been sorely lacking over the past 20 years.

2.7 Other Concerns

Not all of the concerns about biocriteria and bioassessment are being expressed by USEPA and environmental groups. States and the regulated community, while generally in favor of the approach, have also expressed concerns. Cost and resource constraints are most frequently raised by states that are facing an ever increasing burden of mandates without external funding increases. The up-front investment required by biocriteria, while no more expensive than the other tools, represents an added cost. This is why it is important to provide incentives, funding, or both for states to adopt biocriteria. States should also look to capabilities outside of their immediate purview such as sister state or federal agencies that possess bioassessment capabilities. We do not minimize the difficulties of actually accomplishing this type of interjurisdictional cooperation, but examples do exist and this will be strongly encouraged in the future.

The regulated community is concerned about the potential for more stringent permit limits and other restrictions that may be leveraged by biocriteria. There is little doubt that biocriteria enhance the ability to detect degradation. However, this does not necessarily translate into significantly more stringent limitations. Pifher (1991) leaves us with the notion that as waters improve, biocriteria will become more stringent, leaving the regulated community on a "never ending merry-go-round" of increasingly stringent requirements. We disagree with this position because it presumes that the biota will continue to improve as pollutant concentrations are reduced. As we have previously pointed out the biota are more threshold-response oriented and there is a point beyond which additional pollutant removal will have little or no beneficial effect on the biota. In fact, if a weight-of-evidence approach is employed a regulated entity is more likely to know when to "get off" rather than continue to be subject to the uncertainties of independent application. Another concern is that an entity may be in full compliance with an NPDES permit, yet degradation is detected downstream from the discharge. In this case it would seem, as we have observed in Ohio, that either the permit limitations are not sufficient, the entity self-compliance monitoring is inadequate, there are undetected or unreported violations, or there are other pollutant releases not covered by the permit. While an entity may be reluctant to have these facts revealed, the lack of prior knowledge should not be a license to continue with the status quo.

The regulated community is also concerned about taking on responsibility for conducting the ambient monitoring required to implement biocriteria. We strongly advocate that states have primacy since this is necessary to develop the appropriate expertise in maintaining the biocriteria and in conducting the ambient bioassessments. States are in a much better position to implement a comprehensive program that includes issue beyond point sources such as integrated watershed management. Regulated entities that have existing bioassessment capabilities can contribute significantly to this process, but they should not be expected to shoulder the burden for the entire program. Additionally, this is not an area for volunteer programs either as these rarely, if ever, have the expertise and resources required to operate the level of bioassessment necessary for a credible biocriteria approach (see Table 1). As we have already acknowledged, this will be a difficult area for some states primarily because of the start-up costs. This is an area where USEPA must examine the trade-offs between not having an adequate bioassessment capability and having existing impairment remain undetected or underrated. Based on the overall water program costs incurred by Ohio EPA, this would constitute a shift of approximately 5 to 15% of water program resources depending on what bioassessment capabilities already exist.

3.0 AQUATIC RESOURCES AT RISK — THE CONSEQUENCES OF INACTION

There is little question that aquatic resources have been and continue to be degraded by a myriad of land use and resource use activities. Benke (1990) summarized the status of the nation's high-quality rivers and streams concluding that fewer than 2% remain in this category. Judy et al. (1984) indicate that the declining status of surface waters across the United States is largely the result of nonpoint source impacts. A continued reliance on technology-based and even water quality-based solutions to these problems will simply be insufficient. Water resources in Ohio and elsewhere have historically been and will continue to be impacted by human activities beyond those targeted by the NPDES permit process. These remaining problems are comparatively more complex and subtle, but are no less important or real. In fact, it is these more subtle and diffuse impacts that imperil aquatic resources to the point where additional species are declining in distribution and abundance; this in addition to those already declared as rare, threatened, or endangered (Ohio EPA 1992b).

A monitoring approach, integrating biosurvey data that reflects the integrity of the water resource directly, with water chemistry, physical habitat, bioassay, and other monitoring and source information, must be central to accurately defining these varied and complex problems. Such information must also be used in tracking the progress of efforts to protect and rehabilitate water resources. The arbiter of the success of water resource management programs must shift from a heavy reliance on administrative activity accounting (numbers of permits issued, dollars spent, or management practices installed) and a preoccupation with chemical water quality *alone* to more integrated and holistic measurements with overall water resource integrity as a goal. Biocriteria seems an essential component in making this shift.

Emphasizing aquatic life use attainment is important because: (1) aquatic life criteria oftentimes result in the most stringent requirements compared to those for the other use categories, (i.e., protection for the aquatic life use criteria should assure the protection of other uses); (2) aquatic life uses apply to virtually all waterbody types and the diverse criteria (i.e., includes conventionals, nutrients, toxics, habitat, physical and biological factors, etc.) apply to all water resource management issues; and (3) aquatic life uses and the accompanying chemical, physical, and biological criteria provide a comprehensive and accurate ecosystem perspective towards water resource management that promotes the protection of ecological integrity. The need for an ecological perspective in water resource management is especially evident in the following:

- The assessment and control of wet weather flows (stormwater and combined sewers)
- Nonpoint source assessment and watershed management
- Site-specific criteria modifications
- Regulation of activities that directly impact aquatic habitat

Finally, biocriteria can greatly aid the visualization of aquatic resource values and attributes. This is a critical need if we are to change the prevailing view of watersheds and streams as merely catchments and conveyances for municipal and industrial wastes, excess surface and subsurface drainage, or as obstacles to further land developments. In an effort to stem the virtually unabated loss of riparian habitat and watershed integrity Ohio EPA has proposed a stream protection policy that sets forth guidelines under which various activities will need to be conducted in order to conserve biological integrity. Without biocriteria and the case examples developed over the past 15 years this would not have been possible and any opportunity to affect these degrading influences would have been lost.

ACKNOWLEDGMENTS

This chapter would not have been possible without the many years put into fieldwork, laboratory analysis, and data assessment and interpretation by members (past and present) of the Ohio EPA, Ecological Assessment Section. This includes the following: Dave Altfater, Mike Bolton, Chuck Boucher, Bernie Counts, Jeff DeShon, Jack Freda, Marty Knapp, Chuck McKnight, Dennis Mishne, Randy Sanders, Marc Smith, and Roger Thoma. The insightful comments of Ed Rankin and Dan Dudley are appreciated.

## CHAPTER 22

Filling the Gaps in Water Quality Standards: Legal Perspectives on Biocriteria

Robert W. Adler

1.0 INTRODUCTION

Virtually no major aspect of the Clean Water Act (CWA) regulatory scheme has been immune from challenge, as witnessed by the exhaustive, multiparty challenge to EPA's overall permitting system (the National Pollutant Discharge Elimination System, NPDES) (*NRDC v. USEPA* 1988). While biocriteria remain relatively untouched by lawyers and judges thus far, this innocence likely will be lost as biocriteria move from the realm of science to the reality of regulation. Therefore, program managers would be wise to anticipate and to insulate their biocriteria programs from such legal challenges.

To assist in this effort, it is useful to include in this largely scientific compendium some perspective on legal issues relevant to the use of biocriteria. Specifically, this chapter will evaluate the legal basis for the use of biocriteria in water quality standards, as well as the application of those criteria in various CWA programs. Included as well is discussion of various legal issues and problems that may help program managers insulate biocriteria from challenge.

The chapter begins with a brief legal history of water quality standards. Next, it describes the potential legal bases for biocriteria, beginning with the statute and its legislative history, and proceeding to U.S. Environmental Protection Agency (USEPA) regulations and to potentially relevant case law. These legal principles are then used to analyze the various existing and potential applications of biocriteria in clean water programs.

2.0 A BRIEF LEGAL HISTORY OF WATER QUALITY STANDARDS

While used in various forms for decades, the first effort to integrate water quality standards into a comprehensive federal water pollution control program came in Section 5 of the Water Quality Act of 1965, which required states to develop water quality standards for interstate waters. Water quality standards were defined in section 10(c)(1) of the law to include (1) designated uses of waters; (2) numeric or narrative criteria to protect those uses; and (3) an implementation plan.

Narrative criteria are verbal descriptions of water quality, such as "no toxics in toxic amount," "no floatable wastes," or "no putrescible wastes." As such, narrative criteria could address chemical, physical, or biological conditions. Although the use of purely narrative criteria is well accepted (*EDF v. Costle* 1981), cases involving water quality standards reflect the pervasive bias that criteria define numeric limits on chemical pollutants rather than broader definitions of impairment, much less a positive statement describing biological well-being. For example, in 1990 two U.S. Courts of Appeals described water quality criteria as "the maximum concentrations of pollutants that could occur without jeopardizing the

0-87371-894-1/95/$0.00+$.50
© 1995 by CRC Press, Inc.

use" (*NRDC v. USEPA* 1990, Ninth Circuit) and as the "amount of various pollutants" that may be present (*Westvaco v. USEPA* 1990, Fourth Circuit).

In 1972, Congress shifted efforts to control water pollution in general, and point source pollution in particular, away from a focus on receiving water quality in favor of technology-based reductions in pollutant discharges. Under the 1972 CWA, all point sources were subject to "best technology" requirements irrespective of receiving water quality. In theory, these requirements were designed to become increasingly stringent until the ultimate goal of eliminating the discharge of pollutants into receiving waters is achieved (CWA Sections 101(a)(1), 301, 402). Water quality standards, however, retained important functions. States were required to develop water quality standards for intrastate as well as interstate waters; to identify all waters not meeting these standards; to calculate the additional pollutant reductions needed to achieve the standards (known as "total maximum daily loads," or TMDLs); and to impose these requirements through wasteload allocations (WLAs), NPDES permits, and best management practices for nonpoint sources (CWA Sections 208, 302, 303, 402). If a state failed to perform any of these functions, USEPA was obligated to do so instead. Specific requirements are defined further in USEPA regulations (e.g., 40 CFR Parts 130, 131) and guidance (e.g., USEPA 1991f).

States are given the first opportunity to adopt water quality standards (CWA Section 303), reflecting Congress' recognition that biological conditions vary from state to state (*State of Alabama v. USEPA* 1977). However, USEPA maintains critical guidance and oversight functions, and retains the authority to promulgate federal water quality standards for any state that fails to issue standards adequate to achieve the goals and purposes of the act. Under Section 304(a), USEPA issues water quality criteria to guide states in their adoption of enforceable water quality standards. Under Section 303(c), USEPA must review state standards to ensure that they meet the requirements of the act. States are afforded a second chance to address any deficiencies identified by USEPA, but failing such corrections, USEPA is *required* to adopt federal standards under Section 303(d) to ensure full compliance with the law (*Mississippi Commission on Natural Resources v. USEPA* 1980).

The shift to technology-based controls cannot be attributed to a single factor, as the 1972 Senate Report characterized previous water pollution control efforts as "inadequate in every vital aspect" (1972). One major problem, however, was the difficulty, given existing monitoring and modeling techniques, in working backwards from receiving water quality standards to individual discharge requirements. "Regulators had to work backwards from an over-polluted body of water and determine which entities were responsible; proving cause and effect was not always easy" (*NRDC v. USEPA* 1990). In part, the U.S. Supreme Court blamed water quality standards per se: "The problems stemmed from the character of the standards themselves, which focused on the tolerable effects rather than the preventable causes of water pollution..." (*USEPA v. California* 1976).

From the perspective of biocriteria proponents, this critique was insightful and prophetic. Chemical-specific water quality criteria that define "tolerable" amounts of chemical pollution were viewed more as "negative" statements of permissible pollution rather than positive statements of water quality or aquatic ecosystem goals: How much pollution can be allowed without interfering with designated uses? The same is true of whole effluent toxicity (WET) criteria: How much toxicity in effluent and in receiving waters is permissible without impairing uses? Even traditional narrative standards (free from floatables, odors, or other undesirable characteristics) typically are stated as negatives rather than positives. These general rules, of course, are subject to exceptions. For example, does a water quality criterion for dissolved oxygen (DO) establish a positive view of a healthy waterbody or a permissible level of DO reduction?

Biocriteria *could* be viewed in the same way: How much deviation from reference conditions is acceptable without concluding that uses are impaired? More properly, however, biocriteria establish *positive* statements of the desired biological conditions of waterbodies. Rather than defining the "tolerable effects...of pollution," such criteria establish an affirmative statement of desirable ecological attributes. They are aims to achieve, not ills to avoid. It is interesting then, that the Supreme Court in 1992 stated that the purpose of water quality standards is to "establish the desired condition of a waterway" (*Arkansas v. Oklahoma* 1992), and not, as it and other courts had previously, to define tolerable levels of pollution. Given the facts of the *Arkansas* case, it is doubtful that this shift in legal description was intentional; but the new language mirrors the gradual addition of positive goals to a system of water quality standards traditionally stated in negative terms.

Now, USEPA is beginning to view water quality standards as a holistic, affirmative statement of the overall integrity of a waterbody (USEPA 1990a, 1991f). A combination of impermissible negatives (chemical-specific, condition-specific, and WET) and desired positives (biocriteria) can be used to determine whether uses are attained or impaired, and to guide future assessment and regulatory efforts. While USEPA and state scientists and other water quality officials have been making this shift for about a decade, few hard regulatory requirements have been imposed as a result. Thus, little legal attention has been paid to this trend. As biocriteria evolve from assessment to regulatory tools, this will likely change.

3.0 THE LEGAL BASIS FOR BIOCRITERIA

The objective of the CWA, stated in Section 101(a), is "to restore and maintain the chemical, physical, and biological integrity of the Nation's waters." Relatedly, "pollution" in Section 502(19) (as distinct from the narrower definition of "pollutant" in Section 502(6)) "means the man-made or man-induced alteration of the chemical, physical, biological, and radiological integrity of water." Logically, then, water quality standards must address biological as well as chemical and physical characteristics of waterbodies (Karr 1990).

During the first two decades of the CWA, however, regulatory agencies paid far more attention to chemical than to biological integrity of the nation's waters (Adler et al. 1993). These omissions were not due to lack of understanding. Over a decade ago, for example, USEPA and the U.S. Fish and Wildlife Service (FWS) commented that "attempts to monitor the condition of the nation's waters have focused only on the physical and chemical characteristics of the water, while the components of the biological communities were largely ignored" (Judy et al. 1984). Similar conclusions had been reached by state and independent researchers even earlier (Karr 1981; Karr and Dudley 1981). In a seminal article published in 1981, Karr and Dudley wrote:

> Water quality is traditionally interpreted as the physical/chemical properties of water, a fact that greatly limits the scope of the goal [of the Act].... A comprehensive evaluation of both physical/chemical and biological data is a better determination of whether or not fishable/swimmable conditions are being achieved. [The authors defined ecosystem integrity as] the capability of supporting and maintaining a balanced, integrated, adaptive community of organisms having a species composition, diversity, and functional organization comparable to that of natural habitat of the region. (Karr and Dudley 1981)

As a basic principle of statutory construction, general statements of statutory goals and objectives have no legal force and effect, absent specific operative provisions in the law. However laudable the goals, implementation of specific programs requires adequate statutory authority, especially when the legal obligations of third parties are affected. Support for the issuance and implementation of biocriteria, therefore, should rest on a more rigorous legal analysis. This evaluation will proceed from the operative language of the CWA and relevant legislative history, to USEPA and state regulatory interpretations, to relevant case law (of which there is little at this stage). Two distinct questions will be addressed: first, in Section 3, whether biocriteria issued and applied by the states are legally defensible; and second, in Section 4, whether USEPA has the authority to require states to adopt biocriteria in some form.

3.1 The Clean Water Act and its Legislative History

As noted above, the driving force behind biocriteria is the basic objective of the law to restore and maintain the biological as well as the chemical and physical integrity of the nation's waters. But precise application of this principle requires operative language. This authority stems primarily from Sections 303 and 304, which dictate the manner in which water quality standards are adopted, and Sections 301, 303, and 402, which require that those standards be implemented, and describe how.

Section 303 of the 1972 act recodified and expanded the water quality standards provisions of the 1965 law, but there were no fundamental changes — water quality standards consist of designated uses of waters and water quality criteria based on those uses. "Such standards shall be such as to protect the public health or welfare, enhance the quality of water *and serve the purposes of this Act*" (CWA Section 303(c)(2)(A); emphasis added). The italicized language, while seemingly innocuous, in fact makes fully operative the objectives stated in Section 101. To "serve the purposes of th[e] Act," water quality standards must address chemical, physical, and biological integrity of the nation's waters.

Additional statutory authority to support the use of biocriteria is included in Section 304(a), under which USEPA develops and publishes water quality criteria used by states in developing their water quality standards. Section 304(a) criteria do not have independent legal effect; if a state fails to adopt adequate water quality standards, USEPA must do so through separate rule making under Section 303(c). However, since Congress intended that state water quality standards would be based on the USEPA criteria, logically state water quality standards can be at least as broad as suggested by the wording of Section 304(a). The basic requirements for USEPA criteria are set forth in Subsections 304(a)(1) and (2). In relevant part, Subsection (a)(1) reads:

> The Administrator...shall develop and publish...criteria for water quality accurately reflecting the latest scientific knowledge (A) of the kind and extent of all identifiable effects on health and welfare including, but not limited to, plankton, fish, shellfish, wildlife, plant life, shorelines, beaches, esthetics, and recreation which may be expected from the presence of pollutants in any body of water, including ground water; (B) on the concentration and dispersal of pollutants, or their byproducts, through biological, physical, and chemical processes; and (C) on the effects of pollutants on biological community diversity, productivity, and stability.

The breadth of this language, encompassing "all identifiable effects" on health and welfare, supports USEPA and state authority to issue water quality criteria that address biological as well as chemical impairment. Particularly notable is the authority to address "biological community diversity, productivity, and stability," the very types of metrics used to develop biocriteria. However, Subsection (a)(1) is also notable in its focus on the effects of *pollutants* on the various factors to be addressed, as opposed to broader sources of impairment. In this respect, the distinction between the definitions of "pollutant" and "pollution" in Section 502, discussed above, becomes a two-edged sword, used in this case to slice away at the breadth of Section 304(a)(1). Undoubtedly, challenges to biocriteria that address impairment caused by factors other than the release of chemical pollutants will cite this distinction.

Any doubts raised by this limitation, however, appear to be resolved by the broader context of Subsection (a)(2):

> The Administrator...shall develop and publish...information (A) on the factors necessary to restore and maintain the chemical, physical, and biological integrity of all navigable waters, ground waters, waters of the contiguous zone, and the oceans; (B) on the factors necessary for the protection and propagation of shellfish, fish, and wildlife for classes and categories of receiving waters...(C) on the measurement and classification of water quality; and (D) for the purpose of section 303, on the identification of pollutants suitable for maximum daily load measurement correlated with the achievement of water quality objectives.

[For historical reasons not important to this analysis, part of this language is repeated in Subsection 304(a)(5).]

With the exception of Subsection (D) regarding TMDLs (discussed further in the following section on implementation of biocriteria), this subsection is not limited to effects caused by the release of chemical pollutants, but rather may address effects from any cause of impairment. Stated differently, criteria developed under Section 304(a)(1) could be viewed as criteria to address the effects of *pollutants*, while criteria under Section 304(a)(2) address the broader effects of *pollution*. While in practice USEPA has never made this distinction, as a matter of statutory interpretation the difference is important, and supports USEPA's authority to issue broader biological water quality criteria not necessarily related to releases of specific pollutants. Subsection 304(a)(2) also supports development of biocriteria through its direct incorporation of the requirement to protect chemical, physical, *and* biological integrity.

Indeed, the 1972 Senate Committee Report describes water quality criteria under Section 304(a) in language that is remarkably prescient of the types of biocriteria that have been adopted widely by state agencies only in the past several years:

> Criteria establish the effects of pollutants on health and welfare, including the effects of pollutants on receiving water ecosystems and man, and identify the natural chemical, physical and biological integrity of the Nation's waters. The concentration and dispersal of pollutants and their by-products

through biological, physical and chemical processes and any related changes in the diversity, productivity, or stability of receiving waters should be part of the information provided.

The "natural integrity" of the waters may be determined partially by consultation of historical records on species composition, partially from ecological studies in the area or comparable habitats; partially from modeling studies which make estimations of the balanced natural ecosystem based on the information available. (CRS 1972)

USEPA and state authority to issue biological criteria was given an additional boost in the 1987 Water Quality Act, which added Sections 303(c)(2)(B) and 304(a)(8). Section 303(c)(2)(B) was added to force delinquent states to adopt water quality criteria to address priority toxic pollutants present in their waters at levels that might interfere with uses. Numeric criteria were required wherever available. Where such criteria were not available, however, Congress expressly directed that:

...such State shall adopt criteria based on biological monitoring or assessment methods consistent with information published pursuant to section 304(a)(8). Nothing in this section shall be construed to limit or delay the use of effluent limitations or other permit conditions based on or involving biological monitoring or assessment methods or previously adopted numerical criteria.

With this language, Congress simultaneously authorized the use of biocriteria prospectively and ratified their previous use. In Section 304(a)(8), Congress directed USEPA to "develop and publish information on methods for establishing and measuring water quality criteria for toxic pollutants on other bases than pollutant-by-pollutant criteria, including biological monitoring and assessment methods."

The breadth of this additional authority will be questioned on two grounds. First, because the provisions were added as part of Congress' program to require states to address toxic priority pollutants, it will be argued that Congress only intended to authorize biological monitoring methods in this limited context. The force of this argument is weakened substantially, however, because in Section 303(c)(2)(B) Congress expressly recognized the validity of previous biocriteria adopted under more general authority. USEPA agrees in its biocriteria guidance document, noting that "[t]hese specific directives do not serve to restrict the use of biological criteria in other settings where they may be helpful" (USEPA 1990a).

Second, substitutes for chemical-specific numeric criteria for toxicants generally used and understood in 1987 were WET testing, not biocriteria. Thus, it may be argued that Congress' use of the term "biological monitoring and assessment" was an unintentionally broad reference to WET. Neither the Conference Report nor the Senate Report (this provision derived from the Senate bill), however, shed additional light on Congress' intended use of these terms. Moreover, Section 304(a)(8) refers to bases other than pollutant-specific criteria, *including* biological monitoring and assessment methods, and grants USEPA broad discretion to explicate the otherwise undefined terms through guidance. Thus, there is no clear evidence that Congress intended to restrict this terminology to any single methodology. Given judicial deference to reasonable USEPA interpretations of ambiguous statutory provisions (*Chevron v. NRDC* 1984), however, USEPA's view that this provision authorizes biocriteria as well as WET criteria is likely to be upheld.

One additional provision of the law merits attention. The CWA establishes only *minimum* federal requirements for state water quality standards and other aspects of state programs (Section 4.0 *infra*). Under Section 510, states are free to establish requirements stricter than those required by the Act (*PCMA v. Watt* 1983; 40 CFR. Section 131.4). However, once state criteria are adopted and approved by USEPA, they become "part of the federal law of pollution control" (*Arkansas v. Oklahoma* 1992). Thus, so long as biocriteria are authorized by state law, they are not prohibited by the CWA; and once approved by USEPA, they become enforceable under federal law as well with respect to any activity that affects receiving waters in the state that issued the standards. An analysis of additional authority for biocriteria in 50 states obviously is beyond the scope of this analysis. As one example, however, in Maine, water quality standards are adopted by statute rather than regulation (Maine Water Pollution Law, Article. 4-A). Thus, Maine's narrative biocriteria are insulated from attack under the CWA itself.

Given the far-reaching implications of the adoption and application of biocriteria, more specific statutory authority is at least desirable. Indeed, entities most likely to feel the regulatory weight of biocriteria undoubtedly will argue that more direct Congressional attention to this issue is essential. But unless and until Congress addresses biocriteria in the next CWA reauthorization, as proposed, for example, in S.1114 introduced by Senators Baucus and Chafee in 1993, legal acceptance of biocriteria will turn on existing statutory language. The manner in which the existing law will be interpreted is guided by USEPA regulations and judicial precedent, discussed in the following two sections.

3.2 USEPA Regulations

USEPA's water quality standards regulations are contained in Title 40, Part 131 of the Code of Federal Regulations (40 CFR Part 131). Related rules governing water quality planning and management are included in 40 CFR Part 130.

Most aspects of the above statutory analysis are supported or paralleled by USEPA regulations. In some respects, however, because many of USEPA's rules predate the development of biocriteria, they contain language that reflects the earlier, narrower view of water quality standards as limits on concentrations of chemical pollutants in waterbodies. While none of this language is fatal to USEPA or state development of biocriteria, especially given the strong statutory support outlined above, USEPA should broaden this language when it revises the relevant regulations.

USEPA regulations confirm that, to "serve the purposes of the Act," water quality standards must address the basic goals set forth in Section 101(a), including "protection and propagation of fish, shellfish and wildlife" (40 CFR. Sections 130.3, 131.2). As shown above, Section 304(a) and its legislative history clarify that Congress intended the definition of a healthy aquatic ecosystem to reflect a balanced, indigenous population of fish, shellfish, and wildlife, based on historical data, reference streams, or modeling based on available data. This matches almost precisely the methods approved by USEPA's biocriteria guidance document (USEPA 1990a).

This interpretation is supported further by USEPA's definition of "use attainability analysis" as a "structured scientific assessment of the factors affecting the attainment of the use which may include physical, chemical, biological, and economic factors..." (40 CFR. Section 131.3(g)). A use attainability analysis is required before water uses can be downgraded below those needed for protection and propagation of fish, shellfish, and wildlife (as well as human contact recreation). Clearly, this requires far more than analysis of chemical water quality.

Moreover, USEPA's water quality standards regulations highlight the desired end goal, protection of designated uses, as the primary concern, as opposed to chemical water quality goals alone. Water quality standards must include uses consistent with Sections 101(a) and 303(c)(2) (40 C F R. Section 131.6(a)), and water quality criteria must be sufficient to protect those uses (40 CFR. Section 131.6(c), 131.11).

USEPA's recognition of use protection as distinct from pure water quality goals is perhaps clearest in the antidegradation rule, which provides that "[e]xisting instream water uses *and* the level of water quality necessary to protect the existing uses shall be maintained and protected" (40 CFR. Section 131.12(a)(1); emphasis added). At least in the context of antidegradation, USEPA recognized that human and ecological use protection involves more than maintenance of chemical water quality, or the dual focus of this rule would be redundant. Of course, application of biocriteria in the antidegradation program will raise interesting questions. For example, what are the antidegradation ramifications where water quality exceeds levels necessary to protect existing uses, but uses do not exist, or vice versa? When permission for additional degradation is requested under the "social and economic" need test for high-quality waters, what degree of biological use impairment will be allowed? One possible strategy is to allow some chemical degradation, where fully justified on social and economic grounds, but only if no biological impairment is shown through the use of biocriteria, which are more sensitive to subtle impairment than traditional indicators of use attainment. These questions are complex, although not unique to biocriteria, and will require additional policy development efforts by USEPA and states.

Finally, USEPA's regulations allow states to adopt water quality criteria for toxic pollutants based on "biomonitoring methods where numerical criteria cannot be established or to supplement numerical criteria" (40 CFR. Section 131.11(b)(2)). As with Section 303(c)(2)(B), which undoubtedly drew its reference to "biomonitoring" from this existing regulation, USEPA was referring largely or exclusively to whole effluent toxicity monitoring. Nonetheless, USEPA recognized even in 1983, when this rule was issued, that water quality standards needed to extend beyond simple numerical chemical criteria to address the full range of impacts to aquatic species and ecosystems.

In other respects, however, USEPA's regulations retain the narrow notion that it is water quality, as opposed to broader notions of ecosystem health, that are primarily at issue. For example, the rules define water quality criteria as "elements of State water quality standards, expressed as constituent concentrations, levels, or narrative statements, representing a *quality of water* that supports a particular use" (40 CFR. Section 131.3(b); emphasis added). This definition should be expanded to cover biological and physical, as well as chemical integrity of waterbodies.

3.3 Judicial Interpretations

Because biocriteria have been developed only relatively recently, and have been used *primarily* for monitoring and assessment rather than as a permitting tool, there is little judicial precedent on the subject — and to date no federal cases have addressed the validity of biocriteria directly. Other cases on water quality standards in general, however, especially cases dealing with whole effluent toxicity, suggest that the development and implementation of biocriteria will be upheld by federal courts as a legitimate exercise of USEPA's authority to achieve the overall purposes of the CWA.

Courts have recognized for many years that water quality standards are not limited to numeric water quality parameters (*EDF v. Costle* 1981) ("Water quality criteria may be, and often are, totally narrative.") Purely narrative criteria, of course, have serious limitations when the time comes to translate them into regulatory action:

> ...the criteria listed by the states, particularly for toxic pollutants, were often vague narrative or descriptive criteria as opposed to specific numerical criteria. These descriptive criteria were difficult to translate into enforceable limits on discharges from individual polluters. (*NRDC v. USEPA* 1990)

Moreover, until the Supreme Court described water quality standards as affirmative statements of the "desired condition of the waterway" (*Arkansas v. Oklahoma* 1992), courts generally have viewed water quality criteria as statements — whether narrative or numeric — of acceptable levels of pollution. This does not mean, however, that the more recent addition of biological criteria that describe desired affirmative values rather than acceptable negative conditions will be rejected by the courts. Indeed, judicial acceptance of biocriteria is suggested by analogy to cases that have addressed the use of another type of innovative water quality criteria designed to supplement chemical-specific numeric water quality criteria — whole effluent toxicity (WET) criteria.

WET criteria are designed to address the cumulative and synergistic effects of pollutants in a wastestream. This is done by measuring the impact of the entire wastestream — as opposed to a single pollutant — on test organisms such as minnows or daphnia. The more toxic the whole effluent, for example, the more test organisms exposed over a fixed period of time will die (USEPA 1991f). To date, courts have upheld USEPA's innovations in regulating toxicity as opposed to individual chemicals. USEPA's regulatory approval of WET limits was challenged first by various industry groups as part of the multi-party litigation over USEPA's programmatic NPDES regulations (*NRDC v. USEPA* 1988). Industry argued that "toxicity" was not a "pollutant" as defined by the law; therefore, USEPA had no authority to regulate toxicity in NPDES permits. The U.S. Court of Appeals for the District of Columbia, however, summarily rejected this argument:

> While "toxicity" appears to be an attribute of pollutants rather than a pollutant itself, we see no reason why this should preclude the agency from using it as a measure to regulate effluents that are pollutants.

The Court reasoned further that because the statutory definition of "pollutant" includes all "industrial, municipal and agricultural waste" (CWA Section 502(6)), any discharge to which a toxicity limit could be applied "would seem to fall within this broad definition."

More recently, in response to the "toxic hotspots" provisions in Section 304(l) of the 1987 Water Quality Act, the paper industry challenged USEPA's regulation allowing states to translate narrative "no toxics in toxic amounts" criteria into numeric NPDES permit limits through one of three methods. Once again, the U.S. Court of Appeals for the District of Columbia read USEPA's authority expansively, in order to serve the ultimate purposes of the law. Because the court recognized that water quality standards can have no effect on pollution — at least from point sources — unless translated into effluent limits in NPDES permits, it found USEPA's regulation to be a "preeminent example of gap-filling in the interest of a continuous and cohesive regulatory regime" (*API v. USEPA* 1993). This decision appears to resolve any residual doubts about the legality of using WET to interpret and apply narrative water quality standards (i.e., "no toxics in toxic amounts") to complement or address the absence of chemical-specific criteria.

Traditionally, federal courts defer to reasonable administrative agency interpretations of federal statutes, so long as those readings are not foreclosed by the language or legislative history of the statute (*Chevron v. NRDC* 1984). This is true in particular when agencies exercise their technical expertise. There is no reason why these same principles would not apply here. As discussed above, protection of biological as well as chemical and physical integrity is one of the principal goals of the statute; and there is considerable support in the language of the CWA and its legislative history to support the use of biocriteria. In fact, the absence of water quality criteria that address biological integrity leaves a major void in USEPA and state water quality programs. It is likely, then, that federal courts will uphold the development and implementation of biocriteria as another legitimate example of "gap-filling in the interest of a continuous and cohesive regulatory regime."

State courts have not had the opportunity to review the adoption or implementation of biocriteria. In Ohio, which has perhaps the most extensive experience with biocriteria to date (see Yoder and Rankin, Chapter 9), courts have not yet ruled on the validity of biocriteria *per se* or as applied. In one case, a use designation on the Cuyahoga River was challenged, where that designation was based in part on the use of not-yet-promulgated biocriteria (*NEORSD v. Shank* 1991). However, the plaintiffs in this case did not challenge, and the courts did not address, the use of biocriteria in the decision. Ohio's actual adoption of biocriteria has been challenged before the state's Environmental Review Board, but that case has not yet been heard (Yoder and Rankin, Chapter 9).

4.0 USEPA AUTHORITY TO REQUIRE STATE BIOCRITERIA

The previous section discussed the legal authority to support USEPA issuance of biocriteria under Section 304(a), at least in the form of guidance, and state adoption of legally enforceable biocriteria. A separate question, however, is whether USEPA has the authority — or even the duty — to *require* states to adopt biocriteria in some form, or to issue federal biocriteria applicable in a state that fails to do so. Obviously, biocriteria must be adopted as state or federal water quality standards before being used to support regulatory action, as opposed to monitoring and assessment. This legal question is somewhat premature, since thus far USEPA has addressed biocriteria only in the form of guidance. Thus, there is no case law directly on point, and no regulations that apply in more than a general way, and the issue can be addressed only briefly.

As discussed in Section 2.0, under Section 303 of the act, states have the initial role in adopting water quality standards. USEPA issues criteria *guidance* under Section 304(a), but issues enforceable water quality standards only when it finds that the standards in a given state are deficient, and the state fails to correct them. The question, then, is whether USEPA can find state water quality standards deficient for failure to address biological integrity adequately, and issue federal biocriteria if the state fails to do so.

Advocates of states' rights will argue that Congress intended USEPA's role in approving state water quality standards to be narrow. Section 101(b) of the act states:

> the policy of Congress to recognize, preserve, and protect the primary responsibilities and rights of states to prevent, reduce, and eliminate pollution, to plan the development and use (including restoration, preservation, and enhancement) of land and water resources...

Nothing in this statement of policy, however, defines the breadth of USEPA's authority or responsibility in approving state water quality standards. Rather, the controlling statutory language appears in Section 303(c)(3) and (4). Under Section 303(c)(3), state water quality standards may take effect only after USEPA "determines that such standard meets the requirement of th[e] Act." If USEPA finds otherwise, it must inform the state, which has 90 days to correct the deficiency; thereafter, USEPA must issue federal standards under Section 303(c)(4) to do so.

To some extent, this analysis begs the question. Clearly, Section 303(c)(3) gives USEPA the authority to disapprove a specific water quality standard adopted by the state and submitted to USEPA for approval, if that standard inadequately protects state waters. For example, if the state adopts a standard for dissolved oxygen that will cause harm to resident fish or other species, USEPA may disapprove the standard (*Mississippi Commission on Natural Resources v. Costle* 1980). Courts have given USEPA considerable latitude in making such scientific judgments. Thus, courts have rejected both challenges brought by states arguing that USEPA was too strict (ibid.), and those by environmental groups that USEPA was too lenient (*EDF v. Costle* 1981).

But what if the state simply fails to adopt a standard to address a particular pollutant, or more to the point, to address biological in addition to chemical integrity of its waters? From a policy perspective, it is illogical to assume that states may underprotect their waters by omission but not by commission. Viewed solely from the perspective of chemical pollutants, this would mean that USEPA could disapprove a standard for dissolved oxygen that was too weak, but not the absence of a standard altogether. This view is supported by the language of Section 304(c)(4), which requires USEPA to issue federal water quality standards not only when a revised or new state standard is deemed inadequate (Section 303(c)(4)(A)), but also "in any case where [USEPA] determines that a revised or new standard is necessary to meet the requirements of this Act" (section 303(c)(4)(B)). Thus, Section 303(c)(2)(B), added in 1987 (see Section 3.1 supra), clarified that states may not simply omit criteria for toxic pollutants "the discharge or presence of which in the affected waters could reasonably be expected to interfere with…designated uses…"

There is no logical reason to reach a different conclusion with respect to the adequacy of state water quality standards to protect the biological as well as the chemical integrity of state waters. USEPA's duty is to ensure that state standards meet the requirements of the act. The most fundamental requirement of the act is to restore and maintain the chemical, physical, and biological integrity of the nation's waters. State water quality standards that fail to address biological integrity, therefore, do not meet all of the requirements of the act, and USEPA has both the authority and the duty to fill this gap if the state, on notice, fails to do so.

The fact that USEPA may (or must) insist that states adequately address biological as well as chemical integrity, however, does not necessarily mean that USEPA may dictate the precise manner in which biological integrity is protected. Under the principles discussed above, USEPA may disapprove a state standard only if it does not meet the requirements of the act — not if it fails to do so in the same way as USEPA would choose. For example, with respect to water quality standards for toxic pollutants, USEPA will approve standards based on USEPA criteria, USEPA criteria modified to reflect site-specific circumstances, or criteria based on other "scientifically defensible" methods (40 CFR. Section 131.11). Of the many variations on biocriteria discussed elsewhere in this book, then, states may choose the method(s) best suited to its conditions, or develop its own variations, so long as they are scientifically sound, and so long as USEPA finds them to be adequately protective.

It is likely that some states will continue to resist the adoption and application of biocriteria as defined in this book, just as many states have resisted the adoption and use of WET criteria, and the adoption and application of many chemical-specific standards. Nothing in the CWA mandates state adoption of biocriteria per se. Thus, in principle, states can attempt to prove to USEPA that their other water quality standards, taken as a whole, are sufficient to protect the biological integrity of their waters. This is primarily a scientific rather than a legal judgment. But as evidence continues to mount that biological impairment is detected through biocriteria even when chemical-specific criteria are met, and as the relative focus of water quality programs shifts from chemical releases from point sources to physical and biological impairments caused by polluted runoff, physical habitat alterations, and other nonpoint sources, states will be increasingly hard-pressed to make this case.

5.0 IMPLEMENTATION OF BIOCRITERIA

Having established the legal basis for issuing biocriteria, the next, arguably harder inquiry is for what they may legally be used. Legal challenges are more likely as biocriteria are relied on for harder regulatory purposes, such as the imposition of stricter permit conditions on point sources or more rigorous management practices on nonpoint sources.

In its biocriteria guidance, USEPA lists a range of potential applications of biocriteria, for purposes of the CWA and other federal statutes (USEPA 1990a). These range from monitoring and assessment, to program planning and management, to the imposition of regulatory controls. Curiously (perhaps out of excess caution), USEPA omits reference to NPDES permits under Section 402, the principal legal tool for preventing violation of *any* water quality standards due to point source discharges. Nor does USEPA cite enforcement applications, which have been initiated more recently by USEPA. The legal basis for each of these applications of biocriteria will be addressed in turn.

5.1 Monitoring and Assessment

The various monitoring and assessment provisions of the CWA provide firm support for the use of biocriteria to assess the biological health of waterbodies. In addition to the triad of chemical, physical, and biological integrity established in Section 101(a), discussed earlier, specific monitoring provisions require biological as well as chemical and physical monitoring. Section 305(b) of the law requires each state to monitor and assess its waters every two years, and to report the results to USEPA. In addition to assessing chemical pollutants and their impacts, states must analyze "the extent to which all waters of such State provide for the protection and propagation of a balanced population of shellfish, fish, and wildlife." (CWA Section 305(b)(1)(B)). Similar or identical language appears in a long list of other monitoring provisions in the act, including Sections 303(d) (priority lists of impaired waters), 304(l) (toxic hotspots), 319(a) (nonpoint source assessments), and 320(b) (National Estuary Program assessments). Moreover, USEPA notes that biocriteria can be used to conduct risk assessments under other federal statutes, including the Toxic Substances Control Act; the Resource Conservation and Recovery Act; Superfund; the Federal Insecticide, Fungicide and Rodenticide Act; the National Environmental Policy Act; the Federal Lands Policy and Management Act; the Fish and Wildlife Conservation Act; the Marine Protection, Research and Sanctuaries Act; the Coastal Zone Management Act; and the Fish and Wildlife Coordination Act (USEPA 1990a).

USEPA regulations require state monitoring programs to include physical, chemical and biological data (40 CFR. Section 130.3(b)). Until the advent of biocriteria, however, this requirement had little real meaning. In its official guidelines for the 1994 cycle of state assessments under Section 305(b), USEPA "strongly recommends" that states begin to use biocriteria to supplement existing methods of determining use attainment or impairment, and believes that such use will improve the national consistency of 305(b) reports (USEPA 1993b). Appendix B of the Guidance provides detail on how to translate biocriteria assessments to the 305(b) existing categories of "fully supporting", "partially supporting", and "not supporting" the designated use. To change this recommendation to a requirement, USEPA would either have to amend its existing regulations, or interpret them to require the use of biocriteria to identify impairment.

Identification of a waterbody as "not supporting" or "partially supporting" a designated use has specific legal as well as scientific significance, for example, under CWA Sections 303(c) and (d), 304(l), 319, and 402. As discussed further below, such classification may result in the imposition of additional controls on individual sources of pollution. Thus, the use of biocriteria to supplement existing narrative and chemical-specific criteria will raise two related legal questions.

First, the potential subjects of additional controls may question the validity of classifying a waterbody as impaired under the CWA when all applicable chemical water quality criteria are met. In Ohio's 1988 assessment of 431 sites, for example, impairment was detected using biological but not chemical criteria at 36% of the sites assessed (USEPA 1990a). However, the law clearly is aimed at biological as well as chemical integrity. Moreover, it is hard to quarrel with data showing actual impairment of the biological community as measured by such factors as species abundance, diversity, or community structure. So long as the science is sound and well supported, the findings likely will be upheld in the courts. A more likely (and potentially more fruitful) line of attack, discussed in more detail in the following sections, is what types of controls can be imposed, and on whom, when chemical but not biological criteria are met.

Second, and conversely, parties may challenge a finding of impairment based on the exceedence of chemical criteria when no biological impairment is detected through the use of biocriteria. Obviously, this will be at issue only with respect to chemical criteria adopted to protect the biological use, and not when the criteria protect against human health or other adverse effects. Either the legal ramifications of the finding will be assailed (such as stricter water quality-based effluent limits), or the validity of the chemical criteria will be questioned. Because the chemical-specific criteria were adopted to protect the use, parties will argue that if the criteria are violated but the use is not impaired, the criteria must be unduly conservative (Pifher 1991). Absent adequate explanation of the apparent anomaly, this simple logic may be appealing to some judges.

From an empirical perspective, to date this has not caused a major problem. In the same Ohio results cited above, chemical and biosurvey results agreed 58% of the time, and impairment was indicated by chemical but not biological criteria at only 6% of the sites (USEPA 1990a). Moreover, USEPA has taken the firm position that violation of any one of the three types of water quality criteria — chemical-specific, toxicity, or biological — should be interpreted as suggesting impairment, because each type of standard has strengths and limitations:

> Since each type of criteria…has different sensitivities and purposes, a criterion may fail to detect real impairments when used alone. As a result, these methods should be used together…If any one type of criteria indicates impairment of the surface water, regulatory action can be taken to improve water quality. However, no one type of criteria can be used to confirm attainment of a use if another form of criteria indicates nonattainment. (USEPA 1990a)

A more specific list of the capabilities and limitations of each type of water quality standard appears in USEPA's *Technical Support Document for Water Quality-Based Toxicity Control* (USEPA 1991f). As one example, USEPA notes that "chemical criteria may provide earlier indications of impairment from a bioaccumulative chemical because aquatic communities require exposure over time to incur the full effect" (USEPA 1990a). Since the goal of the law is to *maintain* as well as to restore waterbody integrity (CWA Section 101(a)), that is, to *prevent* adverse effects before they occur, it would be foolhardy (and illegal) to ignore chemical warning signals simply because biological ramifications have not yet appeared. Nonetheless, to guard against potential legal attacks on water quality criteria based on chemistry or toxicity, agencies should be prepared to explain fully the limitations as well as the capabilities of biocriteria. USEPA has evaluated the capabilities and limitations of all three major forms of water quality criteria — chemical-specific, WET, and biocriteria (USEPA 1991f).

5.2 Program Planning and Management

USEPA also suggests that biocriteria be used to guide water quality planning and management (USEPA 1990a). To the extent that biocriteria are used simply to set agency priorities and to develop plans, they are not likely to be challenged. Many of these planning activities, however, have real-world implications for point sources, nonpoint sources and other sources of waterbody impairment.

The 1972 act specifically required states not only to identify impaired waters, but to rank them in order of priority for cleanup and restoration. For example, under Section 303(d) states were to identify any waters for which technology-based effluent limitations on point sources were not strict enough to implement "any water quality standard," and to "establish a priority ranking for such waters, taking into account the severity of the pollution and the uses to be made of such waters" (CWA Section 303(d)(1)(A)). These lists were to drive the state's continuing planning process under Section 303(e). The phrase "any water quality standard" obviously is broad enough to encompass biocriteria, but more important, the relative degree of biological use impairment can be ranked more precisely through numerical biocriteria than through chemical criteria alone. Thus, biocriteria are useful to identify which waters require the most immediate and the most forceful remedial action, and which waters retain biological attributes worthy of protection through antidegradation programs. However, states are not likely to use biocriteria in preparing their lists of priority waters until biocriteria are adopted as part of their water quality standards.

Because the 1972 law set a 1977 deadline for compliance with all water quality standards (CWA Section 301(b)(1)(C)), the notion of "priority lists" under Section 303(d), as opposed to a plan to address all polluted waters, was of questionable validity after 1977. Nevertheless, recognizing the sketchy implementation of these requirements over the law's first 15 years, in 1987 Congress established even more focused assessment requirements, with specific deadlines for implementation.

In section 304(l), for example, states were required to identify all waters not expected to meet minimum statutory uses due to any cause (CWA Section 304(l)(1)(A)(ii)); waters not expected to meet water quality standards for toxic pollutants due to any cause (CWA Section 304(l)(1)(A)(ii)); and waters not expected to meet water quality standards "due entirely or substantially to discharges from point sources of [toxic] pollutants" (CWA Section 303(l)(1)(B)). Waters on at least some of these lists were to be subject to "individual control strategies." USEPA's first limited interpretation of this requirement was rejected by the U.S. Court of Appeals for the Ninth Circuit (*NRDC v. USEPA* 1990). The full scope of the provision has not been finally revisited by USEPA on remand, although USEPA's draft regulation would decline to expand the program (USEPA, 57 Fed. Reg. 33051 et seq. 1992). Because over 17,000 waterbodies were identified in the states' first effort under Section 304(l), with most impairment due largely to nonpoint sources, biocriteria could be used to focus efforts and establish cleanup priorities under this program.

Similarly, impatient with the virtual absence of serious efforts to abate nonpoint source pollution under the planning and management provisions of Section 208, Congress added a new nonpoint source program in 1987 (CWA Section 319). The first step in this program was the identification of waters that would not meet water quality standards without additional nonpoint source controls (CWA Section 319(a)). Biocriteria are particularly useful in identifying impairment from polluted runoff and other nonpoint sources, because many of the impacts from those sources, such as smothering of gravel spawning habitat or alteration of flow velocities and bank morphology due to alteration of hydrology, are not detected through chemical or toxicity tests (USEPA 1990a, 1991f). Moreover, while the existing Section 319 is extremely vague on the question of which waters must be addressed in which sequence, proposals to modify Section 319 would require states to target the most threatened or impaired waters first (Adler et al. 1993). Biocriteria would be useful in identifying and establishing priorities under such a scheme.

Increasingly, a consensus is developing that waterbodies should be restored and protected on a watershed basis. While definitions of this concept vary considerably, the underlying principles are that all sources of impairment of a water should be identified rather than focusing exclusively on individual point sources; and that restoration efforts should focus on physical habitat and biological integrity as well as preventing chemical pollution (Water Quality 2000 1992). Biocriteria can be extremely useful in focusing watershed protection and restoration efforts, because biological assessments can help to pinpoint the location and nature of impairment. Additional efforts are then needed, of course, to isolate causes and identify solutions. USEPA recognizes this utility in the context of the National Estuary Program (CWA Section 320) (USEPA 1990a).

Antidegradation programs, which have been seriously underutilized to date (NWF 1992), also could profit from the use of biocriteria. Waterbodies, or portions thereof, with healthy identified biological communities can be targeted for special protection for purposes of either Tier II or Tier III antidegradation (40 CFR. Section 131.12). Biological assessments are also useful in determining whether antidegradation programs are successful in fully preserving existing uses.

5.3 Permits and Other Regulatory Requirements

The most controversial use of biocriteria, of course, will be to strengthen or to impose new regulatory requirements on individual dischargers. It is here where USEPA or a state agency must be most careful about documenting cause and effect, and to demonstrate why the chosen regulatory action is likely to reduce or eliminate the harm. USEPA's biocriteria guidance document identifies only two programs for which biocriteria may be used for strict regulatory purposes (although by its terms the list is not intended to be exclusive) (USEPA 1990a). Under Section 403(c), USEPA is prohibited from issuing a permit for discharges to ocean and certain near-coastal waters if such discharges will result in degradation of those waters pursuant to guidelines issued by USEPA, which must consider a range of impacts on biota. In this circumstance, of course, biocriteria would have to be used as a predictive tool, for which there is no precedent (USEPA 1991f; however, see Section 2.2.8 in Yoder and Rankin, Chapter 21). Similarly, to determine acceptable sites for the disposal of dredge and fill material under Section 404, biocriteria would have to be used predictively. Since biocriteria are designed to measure status rather than to predict impacts, more likely they will be used in these programs to assess — and perhaps to modify — regulatory decisions made on other grounds.

More controversy, however, is likely to surround the use of biocriteria in NPDES permits, an issue apparently not addressed by USEPA in its existing guidance documents. From a purely legal perspective, in principle there should be no reason why biocriteria — like whole effluent toxicity requirements — cannot be used to support additional requirements in NPDES permits, at least to the extent that detectable harm is caused by the discharge of pollutants from a point source. In practice, however, the use of biocriteria and biological assessments to establish new permit requirements raises interesting and challenging questions.

The language of the statute, USEPA regulations, and relevant case law all suggest that biological water quality criteria can be, or must be, enforced through appropriate NPDES conditions or limitations.

A fundamental requirement of the CWA is that NPDES permits must be designed to ensure compliance with *all* water quality standards. Section 301(b)(1)(C) required effluent limits adequate to meet applicable water quality standards by July 1, 1977. Under Section 302(a), USEPA can require the use of stricter permit requirements where needed to ensure the "protection and propagation of a balanced population of fish, shellfish and wildlife," although permittees can seek to override this requirement based on excessive "social and economic costs." "Effluent limitations" are defined by the statute (CWA Section 502(11)) as "any restriction established by a State or [USEPA] on quantities, rates, and concentrations of chemical, physical, biological, and other constituents…" And if all else fails, Section 402(a)(2) allows the USEPA Administrator to impose in NPDES permits "such other conditions as he deems appropriate." Such "other conditions" are likely to be upheld if they promote the basic goal of the law to protect or restore the biological integrity of a waterbody. Interpreting these provisions, USEPA regulations confirm that NPDES permits *must* ensure compliance with all applicable water quality standards (40 C F R. Sections 122.4, 122.43, and 122.44). Thus, so long as biocriteria are valid elements of water quality standards, as shown in the previous section, they must be enforced in NPDES permits.

The U.S. Supreme Court and lower courts have indicated repeatedly that water quality standards must be enforced through NPDES permits, and in fact, that NPDES permits are the primary means of translating water quality standards into enforceable requirements. "An NPDES permit serves to transform generally applicable effluent limitations and other standards — including those based on water quality — into the obligations of the individual discharger…" (*California v. USEPA* 1981; see also *Trustees for Alaska v. USEPA 1984*; *Arkansas v. Oklahoma* 1992; *NRDC v. USEPA* 1990; *Westvaco Corp. v. USEPA* 1990; *API v. USEPA* 1993).

These principles were used by the U.S. Court of Appeals for the District of Columbia to approve in two separate cases, at least in concept, the application of whole effluent toxicity in NPDES permits (*NRDC v. USEPA* 1988; *API v. USEPA* 1993). Notably, in the first case (the multiparty challenge to USEPA's general NPDES program rules in which industry groups challenged USEPA's toxicity rule), the court refused to address the technical feasibility of applying WET limits in a permit, expressly so that individual permit writers can develop an appropriate record (scientific basis) to support a particular application. The same should be true with respect to biocriteria. In concept, there is no legal reason why NPDES permits cannot be used to implement biocriteria. Judicial review of the technical feasibility and appropriate implementation of that principle, however, should await a specific permit, to enable USEPA or a state agency to develop the appropriate technical record.

Nevertheless, courts have noted that the implementation of water quality criteria that are *not* based on chemical-specific numeric criteria can be difficult as a matter of practice. The Ninth Circuit Court of Appeals, for example, explained that narrative criteria "were difficult to translate into enforceable limits on discharges from individual polluters" (*NRDC v. USEPA* 1990). The harder it is to show a cause-and-effect relationship between a particular discharge and the condition of a waterbody, the more the agency must rely on expert testimony and judgment rather than simple mathematic calculations, as is true for numeric water quality criteria.

The logical question, then, is how a violation of biocriteria (or an indication of impairment based on a biological survey) can be translated into an enforceable requirement of an NPDES permit. Clearly, there is no way to translate biocriteria *directly* into water quality-based effluent limits. A number of results are plausible, however. If the impairment is caused by a pollutant that was not previously detected, for example, the biological assessment may precipitate additional effluent monitoring and appropriate chemical effluent limits, or appropriate new WET limits. The impairment may also result from the timing, as opposed to the raw concentration, of pollutant releases (Courtemanch, Chapter 20). Of course, as the number of sources of impairment of a waterbody increase (point and nonpoint), the more difficult it will be to establish the cause-and-effect relationship needed to use biocriteria to impose specific permit requirements. The full range of potential applications must await actual implementation of biocriteria in NPDES permits.

This does not mean, however, that the use of biocriteria in NPDES permits is unlimited. For example, an NPDES permit will not be required unless there is a release of one cr more pollutants from a point source (*NWF v. Gorsuch* 1982). In fact, in finding that dams were not required to have NPDES permits, one court ruled that the fact that dams may cause "pollution" does not necessitate an NPDES permit where there is no addition of "pollutants." States may still implement their water quality standards, of course, for nonpoint sources, including dams, either through their own programs under Section 319, or through water quality certifications issued for federal licenses and permits under section 401. Even under Section

401, however, some courts have rejected conditions based on biological as opposed to purely chemical impairment (Adler et al. 1993). However, in 1994 the U.S. Supreme Court verified that states may use Section 401 to protect designated uses through maintenance of biological habitat, including minimum instream flows (PUD, No. 1 of Jefferson County V. Washington Dept. Ecology, 1994.)

Finally, biocriteria may be useful as well under CWA Section 319, to determine the appropriate best management practices to control nonpoint sources of pollution, and to assess their relative success. Since, as discussed above, traditional chemical water quality criteria do not adequately identify impairment from nonpoint sources, biocriteria will be useful to pinpoint the areas of greatest impairment, potential causes, and appropriate remedial actions.

5.4 Enforcement

Biological assessments based on biocriteria are beginning to be used to support CWA enforcement actions (Yoder and Rankin, Chapters 9 and 17). This application can be useful in a number of ways. First, biological assessments can be used to identify potential violators. If a multisite assessment indicates that impairment occurs below but not above a particular facility, that might be an indication — although probably not sufficient proof without additional evidence — of undetected permit violations (or, as discussed above, that the permit should be strengthened).

Second, biological assessments can be used to prove the harm caused by an identified violation. For example, in determining appropriate administrative or civil penalties, USEPA and federal courts respectively must consider, among other factors, the severity of the violation (CWA Sections 309(d) and 309(g)(3)). Courts have reduced the amount of violations where the plaintiff was not able to show serious harm to the environment (Atlantic States Legal Foundation 1992). Biocriteria could be used not only to demonstrate the degree of impairment downstream of a violator, but also to help prove the likely source of the harm through comparative analysis. Again, more direct independent evidence of cause and effect likely will be needed once harm is detected.

Biocriteria are also potentially useful to determine natural resource damages under CWA Section 311 (or analogous provisions of other statutes, such as Superfund) (*Ohio v. Department of Interior* 1990), or for purposes of criminal penalties which may include fines to account for environmental harm (U.S. Sentencing Commission 1990). Under Section 311, for example, the federal or state governments may collect costs incurred "in the restoration or replacement of natural resources destroyed or damaged as a result of a discharge of oil or a hazardous substance..." Since the cause and extent of damage may be difficult to prove, biological assessments based on biocriteria may be useful additional ammunition. This is particularly true if baseline data are available for the waterbody in question.

Finally, biocriteria could be useful in fashioning and justifying injunctive relief. Since one of the standards for a preliminary or permanent injunction is the balance of harm to the plaintiff and defendant, as well as the public interest, the ability to prove harm may be important either to a prohibitive injunction (to stop polluting), or a mandatory injunction (to remediate past or existing harm).

6.0 CONCLUSION

However well founded the legal basis for biocriteria and their implementation, it is almost certain that they will be challenged in various ways. Already, several avenues of legal attack have been suggested. One legal critic, for example, questioned whether biocriteria are reliable, scientifically repeatable, provide dischargers with adequate notice of acceptable conduct, and can reliably determine cause-and-effect relationships (Pifher 1991).

Therefore, program managers would be wise to anticipate and to insulate their biocriteria programs, from development to implementation, from such legal challenges. Ample authority exists under the CWA to support the development and use of biocriteria for various purposes. Moreover, because existing water quality criteria have not provided adequate protection for biological resources, USEPA and state agencies have a strong policy argument that biocriteria are not only appropriate, but essential, to fulfill the underlying goals of the CWA. The degree to which biocriteria will be upheld turns as well, however, on the sound scientific basis for specific criteria and the application of those criteria. As with most aspects of CWA implementation, good science will ensure good law.

REFERENCES

Abe, J., W. Davis, T. Flanigan, A. Schwarz, and M. McCarthy. 1992. *Environmental Indicators for Surface Water Quality Programs — Pilot Study.* EPA-905-R-92–001. U.S. Environmental Protection Agency, Region 5, Chicago, Illinois and Office of Policy, Planning and Evaluation, Washington, D.C.

Adams, S.M., L.R. Shugart, G.R. Southworth, and D.E. Hinton. 1990. Application of bioindicators in assessing the health of fish populations experiencing contaminant stress. Pages 333–353 in J.F. McCarthy and L.R. Shugart (editors). *Biomarkers of Environmental Contamination.* CRC Press, Inc., Boca Raton, Florida.

Adler, R.W., J.C. Landman, and D. Cameron. 1993. *The Clean Water Act 20 Years Later.* Natural Resources Defense Council. Island Press, Covelo, California.

Alexander, G. 1876. The *Law as to the Pollution of Rivers: the Rivers Pollution Prevention Act, 1876.* Knight and Co., Local Government Publishers, London.

Allan, J.D. and A.S. Flecker. 1993. Biological diversity conservation in running waters. *BioScience* 43: 32–43.

Allen, R.K. and G.F. Edmunds. 1962. A revision of the genus *Ephemerella* (Ephemeroptera: Ephemerellidae). IV. The subgenus *Danella. Journal of the Kansas Entomological Society* 35: 333–338.

Allen, R.K. and G.F. Edmunds. 1963a. A revision of the genus *Ephemerella* (Ephemeroptera: Ephemerellidae). VII. The subgenus *Eurylophella. The Canadian Entomologist* 95: 597–623.

Allen, R.K. and G.F. Edmunds. 1963b. A revision of the genus *Ephemerella* (Ephemeroptera: Ephemerellidae). VI. The subgenus *Serratella* in North America. *Annals of the Entomological Society of America* 56: 583–600.

Allen, R.K. and G.F. Edmunds. 1965. A revision of the genus *Ephemerella* (Ephemeroptera: Ephemerellidae). VIII. The subgenus *Ephemerella* in North America. *Miscellaneous Publications of the Entomological Society of America* 4: 243–282.

Aloi, J.E. 1990. A critical review of recent freshwater periphyton field methods. *Canadian Journal of Fisheries and Aquatic Sciences* 47: 656–670.

APHA (American Public Health Association), American Water Works Association, and Water Pollution Control Federation. 1985. *Standard Methods for Examination of Water and Wastewater.* 16th Edition. Washington, D.C.

APHA (American Public Health Association), American Water Works Association, and Water Environment Federation. 1993. *Standard Methods for the Examination of Water and Wastewater.* 18th edition. Washington, D.C.

Anderson, K., A. Boulanger, H. Gish, J. Kelly, J. Morrill, and L. Davis. 1991. Using machine learning techniques to visualize and refine criteria for biological integrity. Pages 123–128 in *Biological Criteria: Research and Regulation, Proceedings of a Symposium.* EPA-440–5-91–005. U.S. Environmental Protection Agency, Office of Water, Washington, D.C.

Anderson, R.L., C.T. Walbridge, and J.T. Fiandt. 1980. Survival and growth of *Tanytarsus dissimilis* (Chironomidae) exposed to copper, cadmium, zinc, and lead. *Archives of Environmental Contamination and Toxicology* 9: 329–335.

Andrus, C.W., B.A. Long, and H.A. Froehlich. 1988. Woody debris and its contribution to pool formation in a coastal stream 50 years after logging. *Canadian Journal of Fisheries and Aquatic Sciences* 45: 2080–2086.

Angermeier, P.L. and J.R. Karr. 1984. Relationships between woody debris and fish habitat in a small warmwater stream. *Transactions of the American Fisheries Society* 113: 716–726.

Angermeier, P.L. and J.R. Karr. 1986. Applying an index of biotic integrity based on stream-fish communities: considerations in sampling and interpretation. *North American Journal of Fisheries Management* 6: 418–429.

Angermeier, P.L. and I.J. Schlosser. 1987. Assessing biotic integrity of the fish community in a small Illinois stream. *North American Journal of Fisheries Management* 7: 331–338.

Archibald, R.E.M. 1972. Diversity in some South African diatom associations and its relation to water quality. *Water Research* 6: 1229–1238.

Arkansas Department of Pollution Control and Ecology. 1988. *Regulation Establishing Water Quality Standards for Surface Waters of the State of Arkansas.* Arkansas DPCE, Little Rock, Arkansas.

Armantrout, N.B. 1982. Aquatic habitat inventories, the current situation. Pages 7–9 in N.B. Armantrout (editor). *Acquisition and Utilization of Aquatic Habitat Inventory Information*, Western Division, American Fisheries Society, Bethesda, Maryland.

Armour, C.L., K.P. Burnham, and W.S. Platts. 1983. *Field Methods and Statistical Analyses for Monitoring Small Salmonid Streams*. U.S. Fish and Wildlife Service, Washington, D.C.

Bahls, L. 1993. *Periphyton Bioassessment Methods for Montana Streams*. Water Quality Bureau, Department of Health and Environmental Science, Helena, Montana.

Bailey, R.G. 1976. *Ecoregions of the United States*. Map (scale 1:7,500,000). U.S. Dept. of Agriculture, Forest Service, Intermountain Region, Ogden, Utah.

Bailey, R.G. 1989. Explanatory supplement to ecoregions map of the continents. Map (scale 1:30,000,000). *Environmental Conservation* 16: 307–309.

Bailey, R.G. 1991. Design of ecological networks for monitoring global change. *Environmental Conservation* 18: 173–175.

Bailey, R.G. and C.T. Cushwa. 1981. *Ecoregions of North America*. Map (scale 1:12,000,000). FWS/OBS-81/29. *U.S. Fish and Wildlife Service*, Washington, D.C.

Bailey, P.A., J.W. Enblom, S.R. Hanson, P.A. Renard, and K. Schmidt. 1993. *A Fish Community Analysis of the Minnesota River Basin*. Minnesota Pollution Control Agency, St. Paul, Minnesota.

Bain, M.B., J.T. Finn, and H.E. Booke. 1985. Quantifying stream substrate for habitat analysis studies. *North American Journal of Fisheries Management* 5: 499–500.

Bain, M.B., J.T. Finn, and H.E. Booke. 1988. Streamflow regulation and fish community structure. *Ecology* 69: 382–392.

Baker, J.R., G.D. Merritt, and D.W. Sutton. 1993a. *Environmental Monitoring and Assessment Program: Surface Waters-Lakes Field Operations Manual. Volume 1*. U.S. Environmental Protection Agency. Las Vegas, Nevada.

Baker, J.P., W.J. Warren-Hicks, J. Gallagher, and S.W. Christensen. 1993b. Fish population losses from Adirondack lakes: the role of surface water acidity and acidification. *Water Resources Research* 29: 861–874.

Ball, J. 1982. *Stream Classification Guidelines for Wisconsin*. Wisconsin Department of Natural Resources Technical Bulletin, Madison, Wisconsin.

Balon, E.K. 1975. Reproductive guilds of fishes: a proposal and definition. *Journal Fisheries Research Board of Canada* 32: 821–864.

Balon, E.K. 1981. Addition and amendments to the classification of reproductive styles in fishes. *Environmental Biology of Fishes* 6: 377–389.

Barbour, M.T., J.L. Plafkin, B.P. Bradley, C.G. Graves and R.W. Wisseman. 1992. Evaluation of EPA's rapid bioassessment benthic metrics: metric redundancy and variability among reference stream sites. *Journal of Environmental Toxicology and Chemistry* 11: 437–449.

Barbour, M.T. and J.B. Stribling. 1991. Use of habitat assessment in evaluating the biological integrity of stream communities. Pages 25–38 in *Biological Criteria: Research and Regulation, Proceedings of a Symposium, 12–13 December 1990, Arlington, Virginia*. EPA-440–5-91–005. U.S. Environmental Protection Agency, Office of Water, Washington, DC.

Barbour, M.T. and J.B. Stribling. 1994. A technique for assessing stream habitat structure. Pages 156–178 in Proceedings of the conference "Riparian ecosystems in the humid United States: function, values, and management." National Association of Conservation Districts, Washington, D.C.

Barnes, B.V. 1993. The landscape ecosystem approach and conservation of endangered spaces. *Endangered Species UPDATE* 10: 13–19.

Batshelet, E. 1976. *Introduction to Mathematics for Life Sciences*. Second edition. Springer-Verlag, New York.

Bayer, C.W., J.R. Davis, S.R. Twidwell, R. Kleinsasser, G. Linam, K. Mayes and E. Hornig. 1992. Texas Aquatic Ecoregion Project: An Assessment of Least Disturbed Streams. Draft Report. Texas Natural Resources Conservation Commission, Austin, Texas.

Bazata, K. 1991. *Nebraska Stream Classification Study*. Nebraska Department of Environmental Control, Lincoln, Nebraska.

Beak, T.W. 1965. A biotic index of polluted streams and its relationship to fisheries. Pages 191–219 in O. Jagg (editor). *Advances in Water Pollution Research*. Proceedings of the Second International Association on Water Pollution Research Conference, Tokyo, August 1964. Pergamon Press Ltd., London.

Beardsley, D.P. 1992. Using environmental indicators for policy and regulatory decisions. Pages 61–64 in D.H. McKenzie, D.E. Hyatt, and V.J. McDonald (editors). *Ecological Indicators. Proceedings of the International Symposium on Ecological Indicators*, October 16–19, 1990, Ft. Lauderdale, FL. Elsevier Science Publishers Ltd., Essex, England.

Beck, W.M., Jr. 1954. Studies in stream pollution biology. I. A simplified ecological classification of organisms. *Quarterly Journal of the Florida Academy of Sciences* 17: 211–227.

Beck, W.M., Jr. 1955. Suggested method for reporting biotic data. *Sewage and Industrial Wastes* 27: 1193–1197.

Beck, W.M., Jr. 1977. *Environmental Requirements and Pollution Tolerance of Common Freshwater Chironomidae*. EPA-600-4-77-024. United States Environmental Protection Agency, Environmental Monitoring and Support Laboratory, Cincinnati, Ohio.

Bednarik, A.F. and W.P. McCafferty. 1979. Biosystematic revision of the genus *Stenonema* (Ephemeroptera: Heptageniidae). *Canadian Bulletins of Fisheries and Aquatic Sciences* 201: 1–73.

Bencala, K.E. 1993. A perspective on stream-catchment connections. *Journal of the North American Benthological Society* 12: 44–47.

Benke, A.C. 1990a. Ecosystem characteristics and biological productivity of Southeastern Coastal Plain Blackwater Rivers. *Biological Report* 90: 6–7.

Benke, A.C. 1990b. A perspective on America's vanishing streams. *Journal of the North American Benthological Society* 9: 77–88.

Benke, A.C., R.L. Henry, D. M. Gillespie, and R.J. Hunter. 1985. Importance of snag habitat for animal production in southeastern streams. *Fisheries* 10: 8–13.

Bennett, G.W. 1958. Aquatic biology. Pages 163–178 in A Century of Biological Research. *Bulletin of the Illinois Natural History Survey* 27: 1–234.

Benveniste, G. 1977. *The Politics of Expertise*. 2nd edition, Boyd and Fraser, San Francisco, California.

Benveniste, G. 1981. *Regulation and Planning: The Case of Environmental Politics*. Boyd and Fraser, San Francisco, California.

Berg, M.B. and R.A. Hellenthal. 1990. Data variability and the use of chironomids in environmental studies: the standard error of the midge. Pages 1–8 in W.S. Davis (editor). *Proceedings of the 1990 Midwest Pollution Control Biologists Meeting*. EPA-905-9-90-005. United States Environmental Protection Agency, Region 5, Environmental Sciences Division, Chicago, Illinois.

Berkman, H.E. and C.F. Rabeni 1987. Effect of siltation on stream fish communities. *Environmental Biology of Fishes* 18: 285–294.

Berkman, H.E., C.F. Rabeni, and T.P. Boyle. 1986. Biomonitors of stream quality in agriculture areas: fish vs. invertebrates. *Environmental Management* 10: 413–419.

Berthouex, P.M. and I. Hau. 1991. Difficulties related to using extreme percentiles for water quality regulations. *Research Journal of the Water Pollution Control Federation* 63: 873–879.

Bick, H. 1963. A review of Central European methods for the biological estimation of water pollution levels. *Bulletin of the World Health Organization* 29: 401–413.

Bickers, C.A., M.H. Kelly, J.M. Levesque, and R.L. Hite. 1988. *Users Guide to IBI-AIBI- Version 2.01 (A BASIC Program for Computing the Index of Biotic Integrity with the IBM-PC)*. IEPA/WPC/89-007. Illinois Environmental Protection Agency, Division of Water Pollution Control, Planning Section, Springfield, Illinois.

Bickford, C.A., C.E. Mayer, and K.D. Ware. 1963. An efficient sampling design for forest inventory: the northeast forest resurvey. *Journal of Forestry* 61: 826–833.

Biggs, B.J.F., M.J. Duncan, I.G. Jowett, J.M. Quinn, C.W. Hickey, R.J. Davies-Colley, and M.E. Close. 1990. Ecological characterisation, classification, and modelling of New Zealand rivers: an introduction and synthesis. *New Zealand Journal of Marine and Freshwater Research* 24: 277–304.

Bilyard, G.R. and M.B. Brooks-McAuliffe. 1987. *Recommended Biological Indices for 301(h) Programs*. EPA 430-9-86-002. U.S. Environmental Protection Agency, Office of Marine and Estuarine Protection, Washington, D.C.

Binns, N.A. 1982. *Habitat Quality Index Procedures Manual*. Wyoming Game and Fish Department, Laramie, Wyoming.

Binns, N.A. and F.M. Eiserman. 1979. Quantification of fluvial trout habitat in Wyoming. *Transactions of the American Fisheries Society* 108: 215–228.

Bisson, P.A., J.L. Nielson, R.A. Palmason, and L.E. Grove. 1982. A system of naming habitat types in small streams, with examples of habitat utilization by salmonids during low streamflow. Pages 62–73 in N. B. Armantrout. (editor). *Acquisition and Utilization of Aquatic Habitat Inventory Information, Portland, Oregon, 28 October 1981*. Western Division, American Fishery Society, Portland, Oregon.

Blackburn, T.M., J.H. Lawton, and J.N. Perry. 1992. A method of estimating the slope of upper bounds plots of body size and abundance in natural animal assemblages. *Oikos* 65: 107–112.

Blair, W.F. 1950. The biotic provinces of Texas. *Texas Journal of Science* 2: 93–117.

Blaxter, J.H.S. 1974. *The Early Life History of Fish. The Proceedings of an International Symposium Held at Dunstaffnage Marine Research Laboratory of the Scottish Marine Biological Association at Oban, Scotland, from May 17–23, 1973.* Springer-Verlag, New York.

Blockstein, D.E. 1992. An aquatic perspective on U.S. biodiversity policy. *Fisheries* 17(3): 26–30.

Bode, R.W. 1983. Larvae of North American *Eukiefferiella* and *Tvetenia* (Diptera: Chironomidae). *New York State Museum Bulletin* 452: 1–40.

Bode, R.W. 1988a. *Methods for Rapid Biological Assessment of Streams.* New York State Department of Environmental Conservation Report, Albany, New York.

Bode, R.W. 1988b. *Biological Impact Assessment: Skaneateles Creek.* New York State Department of Environmental Conservation report, Albany, New York.

Bode, R.W. and M.A. Novak. 1989. Proposed biological criteria for New York streams. Pages 42–48 in T.P. Simon, L.L. Holst, and L.J. Shepard (editors). *Proceedings of the First National Workshop on Biological Criteria.* EPA 905–9-89–003. U.S. Environmental Protection Agency, Region 5, Environmental Sciences Division, Chicago, Illinois.

Bode, R.W., M.A. Novak, and L.E. Abele. 1986a. *Rapid Biological Stream Assessment: Ramapo River, 1986 Survey.* New York State Department of Environmental Conservation Report, Albany, New York.

Bode, R.W., M.A. Novak, and L.E. Abele. 1986b. *Rapid Biological Stream Assessment: Gooseberry Creek, 1986 Survey.* New York State Department of Environmental Conservation Report, Albany, New York.

Bode, R.W., M.A. Novak, and L.E. Abele. 1990a. *Biological Stream Assessment: Shanty Hollow Brook, 1990 Survey.* New York State Department of Environmental Conservation Report, Albany, New York.

Bode, R.W., M.A. Novak, and L.E. Abele. 1990b. *Biological Stream Assessment: Wawayanda Creek, 1989 Survey.* New York State Department of Environmental Conservation Report, Albany, New York.

Bode, R.W., M.A. Novak, and L.E. Abele. 1990c. *Biological Impairment Criteria for Flowing Waters in New York State.* New York State Department of Environmental Conservation Report, Albany, New York.

Bode, R.W., M.A. Novak and L.E. Abele. 1992. *Biological Stream Assessment: Cayadutta Creek, 1992 Survey.* New York State Department of Environmental Conservation Report, Albany, New York.

Bolton, M.J. *In Preparation.* Guide to the identification of larval Chironomidae (Diptera) in the temperate eastern Nearctic north of Florida. Ohio EPA, Division of Surface Water, Ecological Assessment Section, Columbus, Ohio.

Bovee, K.D. 1982. *A Guide to Stream Habitat Analysis Using the Instream Flow Incremental Methodology.* U.S. Fish and Wildlife Service Flow Information Paper No. 12, FWS/OBS-82/26, Ft. Collins, Colorado.

Bovee, K.D. 1986. *Development and Evaluation of Habitat Suitability Criteria for Use in Instream Flow Incremental Methodology.* U.S. Fish and Wildlife Service, Washington, D.C.

Boyle, T.P., G.M. Smillie, J.C. Anderson, and D.P. Beeson. 1990. A sensitivity analysis of nine diversity and seven similarity indices. *Journal of the Water Pollution Control Federation* 62: 749–762.

Bramblett, R.G. and K.D. Fausch. 1991. Variable fish communities and the Index of Biotic Integrity in a western Great Plains river. *Transactions of the American Fisheries Society* 120: 752–769.

Brinkhurst, R.O. 1965. The biology of the tubificidae with special reference to pollution. Pages 57–66 in C.M. Tarzwell (editor). *Biological Problems in Water Pollution, Third Seminar.* U.S. Department of Health, Education, and Welfare, Public Health Service, Cincinnati, Ohio.

Brinkhurst, R.O. 1969. Discussion. [Comments on the paper by Tümpling, Wolf von. Suggested classification of water quality based on biological characteristics. pp. 279–290.] Pages 287–288 in S.H. Jenkins (editor). *Advances in Water Pollution Research.* Proceedings of the Fourth International Association on Water Pollution Research Conference, Prague, April 21–25, 1969. Pergamon Press Ltd., London.

Brooks, R.P., M.J. Croonquist, E.T. D'Silva, J.E. Gallagher, and D.E. Arnold. 1991. Selection of biological indicators for integrating assessments of wetland, stream, and riparian habitats. Pages 81–89 in *Biological Criteria: Research and Regulation, Proceedings of a Symposium, 12–13 December 1990, Arlington, Virginia.* EPA-440-5-91–005. U.S. Environmental Protection Agency, Office of Water, Washington, DC.

Brown, D.E. and C.H. Lowe. 1980. *Biotic Communities of the Southwest.* Map (scale 1:1,000,000). General Technical Report RM-78, U.S. Department of Agriculture, Forest Service, Fort Collins, Colorado.

Brown, P. 1972. *Aquatic Dryopoid Beetles (Coleoptera) of the United States.* Biota of Freshwater Ecosystems Identification Manual No. 6. U.S. Environmental Protection Agency, Washington, D.C.

Brussock, P.P., A.V. Brown, and J.C. Dixon. 1985. Channel form and stream ecosystem models. *Water Resource Bulletin* 21: 859–866.

Buikema, A.L., Jr. and J.R. Voshell, Jr. 1993. Toxicity studies using freshwater benthic macroinvertebrates. Pages 344–398 in D.M. Rosenberg and V.H. Resh (editors). *Freshwater Biomonitoring and Benthic Macroinvertebrates*. Chapman and Hall, New York.

Burch, J.B. 1982. *Freshwater Snails (Mollusca: Gastropoda) of North America*. EPA-600–3-82–026. U.S. Environmental Protection Agency, Environmental Monitoring and Support Laboratory, Cincinnati, Ohio.

Burgess, R.L. 1980. The national biological monitoring inventory. Pages 153–165 in D.L. Worf (editor). *Biological Monitoring for Environmental Effects*. Lexington Books, Lexington, Massachusetts.

Burks, B.O. 1953. The mayflies or Ephemeroptera of Illinois. *Bulletin of the Illinois Natural History Survey Division* 26: 1–216.

Burr, B.M. and W.L. Warren, Jr. 1986. *A Distributional Atlas of Kentucky Fishes*. Kentucky Nature Preserves Commission Scientific and Technical Series 4. Frankfort, Kentucky.

Burton, T.A., W.H. Clark, G.W. Harvey, and T.R. Maret. 1991. Development of sediment criteria for the protection and propagation of salmonid fishes. Pages 142–144 in *Biological Criteria: Research and Regulation, Proceedings of a Symposium Arlington, Virginia, 12 December 1990*. EPA-440–5-91–005. U.S. Environmental Protection Agency, Office of Water, Washington, D.C.

Butkus, S.R. 1989. Pilot program for water quality data collection by volunteers. Tennessee Valley Authority.

Buzan, D. 1992. The Texas Water Commission in partnership with citizens. Pages 26–27 J. Douherty (editor). *Proceedings of the Third National Citizens' Volunteer Water Monitoring Conference*. EPA 841-R-92–004. U.S. Environmental Protection Agency, Office of Water, Washington, D.C.

Byrne, J. 1992. Fighting a proposed river reclassification. *Volunteer Monitor* 4: 6–7.

Cain, L.P. 1978. *Sanitation Strategy for a Lake Front Metropolis. The Case of Chicago*. Northern Illinois University Press, DeKalb, Illinois.

Cairns, J., Jr. 1974. Indicator species vs. the concept of community structure as an index of pollution. *Water Resources Bulletin* 10: 338–347.

Cairns, J., Jr. 1977. Quantification of biological integrity. Pages 171–187 in R.K. Ballentine and L.J. Guarraia (editors). *The Integrity of Water. Proceedings of a Symposium*. March 10–12, 1975. U.S. Environmental Protection Agency, Washington, D.C.

Cairns, J., Jr. 1986. Freshwater. In *Proceedings of the Workshop on Cumulative Environmental Effects: A Binational Perspective*. CEARC, Ottawa, Ontario and NRC, Washington, D.C.

Cairns, J., Jr., D.W. Albaugh, F. Bussey, and M.D. Chaney. 1968. The sequential comparison index — a simplified method for non-biologists to estimate relative differences in biological diversity in stream pollution studies. *Journal of the Water Pollution Control Federation* 40: 1607–1613.

Cairns, J. Jr. and K.L. Dickson. 1971. A simple method for the biological assessment of the effects of water discharges on aquatic bottom-dwelling organisms. *Journal of the Water Pollution Control Federation* 41: 755–772.

Cairns, J., Jr. and P.V. McCormick. 1992. Developing an ecosystem-based capability for ecological risk assessments. *The Environmental Professional* 14: 186–196.

Cairns, J., Jr., P.V. McCormick, and B.R. Niederlehner. 1993. A proposed framework for developing indicators of ecosystem health. *Hydrobiologia* 263: 1–44.

Cairns, J., Jr. and J.R. Pratt. 1993. A history of biological monitoring using benthic macroinvertebrates. Pages 10–27 in D.M. Rosenberg and V.H. Resh (editors). *Freshwater Biomonitoring and Benthic Macroinvertebrates*. Chapman & Hall, Inc., New York.

Cairns, M.A. and R.T. Lackey. 1992. Biodiversity and management of natural resources: the issues. *Fisheries* 17(3): 6–10.

Carlson, R.E. 1977. A trophic state index for lakes. *Limnology and Oceanography* 22: 361–368.

Carson, R. 1963. *Silent Spring*. Houghton Miflin, Boston, Massachusetts.

Carlson, C.A. and R.T. Muth. 1989. The Colorado River: lifeline of the American Southwest. *Canadian Special Publication of Fisheries and Aquatic Sciences* 106: 220–239.

Cattaneo, A. and G. Roberge. 1991. Efficiency of a brush sampler to measure periphyton in streams and lakes. *Canadian Journal of Fisheries and Aquatic Sciences* 48: 1877–1881.

CEQ (Council on Environmental Quality). 1993. *Incorporating Biodiversity Considerations into Environmental Impact Analysis under the National Environmental Policy Act*. Executive Office of the President, Washington, D.C.

Chadwick, E. 1842. *On an Inquiry into the Sanitary Condition of the Labouring Population of Great Britain. London.* Printed by W. Clowesand Sons, Stamford Street, for Her Majesty's Stationery Office, London.

Chapman, D.W. 1988. Critical review of variables used to define effects of fines in redds of large salmonids. *Transactions of the American Fisheries Society* 117: 1–20.

Chapman, P.M., L.M. Churchland, P.A. Thomson, and E. Michnowsky. 1980. Heavy metal studies with oligochaetes. Pages 477–502 in R.O. Brinkhurst and D.G. Cook (editors). *Aquatic Oligochaete Biology.* Plenum Publishing, New York.

Charles, D.F. and J.P. Smol. 1994. Long-term chemical changes in lakes: Quantitative inferences from biotic remains in sediment record. Pages 3–31 in L. A. Baker (editor). *Environmental Chemistry of Lakes and Reservoirs.* Advances in Chemistry Series 237, American Chemical Society, Washington, D.C.

Chesbrough, E.S. 1858. *Report of the Results of Examinations Made in Relation to Sewerage in Several European Cities, in the Winter of 1856–57.* Report of Chief Engineer. Board of Sewerage Commissioners Office, Chicago, Illinois.

Chessman, B.C. 1986. Diatom flora of an Australian river system: spatial patterns and environmental relationships. *Freshwater Biology* 16: 805–819.

Chutter, F.M. 1972. An empirical biotic index of the quality of water in South African streams and rivers. *Water Research* 6: 19–30.

Clarke, S.E., D. White, and A.L. Schaedel. 1991. Oregon, USA, ecological regions and subregions for water quality management. *Environmental Management* 15: 847–856.

Cleary, E.J. 1955. Aquatic life water quality criteria. First progress report of the aquatic life advisory committee of the Ohio River Valley Sanitation Commission. *Sewage and Industrial Wastes* 27: 321–331.

Clements, F.E. 1916. *Plant Succession: An Analysis of the Development of Vegetation.* Carnegie Institute, Washington, D.C.

Cleve, P.T. 1899. Postglaciala bildningarnas klassifikation pa grund af deras fossila diatomaceer. Pages 59–61 in N.O. Holst (editor). *Bidrag till kännedomen om Östersöns och Bottniska Vikens postglaciala geologi.* Sveriges Geol., Undersökning, C. No. 180.

Clifford, H.F. 1992. *Aquatic Invertebrates of Alberta.* University of Alberta Press, Edmonton, Alberta, Canada.

Cochran, W.G. 1977. *Sampling Techniques,* 3rd ed., John Wiley & Sons, New York.

Cohn, F. 1853. Über lebendige Organismen im Trinkwasser. *Zeitschrift für klinische Medicin von Dr. Günsberg* 4: 229–237.

Colburn, T.E., A. Davidson, S.N. Green, R.A Hodge, C.I. Jackson, and R.A. Liroff. 1990. *Great Lakes, Great Legacy?* The Conservation Foundation, Washington, D.C.

Colby, P.J., P.A. Ryan, D.H. Schupp, and S.L. Serns. 1987. Interactions in north-temperate lake fish communities. *Canadian Journal of Fisheries and Aquatic Sciences* 44 (Supplement 2): 104–128.

Cole, G.A. 1975. *Textbook of Limnology.* Mosby. Saint Louis, Missouri.

Converse, J.M. 1987. *Survey Research in the United States: Roots and Emergence, 1890–1960.* University of California Press, Berkeley, California.

Cooper, S.D. and L. Barmuta. 1993. Field experiments in biomonitoring. Pages 399–441 in D. M. Rosenberg and V. H. Resh (editors). *Freshwater Biomonitoring and Benthic Macroinvertebrates.* Chapman and Hall, New York.

Cotter, J. and J. Nealon. 1987. *Area Frame Design for Agricultural Surveys.* Area Frame Section, Research and Applications Division, National Agricultural Statistics Service, U.S. Department of Agriculture.

Courtemanch, D.L. and S.P. Davies. 1987. A coefficient of community loss to assess detrimental change in aquatic communities. *Water Research* 21: 217–222.

Courtemanch, D.L., S.P. Davies, and E.B. Laverty 1989. Incorporation of biological information in water quality planning. *Environmental Management* 13: 35–41.

Courtney, W.R., Jr. and D.A. Hensley. 1980. Special problems with monitoring exotic species. Pages 281–307 in C.H. Hocutt and J.R. Stauffer, Jr. (editors). *Biological Monitoring of Fish.* Lexington Books, Lexington, Kentucky.

Covich, A., D. Allan, T. Arsuffi, A. Beuke, B. Cushing, C. Dahm, P. Firth, S. Fisher, M. Gurtz, J. Meyer, P. Mulholland, M. Molles, J. Morse, R. Naiman, M. Palmer, L. Poff, S. Reice, A. Sheldon, L. Smock, A. Steinman, B. Wallace, and J. Webster. 1992. Report of the NABS FWI (Freshwater Imperative) working group. *Bulletin of the North American Benthological Society* 9: 266–271.

Cowardin, L.M., V. Carter, F.C. Golet, and E.T. LaRoe. 1979. *Classification of Wetlands and Deepwater Habitats of the United States*. FWS/OBS-79/31. U.S. Fish and Wildlife Service, Department of the Interior, Washington, D.C.

Crowley, J.M. 1967. Biogeography. *Canadian Geographer* 11: 312–326.

Crumby, W.D., M.A. Webb, F.J. Bulow and H.J. Cathey. 1990. Changes in the biotic integrity of a river in north-central Tennessee. *Transactions of the American Fisheries Society* 119: 885–893.

Cumming, H.S. 1916. *Investigation of the Pollution and Sanitary Conditions of the Potomac Watershed (With Special Reference to Self Purification and the Sanitary Condition of Shellfish in the Lower Potomac River)*. Hygienic Laboratory Bulletin No. 104, U.S. Public Health Service, Washington, D.C.

Cummins, K.W. 1962. An evaluation of some techniques for the collection and analysis of benthic samples with special emphasis on lotic waters. *American Midland Naturalist* 67: 477–504.

Cummins, K.W. 1988. Rapid bioassessment using functional analysis of running water invertebrates. Pages 49–54 in T.P. Simon, L.L. Holst, and L.J. Shepard (editors). *Proceedings of the First National Workshop on Biological Criteria*. EPA-905–9-89–003. U.S. Environmental Protection Agency, Region 5, Chicago, Illinois.

Cummins, K.W. and M.A. Wilzbach. 1985. *Field Procedures for Analysis of Functional Feeding Groups of Stream Macroinvertebrates*. Contribution 1611. Appalachian Environmental Laboratory, University of Maryland, Frostburg, Maryland.

Cunningham, P.A., J.M. McCarthy, and D. Zeitlin. 1990. *Results of the 1989 Census of State Fish/Shellfish Consumption Advisory Programs*. Research Institute, Research Triangle Park, North Carolina.

Cusimano, R.F., J.P. Baker, W.J. Warren-Hicks, V. Lesser, W.W. Taylor, M.C. Fabrizio, D.B. Hayes, and B.P. Baldigo. 1989. *Fish Communities in Lakes in Subregion 2B (Upper Peninsula of Michigan) in Relation to Lake Acidity*. EPA-600–3-89–021. U.S. Environmental Protection Agency, Office of Research and Development, Washington, D.C.

Dahl, T.E. and C.E. Johnson. 1991. *Status and Trends of Wetlands in the Conterminous United States, 1970s to 1980s*. U.S. Department of the Interior, Fish and Wildlife Service, Washington, D.C.

Daneke, G.A. 1982. The future of environmental protection and reflection on the difference between planning and regulating. *Public Administration Review* 42: 227–233.

Darwin, C.E. 1859. *On the Origin of Species by Natural Selection, or the Preservation of Favoured Races in the Struggle for Life*. John Murray, London. 502 p. (Reprinted in 1975 by Cambridge University Press, New York.)

Davies, S.P., L. Tsomides, D.L. Courtemanch, and F. Drummond. 1993. *Maine Biological Monitoring and Biocriteria Development Program*. Maine Department of Environmental Protection, Bureau of Water Quality Control, Division of Environmental Evaluation and Lake Studies, Augusta, Maine.

Davis, W.S. (editor). 1990a. *Proceedings of the 1990 Midwest Pollution Control Biologists Meeting*. EPA-905–9-90–005. U.S. Environmental Protection Agency, Region 5, Environmental Sciences Division, Chicago, Illinois.

Davis, W.S. 1990b. Foreword: A historical perspective on regulatory biology. Pages i-xii in W.S. Davis (editor). *Proceedings of the 1990 Midwest Pollution Control Biologists Meeting, Chicago, IL, April 10–13, 1990*. EPA-905–9-90–005. U.S. Environmental Protection Agency, Region 5, Environmental Sciences Division, Chicago, Illinois.

Davis, W.S. and A. Lubin. 1989. Statistical validation of Ohio EPA's Invertebrate Community Index. Pages 23–32 in W.S. Davis and T.P. Simon (editors). *Proceedings of the 1989 Midwest Pollution Control Biologists Meeting, 14–17 February 1989*. EPA-905–9-89–007. U.S. Environmental Protection Agency, Region 5, Environmental Sciences Division, Chicago, Illinois.

Davis, W.S. and T.P. Simon. 1989. Sampling and data evaluation requirements for fish and macroinvertebrate communities. Pages 89–97 in T.P. Simon, L.L. Holst and L.J. Shepard (editors). *Proceedings of the First National Workshop on Biological Criteria, Lincolnwood, Illinois, December 2–4, 1987*. EPA 905–9-89–003. U.S. Environmental Protection Agency, Region 5, Environmental Sciences Division, Chicago, Illinois.

Davis, W.S., C. Weber, D. Duff, and F. Mangum. 1992. River and stream monitoring techniques. Pages 148–151 in J. Douherty (editor). *Proceedings of the Third National Citizens' Volunteer Water Monitoring Conference*. EPA 841-R-92–004. U.S. Environmental Protection Agency, Office of Water, Washington, D.C.

De Pauw, N. and G. Vanhooren. 1983. Method for biological quality assessment of watercourses in Belgium. *Hydrobiologia* 100: 153–168.

Denevan, W.M. 1992. The pristine myth: the landscape of the Americas in 1492. *Annals of the Association of American Geographers* 82: 369–385.

DeNicola, D.M., K.D. Hoagland, and S.C. Roemer. 1992. Influences of canopy cover on spectral irradiance and periphyton assemblages in a prairie stream. *Journal of the North American Benthological Society* 11: 391–404.

Descy, J.P. 1979. A new approach to water quality estimation using diatoms. *Nova Hedwigia* 64: 305–323.

Descy, J.P. and M. Coste. 1990. Utilisation des diatomées benthiques por l'évaluation de la qualité des eaux courantes, Rapport Final Contract CEE B-71–23. Unpublished

DeShon, J.E., D.O. McIntyre, J.T. Freda, C.D. Webster, and J.P. Abrams. 1980. *Volume IV, Biological Evaluations: Ohio 305(b) Report.* Ohio EPA, Division of Surveillance and Water Quality Standards, Columbus, Ohio.

Dice, L.R. 1943. *The Biotic Provinces of North America.* University of Michigan Press, Ann Arbor, Michigan.

Dickson, K.L., W.T. Waller, J.H. Kennedy and L.P. Ammann. 1992. Assessing the relationship between ambient toxicity and instream biological response. *Environmental Toxicology and Chemistry* 11: 1307–1322.

Dilley, M.A. 1992. A comparison of the results of a volunteer stream quality monitoring program and the Ohio EPA's biological indices. Pages 61–72 in T.P. Simon and W.S. Davis (editors). *Proceedings of the 1991 Midwest Pollution Control Biologists Meeting: Environmental Indicators: Measurement and Assessment Endpoints.* EPA-905-R-92–003, U.S. Environmental Protection Agency, Region 5, Chicago, Illinois.

Dionne, M. and J. R. Karr. 1992. Ecological monitoring of fish assemblages in Tennessee River reservoirs. Pages 259–281 in D.H. McKenzie, D.E. Hyatt, and V.J. McDonald (editors). *Ecological Indicators, Volume 1.* Elsevier Applied Science, New York.

Dixit, S.S., J.P. Smol, J.C. Kingston, and D.F. Charles. 1992. Diatoms: powerful indicators of environmental change. *Environmental Science and Technology* 26: 23–33.

DNREC (Delaware Department of Natural Resources). 1992a. Appendix C [Delaware 1992 305(b) Report], *Biological Integrity and Habitat Quality of Nontidal Streams of Kent and Sussex Counties, Delaware.* Delaware Department of Natural Resources and Environmental Control, Dover, Delaware.

DNREC (Delaware Department of Natural Resources). 1992b. *Regional Reference Site Selection for Classifying Nontidal Coastal Plain Streams in Delaware.* Delaware Department of Natural Resources and Environmental Control, Dover, Delaware.

Doppelt, B., M. Scurlock, C. Frissell, and J. Karr. 1993. *Entering the Watershed: A New Approach to Save America's River Ecosystems.* Island Press, Washington, D.C.

Doudoroff, P. and C.E. Warren. 1957. Biological indices of water pollution with particular reference to fish populations. Pages 144–163 in C.M. Tarzwell (editor). *Biological Problems in Water Pollution, Transactions of the 1956 Seminar.* U.S. Public Health Service, Cincinnati, Ohio.

Duff, D.A., F.A. Mangum, and R. Maw (editors). 1989. *Fisheries Survey Handbook.* U.S. Department of Agriculture — Forest Service, Intermountain Region, Ogden, Utah.

Dunham, D. and A. Collotzi. 1975. *The Transect Method of Stream Survey.* U.S. Department of Agriculture — Forest Service, Intermountain Region, Ogden, Utah.

Dycus, D.L. and D.L. Meinert. 1993. *Monitoring and Evaluation of Aquatic Resource Health and Use Suitability in Tennessee Valley Authority Reservoirs.* Tennessee Valley Authority, Resource Group, Water Management, Chattanooga, Tennessee.

EA Engineering, Science, and Technology. 1993. 1991 Ohio River ecological research program. Prepared for American Electric Power Company, Ohio Edison Company, OhioValley Electric Corporation, Cincinnati Gas & Electric Compay, and Tennessee Valley Authority. Deerfield, Illinois.

Eagleson, K.W., Lenat, D.L., Rusley, L.W. and Winborne, R.B. 1990. Comparison of measured instream biological responses and responses predicted using the *Ceriodaphnia dubia* chronic toxicity test. *Environmental Toxicology and Chemistry* 9: 1019–1028.

Eaton, L.E. and D.R. Lenat. 1991. Comparison of a rapid bioassessment method with North Carolina's qualitative macroinvertebrate collection method. *Journal of the North American Benthological Society* 10: 335–338.

Ebel, W.J., C.D. Becker, J.W. Mullan, and H.L. Raymond. 1989. The Columbia River — toward a holistic understanding. Pages 205–219 in D.P. Dodge (editor). *Proceedings of the Large River Symposium.* Canadian Special Publication Fisheries and Aquatic Sciences 106.

Edmunds, G.F., S.L. Jensen and L. Berner. 1976. *The Mayflies of North and Central America.* University of Minnesota Press, Minneapolis, Minnesota.

Einstein, H.A. 1972. Sedimentation (suspended solids). Pages 309–318 in R.T. Oglesby, C.A. Carlson, and J. McCann (editors). *River Ecology and Man*. Academic Press, New York.

Ellett, K.K. 1988. *An Introduction to Water Quality Monitoring Using Volunteers: A Handbook for Coordinators*. Alliance for the Chesapeake Bay, Baltimore, Maryland.

Ely, E. 1991. Benthic macroinvertebrate monitoring in estuaries. *Volunteer Monitor* 3: 5.

Emerson, J.W. 1971. Channelization, a case study. *Science* 325–326.

Epler, J.H. 1987. Revision of the Nearctic *Dicrotendipes* Kieffer, 1913 (Diptera: Chironomidae). *Evolutionary Monographs* 9: 1–102.

Evenson, W.E., S.R. Rushforth, J.D. Brotherson, and N. Fungladda. 1981. The effects of selected physical and chemical factors on attached diatoms in the Uintah Basin of Utah, U.S.A. *Hydrobiologia* 83: 325–330.

Everest, F.H., F.B. Lotspeich, and W.R. Meehan. 1981. New perspectives on sampling, analysis, and interpretation of spawning gravel quality. Pages 325–333 in N.B. Armantrout (editor). *Acquisition and Utilization of Aquatic Habitat Inventory Information*. Portland, Oregon, 28 October 1981. Western Division, American Fisheries Society, Bethesda, Maryland.

Fajen, O.F. and R.E. Wehnes. 1982. Missouri's method of evaluating stream habitat. Pages 117–123 in N. B. Armantrout (editor). *Acquisition and Utilization of Aquatic Habitat Inventory Information*. Western Division, American Fisheries Society, Bethesda, Maryland.

Fausch, K.D., J.R. Karr, and P.R. Yant. 1984. Regional application of an index of biotic integrity based on stream fish communities. *Transactions of the American Fisheries Society* 113: 39–55.

Fausch, K.D., J. Lyons, J.R. Karr, and P.L. Angermeier. 1990. Fish communities as indicators of environmental degradation. Pages 123–144 in S.M. Adams (editor). *Biological Indicators of Stress in Fish*. American Fisheries Society Symposium 8, Bethesda, Maryland.

Fenneman, N.M. 1946. *Physical Divisions of the United States*. Map (scale 1:7,000,000). U.S. Geological Survey, Reston, Virginia.

Firehock, K. 1992. Using volunteers for statewide networks. Pages 78–80 in J. Douherty (editor). *Proceedings of the Third National Citizens' Volunteer Water Monitoring Conference*. EPA 841-R-92-004. U.S. Environmental Protection Agency, Office of Water, Washington, D.C.

Fisher, T.R. 1989. *Application and Testing of Indices of Biotic Integrity in Northern and Central Idaho Headwater Streams*. Master of Science Thesis, University of Idaho, Moscow, Idaho.

Fitter, R.S.R. 1945. London's natural history. *New Naturalist (London)* 3: 1–282.

Fjerdingstad, E. 1960. Water pollution estimated by biological measures. *Nordisk Hyg. Tidskr* 41: 149–196. [Cited in Wilber, Charles G. 1969. *The Biological Aspects of Water Pollution*. Charles C. Thomas Publisher, Springfield, Illinois.]

Fjerdingstad, E. 1964. Pollution of streams estimated by benthal phytomicroorganisms. I. *International Revue der Gesamten Hydrobiologie* 49: 63–131.

Fjerdingstad, E. 1965. Some remarks on a new saprobic system. Pages 232–235 in C.M. Tarzwell (editor). *Third Seminar on Biological Problems in Water Pollution*. U.S. Public Health Service, Cincinnati, Ohio.

Flinn, D.W. (editor). 1965. *Report on the Sanitary Condition of the Laboring Population of Gt. Britain, 1842, by Edwin Chadwick*. Edinburgh at the University Press.

Flint, O.S. 1962. Larvae of the caddisfly genus *Rhyacophila* in eastern North America (Trichoptera: Rhyacophilidae). *Proceedings of the United States National Museum* 113: 465–493.

Forbes, S.A. 1887. The lake as a microcosm. *Bulletin of the Peoria Scientific Association*. (Reprinted 1925. *Illinois State Laboratory of Natural History Bulletin* 15: 537–550).

Forbes, S.A. 1895. Illinois State Laboratory of Natural History, biennial report of the director. 1893–1895. *Illinois Fish Commissioners Report for 1892–1894*: 39–52.

Forbes, S.A. 1928. The biological survey of a river system — its objects, methods and results. *Illinois Natural History Survey Bulletin* 17: 277–284.

Forbes, S.A. and R.E. Richardson. 1908. *The Fishes of Illinois*. Illinois State Laboratory of Natural History, Urbana, Illinois.

Forbes, S.A. and R.E. Richardson. 1913. Studies on the biology of the upper Illinois River. *Illinois State Laboratory of Natural History Bulletin* 9: 1–48.

Ford, J. 1989. The effects of chemical stress on aquatic species composition and community structure. Pages 99–144 in S.A. Levin, M.A. Harwell, J.R. Kelly, and K.D. Kimball (editors). *Ecotoxicology: Problems and Approaches*. Springer-Verlag, New York.

Fore, L.S., J. R. Karr and L. L. Conquest. 1994. Statistical properties of an index of biotic integrity used to evaluate water resources. *Canadian Journal of Fisheries and Aquatic Sciences* 51: 1077–1087.

Freedman, D., R. Pisan, R. Purves, and A. Adhikari. 1991. *Statistics*, Second Edition. W.W. Norton & Company, Inc., New York.

Frey, D.G. 1977. Biological integrity of water — an historical approach. Pages 127–140 in R.K. Ballantine and L.J. Guarraia (editors). *The Integrity of Water. Proceedings of a Symposium, March 10–12, 1975.* U.S. Environmental Protection Agency, Washington, D.C.

Friedrich, G. 1990. Eine Revision der Saprobiensystem (Revision of the saprobien system). *Zeitschrift für Wasser und Abwasser Forschung* 23: 141–52.

Friedrich, G., D. Chapman, and A. Beim. 1992. The use of biological material. Pages 171–238 in D. Chapman (editor). *Water Quality Assessments: A Guide to the Use of Biota, Sediments and Water in Environmental Monitoring.* Chapman and Hall, London.

Frissell, C.A. 1993. Topology of extinction and endangerment of native fishes in the Pacific Northwest and California (USA). *Conservation Biology* 7: 342–354.

Frissell, C.A., W.L. Liss, C.E. Warren, and M.C. Hurley. 1986. A hierarchial framework for stream habitat classification, viewing streams in a watershed context. *Environmental Management* 10: 199–214.

Fritz, S.C. 1990. Twentieth century salinity and water level fluctuations in Devils Lake, North Dakota: a test of a diatom based transfer function. *Limnology and Oceanography* 35: 1771–1781.

Frost, S.L. and W.J. Mitsch. 1989. Resource development and conservation history along the Ohio River. *Ohio Journal of Science* 89: 143–152.

Funk, J.L. 1954. Movement of stream fishes in Missouri. *Transactions of the American Fishery Society* 85: 39–57.

Gallant, A.L., T.R. Whittier, D.P. Larsen, J.M. Omernik, and R.M. Hughes. 1989. *Regionalization as a Tool for Managing Environmental Resources.* EPA-600–3-89–060. U.S. Environmental Protection Agency, Environmental Research Laboratory, Corvallis, Oregon.

Gammon, J.R. 1976. *The Fish Population of the Middle 340 km of the Wabash River.* Purdue University Water Resources Research Center Technical Report 86, LaFayette, Indiana.

Gammon, J.R., C.W. Gammon, and M.K. Schmid. 1990. Land use influences on fish communities in central Indiana streams. Pages 111–120 in W.S. Davis (editor). *Proceedings of the 1990 Midwest Pollution Control Biologists Meting.* EPA 905–90–005. U.S. Environmental Protection Agency, Region 5, Environmental Sciences Division, Chicago, Illinois.

Gammon, J.R., M.D. Johnson, C.E. Mays, D.A. Schiappa, W.L. Fisher, and B.L. Pearman. 1983. *Effects of Agriculture on Stream Fauna in Central Indiana.* EPA-600-S3–83–020. U.S. Environmental Protection Agency, Office of Research and Development, Washington, D.C.

Gammon, J.R., A. Spacie, J.L. Hamelink, and R.L. Kaesler. 1981. Role of electrofishing in assessing environmental quality of the Wabash River. Pages 307–324 in J. Bates and C. I. Weber (editors). *Ecological Assessments of Effluent Impacts on Communities of Indigenous Aquatic Organisms.* ASTM STP 730. Philadelphia, Pennsylvania.

Geise, J.W. and W.E. Keith. 1989. The use of fish communities in ecoregion reference streams to characterize the stream biota in Arkansas waters. Pages 26–41 in T.P. Simon, L.L. Holst, and L.J. Shepard (editors). *Proceedings of the First National Workshop on Biological Criteria.* EPA 905–9-89–003. U.S. Environmental Protection Agency, Region 5, Environmental Sciences Division, Chicago, Illinois.

Gerking, S.D. 1953. Evidence for the concept of a home range and territory in stream fishes. *Ecology* 34: 347–365.

Gerking, S.D. 1959. The restricted movement of fish populations. *Biological Review* 34: 221–242.

Gerritsen, J., M.T. Barbour, G.R. Gibson, E.H. Liu, P.J. Peak, J.B. Stribling, and J.S. White. 1994a. Lake and reservoir bioassessment and biocriteria. (Internal Review Draft). U.S. Environmental Protection Agency, Office of Wetlands, Oceans, and Watersheds and Office of Science & Technology, Washington, D.C.

Gerritson, J. and M.L. Bowman. 1994. *Periphytic Diatom Communities of High Elevation Rocky Mountain Lakes: Characterization of Reference Conditions and Gradient Analysis.* Prepared for U.S. Environmental Protection Agency, Office of Water, Washington, D.C. and Montana Department of Health and Environmental Sciences, Helena, Montana. Tetra Tech, Inc., Owings Mills, Maryland.

Gerritsen, J., J. Green, and R. Preston. 1994b. Establishment of regional reference conditions for stream biological assessment and watershed management. Pages 797–801 in *Proceedings: Watershed 93. A National Conference on Watershed Management.* EPA-840-R-94-002. U.S. Environmental Protection Agency, Washington, D.C.

Ghetti, P.F. and O. Ravera. 1994. European perspective on biological monitoring. Pages 31–46 in S.L. Loeb and A. Spacie (editors). *Biological Monitoring of Aquatic Systems*. Lewis Publishers, Boca Raton, Florida.

Gibson, G.R. (editor). 1994. *Biological Criteria: Technical Guidance for Streams and Small Rivers*. EPA-822-B-94-001. U.S. Environmental Protection Agency, Office of Science & Technology, Washington, D.C.

Gleason, H.A. 1926. The individualistic concept of the plant association. *Bulletin of the Torrey Botanical Club* 53: 7–26.

Goede, R.W. 1988. *Fish Health/Condition Assessment Procedures*. Utah Division of Wildlife Resources and Fisheries Publication, Experiment Station, Logan, Utah.

Goldstein, R.M., T.P. Simon, P.A. Bailey, M. Ell, K. Schmidt and J. W. Emblom. 1994. Proposed metrics for the index of biotic integrity for the streams of the Red River of the North basin. *Transactions of the North Dakota Academy of Science* (in press).

Golledge, R.G., et al., 1982. Commentary on "The highest form of the geographers's art." *Annals of the Association of American Geographers* 7: 557–558.

Gordon, N.D., T.A. McMahon, and B.L. Finlayson. 1992. *Stream Hydrology: An Introduction for Ecologists*. John Wiley and Sons, Ltd, West Sussex, England.

Gorman, O.T. 1986. Assemblage organization of stream fishes: the effect of rivers on adventitious streams. *American Naturalist* 128: 611–616.

Gorman, O.T. 1987a. Fishes and aquatic insects of Nippersink Creek, McHenry County, Illinois. *Transactions of the Illinois Academy of Science* 80: 233–254.

Gorman, O.T. 1987b. A survey of the fishes and macroinvertebrates of some small streams in Cook, Lake, and DuPage Counties, Illinois. *Transactions of the Illinois Academy of Science* 80: 253–272.

Gorman, O.T. and J.R. Karr. 1978. Habitat structure and stream fish communities. *Ecology* 59: 507–515.

Gottfried, J. and V.H. Resh. 1979. Developing modules for field exercises in aquatic entomology. Pages 81–93 in V.H. Resh and D.M. Rosenberg (editors). *Innovative Teaching in Aquatic Entomology*. Canadian Special Publications in Fisheries and Aquatic Sciences 43.

Government of Canada 1991. *The State of Canada's Environment*. Canada Communication Group — Publishing. Ottawa, Canada.

Gray, J.S. 1989. Effects of environmental stress on species rich assemblages. *Biological Journal of the Linnean Society* 37: 19–32.

Green, R.H. 1979. *Sampling Design and Statistical Methods for Environmental Biologists*. John Wiley & Sons, New York.

Greenfield, D.W. and J.D. Rogner. 1984. An assessment of the fish fauna of Lake Calumet and adjacent wetlands, Chicago, Illinois: past, present, and future. *Transactions of the Illinois Academy of Science* 77: 77–93.

Gregory, S.V., F.J. Swanson, W.A. McKee, and K.W. Cummins. 1991. An ecosystem perspective of riparian zones. *Bioscience* 41: 540–551.

Griffith, G.E. and J.M. Omernik. 1991. *Alabama/Mississippi Project: Geographic Research Component: Ecoregion/Subregion Delineation and Reference Site Selection*. U.S. Environmental Protection Agency, Environmental Research Laboratory, Corvallis, Oregon.

Griffith, G.E., J.M. Omernik, S.M. Pierson, and C.W. Kiilsgaard. 1993a. *Massachusetts Regionalization Project*. U.S. Environmental Protection Agency, Environmental Research Laboratory, Corvallis, Oregon.

Griffith, G.E., J.M. Omernik, C. Rohm, and S.M. Pierson. 1993b. *Florida Regionalization Project*. U.S. Environmental Protection Agency, Environmental Research Laboratory, Corvallis, Oregon.

Grigg, D.B. 1967. Regions, models and classes. Pages 461–509 in R.J. Choley and P. Hagget (editors). *Models in Geography*. Methven and Company, London.

Grodhaus, G. 1987. *Endochironmus* Kieffer, *Tribelos* Townes, *Synendotendipes* new genus, and *Endotribelos* new genus (Diptera: Chironomidae) of the Nearctic region. *Journal of the Kansas Entomological Society* 60: 167–247.

Guhl, W. 1986. Bemerkungen ueber den Saprobitaetsgrad (Remarks on the saprobity index). *Limnologica* 17: 119–126.

Gurtz, M.E. 1994. Design considerations for biological components of the national water quality assessment (NAWQA) program. Pages 323–354 in S.L. Loeb and A. Spacie (editors). *Biological Monitoring of Aquatic Systems*. Lewis Publishers, Boca Raton, Florida.

Haddock, J.D. 1977. The biosystematics of the caddisfly genus *Nectopsyche* in North American with emphasis on the aquatic stages. *The American Midland Naturalist* 98: 382–421.

Hammond, E.H. 1970. *Classes of land surface form. Map (scale 1:7,500,000)*. Pages 62–63 in *The National Atlas of the United State of America*. U.S. Geological Survey, Washington, D.C.

Hanna, G.D. 1933. Diatoms of the Florida peat deposits. *Florida State Geological Survey 23rd-24th Annual Report* 1930–1932: 65–120.

Harris, T.L. and T.M. Lawrence. 1978. *Environmental Requirements and Pollution Tolerance of Trichoptera*. EPA-600–4-78–063. U.S. Environmental Protection Agency, Environmental Monitoring and Support Laboratory, Cincinnati, Ohio.

Hart, C.W., Jr. and S.L.H. Fuller (editors). 1974. *Pollution Ecology of Freshwater Invertebrates*. Academic Press, New York.

Hart, J.F. 1982a. The highest form of the geographer's art. *Annals of the Association of American Geographers* 72: 1–29.

Hart, J. F. 1982b. Comment in reply. *Annals of the Association of American Geographers* 72: 559.

Hart, J.F. 1983. More gnashing of false teeth. *Annals of the Association of American Geographers* 3: 441–443.

Hartig, J.H., and M.A. Zarull. 1992. Towards defining aquatic ecosystem health for the Great Lakes. *Journal of Aquatic Ecosystem Health* 1: 97–107.

Hartman, W.L. 1972. Lake Erie: effects of exploitation, environmental changes and new species on the fishery resources. *Journal of the Fisheries Research Board of Canada* 29: 899–912.

Hassall, A.H. 1850. *A Microscopic Examination of the Water Supplied to the Inhabitants of London and the Suburban Districts*. [Cited in Ingram, W.M., K.M. Mackenthun, and A.F. Bartsch. 1966. *Biological Field Investigative Data for Water Pollution Surveys*. Federal Water Pollution Control Administration, Washington, D.C.]

Hawkes, C.L., D.L. Miller and W.G. Layher. 1986. Fish ecoregions of Kansas: stream fish assemblage patterns and associated environmental correlates. *Environmental Biology of Fishes* 17: 267–279.

Hawkes, H.A. 1957. Biological aspects of river pollution. Pages 191–239 in L. Klein (editor). *Aspects of River Pollution*. Academic Press, New York.

Hawkes, H.A. 1975. River zonation and classification. Pages 312–374 in B.A. Whitton (editor). *River Ecology*. University of California Press, Berkeley, California.

Hawkins, C.P. et al., 1993. A hierarchical approach to classifying stream habitat features. *Fisheries* 18: 3–12.

Hayslip, G.A. 1993. *EPA Region 10 In-Stream Biological Monitoring Handbook (for Wadable Streams in the Pacific Northwest)*. EPA-910–9-92–013. U.S. Environmental Protection Agency-Region 10, Environmental Services Division, Seattle, Washington.

Hazard, J.W. and B.E. Law. 1989. *Forest Survey Methods Used in the USDA Forest Service*. EPA-600–3-89–065, NTIS PB89 220 594/AS. U.S. Environmental Protection Agency, Environmental Research Laboratory, Corvallis, Oregon.

Healey, R.G. 1983. Regional geography in the computer age: A further commentary on "The highest form of the geographer's art." *Annals of the Association of American Geographers* 73: 439–441.

Heede, B.H. and J.N. Rinne. 1990. Hydrodynamic and fluvial morphologic processes, implications for fisheries management and research. *North American Journal of Fisheries Management* 10: 249–268.

Heiskary, S.A. and C.B. Wilson. 1989. The regional nature of lake water quality across Minnesota: an analysis for improving resource management. *Journal of the Minnesota Academy of Sciences* 55: 71–77.

Helm, W.T. (editor). 1985. *Aquatic Habitat Inventory: Glossary of Stream Habitat Terms*. Western Division, Habitat Inventory Committee, American Fisheries Society, Bethesda, Maryland.

Henjum, M.G., J.R. Karr, D.L. Bottom, D.A. Perry, J.C. Bednarz, S.G. Wright, S.A. Beckwitt, and E. Beckwitt. 1994. Interim Protection for Late-Successional Forests, Fisheries, and Watersheds: National Forests East of the Cascade Crest, Oregon and Washington. The Wildlife Society. Bethesda, Maryland.

Herlihy, A.T., P.R. Kaufmann, and M.E. Mitch. 1991. Stream chemistry in the Eastern United States. 2. Current sources of acidity in acidic and low acid neutralizing capacity streams. *Water Resources Research* 27: 629–642.

Herricks. E.E. and D.J. Schaeffer. 1985. Can we optimize biomonitoring? *Environmental Management* 9: 487–492.

Herz, M. 1992. Compliance monitoring. *Volunteer Monitor* 4:8.

Herz, M., M. Podlich, and J. Tiedmann. 1992. Enforcement and compliance monitoring. Pages 88–92 in J. Douherty (editor). *Proceedings of the Third National Citizens' Volunteer Water Monitoring Conference*. EPA 841-R-92–004. U.S. Environmental Protection Agency, Office of Water, Washington, D.C.

Hesse, L. W., J.C. Schmulbach, J.M. Carr, K.D. Keenlyne, D.G. Unkenholz, J.W. Robinson, and G.E. Mestl. 1989. Missouri River fishery resources in relation to past, present, and future status. *Canadian Special Publication of Fisheries and Aquatic Sciences* 106: 352–371.

Hester, F.E. and J.S. Dendy. 1962. A multiple-plate sampler for aquatic macroinvertebrates. *Transactions of the American Fisheries Society* 91: 420–421.

Hildreth, S.P. 1848. *Pioneer History: Being an Account of the First Examinations of the Ohio Valley, and the Early Settlement of the Northwest Territory. Historical Society of Cincinnati.* H.W. Derby and Co., Cincinnati. [Cited on p. 19 in Trautman, M.B. 1981. *The Fishes of Ohio.* (Revised edition). Ohio State University Press, Columbus, Ohio.]

Hill, M.T., W.S. Platts, and R.L. Beschta. 1991. Ecological and geomorphological concepts for instream and out-of-channel flow requirements. *Rivers* 2: 198–210.

Hilsenhoff, W.L. 1977. *Use of Arthropods to Evaluate Water Quality of Streams.* Technical Bulletin No. 100, Wisconsin Department of Natural Resources, Madison, Wisconsin.

Hilsenhoff, W.L. 1982a. *Using a Biotic Index to Evaluate Water Quality in Streams.* Technical Bulletin No. 132, Wisconsin Department of Natural Resources, Madison, Wisconsin.

Hilsenhoff, W.L. 1982b. Aquatic insects of Wisconsin. Keys to Wisconsin genera and notes on biology, distribution and species. University of Wisconsin - Madison. *Publication of the Natural History Council* 2: 1–60.

Hilsenhoff, W.L. 1987. An improved biotic index of organic stream pollution. *The Great Lakes Entomologist* 20: 31–39.

Hilsenhoff, W.L. 1988. Rapid field assessment of organic pollution with a family-level biotic index. *Journal of the North American Benthological Society* 7: 65–68.

Hitchcock, S.W. 1974. Guide to the insects of Connecticut. Part VII. The Plecoptera or stoneflies of Connecticut. *State Geological and Natural History Survey of Connecticut Bulletin* 107: 1–262.

Hite, R. 1988. Overview of stream quality assessments and stream classification in Illinois. Pages 98–120 in T.P. Simon, L.L. Holst, and L.J. Shepard (editors). *Proceedings of the First National Workshop on Biological Criteria, Lincolnwood, Illinois.* EPA 905–9-89–003. U.S. Environmental Protection Agency, Region 5, Chicago, Illinois.

Hite, R.L. and B.A. Bertrand. 1989. *Biological Stream Characterization (BSC): A Biological Assessment of Illinois Stream Quality.* IEPA/WPC/89–275. Special Report Number 13, Illinois Environmental Protection Agency, Illinois State Water Plan Task Force, Springfield, Illinois.

Hocutt, C.H. and E.O. Wiley (editors). 1986. *The Zoogeography of North American Freshwater Fishes.* John Wiley and Sons, New York.

Hoefs, N.J., and T.P. Boyle. 1992. Contribution of fish community metrics to the index of biotic integrity in two Ozark rivers. Pages 283–303 in D.H. McKenzie, D.E. Hyatt, and V.J. McDonald (editors). *Ecological Indicators, Volume 1.* Elsevier Applied Science, New York.

Holdridge, L.R. 1959. *Mapa ecologico de Central America.* Segun Publicaciones de Unidad de Recursos Naturalos, Dept. de Asuntos Ecconmicos, Union Panamericana.

Holland, A.F. (editor). 1990. *Near Coastal Program Plan for 1990: Estuaries.* EPA-600–4-40–033. U.S. Environmental Protection Agency, Environmental Research Laboratory, Office of Research and Development, Narragansett, Rhode Island.

Holsinger, J.R. 1972. *The Freshwater Amphipod Crustaceans (Gammaridae) of North America.* Biota of Freshwater Ecosystems Identification Manual No. 5. U.S. Environmental Protection Agency, Washington, D.C.

Hoorman, J.J., L.C. Brown, and A.D. Ward. 1992. *Understanding Agricultural Drainage.* Ohio Cooperative Extensive Service, Columbus, Ohio.

Horton, R.E. 1945. Erosional developments of streams and their drainage basins: hydrophysical approach to quantitative morphology. *Bulletin of the Geological Society of America* 56: 275–370.

Hubbard, M.D. and W.L. Peters. 1978. *Environmental Requirements and Pollution Tolerance of Ephemeroptera.* EPA-600–4-78–061. U.S. Environmental Protection Agency, Environmental Monitoring and Support Laboratory, Cincinnati, Ohio.

Hubbs, C.L. 1933. Sewage treatment and fish life. *Sewage Works Journal* 5: 1033–1040.

Hubbs, C. 1982. *A Checklist of Texas Freshwater Fishes.* Tech. Series II. Texas Parks and Wildlife Department, Austin, Texas.

Hughes, R.M. 1985. Use of watershed characteristics to select control streams for estimating effects of metal mining wastes on extensively disturbed streams. *Environmental Management* 9: 253–262.

Hughes, R.M., J.H. Gakstatter, M.A. Shirazi, and J.M. Omernik. 1982. An approach for determining biological integrity in flowing waters. Pages 877–888 in T.B. Brann, L.O. House, and H.G. Lund (editors). *Inplace Resource Inventories: Principles and Practices, A National Workshop*. Society of American Foresters, Bethesda, Maryland.

Hughes, R.M. and J.R. Gammon. 1987. Longitudinal changes in fish assemblages and water quality in the Willamette River, Oregon. *Transactions of the American Fisheries Society* 116: 196–209.

Hughes, R.M., S.A. Heiskary, W.J. Matthews, and C.O. Yoder. 1994. Use of ecoregions in biological monitoring. Pages 125–151 in S.L. Loeb and A. Spacie (editors). *Biological Monitoring of Aquatic Systems*. Lewis Publishers, Chelsea, Michigan.

Hughes, R.M., C. B. Johnson, S.S. Dixit, A.T. Herlihy, P.R. Kaufmann, W.L. Kinney, D.P. Larsen, P.A. Lewis, D.M. McMullen, A.K. Moors, R.J. O'Connor, S.G. Paulsen, R.S. Stemberger, S.A. Thiele, T.R. Whittier, and D.L. Kugler. 1993. Development of lake condition indicators for EMAP — 1991 pilot. Pages 7–90 in D.P. Larsen and S.J. Christie (editors). *EMAP — Surface Waters 1991 Pilot Report*. EPA-620-R-93–003. U.S.Environmental Protection Agency, Office of Research and Development, Corvallis, Oregon.

Hughes, R.M. and D.P. Larsen. 1988. Ecoregions: an approach to surface water protection. *Journal of the Water Pollution Control Federation* 60: 486–493.

Hughes, R.M., D.P. Larsen, and J.M. Omernik. 1986. Regional reference sites: a method for assessing stream potential. *Environmental Management* 10: 629–635.

Hughes, R.M. and J.M. Omernik. 1981. Use and misuse of the terms watershed and stream order. Pages 320–326 in L.A. Krumholz (editor). *Proceedings of the Warmwater Streams Symposium*. American Fisheries Society, Bethesda, Maryland.

Hughes, R.M. and J.M. Omernik. 1982. A proposed approach to determine regional patterns in aquatic ecosystems. Pages 92–102 in N. B. Armantrout (editor). *Acquisition and Utilization of Aquatic Habitat Inventory Information*. American Fisheries Society, Western Division, Bethesda, Maryland.

Hughes, R.M. and J.M. Omernik. 1983. An alternative for characterizing stream size. Pages 87–101 in T.D. Fontaine and S.M. Bartell (editors). *Dynamics of Lotic Ecosystems*. Ann Arbor Science Publishing, Ann Arbor, Michigan.

Hughes, R.M. and R.F. Noss. 1992. Biological diversity and biological integrity: current concerns for lakes and streams. *Fisheries* 17(3): 11–19.

Hughes, R.M., E. Rexstad, and C.E. Bond. 1987. The relationship of aquatic ecoregions, river basins, and physiographic provinces to the ichthyogeographic regions of Oregon. *Copeia* 1987: 423–432.

Hughes, R.M., T.R. Whittier, C.M. Rohm, and D.P. Larsen. 1990. A regional framework for establishing recovery criteria. *Environmental Management* 14: 673–683.

Hughes, R.M., T.R. Whittier, S.A. Thiele, J.E. Pollard, D.V. Peck, S.G. Paulsen, D. McMullen, J. Lazorchak, D.P. Larsen, W.L. Kinney, P.R. Kaufman, S. Hedtke, S.S. Dixit, G.B. Collins, and J.R. Baker. 1992. Lake and stream indicators for the United States Environmental Protection Agency's environmental monitoring and assessment program. Pages 305–335 in D.H. McKenzie, D.E. Hyatt and V.J. McDonald (editors). *Ecological Indicators. Volume 1*. Elsevier Applied Science, New York.

Hunsaker, C.T. and D.E. Carpenter. 1990. *Environmental Monitoring and Assessment Program: Ecological Indicators*. EPA-600–3-90–060. U.S. Environmental Protection Agency, Office of Research and Development, Research Triangle Park, North Carolina.

Hunt, C.E. 1992. *Down by the River: The Impact of Federal Water Projects and Policies on Biological Diversity*. Island Press, Washington, D.C.

Hurlbert, S.H. 1984. Pseudoreplication and the design of ecological field experiments. *Ecological Monographs* 54: 187–211.

Hurley, A.C. 1981. *Review of the Environmental Protection Agency's Attempts to Define Biological Integrity and Comments on Water Monitoring Strategies*. A draft report prepared as a product of the Environmental Science Engineer Fellowship Coordinated by the AAAS and Sponsored by the U.S. Environmental Protection Agency. Washington, D.C.

Hynes, H.B.N. 1960. *The Biology of Polluted Waters*. Liverpool University Press, Liverpool, England.

Hynes, H.B.N. 1965. The significance of macroinvertebrates in the study of mild river pollution. Pages 235–240 in C.M. Tarzwell (editor). *Biological Problems in Water Pollution. Third Seminar-1962*. U.S. Public Health Service, Cincinnati, Ohio.

Hynes, H.B.N. 1966. *The Biology of Polluted Waters.* Liverpool University Press, Liverpool, England.

Hynes, H.B.N. 1970. *The Ecology of Running Waters.* University of Toronto Press, Toronto, Canada.

Hynes, H.B.N. 1975. The stream and its valley. *Verhandlungen Internationale Vereinigung für Theoretische und Angewandte Limnologie* 19: 1–15.

Illinois EPA. 1989. *Biological Stream Characterization (BSC), A Biological Assessment of Illinois Stream Quality.* Special Report 13 of Illinois State Water Plan Task Force, Illinois EPA, Division of Water Pollution Control, Springfield, Illinois.

Ingram, W.M., K.M. Mackenthun, and A.F. Bartsch. 1966. *Biological Field Investigative Data for Water Pollution Surveys.* Federal Water Pollution Control Administration, Washington, D.C.

International Union for Conservation of Nature and Natural Resources. 1974. Biotic provinces of the world — further development of a system for defining and classifying natural regions for purposes of conservation. *Occasional Paper* 9: 1–57.

ISO (International Organization for Standardization). 1984. *Water Quality — Assessment of the Water and Habitat Quality of Rivers by a Micro-Invertebrate "Score."* ISO/T 147/SC 5/WG 6 N 40. Draft Proposal ISO/DP 8689.

ITFM (Intergovernmental Task Force for Monitoring Water Quality). 1992. *Ambient Water-Quality Monitoring in the United States. First Year Review, Evaluation, and Recommendations.* Interagency Advisory Committee on Water Data, U.S. Geological Survey, Reston, Virginia.

ITFM. 1994. Water Quality Monitoring in the United States. 1993 Report of the Intergovernmental Task Force on Monitoring Water Quality. Interagency Advisory Committee on Water Data. U.S. Geological Survey, Reston, Virginia.

Iudicello, S. 1992. Reaching the decision makers. *Volunteer Monitor* 4:15.

Jaccard, P. 1912. The distribution of flora in the alpine zone. *New Phytologist* 11: 37–50.

Jackson, G.A. 1977. Nearctic and Palearctic *Paracladopelma* Harnisch and *Saetheria* n.gen. (Diptera: Chironomidae). *Journal of the Fisheries Research Board of Canada* 34: 1321–1349.

Jackson, S. 1992. Re-examining independent applicability: agency policy and current issues. Pages 135–138 in K. Swetlow (editor). *Water Quality Standards for the 21st Century, Proceedings of the Third National Conference.* EPA 823-R-92–009. U.S. Environmental Protection Agency, Office of Science and Technology, Washington, D.C.

Jacobson, J.L., S.W. Jacobson, and H.E.B. Humphrey. 1990. Effects of *in utero* exposure to polychlorinated biphenyls and related contaminants on cognitive functioning in young children. *Journal of Pediatrics* 116: 38–45.

Jenkins, R.E. and N.M. Burkhead. 1994. *The Freshwater Fishes of Virginia.* American Fisheries Society, Bethesda, Maryland.

Jezerinac, R.F. 1978. Key to the first form male *Procambarus* and *Orconectes* (Decapoda: Cambaridae) of Ohio. Unpublished.

Jezerinac, R.F. and R.F. Thoma. 1984. An illustrated key to the Ohio *Cambarus* and *Fallicambarus* (Decapoda: Cambaridae) with comments and a new subspecies record. *Ohio Journal of Science* 84: 120–125.

Johannsen, O.A. 1935. Aquatic diptera. Part II. Orthorrphapha - Brachycera and Cyclorrhapha. *Cornell University Agricultural Experiment Station Memoir* 177: 1–62.

Johnson, W.C. 1992. Dams and riparian forests, case study from the Missouri River. *Rivers* 3: 229–242.

Johnson, R.K., T. Wiederholm, and D.M. Rosenberg. 1993. Freshwater biomonitoring using individual organisms, populations, and species assemblages of benthic macroinvertebrates. Pages 40–158 in D.M. Rosenberg and V.H. Resh (editors). *Freshwater Biomonitoring and Benthic Macroinvertebrates.* Chapman and Hall, New York.

Jones, J.G. 1978. Spatial variation in epilithic algae in a stony stream (Wilfin Beck) with special reference to *Cocconeis placentula. Freshwater Biology* 8: 539–546.

Jorling, T. 1977. Incorporating ecological interpretation into basic statutes. Pages 9–14 in R.K. Ballentine and L.J. Guarraia (editors). *The Integrity of Water: A Symposium.* U.S. Environmental Protection Agency, Washington, D.C.

Judy, R.D., Jr, P.N. Seeley, T.M. Murray, S.C. Svirsky, M.R. Whitworth, and L.S. Ischinger. 1984. *1982 National Fisheries Survey. Volume I.* Technical Report Initial Findings. FWS/OBS-84/06. U.S. Fish and Wildlife Service, Washington, D.C.

Kämäri, J., M. Forsius and L. Kauppi, L., 1990. Statistically based lake survey: A representative picture of nutrient status in Finland. *Internationale Vereinigung für Theoretische und Angewandte Limnologie Verhandlungen* 24: 663–666.

Karr, J. R. 1981. Assessment of biotic integrity using fish communities. *Fisheries* 6: 21–27.

Karr, J.R. 1987. Biological monitoring and environmental assessment: a conceptual framework. *Environmental Management* 11: 249–256.

Karr, J.R. 1990. Biological integrity and the goal of environmental legislation: lessons for conservation biology. *Conservation Biology* 4: 244–50.

Karr, J.R. 1991. Biological integrity: a long-neglected aspect of water resource management. *Ecological Applications* 1: 66–84.

Karr, J.R. 1993a. Defining and assessing ecological integrity: beyond water quality. *Environmental Toxicology and Chemistry* 12: 1521–1531.

Karr, J.R. 1993b. Measuring biological integrity: lessons from streams. Pages 83–104 in S. Woodley, J. Kay, and G. Francis (editors). *Ecological Integrity and the Management of Ecosystems*. St. Lucie Press, Delray Beach, Florida.

Karr, J.R. and B.L. Kerans. 1991. Components of biological integrity: their definition and use in development of an invertebrate IBI. Pages 1–16 in W.S. Davis and T.P. Simon (editors). *Proceedings of the 1991 Midwest Pollution Control Biologists Meeting*. EPA 905-R-92-003. U.S. Environmental Protection Agency, Environmental Sciences Division, Chicago, Illinois.

Karr, J.R., R.C. Heidinger, and E.H. Helmer. 1985a. Sensitivity of the index of biotic integrity to changes in chlorine and ammonia levels from wastewater treatment facilities. *Journal of the Water Pollution Control Federation* 57: 912–915.

Karr, J.R., P.R. Yant, K.D. Fausch, and I.J. Schlosser. 1987. Spatial and temporal variability of the Index of Biotic Integrity in three midwestern streams. *Transactions of the American Fisheries Society* 116: 1–11.

Karr, J.R., L.A. Toth, and D.R. Dudley. 1985b. Fish communities of midwestern rivers: a history of degradation. *BioScience* 35: 90–95.

Karr, J.R. and I.J. Schlosser. 1977. *Impact of Nearstream Vegetation and Stream Morphology on Water Quality and Stream Biota*. EPA-600-3-77-097. U.S. Environmental Protection Agency, Environmental Research Laboratory, Athens, Georgia.

Karr, J.R. and I.J. Schlosser. 1978. Water resources and the land-water interface. *Science* 201: 229–234.

Karr, J.R. and D.R. Dudley 1981. Ecological perspective on water quality goals. *Environmental Management* 5: 55–68.

Karr, J.R., K.D. Fausch, P.L. Angermeier, P.R. Yant, and I.J. Schlosser. 1986. *Assessing Biological Integrity in Running Waters: A Method and its Rationale*. Illinois Natural History Survey Special Publication 5, Champaign, Illinois.

Karr, J.R. and M. Dionne. 1991. Designing surveys to assess biological integrity in lakes and reservoirs. Pages 62–72 in *Biological Criteria: Research and Regulation, Proceedings of a Symposium, 12–13 December 1990, Arlington, Virginia*. EPA-440-5-91-005. U.S. Environmental Protection Agency, Office of Water, Washington, D.C.

Kaufmann, P.R. 1993. Physical habitat. Pages 59–69 in R. M. Hughes (editor). *Stream Indicator and Design Workshop*. EPA 600-R-93-138. U.S. Environmental Protection Agency, Environmental Research Laboratory, Corvallis, Oregon.

Kaufmann, P.R., A.T. Herlihy, J.W. Elwood, M.E. Mitch, W.S. Overton, M.J. Sale, J.J. Messer, K.A. Cougan, D.V. Peck, K.H. Reckhow, A.J. Kinney, S.J. Christie, D.D. Brown, C.A. Hagley, and H.I. Jager. 1988. *Chemical Characteristics of Streams in the Mid-Atlantic and Southeastern United States. Volume I: Population Descriptions and Physico-Chemical Relationships*. EPA-600-3-88-021a. U.S. Environmental Protection Agency, Office of Research and Development, Washington, D.C.

Kaufmann, P.R., A.T. Herlihy, M.E. Mitch, and J.J. Messer. 1991. Stream chemistry in the Eastern United States. 1. Synoptic survey design, acid-base status, and regional patterns. *Water Resources Research* 27: 611–627.

Kay, J.J. 1990. A non-equilibrium thermodynamic framework for discussing ecosystem integrity. Pages 209–37 in C.J. Edwards and H.A. Regier (editors). *An Ecosystem Approach to the Integrity of the Great Lakes in Turbulent Times*. Special Publication 90-4. Great Lakes Fisheries Committee, Ann Arbor, Michigan.

KDEP (Kentucky Department of Environmental Protection). 1992. *Kentucky Report to Congress*. Division of Water, Frankfort, Kentucky.

Kellogg, L. and W.S. Davis. 1992. Meeting scientific standards for biological monitoring. Pages 108–113 in J. Douherty (editor). *Proceedings of the Third National Citizens' Volunteer Water Monitoring Conference*. EPA 841-R-92-004. U.S. Environmental Protection Agency, Office of Water, Washington, D.C.

Kelly, J.R. and M.A. Harwell. 1990. Indicators of ecosystem recovery. *Environmental Management* 14: 527–545.

Kentucky Division of Water. 1993. *Methods for Assessing Biological Integrity of Surface Waters.* Kentucky Dept. of Environmental Protection, Frankfort, Kentucky.

Kerans, B.L. and J.R. Karr. 1994. Development and testing of a benthic index of biotic integrity (B-IBI) for rivers of the Tennessee Valley. *Ecological Applications* 4(4).

Kerans, B.L, J.R. Karr, and S.A. Ahlstedt. 1992. Aquatic invertebrate assemblages: spatial and temporal differences among sampling protocols. *Journal of the North American Benthological Society* 11: 377–390.

Kingdon, J.W. 1984. *Agendas, Alternatives and Public Policies.* Little, Brown and Co., Boston, Massachusetts.

Kirtland, J.P. 1838. Report on the zoology of Ohio. *Annual Report of the Geological Survey. State of Ohio* 2: 157–197.

Kirtland, J.P. 1850. Fragments of natural history. *The Family Visitor. Cleveland, Ohio* 1:1.

Kish, L. 1967. *Survey Sampling.* John Wiley & Sons, New York.

Kish, L. 1987. *Statistical Design for Research.* John Wiley & Sons, New York.

Klein, L. 1957. *Aspects of River Pollution.* Academic Press,, New York.

Kleinsasser, R. and G. Linam. 1990. Water quality and fish assemblages in the Trinity River, Texas, between Ft. Worth and Lake Livingston. Texas Parks and Wildlife Department, Austin, Texas.

Klemm, D.J. 1982. Leeches *(Annelida: Hirudinea) of North America.* EPA-600-3-82-025. U.S. Environmental Protection Agency, Environmental Monitoring and Support Laboratory, Cincinnati, Ohio.

Klemm, D.J., P.A. Lewis, F. Fulk, and J.M. Lazorchak. 1990. *Macroinvertebrate Field and Laboratory Methods for Evaluating the Biological Integrity of Surface Waters.* EPA-600-4-90-030. U.S. Environmental Protection Agency, Environmental Monitoring and Support Laboratory, Cincinnati, Ohio.

Kofoid, C.A. 1903. The plankton of the Illinois River, 1894–1899, with introductory notes upon the hydrography of the Illinois River and its basin. Part I. Quantitative investigations and general results. *Bulletin of the Illinois State Laboratory of Natural History* 6: 1–535.

Kofoid, C.A. 1908. The plankton of the Illinois River, 1894–1899, with introductory notes upon the hydrography of the Illinois River and its basin. Part II. Constituent organisms and their seasonal distribution. *Bulletin of the Illinois State Laboratory of Natural History* 8: 1–360.

Kolbe, R.W. 1927. Zur Ökologie, Morphologie, und Systematik der Brackwasser-Diatomeen. *Pflanzenforschung* 7: 1–146.

Kolkwitz, R. 1950. *Öekologie der Saprobien.* (Ecology of the saprobien). Schriftenreihe des Vereins für Wasserhygiene. No. 4 Piscutor-Verlag, Stuttgart.

Kolkwitz, R. and M. Marsson. 1902. Grundsätze für die biologische Beurteilung des Wassers nach seiner Flora und Fauna. (Principles for the biological assessment of waterbodies according to their flora and fauna). *Mittheilungen der Kgl. Prüfungsanstalt für Wasserversorgung und Abwässerbeseitigung Berlin-Dahlem.* 1: 33–72.

Kolkwitz, R. and M. Marsson. 1908. Ökologie der pflanzlichen Saprobien. *Berichte der Deutshcen botanischen Gesellschaft* 26a:505–519. [Translated 1967. Ecology of plant saprobia. Pages 47–52. in L.E. Keup, W.M. Ingram, and K.M. Mackenthum (editors). *Biology of Water Pollution.* Federal Water Pollution Control Administration, Washington, D.C.]

Kolkwitz, R. and M. Marsson. 1909. Ökologie der tierischen Saprobien. Beiträge zur Lehre von der biologischen Gewässerbeurteilung. *Internationale Revue der Gesamten Hydrobiologie und Hydrogeographie* 2: 126–152. [Translated 1967. Ecology of animal saprobia. Pages 485–95 in L.E. Keup, W.M. Ingram and K.M. Mackenthum (editors). *Biology of Water Pollution.* Federal Water Pollution Control Administration, Washington, D.C.]

Kondratieff, B.C., R.F. Kirchner, and K.W. Stewart. 1988. A review of *Perlinella* Banks (Plecoptera: Perlidae). *Annals of the Entomological Society of America* 81: 19–27.

Kopec, J.S. 1989. The Ohio Scenic Rivers Program: citizens in action. Pages 123–127 in W.S. Davis and T.P. Simon (editors). *Proceedings of the 1989 Midwest Pollution Control Biologists Meeting.* EPA-905–9-89–007. U.S. Environmental Protection Agency, Region 5, Environmental Sciences Division, Chicago, Illinois.

Kopec, J.S. 1992. The advantages of integrating various monitoring programs. Pages 72–73 in J. Douherty (editor). *Proceedings of the Third National Citizens' Volunteer Water Monitoring Conference.* EPA 841-R-92–004. U.S. Environmental Protection Agency, Office of Water, Washington, D.C.

Kopec, J.S. and S. Lewis. 1983. *Stream Quality Monitoring*. Ohio Department of Natural Resources, Division of Natural Areas and Preserves, Scenic Rivers Program, Columbus, Ohio.

Kremen, C. 1992. Assessing the indicator properties of species assemblages for natural areas monitoring. *Ecological Applications* 2: 203–217.

Krueger, H.O., J.P. Ward, and S.H. Anderson. 1988. *A Resource Managers Guide for Using Aquatic Organisms to Assess Water Quality for Evaluation of Contaminants*. Biological Report 88(20), U.S. Fish and Wildlife Service, Fort Collins, Colorado.

Krumholz, L.A. 1981. *The Warmwater Streams Symposium, A National Symposium on Fisheries Aspects of Warmwater Streams, Knoxville, Tennessee, 9 March 1980*. Lawrence, Kansas, Southern Division, American Fisheries Society, Bethesda, Maryland.

Küchler, A.W. 1964. *Potential Natural Vegetation of the Conterminous United States*. American Geographical Society, Special Publication No. 36. A. Hoen & Co., Baltimore, Maryland.

Küchler, A.W. 1970. Potential natural vegetation. Map (scale 1:7,500,000). Plates 90–91 in *The National Atlas of the United States of America*. U.S. Geological Survey, Washington, D.C.

Kuehne, R.A. and R.W. Barbour. 1983. *The American Darters*. University Press of Kentucky, Lexington, Kentucky.

Kurtenbach, J. 1991. A method for rapid bioassessment of streams in New Jersey using benthic macroinvertebrates. Page 138 in *Biological Criteria: Research and Regulation, Proceedings of a Symposium, 12–13 December 1990, Arlington, Virginia*. EPA-440–5-91–005. U.S. Environmental Protection Agency, Office of Water, Washington, D.C.

Lampkin, A.J. III and M.R. Sommerfeld. 1982. Algal distribution in a small intermittent stream receiving acid mine-drainage. *Journal of Phycology* 18: 196–199.

Landers, D.H., J.M. Eilers, D.F. Brakke, W.S. Overton, P.E. Kellar, M.E. Silverstein, R.D. Schonbrod, R.E. Crowe, R.A. Linthurst, J.M. Omernik, S.A. Teague, and E.P. Meier. 1987. *Characteristics of Lakes in the Western United States. Volume I: Population Descriptions and Physico-Chemical Relationships*. EPA-600–3-86–054a. U.S. Environmental Protection Agency, Office of Research and Development, Washington, D.C.

Landers, D.H., W.S. Overton, R.A. Linthurst, and D.F. Brakke. 1988. Eastern Lake Survey: regional estimates of lake chemistry. *Environmental Science and Technology* 22: 128–135.

Langdon, R. 1989. The development of fish population-based biocriteria in Vermont. Pages 12–25 in T.P. Simon, L.L. Holst, and L.J. Shepard (editors). *Proceedings of the First National Workshop on Biological Criteria*. EPA 905–9-89–003. U.S. Environmental Protection Agency, Region 5, Environmental Sciences Division, Chicago, Illinois.

Lange-Bertalot, H. 1979. Pollution tolerance as a criterion for water quality estimation. *Nova Hedwigia* 64: 285–304.

LaPoint, T.W., S.M. Melacon, and M.K. Morris. 1984. Relationships among observed metal concentrations, criteria and benthic community structural responses in 15 streams. *Journal of the Water Pollution Control Federation* 56: 1030–1038.

Larimore, R.W., W.S. Childers, and C. Heckrotte. 1957. Destruction and reestablishment of stream fish and invertebrates affected by drought. *Transactions of the American Fisheries Society* 88: 261–269.

Larimore, R.W. and P.W. Smith. 1963. The fishes of Champaign County, Illinois, as affected by 60 years of stream changes. *Illinois Natural History Survey Bulletin* 28: 299–382.

Larsen, D.P. and S.J. Christie. 1993. *EMAP-Surface Waters 1991 Pilot Report*. EPA-620-R-93–003. U.S. EPA, Environmental Research Laboratory, Corvallis, Oregon.

Larsen, D.P., D.R. Dudley, and R.M. Hughes. 1988. A regional approach for assessing attainable surface water quality: Ohio as a case study. *Journal of Soil and Water Conservation* 43: 171–176.

Larsen, D.P., J.M. Omernik, R.M. Hughes, C.M. Rohm, T.R. Whittier, A.J. Kinney, A.L. Gallant, and D.R. Dudley. 1986. The correspondence between spatial patterns in fish assemblages in Ohio streams and aquatic ecoregions. *Environmental Management* 10: 815–828.

Larson, R. 1993. Interior's 9th bureau set to survey life. *The Washington Times*. September 24.

Lathrop, J.E. 1989. A naturalists key to stream macroinvertebrates for citizen monitoring programs in the midwest. Pages 107–118 in W.S. Davis and T.P. Simon (editors). *Proceedings of the 1989 Midwest Pollution Control Biologists Meeting, Chicago, Illinois*. EPA 905–9-89–007. USEPA Region V, Instream Biocriteria and Ecological Assessment Committee, Chicago, Illinois.

Lauterborn, R. 1901. Die sapropelische Lebewelt. *Zoologischer Anzeiger* 24: 50–55.

Layher, W.G. and K.L. Brunson. 1992. A modification of the habitat evaluation procedures for determining instream flow requirements for warmwater streams. *North American Journal of Fisheries Management* 12: 47–54.

Layher, W.G. and O.E. Maughan. 1985. Relations between habitat variables and channel catfish populations in prairie streams. *Transactions of the American Fisheries Society* 114: 771–781.

Leclercq, L. and E. Depiereux. 1987. Typologie des rivières oligotrophes du massif Ardennais (Belgique) par l'analyse multivariée de relevés de diatomées benthiques. *Hydrobiologia* 153: 175–192.

Leclercq, L. and B. Maquet. 1987. Deux nouveaux indices chimique et diatomique de qualité d'eau courante, application au Samson et a des affluents, comparaison avec d'autrtes indices chimiques, biocenotiques et diatomiques. *Institut Royal des Sciences Naturelles de Belgique Documents de travail* 38.

Lee, K.N. 1993. *Compass and Gyroscope: Integrating Science and Politics for the Environment.* Island Press, Washington, D.C.

Lehmkuhl, D.M. 1979. *How to Know the Aquatic Insects.* William C. Brown Company Publishers, Dubuque, Iowa.

Leighton, M.O. 1907. *Pollution of Illinois and Mississippi Rivers by Chicago Sewage: A Digest of the Testimony Taken in Case of the State of Missouri v. the State of Illinois of the Sanitary District of Chicago.* Water Supply and Irrigation Paper No, 194, Series L, Quality of Water, 20. U.S. Geological Survey, Washington, D.C.

Lenat, D.R. 1987. *Water Quality Assessment Using a New Qualitative Collection Method for Freshwater Benthic Macroinvertebrates.* North Carolina DEM Technical Report, Raleigh, North Carolina.

Lenat, D.R. 1988. Water quality assessment of streams using a qualitative collection method for benthic macroinvertebrates. *Journal of the North American Benthological Society* 7: 222–233.

Lenat, D.R. 1990. Reducing variability in freshwater macroinvertebrate data. Pages 19–32 in W. S. Davis (editor). *Proceedings of the 1990 Midwest Pollution Control Biologists Meeting. EPA-905-9-90-005.* U.S. Environmental Protection Agency, Region 5, Environmental Sciences Division, Chicago, Illinois.

Lenat, D.R. 1993. A biotic index for the southeastern United States: derivation and list of tolerance values, with criteria for assigning water quality ratings. *Journal of the North American Benthological Society* 12: 279–290.

Leonard, P.M. and D.J. Orth. 1986. Application and testing of an index of biotic integrity in small, coolwater streams. *Transactions of the American Fisheries Society* 115: 401–414.

Leopold, A. 1949. *A Sand County Almanac and Sketches Here and There.* Oxford Press, New York.

Leopold, L.B., M.G. Wolman, and J.P. Miller. 1964. *Fluvial Processes in Geomorphology.* W. H. Freeman and Company, San Francisco, California.

LeSueur, C.A. 1827. *American Ichthyology or, Natural History of the Fishes of North America: With Colored Figures from Drawings Executed from Nature.* New Harmony, Indiana.

Levin, S.A. 1992. The problem of pattern and scale in ecology. *Ecology* 73: 1943–1967.

Liebmann, H. 1951. *Handbuch der Frischwasser und Abwasserbiologie.* Vol.1 Verlag R. Oldenbourg, München.

Liebmann, H. 1962. *Handbuch der Frischwasser und Abwasserbiologie.* Vol.1. 2nd edition. Verlag R. Oldenbourg, München.

Likens, G.E. and F.H. Bormann. 1974. Linkages between terrestrial and aquatic ecosystems. *BioScience* 24: 447–456.

Linam, G. and R. Kleinsasser. 1987. *Fisheries Use Attainability Study for Bosque River.* Report Number 0-265A-11/09/87. Texas Parks and Wildlife Department, San Marcos, Texas.

Linthurst, R.A., D.H. Landers, J.M. Eilers, D.F. Brakke, W.S. Overton, E.P. Meier, and R.E. Crowe. 1986. *Characteristics of Lakes in the Eastern United States. Volume I: Population Descriptions and Physico-Chemical Relationships.* EPA-600-4-86-007a. U.S. Environmental Protection Agency, Office of Research and Development, Washington, D.C.

Lobb, M.D.I. and D.J. Orth. 1991. Habitat use by an assemblage of fish in a large warmwater stream. *Transactions of the American Fisheries Society* 120: 65–78.

Loftis, J.C., R.C. Ward, D. Phillips, and C.H. Taylor. 1989. *An Evaluation of Trend Detection Techniques for Use in Water Quality Monitoring Programs.* EPA-600-3-89-037. U.S. Environmental Protection Agency, Environmental Research Laboratory, Corvallis, Oregon.

Loftis, J.C., R.C. Ward, and G.M. Smillie. 1983. Statistical models for water quality regulation. *Journal of the Water Pollution Control Federation* 55: 1098–1104.

Loveland, T.R., J.W. Merchant, D.O. Ohlen, and J.F. Brown. 1991. Development of a land-cover character-istics database for the conterminous U.S. *Photogrammetric Engineering and Remote Sensing* 57: 1453–1463.

Lowe, R.L. 1974. *Environmental Requirements and Pollution Tolerance of Freshwater Diatoms.* U.S. Envi-ronmental Protection Agency, Environmental Monitoring Series, Cincinnati, Ohio.

Lowrance, R.R., R. Todel, J. Fail, O. Hendrickson, Jr., O. R. Leonard, and L. Asmussen. 1984. Riparian forests as nutrient filters in agricultural watersheds. *BioScience* 34: 374–377.

Luey, J.B., and I.R. Adelman. 1980. Downstream natural areas as refuges for fish in drainage development watersheds. *Transactions of the American Fisheries Society* 109: 332–335.

Lyons, J. 1989. Correspondence between the distribution of fish assemblages in Wisconsin streams and Omernik's ecoregions. *American Midland Naturalist* 122: 163–182.

Lyons, J. 1992a. *Using the Index of Biotic Integrity (IBI) to Measure Environmental Quality in Warmwater Streams of Wisconsin.* General Technical Report NC-149. North Central Forest Experiment Station, U.S. Department of Agriculture, St. Paul, Minnesota.

Lyons, J. 1992b. The length of stream to sample with a towed electrofishing unit when fish species richness is estimated. *North American Journal of Fisheries Management* 12: 198–203.

Lyons, J. and C.C. Courtney. 1990. *A Review of Fisheries Habitat Improvement Projects in Warmwater Streams, with Recommendations for Wisconsin.* Wisconsin Department of Natural Resources, Madison, Wisconsin.

Lyons, J., A.M. Forbes, and M.D. Staggs. 1988. *Fish Species Assemblages in Southwestern Wisconsin Streams with Implications for Smallmouth Management.* Wisconsin Department of Natural Resources, Madison, Wisconsin.

Maas, R., D. Kucken, and P.F. Gregutt. 1991. Developing a rigorous water quality database through volunteer a monitoring network. *Lake Reservoir Management* 7: 123–124.

Mackenthun, K.M and W.M. Ingram. 1967. *Biological Associated Problems in Freshwater Environments: Their Identification, Investigation and Control.* Federal Water Pollution Control Administration. Wash-ington, D.C.

Maine Department of Environmental Protection. 1987. *Methods for Biological Sampling and Analysis of Maine's Waters.* Maine Department of Environmental Protection, Augusta, Maine.

Malley, D.F. and K.H. Mills. 1992. Whole-lake experimentation as a tool to assess ecosystem health, responses to stress and recovery: the Experimental Lakes Area experience. *Journal of Aquatic Ecosystem Health* 1: 159–174.

Mangum, F.A. 1986a. *Use of Aquatic Macroinvertebrates in Land Management.* U.S. Department of Agricul-ture — Forest Service, Wildlife and Fish Ecology Unit, Fort Collins, Colorado.

Mangum, F.A. 1986b. Macroinvertebrates. Chapter 32 in A.Y. Cooperrider, et al. (editors). *Inventory and Monitoring of Wildlife Habitat.* U.S. Department of the Interior, Bureau of Land Management, Denver, Colorado.

Mangum, F.A. 1990. *Use of Aquatic Macroinvertebrates in Evaluations of Environmental Quality in Aquatic Ecosystems and Associated Terrestrial Habitats.* U.S. Department of Agriculture, Forest Service, Provo, Utah.

Marcus, M.D. and L.L. McDonald. 1992. Evaluating the statistical bases for relating receiving water impacts to effluent and ambient toxicities. *Environmental Toxicology and Chemistry* 11: 1389–1402.

Maret, T. 1986. *Nebraska Stream Habitat Quality Index.* Nebraska Department of Environmental Control, Lincoln, Nebraska.

Maret, T. 1988. A stream inventory process to classify use support and develop biological standards in Nebraska. Pages 55–66 in T.P. Simon, L.L. Holst, and L.J. Shepard (editors). *Proceedings of the First National Workshop on Biocriteria, Lincolnwood, Illinois.* EPA 905–9-89–003. U.S. Environmental Pro-tection Agency, Region 5, Chicago, Illinois.

Margalef, R. 1951. Diversidad de especies en las comunidades naturales. *Publicaciones del Instituto de Biologia Aplicada Barcelona* 6: 59–72.

Marsh, P.C. and J.E. Luey. 1982. Oases for aquatic life within agricultural watersheds. *Fisheries* 7(6): 16–24.

Martin, G.L., T.J. Balduf, D.O. McIntyre, and J.P. Abrams. 1979. *Water Quality Study of the Ottawa River, Allen and Putnam Counties, Ohio.* Ohio EPA Water Quality Survey 79/1, Office of Wastewater, Colum-bus, Ohio.

Martin, P.S. 1967. Pleistocene overkill. *Natural History* 76: 32–38.

Maschwitz, D.E. 1976. Revision of the Nearctic species of the subgenus *Polypedilum* (Chironomidae: Diptera). Doctoral Dissertation, University of Minnesota.

Mason, W.T., Jr. 1979. A rapid procedure for assessment of surface mining impacts to aquatic life. Pages 310–323 in *Coal Conference and Expo V, Proceedings of a Symposium, Louisville, Kentucky, October 23–25, 1979*. McGraw-Hill, New York.

Master, L. 1990. The imperiled status of North American aquatic animals. *Biodiversity Network News* (Nature Conservancy) 3: 1–2, 7–8.

Masters, A.E. 1992. *Reservoir Vital Signs Monitoring, 1991 — Benthic Macroinvertebrate Community Results*. TVA-WR-92-3. Tennessee Valley Authority, Chattanooga, Tennessee.

Matthews, W.J. 1986. Fish fauna structure in an Ozark stream, stability, peristence, and a catastrophic flood. *Copeia* 1986: 388–397.

Matthews, W.J. 1990. Fish community structure and stability in warmwater midwestern streams. *Biological Report* 90: 16–17.

Matthews, W.J. and H.W. Robison. 1988. The distribution of the fishes of Arkansas: a multivariate approach. *Copeia* 1988: 358–373.

McBride, B.B., J.C. Loftis, and N.C. Adkins. 1993. What do significance tests really tell us about the environment? *Environmental Management* 17: 423–432.

McCafferty, W.P. 1975. The burrowing mayflies (Ephemeroptera: Ephmeroidea) of the United States. *Transactions of the American Entomological Society* 101: 447–504

McCafferty, W.P. 1981. *Aquatic Entomology. The Fisherman's and Ecologists' Illustrated Guide to Insects and their Relatives*. Science Books International, Boston, Massachusetts.

McCafferty, W.P. and R.D. Waltz. 1990. Revisionary synopsis of the Baetidae (Ephemeroptera) of North and Middle America. *Transactions of the American Entomological Society* 116: 769–799.

McCarthy, J.F. 1990. Concluding remarks: implementation of a biomarker-based environmental monitoring program. Pages 429–439 in J. F. McCarthy and L.R. Shugart (editors). *Biomarkers of Environmental Contamination*. Lewis Publishers, Boca Raton, Florida.

McCormick, P.V. and R.J. Stevenson. 1992. Mechanism of benthic algal succession in lotic environments. *Ecology* 72: 1835–1858.

McKenzie, D.H., D.E. Hyatt, and V.J. McDonald (editors). 1992. *Ecological Indicators. Proceedings of the International Symposium on Ecological Indicators, October 16–19, 1990, Ft. Lauderdale, FL*. Elsevier Science Publishers Ltd., Essex, England.

Meador, M.R., T.F. Cuffney, and M.E. Gurtz. 1993. *Methods for Sampling Fish Communities as Part of the National Water-Quality Assessment Program*. Open-File Report 93–104. U.S. Geological Survey, Raleigh, North Carolina.

Meador, M.R., C.R. Hupp, T.F. Cuffney, and M.E. Gurtz. 1993. *Methods for Characterizing Stream Habitat as Part of the National Water-Quality Assessment Program*. Open File Report 93–408. U.S. Geological Survey, Raleigh, North Carolina.

Mellanby, K. 1974. A water pollution survey, mainly by British schoolchildren. *Environmental Pollution* 6: 161–173.

Menhinick, E.F. 1964. A comparison of some species-individuals diversity indices applied to samples of field insect. *Ecology* 45: 859–861.

Merritt, R.W. and K.W. Cummins (editors). 1984. *An Introduction to the Aquatic Insects of North America*. 2nd edition. Kendall/Hunt Publishing Company, Dubuque, Iowa.

Messer, J.J., C.W. Ariss, J.R. Baker, S.K. Drousé, K.N. Eshleman, P.R. Kaufmann, R.A. Linthurst, J.M. Omernik, W.S. Overton, M.J. Sale, R.D. Schonbrod, S.M. Stambaugh, and J.R. Tuschall, Jr. 1986. *National Surface Water Survey: National Stream Survey, Phase I — Pilot Survey*. EPA-600–4-86-026. U.S. Environmental Protection Agency, Office of Research and Development, Washington, D.C.

Messer, J.J., R.A. Linthurst, and W.S. Overton. 1991. An EPA program for monitoring ecological status and trends. *Environmental Monitoring and Assessment* 17: 67–78.

Metcalfe, J.L. 1989. Biological water quality assessment of running waters based on macroinvertebrate communities: history and present status in Europe. *Environmental Pollution* 60: 101–139.

Miller, D.L., P.M. Leonard, R.M. Hughes, J.R. Karr, P.B. Moyle, L.H. Schrader, B.A. Thompson, R.A. Daniel, K.D. Fausch, G.A. Fitzhugh, J.R. Gammon, D.B. Halliwell, P.L. Angermeier, and D.J. Orth. 1988. Regional applications of an Index of Biotic Integrity for use in water resource management. *Fisheries* 13(5): 12–20.

Miller, R.R., J.D. Williams, and J.E. Williams. 1989. Extinctions of North American fishes during the past century. *Fisheries* 14: 22–38.

Mills, H.B., W.C. Starrett, and F.C. Bellrose. 1966. Man's effect on the fish and wildlife of the Illinois River. *Illinois Natural History Survey Biological Notes* 57.

Miner, R. and D. Borton. 1991. Considerations in the development and implementation of biocriteria. Pages 115–119 in G.H. Flock (editor). *Water Quality Standards for the 21st Century, Proceedings of a Conference.* U.S. Environmental Protection Agency, Office Science and Technology, Washington, D.C.

Minshall, G.W., K.W. Cummins, R.C. Petersen, C.E. Cushing, D.A. Bruns, J.R. Sedell, and R.L. Vannote. 1985. Developments in stream ecosystem theory. *Canadian Journal of Fisheries and Aquatic Sciences* 42: 1045–1055.

Minshall, G.W., R.C. Petersen, K.W. Cummins, T.L. Bott, J.R. Sedell, C.E. Cushing, and R.L. Vannote. 1983. Interbiome comparison of stream ecosystem dynamics. *Ecological Monographs* 53: 1–25.

Morihara, D.K. and W.P. McCafferty. 1979. The *Baetis* larvae of North America (Ephemeroptera: Baetidae). *Transactions of the American Entomological Society* 105: 139–221.

Morisawa, M. 1968. *Streams, Their Dynamics and Morphology.* McGraw-Hill, New York.

Moser, H.G., W.J. Richards, D.M. Cohen, M.P. Fahay, A.W. Kendall, Jr., and S.L. Richardson. 1984. *Ontogeny and Systematics of Fishes.* American Society of Ichthyologists and Herpetologists, Special Publication 1. Allen Press, Lawrence, Kansas.

Mount, D.I. 1987. Comparison of test precision of effluent toxicity tests with chemical analyses. U.S. Environmental Protection Agency, Environmental Research Laboratory, Duluth, Minnesota. Unpublished.

Moyle, P.B., L.R. Brown, and B. Herbold. 1986. *Final Report on Development and Preliminary Tests of Indices of Biotic Integrity for California.* Final Project Report submitted to the U.S. Environmental Protection Agency, Corvallis, Oregon.

Moyle, P.B. and R.A. Leidy. 1992. Loss of aquatic ecosystems: evidence from fish faunas. Pages 127–169 in P.L. Fielder and S.K. Jain (editors). *Conservation Biology: The Theory and Practice of Nature Conservation, Preservation and Management.* Chapman and Hall, New York.

Moyle, P.B. and J.E. Williams. 1990. Biodiversity loss in the temperate zone: decline of the native fish fauna of California. *Conservation Biology* 4: 275–284.

Mueller-Dombois, D. and H. Ellenberg. 1974. *Aims and methods of vegetation ecology.* Wiley, New York.

Mulholland, P.J., J.W. Elwood, and A.V. Palumbo. 1986. Effect of stream acidification on periphyton composition, chlorophyll, and productivity. *Canadian Journal of Fisheries and Aquatic Sciences* 43: 1846–1858.

Mundie, J.H., K.S. Simpson, and C.J. Perrin. 1991. Responses of stream periphyton and benthic insects to increases in dissolved inorganic phosphorus in a mesocosm. *Canadian Journal of Fisheries and Aquatic Sciences* 48: 2061–2072.

Murphy, M.L. and K.V. Koski. 1989. Input and depletion of woody debris in Alaska streams and implications for streamside management. *North American Journal of Fisheries Managment* 9: 427–436.

Myers, N. 1993. *Ultimate Security: The Environmental Basis of Political Stability.* Norton, New York.

Myslinski, E. and W. Ginsburg. 1977. Macroinvertebrates as indicators of pollution. *Journal of the American Water Works Association* 69: 538–544.

Naiman, R.J., D.G. Lonzarich, T.J. Beechie, and S.C. Ralph. 1992. General principles of classification and the assessment of conservation potential in rivers. Pages 93–123 in P.J. Boon, P. Calow, and G.E. Petts (editors). *River Conservation and Management.* Wiley, New York.

NAS (National Academy of Sciences). 1986. *Ecological Knowledge and Environmental Problem-Solving.* National Academy Press, Washington D.C.

National Wetlands Working Group. 1986. *Canada Wetland Regions* (maps). Canada Map Office, Ottawa, Ontario.

Needham, J.G. and M.J. Westfall, Jr. 1955. *A Manual of the Dragonflies of North America (Anisoptera) Including the Greater Antilles and the Provinces of the Mexican Border.* University of California Press, Berkeley, California.

Needham, P.R. and R.L. Usinger. 1956. Variability in the macrofauna of a single riffle in Prosser Creek, California, as indicated by the Surber sampler. *Hilgardia* 24: 383–409.

Nehlsen, W., J.E. Williams, and J.A. Lichatowich. 1991. Pacific salmon at the crossroads: stocks at risk from California, Oregon, Idaho, and Washington. *Fisheries* (Bethesda) 16: 4–21.

Neunzig, H.H. 1966. Larvae of the genus *Nigronia* Banks (Neuroptera: Corydalidae). *Proceedings of the Entomological Society of Washington* 68: 11–16.

Newton, B. 1992. The U.S. Clean Water Act: programmatic needs and experiences. Pages 1211–1217 in D.H. McKenzie, D.E. Hyatt, and V.J. McDonald (editors). *Ecological Indicators. Volume 1.* Elsevier Applied Science, New York.

Nichols, A.B. 1992. It's clear, U.S. waters have improved. *Water Environment and Technology* Oct. 1992: 44–50.

Norris, G. 1992. Monitoring data lead to stream protection order. *Volunteer Monitor* 4:4–5.

Norris, G., M. Kelly, and J. Schloss. 1992. Deciding data objectives. Pages 43–50 in J. Douherty (editor). *Proceedings of the Third National Citizens' Volunteer Water Monitoring Conference.* EPA 841-R-92–004. U.S. Environmental Protection Agency, Office of Water, Washington, D.C.

Norris, R.H. and A. Georges. 1993. Analysis and interpretation of benthic macroinvertebrate surveys. Pages 234–285 in D.M. Rosenberg and V.H. Resh (editors). *Freshwater Biomonitoring and Benthic Macroinvertebrates.* Chapman and Hall, New York.

Norris, R.H., E.P. McElravy, and V.H. Resh. 1992. The sampling problem. Pages 282–306 in J. Petts and P. Calow (editors). *The Rivers Handbook.* Volume 1. Blackwell Scientific Publishers, Oxford, England.

Noss, R.F. 1990. Indicators for monitoring biodiversity: A hierarchical approach. *Conservation Biology* 4: 355–64.

Novak, M.A. and R.W. Bode. 1992. Percent model affinity, a new measure of macroinvertebrate community composition. *Journal of the North American Benthological Society* 11: 80–85.

NRC (National Research Council). 1992. *Restoration of Aquatic Ecosystems: Science, Technology, and Public Policy.* National Academy Press, Washington, D.C.

NRC (National Research Council). 1993. *A Biological Survey for the Nation.* National Committee on the Formation of the National Biological Survey. National Academy Press, Washington, D.C.

NWF (National Wildlife Federation). 1992. *Waters at Risk: Keeping Clean Waters Clean.* Washington, D.C.

Oberdorff, T. and R.M. Hughes. 1992. Modification of an Index of Biotic Integrity based on fish assemblages to characterize rivers of the Seine-Normandie basin, France. *Hydrobiologia* 228: 117–130.

Odum, E.P., J.T. Finn and E.H. Franz 1979. Perturbation theory and the subsidy-stress gradient. *Bioscience* 29: 349–352.

Ohio DNR (Department of Natural Resources). 1990. Ohio's nonpoint source management plan. Ohio DNR, Division of Soil and Water Conservation, Columbus, Ohio.

Ohio EPA. 1983. *Comprehensive Water Quality Report for Big Darby Creek (Plain City and Ranco, Inc.), Madison and Union Counties, Ohio.* Ohio EPA, Division of Wastewater Pollution Control, Columbus, Ohio.

Ohio EPA. 1985a. *Comprehensive Water Quality Report for the Upper Hocking River. Amendments to the Hocking River Basin Water Quality Management Plan.* Ohio EPA, Division of Water Pollution Control, Columbus, Ohio.

Ohio EPA. 1985b. *Lower Black River Comprehensive Water Quality Report.* Ohio EPA, Division of Water Pollution Control, Columbus, Ohio.

Ohio EPA. 1987a. *Biological Criteria for the Protection of Aquatic Life. Volume I: The Role of Biological Data in Water Quality Assessment.* Ohio EPA, Division of Water Quality Monitoring and Assessment, Surface Water Section, Columbus, Ohio.

Ohio EPA 1987b. *Biological Criteria for the Protection of Aquatic Life. Volume II: Users Manual for Biological Field Assessment of Ohio Surface Waters.* Ohio EPA, Division of Water Quality Monitoring and Assessment, Surface Water Section, Columbus, Ohio.

Ohio EPA. 1988. *Ohio Water Quality Inventory,* 1988 305(b) Report, Volume I, and Executive Summary. E.T. Rankin, C.O. Yoder and D.A. Mishne (editors). Ohio EPA, Division of Water Quality Monitoring and Assessment, Columbus, Ohio.

Ohio EPA 1989a. *Biological Criteria for the Protection of Aquatic Life. Volume III: Standarized Biological Field Sampling and Laboratory Methods for Assessing Fish and Macroinvertebrate Communities.* Ohio EPA, Division of Water Quality Planning and Assessment, Ecological Assessment Section, Columbus, Ohio.

Ohio EPA. 1989b. *Addendum to Biological Criteria for the Protection of Aquatic Life, Volume II: Users Manual for Biological Field Assessment of Ohio Surface Water.* Ohio EPA, Division of Water Quality Planning and Assessment, Ecological Assessment Section, Columbus, Ohio.

Ohio EPA. 1990a. *The Use of Biocriteria in the Ohio EPA Surface Water Monitoring and Assessment Program.* Ohio Environmental Protection Agency, Division of Water Quality Planning and Assessment, Columbus, Ohio.

Ohio EPA. 1990b. *Ohio Water Resource Inventory, Volume I, Summary, Status, and Trends.* E. T. Rankin, C.O. Yoder, and D.A. Mishne (editors). Ohio Environmental Protection Agency, Division of Water Quality Planning and Assessment, Columbus, Ohio.

Ohio EPA. 1990c. *Ohio's Nonpoint Source Pollution Assessment.* Ohio EPA, Division of Water Quality Planning and Assessment. Columbus, Ohio.

Ohio EPA. 1991. *Biological and Water Quality Study of the Hocking River Mainstem and Selected Tributaries.* Ohio EPA Technical Report EAS/1991–10–6. Ohio EPA, Division of Water Quality Planning and Assessment, Columbus, Ohio.

Ohio EPA. 1992a. *Biological and Water Quality Study of the Ottawa River, Allen, Auglaize, and Putnam Counties, Ohio.* Ohio EPA Technical Report EAS/1992–9-2. Division of Water Quality Planning and Assessment, Columbus, Ohio.

Ohio EPA. 1992b. *Ohio Water Resource Inventory. Volume 1, Status and Trends.* E. T. Rankin, C.O. Yoder, and D.A. Mishne (editors). Ohio Environmental Protection Agency, Division of Water Quality Planning and Assessment, Columbus, Ohio.

Ohio EPA. 1992c. *Biological and Habitat Investigation of Greater Cincinnati Area Streams (Hamilton and Clermont Counties, Ohio).* Ohio EPA Technical Report EAS/1992–5-1. Division of Water Quality Planning and Assessment, Columbus, Ohio.

Ohio EPA. 1992d. *Biological and Water Quality Study of the Cuyahoga River and Selected Tributaries.* Ohio EPA Technical Report EAS/1992-12-11. Division of Surface Water, Columbus, Ohio.

Oklahoma Conservation Commission. 1993. *Development of Rapid Bioassessment Protocols for Oklahoma Utilizing Characteristics of the Diatom Community.* Oklahoma Conservation Commission.

Oliver D.R. and M.E. Roussel. 1983. The insects and arachnids of Canada. Part II. The genera of larval midges of Canada (Diptera: Chironomidae). *Agriculture Canada Publication* 1746: 1-263.

Olson-Rutz, K.M. and C.B. Marlow. 1988. Analysis and interpretation of stream channel cross-sectional data. *North American Journal of Fisheries Management* 12: 55–61.

Omernik, J.M. 1977. *Nonpoint Source — Stream Nutrient Level Relationships: A Nationwide Study.* EPA-600–3-77–105. U.S. Environmental Protection Agency, Environmental Research Laboratory, Corvallis, Oregon.

Omernik, J.M. 1987. Ecoregions of the conterminous United States. *Annals of the Association of American Geographers* 77: 118–125.

Omernik, J.M. and A.L. Gallant. 1987. *Ecoregions of the South-Central States.* EPA-600-D-87–315. U.S. Environmental Protection Agency, Corvallis, Oregon.

Omernik, J.M. and A.L. Gallant. 1988. *Ecoregions of the Upper Midwest States.* EPA-600–3-88–037. U.S. Environmental Protection Agency, Environmental Research Laboratory, Corvallis, Oregon.

Omernik, J.M., and A.L. Gallant. 1990. Defining regions for evaluating environmental resources. Pages 936–947 in H.G. Lund and G. Preto (coordinators) *Global Natural Resource Monitoring and Assessments: Preparing for the 21st Century.* American Society of Photogrammetry and Remote Sensing, Bethesda, Maryland.

Omernik, J.M. and G.E. Griffith. 1986. Total alkalinity of surface waters: a map of the Upper Midwest region. Map (scale 1:2,500,000). *Environmental Management* 10: 829–839.

Omernik, J.M. and G.E. Griffith. 1991. Ecological regions vs. hydrologic units: frameworks for managing water quality. *Journal of Soil and Water Conservation* 46: 334–340.

Omernik, J.M. and C.F. Powers. 1983. Total alkalinity of surface waters — a national map. *Annals of the Association of American Geographers* 73: 133–136.

Omernik, J.M., C.M. Rohm, S.E. Clarke, and D.P. Larsen. 1988. Summer total phosphorus in lakes: a map of Minnesota, Wisconsin, and Michigan. *Environmental Management* 12: 815–825.

Omernik, J.M., M.A. Shirazi, and R.M. Hughes. 1982. A synoptic approach for regionalizing aquatic ecosystems. Pages 199–218 in *In-Place Resource Inventories: Principles and Practices: Proceedings of a National Workshop, August 9–14, 1981.* Society of American Foresters, University of Maine, Orono.

ORSANCO (Ohio River Valley Sanitation Commission). 1991. *Assessment of ORSANCO Fish Population Data using the Modified Index of Well-Being MIwb.* Ohio River Valley Sanitation Commission, Cincinnati, Ohio.

Osborne, L.L., B. Dickson, M. Ebbers, R. Ford, J. Lyons, D. Kline, E. Rankin, D. Ross, R. Sauer, P. Seelbach, C. Soeas, T. Stefanavage, J. Wait, and S. Walker. 1991. Stream habitat assessment programs in North Central Division States of the AFS. *Fisheries* 16: 28–35.

Osborne, L.L., S.L. Kohler, P.B. Bayley, D.M. Day, W.A. Bertrand, M.J. Wiley, and R. Sauer. 1992. Influence of stream location in a drainage network on the Index of Biotic Integrity. *Transactions of the American Fisheries Society* 121: 635–643.

Overton, J. 1991. Utilization of biological information in North Carolina's water quality regulatory program. Pages 19–24 in *Biological Criteria: Research and Regulation. Proceedings of a Symposium.* EPA-440–5-91–005. U.S. Environmental Protection Agency, Office of Water, Washington, D.C.

Overton, W.S., D. White, and D.L. Stevens, Jr. 1991. *Design Report for EMAP, The Environmental Monitoring and Assessment Program.* EPA-600–3-91–053. U.S. Environmental Protection Agency, Office of Research and Development, Washington, D.C.

Palmer, C.M. 1969. A composite rating of algae tolerating organic pollution. *Journal of Phycology* 5: 78–82.

Pantle, R. and H. Buck. 1955. Die biologische Überwachung der Gewässer und die Daistellung der Ergebnisse. (Biological monitoring of waterbodies and the presentation of results). *Gas und Wasserfach* 96: 604.

Patil, G.P. and C. Taillie. 1976. Ecological diversity: concepts, indices and applications. *Proceedings of the International Biometrics Conference* 9: 383–411.

Patrick, R. 1950. Biological measure of stream conditions. *Sewage and Industrial Wastes* 22: 926–938.

Patrick, R. 1968. The structure of diatom communities in similar ecological conditions. *American Naturalist* 102: 173–183.

Patrick, R. 1977. Identifying integrity through ecosystem study. Pages 155–172 in R.K. Ballantine and L.J. Guarraia (editors). 1977. *The Integrity of Water. Proceedings of a Symposium, March 10–12, 1975.* U.S. Environmental Protection Agency, Washington, D.C.

Patrick, R. 1992. *Surface Water Quality: Have the Laws Been Successful?* Princeton University Press, Princeton, New Jersey.

Patrick, R. and N.A. Roberts. 1979. Diatom communities in the Middle Atlantic States, U.S.A. Some factors that are important to their structure. *Nova Hedwigia* 64: 265–283.

Paulsen, S.G. 1991. *EMAP — Surface Waters Monitoring and Research Strategy, Fiscal Year 1991.* EPA 600–3-91–022. U.S. Environmental Protection Agency, Environmental Research Laboratory, Corvallis, Oregon.

Paulsen, S.G., D.P. Larsen, P.R. Kaufmann, T.R. Whittier, J.R. Baker, D. Peck, J. McGue, R.M. Hughes, D. McMullen, D. Stevens, J.L. Stoddard, J. Lazorchak, W. Kinney, A.R. Selle, and R. Hjort. 1990. *EMAP — Surface Waters Monitoring and Research Strategy, Fiscal Year 1991.* EPA-600–3-91–002. U.S. Environmental Protection Agency, Office of Research and Development, Washington, DC and Environmental Research Laboratory, Corvallis, Oregon.

Paulsen, S.G. and R.A. Linthurst. 1994. Biological monitoring in the environmental monitoring and assessment program. Pages 297–322 in S.L. Loeb and A. Spacie (editors). *Biological Monitoring of Aquatic Systems.* Lewis Publishers, Boca Raton, Florida.

Pearson, W.D. and B.M. Pearson. 1989. Fishes of the Ohio River. *Ohio Journal of Science* 89: 181–187.

Pennak, R.W. 1989. *Fresh-Water Invertebrates of the United States.* 2nd edition. John Wiley & Sons, New York.

Peterjohn, W.T. and D.L. Correll. 1984. Nutrient dynamics in an agricultural watershed, observations on the role of a riparian forest. *Ecology* 65: 1466–1475.

Peterson, C.G. and R.J. Stevenson. 1990. Post-spate development of epilithic algal communities in different current environments. *Canadian Journal of Botany* 68: 2092–2102.

Pflieger, W.L. 1971. A distributional study of Missouri fishes. *University of Kansas Publications, Museum of Natural History* 20: 225–570.

Pflieger, W.L. 1975. *Fishes of Missouri.* Missouri Department of Conservation, Columbia, Missouri.

Pihfer, M.T. 1991. Biocriteria: just when you thought it was safe to go back into the water. *Environment Reporter.* Bureau of National Affairs 0013–9211, August 30, 1991.

Pinkham, C.F.A. and J.G. Pearson. 1976. Applications of a new coefficient of similarity to pollution surveys. *Journal of the Water Pollution Control Federation* 48: 717–723.

Pitzer, D.E. 1989. The original boatload of knowledge down the Ohio River: William Maclure's and Robert Owen's transfer of science and education to the midwest, 1825–1826. *Ohio Journal of Science* 89: 128–142.

Plafkin, J.L., M.T. Barbour, K.D. Porter, S.K. Gross, and R.M. Hughes. 1989. *Rapid Bioassessment Protocols for Use in Streams and Rivers. Benthic Macroinvertebrates and Fish*. EPA 440–4-89–001. Office of Water Regulations and Standards, U.S. Environmental Protection Agency, Washington, D.C.

Planas, D., L. Lapierre, G. Moreau, and M. Allard. 1989. Structural organization and species composition of a lotic periphyton community in response to experimental acidification. *Canadian Journal of Fisheries and Aquatic Sciences* 46: 827–835.

Platts, W.S. 1980. A plea for fishery habitat classification. *Fisheries* 5(1): 2–6.

Platts, W.S. 1982. Stream inventory garbage in — reliable analysis out, only in fairy tales. Pages 75–84 in N.B. Armentrout (editor). *Acquisition and Utilization of Aquatic Habitat Inventory Information, Portland, Oregon, 28 October 1981*. Western Division, American Fisheries Society, Bethesda, Maryland.

Platts, W.S., W.F. Megahan, and G.W. Minshall. 1983. *Methods for Evaluating Stream, Riparian, and Biotic Conditions*. General Technical Report No. INT-138, U.S. Department of Agriculture, Forest Service, Intermountain Forest and Range Experiment Station, Ogden, Utah.

Platts, W.S. and J.N. Rinne. 1985. Riparian and stream enhancement management and research in the Rocky Mountains. *North American Journal of Fisheries Management* 5: 115–125.

Platts, W.S., R.J. Torguemada, M.L. McHenry, and C.K. Graham. 1989. Changes in salmonid spawning and rearing habitat from increased delivery of fine sediment to the South Fork Salmon River, Idaho. *Transactions of the American Fisheries Society* 118: 274–283.

Plotnikoff, R.W. 1992. *Timber/Fish/Wildlife Ecoregion Bioassessment Pilot Project*. Washington Department of Ecology. Olympia, Washington.

Poff, N.L. and J.V. Ward. 1989. Implications of streamflow variability and predictability for lotic community structure, a regional analysis of streamflow patterns. *Canadian Journal of Fisheries and Aquatic Sciences* 46: 1805–1818.

Poten, C. 1992. Monitors add power to grass roots advocacy. *Volunteer Monitor* 4:9.

Price, P.W. 1975. *Insect Ecology*. John Wiley and Sons, New York.

Primrose, N.L, W.L. Butler, and E.S. Friedman. 1991. Assessing biological integrity using EPA rapid bioassessment protocol II: the Maryland experience. Pages 131–132 in *Biological Criteria: Research and Regulation, Proceedings of a Symposium, 12–13 December 1990, Arlington, Virginia*. EPA-440–5-91–005. U.S. Environmental Protection Agency, Office of Water, Washington, D.C.

Pryfogle, P.A. and R.L. Lowe. 1979. Sampling and interpretation of epilithic lotic diatom communities. Pages 77–89 in R.L. Weitzel (editor). *Methods and Measurements of Periphyton Communities: A Review*. ASTM STP 690, American Society for Testing and Materials, Philadelphia, Pennsylvania.

PUD No. 1 of Jefferson County V. Washington Dept. Ecology, 114 S. Ct. 1900 (1994).

Pulliam, H.R. 1988. Sources, sinks, and population regulation. *American Naturalist* 132: 652–661.

Purdy, W.C. 1922. *A Study of the Pollution and Natural Purification of the Ohio River. I. The Plankton and Related Organisms*. Public Health Service Bulletin No. 131. U.S. Public Health Service, Washington, D.C.

Purdy, W.C. 1930. *A Study of the Pollution and Natural Purification of the Illinois River*. Public Health Bulletin No. 198. U.S. Public Health Service, Washington, D.C.

Rabe, B.G. 1986. *Fragmentation and Integration in State Environmental Management*. The Conservation Foundation, Washington, D.C.

Rabe, F.W. 1991. *Streamwalk II: Learning How to Monitor Our Streams*. Idaho Water Resources Research Institute, University of Idaho, Boise, Idaho.

Rafinesque, C.S. 1820. *Ichthyologia Ohioensis, or Natural History of the Fishes Inhabiting the River Ohio and its Tributary Streams, Preceded by a Physical Description of the Ohio and its Branches*. W.G. Hunt, Lexington, Kentucky.

Rahel, F.J. and W.A. Hubert. 1991. Fish assemblages and habitat gradients in a Rocky Mountain-Great Plains stream: biotic zonation and additive patterns of community change. *Transactions of the American Fisheries Society* 120: 319–332.

Randolph, I. 1921. *The Sanitary District of Chicago, and the Chicago Drainage Canal: A Review of Twenty Years of Engineering Work*. The Chicago Sanitary District, Chicago, Illinois.

Rankin, E.T. 1989. *The Qualitative Habitat Evaluation Index (QHEI). Rationale, Methods, and Applications*. Ohio EPA, Division of Water Quality Planning and Assessment, Ecological Analysis Section, Columbus, Ohio.

Rankin, E.T. 1991. *The Use of Biocriteria in the Assessment of Nonpoint and Habitat Impacts in Warmwater Streams.* Ohio Environmental Protection Agency, Ecological Assessment Section, Columbus, Ohio.

Rankin, E.T. and C.O. Yoder. 1990. The nature of sampling variability in the Index of Biotic Integrity in Ohio streams. Pages 9–18 in W.S. Davis (editor). *Proceedings of the 1990 Midwest Pollution Control Biologists Meeting.* EPA 905–9-90–005. U.S. Environmental Protection Agency, Region 5, Environmental Sciences Division, Chicago, Illinois.

Rankin, E.T. and C.O. Yoder. 1992. *Calculation and Uses of the Area of Degradation Value (ADV).* Ohio EPA, Division of Water Quality Planning and Assessment, Columbus, Ohio.

Rawson, D.S. 1939. Some physical and chemical factors in the metabolism of lakes. Pages 9–26 in E. R. Moulton (editor). *Problems of Lake Biology.* Publication 10, American Association of the Advancement of Science, Washington, D.C.

Reash, R.J. 1992. Predicting the similarity of fish assemblages along the Ohio River: The influence of geographic factors. Paper presented at the 1992 Ohio Fish and Wildlife Conference, Columbus.

Reash, R.J. (in press). Factors affecting biological index scores near six Ohio River power plant sites. In T.P. Simon (editor). *Abstracts of the 1993 Midwest Environmental Indicators and Biocriteria Conference.* U.S. Environmental Protection Agency, Region 5, Chicago, Illinois.

Reash, R.J. and J.H. Van Hassel. 1988. Distribution of upper and middle Ohio River fishes, 1973–1985. II. Influence of zoogeographic and physiochemical tolerance factors. *Journal of Freshwater Ecology* 4: 459–476.

Resh. V.H. 1976. The biology and immature stages of the caddisfly genus *Ceraclea* in eastern North America (Trichoptera: Leptoceridae). *Annals of the Entomological Society of America* 69:P 1039–1061.

Resh, V.H. and G. Grodhaus. 1983. Aquatic insects in urban environments. Pages 247–276 in G.W. Frankie and C.S. Koehler (editors). *Urban Entomology: Interdisciplinary Perspectives.* Praeger Publishers. New York.

Resh, V.H. and J.K. Jackson. 1993. Rapid assessment approaches to biomonitoring using benthic macroinvertebrates. Pages 195–233 in D.M. Rosenberg and V.H. Resh (editors). *Freshwater Biomonitoring and Benthic Macroinvertebrates.* Chapman and Hall, New York.

Resh, V.H., J.K. Jackson, and E.P. McElravy. 1990. Disturbance, annual variability, and lotic benthos: examples from a California stream influenced by a Mediterranean climate. Pages 309–329 in R. de Bernardi, G. Giussani, and L. Barbanti (editors). *Scientific Perspectives in Theoretical and Applied Limnology.* Memorie dell' Instituto Italiano di Idrobiologia, Pallanza, Italy.

Resh, V.H. and E.P. McElravy. 1993. Contemporary quantitative approaches to biomonitoring using benthic macroinvertebrates. Pages 159–194 in D.M. Rosenberg and V.H. Resh (editors). *Freshwater Biomonitoring and Benthic Macroinvertebrates.* Chapman and Hall, New York.

Resh, V.H. and R.H. Norris. 1994. Design and implementation of rapid assessment approaches using benthic macroinvertebrates for water quality assessments. *Australian Journal of Ecology.* In press.

Resh, V.H. and D.M. Rosenberg. 1984. Introduction. Pages 1–9 in V.H. Resh and D.M. Rosenberg (editors). *The Ecology of Aquatic Insects.* Praeger Publishers, New York.

Resh, V.H. and J.D. Unzicker. 1975. Water quality monitoring and aquatic organisms: the importance of species identification. *Journal of the Water Pollution Control Federation* 49: 9–19.

Reynoldson, T., L. Hampel, and J. Martin. 1986. Biomonitoring networks operated by schoolchildren. *Environmental Pollution (Series A)* 41: 363–380.

Richardson, R.E. 1921a. The small bottom and shore fauna of the middle and lower Illinois River and its connecting lakes, Chillicothe to Grafton: its valuation; its sources of food; and its relation to the fishery. *Illinois Natural History Survey Bulletin* 13: 363–524.

Richardson, R.E. 1921b. Changes in the bottom and shore fauna of the middle and lower Illinois River and its connecting lakes since 1913–1915 as a result of the increase southward of sewage pollution. *Illinois Natural History Survey Bulletin* 14: 33–75.

Richardson, R.E. 1928. The bottom fauna of the Middle Illinois River, 1913–1925: its distribution, abundance, valuation, and index value in the study of stream pollution. *Illinois Natural History Survey Bulletin* 17: 387–472.

River Watch Network. 1993. *Guide to Macroinvertebrate Sampling.* River Watch Network, Montpelier, Vermont.

Roback. S.S. 1977. The immature chironomids of the eastern United States II. Tanypodinae - Tanypodini. *Proceedings of The Accademy of Natural Sciences of Philadelphia* 128: 55–87.

Roback, S.S. 1985. The immature chironomids of the eastern United States VI. Penaneurini - genus *Ablabesmyia*. *Proceedings of The Academy of Natural Sciences of Philadelphia* 137: 153–212.

Roback, S.S. 1987. The immature chironomids of the eastern United States IX. Pentaneurini - genus *Labrundinia*, with the description of some Neotropical material. *Proceedings of The Academy of Natural Sciences of Philadelphia* 139: 159–209.

Robison, H.W. 1986. Zoogeographic implications of the Mississippi River basin. Pages 267–285 in C.H. Hocutt and E.O. Wiley (editors). *The Zoogeography of North American Freshwater Fishes*. John Wiley & Sons, New York.

Rohm, C.M., J.W. Geise, and C.C. Bennett. 1987. Evaluation of an aquatic ecoregion classification of streams in Arkansas. *Journal of Freshwater Ecology* 4: 127–140.

Rosen, B.H. and R.L. Lowe. 1984. Physiological and ultrastructural responses of *Cyclotella meneghiniana* (Bacillariophyta) to light intensity and nutrient limitation. *Journal of Phycology* 20: 173–183.

Rosenberg, D.M. and V.H. Resh. 1982. The use of artificial substrates in the study of freshwater benthic macroinvertebrates. Pages 175–235 in J. Cairns, Jr. (editor). *Artificial Substrates*. Ann Arbor Science Publishers, Ann Arbor, Michigan.

Rosenberg, D. M. and V. H. Resh. 1993. Introduction to freshwater biomonitoring and benthic macroinvertebrates. Pages 1–8 in D.M. Rosenberg, and V.H. Resh (editors). *Freshwater Biomonitoring and Benthic Macroinvertebrates*. Chapman and Hall, New York.

Ross, H. 1944. The caddisflies or Trichoptera of Illinois. *Bulletin of the Illinois Natural History Survey Division* 23: 1–326.

Ross, L.T. and D.A. Jones (editors). 1979. *Biological Aspects of Water Quality in Florida*. Technical Series 4(3), Department of Environmental Regulation, State of Florida, Tallahassee, Florida.

Ross, S.T. 1991. Mechanisms structuring stream fish assemblages: are there lessons from introduced species? *Environmental Biology of Fishes* 30: 359–368.

Round, F.E. 1991. Diatoms in river water-monitoring studies. *Journal of Applied Phycology* 3: 129–145.

Rowe, J.S. 1990. *Home Place: Essays on Ecology*. Canadian Parks and Wilderness Society. Henderson Book Series No. 12. Ne West Publishers Ltd., Edmonton.

Rowe, J.S. 1992. Regionalization of earth space and its ethical implications. Paper presented at the Landscape Ecology Symposium, Corvallis, Oregon, 10 April 1992. New Denver, British Columbia, Canada.

Ruffier, P.J. 1992. Re-examining independent applicability: regulatory policy should reflect a weight of evidence approach. Pages 139–147 in K. Swetlow (editor). *Water Quality Standards for the 21st Century, Proceedings of the Third National Conference*. EPA 823-R-92–009. U.S. Environmental Protection Agency, Office of Science and Technology, Washington, D.C.

Rushforth, S.R., J.D. Brotherson, N. Fungladda, and W.E. Evenson. 1981. The effects of dissolved heavy metals on attached diatoms in the Uintah Basin of Utah, U.S.A. *Hydrobiologia* 83: 313–323.

Ryder, R.A. and C.J. Edwards. 1985. *A Conceptual Approach for the Application of Biological Indicators of Ecosystem Quality in the Great Lakes Basin*. Report to the Great Lakes Science Advisory Board, Windsor, Ontario.

SAB (Science Advisory Board). 1990. *Reducing Risk: Setting Priorities and Strategies for Environmental Protection*. SAB-EC-90–021. U.S. Environmental Protection Agency, Science Advisory Board, Washington, D.C.

SAB (Science Advisory Board). 1991. *Evaluation of the Ecoregion Concept: Report of the Ecoregions Subcommittee of the Ecological Processes and Effects Committee*. SPA-SAB-EPEC-91–003. Science Advisory Board, Washington, D.C.

SAB (Science Advisory Board). 1993. *An SAB Report: Evaluation of Draft Technical Guidance on Biological Criteria for Streams and Small Rivers*. Prepared by the Biological Criteria Subcommittee of the Ecological Processes and Effects Committee. EPA-SAB-EPEC-94–003. Science Advisory Board, Washington, D.C.

Saether, O.A. 1977. Taxonomic studies on Chironomidae: *Nanocladius, Pseudochironomus*, and the *Harnischia* complex. *Bulletin of the Fisheries Research Board of Canada* 196: 1–143.

Saether, O.A. 1985. A review of the genus *Rheocricotopus* Thienemann & Harnisch, 1932, with the description of three new species (Diptera: Chironomidae). *Spixiana Supplement* 11: 59–108.

Sanders, R.E. 1991. *A 1990 Night Electrofishing Survey of the Upper Ohio River Mainstem (RM 40.5 to 270.8) and Recommendations for a Long-Term Monitoring Program*. Ohio EPA, Division of Water Quality Planning and Assessment, Columbus, Ohio.

SAS Institute, Inc. 1985. *SAS Users Guide: Statistics.* Version 5 edition. Statistical Analysis System Institute, North Carolina.

Saylor, C.F. and S.A. Ahlstedt. 1990. *Application of Index of Biotic Integrity (IBI) to Fixed Station Water Quality Monitoring Sites.* Tennessee Valley Authority, Water Resources, Aquatic Biology Department Publication, Norris, Tennessee.

Saylor, C.F., D.M. Hill, S.A. Ahlstedt, and A.M. Brown. 1988. *Middle Fork Holston River Watershed Biological Assessment, Summers of 1986 and 1987.* TVA/ONRED/AWR 88/25. Tennessee Valley Authority, Office of Natural Resources and Economic Development, Division of Air and Water Resources, Norris, Tennessee.

Saylor, C.F. and E.M. Scott, Jr. 1987. *Application of the Index of Biotic Integrity to Existing Tennessee Valley Authority Data.* TVA/ONRED/AWR 87/32. Tennessee Valley Authority, Office of Natural Resources and Economic Development, Water Resources, Aquatic Biology Department Publication, Norris, Tennessee.

Schaefer, J.M. and M.T. Brown. 1992. Designing and protecting river corridors for wildlife. *Rivers* 3: 14–26.

Schelske, C.L. 1984. *In situ* and natural phytoplankton assemblage bioassays. Pages 15–47 in L.E. Shubert (editor). *Algae as Ecological Indicators.* Academic Press, London.

Schindler, D.W. 1987. Detecting ecosystem responses to anthropogenic stress. *Canadian Journal of Fisheries and Aquatic Sciences* 44: 6–25.

Schlosser, I.J. 1982. Fish community structure and function along two habitat gradients in a headwater stream. *Ecological Monographs* 52: 395–414.

Schlosser, I.J. 1985. Flow regime, juvenile abundance, and the assemblage structure of stream fishes. *Ecology* 66: 1484–1490.

Schlosser, I.J. 1990. Environmental variation, life history attributes, and community structure in stream fishes, implications for environmental management and assessment. *Environmental Management* 14: 621–628.

Schlosser, I.J. 1991. Stream fish ecology, a landscape perspective. *BioScience* 41: 704–712.

Schlosser, I.J. and J.R. Karr. 1981. Riparian vegetation and channel morphology impact on spatial patterns of water quality in agricultural watersheds. *Environmental Management* 5: 233–243.

Schmidt, W.A. 1992. Water quality protection requires independent application of criteria. Pages 157–164 in K. Swetlow (editor). *Water Quality Standards for the 21st Century, Proceedings of the Third National Conference.* EPA 823-R-92-009. U.S. Environmental Protection Agency, Office of Science and Technology, Washington, D.C.

Schneider, E.D. 1992. Monitoring for ecological integrity: the state of the art. Pages 1403–1419 in D.H. McKenzie, D.E. Hyatt, and V.J. McDonald (editors). *Ecological Indicators. Proceedings of the International Symposium on Ecological Indicators, October 16–19, 1990, in Ft. Lauderdale, Florida.* Elsevier Science Publishers Ltd., Essex, England.

Schoeman. F.R. 1976. Diatom indicator groups in the assessment of water quality in the Juskei-Crocodile river system (Transvaal, Republic of South Africa). *Journal of the Limnological Society of Southern Africa* 2: 21–24.

Schrader, L.H. 1986. Testing the Index of Biotic Integrity in the South Platte River basin of the Northeastern Colorado. Master of Science thesis, Colorado State University, Fort Collins, Colorado.

Schregardus, D.R. 1992. Re-examining independent applicability: biological criteria are the best measure of the integrity of a waterbody and should control when there is a conflict. Pages 149–156 in K. Swetlow (editor). *Water Quality Standards for the 21st Century, Proceedings of the Third National Conference.* EPA 823-R-92-009. U.S. Environmental Protection Agency, Office of Science and Technology, Washington, D.C.

Schuster, G.A. and D.A. Etnier. 1978. A manual for the identification of the larvae of the caddisfly gerera *Hydropsyche* Pictet and *Symphitopsyche* Ulmer in eastern and central North America (Trichoptera: Hydropsychidae). EPA-600-4-78-060. U.S. Environmental Protection Agency, Environmental Monitoring and Support Laboratory, Cincinnati, Ohio.

Schuster, E.G. and H.R. Zuuring. 1986. Quantifying the unquantifiable: or, have you stopped abusing measurement scale? *Journal of Forestry* 1986: 25–30.

Scott, J.B., C.R. Steward, and Q.J. Stober. 1986. Effects of urban development on fish population dynamics in Kelsey Creek, Washington. *Transactions of the American Fisheries Society* 115: 555–567.

Sedell, J.R. and J.L. Frogatt. 1984. Importance of streamside forests to large rivers: the isolation of the Willamette River, Oregon, USA. from its floodplain by snagging and streamside forest removal. *Internationale Vereinigung für Theoretische und Angewandte Limnologie Verhandlungen* 22: 1828–1834.

Sedell, J.R. and Swanson, F.J. 1984. Ecological characteristics of streams in old-growth forests of the Pacific Northwest. Pages 9–16 in Meehan, W.R., T.W. Merrell, and T.A. Hanley (editors). *Fish and Wildlife Relationships in Old-Growth Forests, Juneau, Alaska, 12 April 1982*. American Institute of Fishery Research Biologists.

Sefton, D. 1992. Volunteers and nonpoint source monitoring. Pages 124–126 in J. Douherty (editor). *Proceedings of the Third National Citizens' Volunteer Water Monitoring Conference*. EPA 841-R-92–004. U.S. Environmental Protection Agency, Office of Water, Washington, D.C.

Shackleford, B. 1988. *Rapid Bioassessments of Lotic Macroinvertebrate Communities: Biocriteria Development, Biomonitoring Section*. Arkansas Department of Pollution Control and Ecology, Little Rock, Arkansas.

Shannon, C.E. 1948. A mathematical theory of communication. *Bell System Technical Journal* 27: 379–423, 623–656.

Shannon, C.E. and W. Weaver. 1949. *The Mathematical Theory of Communication*, pp. 19–27, 82–83, 104–107. The University of Illinois Press, Urbana, Illinois.

Shurrager, P.S. 1932. An ecological study of the fishes of the Hocking River 1931–1932. *Ohio Biological Survey Bulletin* 5: 377–409.

Simon, T.P. 1986. A listing of regional guides, keys, and selected comparative descriptions of freshwater and marine larval fishes. *American Fisheries Society: Early Life History Section Newsletter* 7: 10–15.

Simon, T.P. 1989. Rationale for a family-level ichthyoplankton index for use in evaluating water quality. Pages 41–65 in W.S. Davis and T.P. Simon (editors). *Proceedings of the 1989 Midwest Pollution Control Biologists Meeting*. EPA 905–9-89–007. U.S. Environmental Protection Agency, Environmental Sciences Division, Region 5, Chicago, Illinois.

Simon, T.P. 1990. Instream water quality evaluation of the Upper Illinois River basin using the Index of Biotic Integrity. Pages 124–142 in W.S. Davis (editor). *Proceedings of the 1990 Midwest Pollution Control Biologists Meeting*. EPA 905–9-90–005. U.S. Environmental Protection Agency, Region 5, Environmental Sciences Division, Chicago, Illinois.

Simon, T.P. 1991. *Development of Ecoregion Expectations for the Index of Biotic Integrity. I. Central Corn Belt Plain*. EPA 905–9-91–025. U.S. Environmental Protection Agency, Region 5, Chicago, Illinois.

Simon, T.P. 1992. *Development of Biological Criteria for Large Rivers with an Emphasis on an Assessment of the White River Drainage, Indiana*. EPA 905-R-92–026. U.S. Environmental Protection Agency, Region 5, Chicago, Illinois.

Simon, T.P., G.R. Bright, J. Rudd, and J. Stahl. 1989. Water quality characterization of the Grand Calumet River basin using the Index of Biotic Integrity. *Proceedings of the Indiana Academy of Science* 98: 257–265.

Simon, T.P., L.L. Holst, and L.J. Shepard (editors). 1988. *Proceedings of the First National Workshop on Biological Criteria*. EPA-905–9-89–003. U.S. Environmental Protection Agency, Region 5, Chicago, Illinois.

Simonson, T.D., J. Lyons, and P.D. Kanehl. 1993. *Guidelines for Evaluating Fish Habitat in Wisconsin Streams*. General Technical Report. U.S. Department of Agriculture, Forest Service, North Central Forest Experimental Station, St. Paul, Minnesota.

Simpson, E. H. 1949. Measurement of diversity. *Nature* 136: 688.

Simpson, K.W. and R.W. Bode. 1980. Common larvae of Chironomidae (Diptera) from New York State streams and rivers with particular reference to the fauna of artificial substrates. *New York State Museum Bulletin* 439: 1–105.

Slàdeček V. 1965. The future of the saprobity system. *Hydrobiologia* 25: 518–537.

Slàdeček, V. 1979. Continental systems for the assessment of river water quality. Pages 3.1–3.33 in A. James and L. Evison (editors). *Biological Indicators of Water Quality*. John Wiley and Sons, Chichester, England.

Slàdeček, V. 1985. Scale of saprobity. *Internationale Vereinigung für Theoretische und Angewandte Limnologie Verhandlungen* 22: 2337–2341.

Slàdeček, V. 1988. Conversions on the scale of saprobity. *Internationale Vereinigung für Theoretische und Angewandte Limnologie Verhandlungen* 23: 1559–1562.

Slàdeček, V. 1991. *Atlas of Freshwater Saprobic Organisms*. Hokuryukan Co, Ltd., Tokyo, Japan.

Smith, P.W. 1971. Illinois streams: a classification based on their fishes and an analysis of factors responsible for the disappearance of native species. *Illinois Natural History Survey Biological Notes* 76.

Smith, P.W., A.C. Lopinot, and W.L. Pflieger. 1971. A Distributional Atlas of Upper Mississippi River Fishes. *Illinois Natural History Survey Biological Notes 73.*

Smith, R.L. 1966. *Ecology and Field Biology.* Harper and Row, New York.

Smith, S.H. 1972. The future of salmonid communities in the Laurentian Great Lakes. *Journal of the Fisheries Research Board of Canada* 29: 951–957.

Smol, J.P. and J.R. Glew. 1992. Paleolimnology. Pages 551–564 in W.A. Nierenberg (editor). *Encyclopedia of Earth System Science.* Academic Press, San Diego, California.

Stalnaker, C.B. 1990. Minimum flow is a myth. *Biological Report* 90: 31–33.

Steedman, R.J. 1988. Modification and assessment of an index of biotic integrity to quantify stream quality in southern Ontario. *Canadian Journal of Fisheries and Aquatic Sciences* 45: 492–501.

Steedman, R.J. 1991. Occurrence and environmental correlates of black spot disease in stream fishes near Toronto, Ontario. *Transactions of the American Fisheries Society* 120: 494–499.

Stephan, C.E., D.I. Mount, D.J. Hansen, J.H. Gentile, G.A. Chapman, and W.A. Brungs. 1985. *Guidelines for Deriving Numerical National Water Quality Criteria for the Protection of Aquatic Organisms and Their Uses.* National Technical Information Service #PB85–227049, U.S. Environmental Protection Agency, Office of Research and Development, Washington, D.C.

Stevens, W.K. 1993. River life through U.S. broadly degraded. *New York Times,* Tuesday, January 26, 1993, B5-B8.

Stevens, J.C. and S.W. Szczytko. 1990. The use and variability of the biotic index to monitor changes in an effluent stream following wastewater treatment plant upgrades. Pages 33–46 in W. S. Davis (editor). *Proceedings of the 1990 Midwest Pollution Control Biologists Meeting.* EPA-905-9-90–005. U.S. Environmental Protection Agency, Region 5, Environmental Sciences Division, Chicago, Illinois.

Stevenson, R.J. 1984. Epilithic and epipelic diatoms in the Sandusky River, with emphasis on species diversity and water pollution. *Hydrobiologia* 114: 161–174.

Stevenson, R.J. and S. Hashim. 1986. Variation in diatom community structure among habitats in sandy streams. *Journal of Phycology* 25: 678–686.

Stevenson, R.J. and R.L. Lowe. 1986. Sampling and interpretation of algal patterns for water quality assessments. Pages 118–149 in B.G. Isom (editor). *Rationale for Sampling and Interpretation of Ecological Data in the Assessment of Freshwater Ecosystems.* ASTM STP 894, American Society for Testing and Materials, Philadelphia, Pennsylvania.

Stevenson, R.J., C.G. Peterson, D.B. Kirschtel, C.C. King, and N.C. Tuchman. 1991. Density-dependent growth, ecological strategies, and effects of nutrients and shading on benthic diatom succession in streams. *Journal of Phycology* 27: 59–69.

Stewart, K.W. and B.P. Stark. 1988. Nymphs of North American stonefly genera (Plecoptera). *The Thomas Say Foundation* XII: 1–436.

Stewart-Oaten, A., W.W. Murdoch, and K.R. Parker. 1986. Environmental impact assessment: "pseudoreplication" in time? *Ecology* 67: 929–940.

Stokes, A. 1992. State of the States: Volunteer monitoring expands nationwide. Pages 20–24 in J. Douherty (editor). *Proceedings of the Third National Citizens' Volunteer Water Monitoring Conference.* EPA 841-R-92–004. U.S. Environmental Protection Agency, Office of Water, Washington, D.C.

Strahler, A.N. 1957. Quantitative analysis of watershed geomorphology. *American Geophysical Union Transactions* 38: 913–920.

Sumita, M. and T. Watanabe. 1983. New general estimation of river pollution using new diatom community index (NDCI) as biological indicators based on specific composition of epilithic diatoms communities: Applied to Asano-gawa and Sai-gawa Rivers in Ishikawa Prefecture. *Japanese Journal of Limnology* 44: 329–340.

Suter, G.W., II. 1993. A critique of ecosystem health concepts and indexes. *Environmental Toxicology and Chemistry* 12: 1533–1539.

Swift, B.L. 1984. Status of riparian ecosystems in the United States. *Water Resources Bulletin* 20: 233–238.

Tappel, P.D. and T.C. Bjornn. 1983. A new method of relating size of spawning gravel to salmonid embryo survival. *North American Journal of Fisheries Management* 3: 123–135.

Ter Braak, C.J.F. 1988. *CANOCO - A FORTRAN Program for Canonical Community Ordination by Partial Detrended Correspondence Analysis and Redundancy Analysis.* Technical Report LWA-88-02. Agricultural Mathematics Group, Wageningen, The Netherlands.

Ter Braak, C.J.F. 1990. Update notes: CANOCO version 3.10. Agricultural Mathematics Group, Wageningen, The Netherlands.

Ter Braak, C.J.F. and I.C. Prentice. 1988. A theory of gradient analysis. *Advances in Ecological Research* 18: 271–317.

Terrell, C.R. and P.B. Perfetti. 1989. *Water Quality Indicators Guide: Surface Waters*. SCS-TP-161. U.S. Department of Agriculture, Soil Conservation Service, Washington, D.C.

Terrell, J.W. (editor). 1984. *Proceedings of a Workshop on Fish Habitat Suitability Index Models*. Biological Report 85, U.S. Fish and Wildlife Service, Washington, D.C.

Thoma, R.F. 1990. *A Preliminary Assessment of Ohio's Lake Erie Estuarine Fish Communities*. Ohio Environmental Protection Agency, Division of Water Quality Planning and Assessment, Ecological Assessment Section, Columbus, Ohio.

Thomas, J.W., E.D. Forsman, J.B. Lint, E.C. Meslow, B.R. Noon, and J. Verner. 1990. *A Conservation Strategy for the Northern Spotted Owl*. U.S. Forest Service, Portland, Oregon.

Thompson, B.A. and G.R. Fitzhugh. 1986. *A Use Attainability Study and an Evaluation of Fish and Macroinvertebrate Assemblages of the Lower Calcasieu River, Louisiana*. Report LSU-CFI-29. Louisiana State University, Center for Wetland Resources, Coastal Fisheries Institute, Baton Rouge, Louisiana.

Thompson, D.H. and F.D. Hunt. 1930. The fishes of Champaign County: a study of the distribution and abundance of fishes in small streams. *Illinois Natural History Survey Bulletin* 29: 1–101.

Thorp, J.H. and A.P. Covich. (editors). 1991. *Ecology and Classification of North American Freshwater Invertebrates*. Academic Press, New York.

Tippie, V.K. 1992. Opening remarks (Day 2: forming private partnerships). Pages 55–57 in J. Douherty (editor). *Proceedings of the Third National Citizens' Volunteer Water Monitoring Conference*. EPA 841-R-92–004. U.S. Environmental Protection Agency, Office of Water, Washington, D.C.

Tonn, W.M. 1990. Climate change and fish communities: a conceptual framework. *Transactions of the American Fisheries Society* 119: 337–352.

Tonn, W.M., J.J. Magnuson, M. Rask, and J. Toivonen. 1990. Intercontinental comparison of small-lake fish assemblages: the balance between local and regional processes. *American Naturalist* 136: 345–375.

Trautman, M.B. 1981. *The Fishes of Ohio. Second edition*. Ohio State University Press, Columbus, Ohio.

Trautman, M.B. and D.K. Gartman. 1974. Re-evaluation of the effects of man-made modifications on Gordon Creek between 1887 and 1973 and especially as regards its fish fauna. *Ohio Journal of Science* 74: 162–173.

Trewartha, G.T. 1943. *An Introduction of Weather and Climate*. Second edition. McGraw-Hill, New York.

Tümpling, W. von. 1962. Statistiche Probleme der biologishcen Bewasseruberwachung. *Wasserwirtschaft-Wassertechnik* 12:353–357.

Tümpling, W. von. 1969. Suggested classification of water quality based on biological characteristics. Pages 279–290 in S.H. Jenkins (editor). *Advances in Water Pollution Research*. Proceedings of the Fourth International Association on Water Pollution Research Conference, Prague, April 21–25, 1969. Pergamon Press Ltd., London.

Twidwell, S. and J.R. Davis. 1989. *An Assessment of Six Least Disturbed Unclassified Texas Streams*. Report Number. LP 89-04. Texas Natural Resources Conservation Commission, Austin, Texas.

Udvardy, M.D.F. 1975. *A Classification of the Biogeographical Provinces of the World*. IUCN Occasional Paper 18. IUCN, Morges, Switzerland.

U.S. Department of Agriculture. 1981. *Land Resource Regions and Major Land Resource Areas of the United States*. Agriculture Handbook 296. U.S. Government and Printing Office, Washington, D.C.

USEPA (U.S. Environmental Protection Agency). 1984a. *Technical Guidance Manual for Performing Wasteload Allocations, in Stream Design Flow for Steady-State Modeling*. U.S. Environmental Protection Agency, Monitoring and Data Support Division, Office of Water Regulation and Standards, Office of Water, Washington, D.C.

USEPA (U.S. Environmental Protection Agency). 1984b. *Technical Support Document for Conducting Use Attainability Studies*. U.S. Environmental Protection Agency, Office of Water Regulation and Standards, Office of Water, Washington, D.C.

USEPA (U.S. Environmental Protection Agency). 1985. *Technical Support Document for Water Quality-Based Toxics Control*. EPA-440–4-85–003. U.S. Environmental Protection Agency, Office of Water, Washington, D.C.

USEPA (U.S. Environmental Protection Agency). 1986. *Technical Guidance Manual for Performing Wasteload Allocations, in Stream Design and Flow for Steady-State Modeling*. U.S. Environmental Protection Agency, Office of Water Regulation and Standards, Office of Water, Washington, D.C.

USEPA (U.S. Environmental Protection Agency). 1987a. *Surface Water Monitoring: A Framework for Change*. Office of Water and Office of Policy, Planning and Evaluation, U.S. Environmental Protection Agency, Washington, D.C.

USEPA (U.S. Environmental Protection Agency). 1987b. *Report on the National Workshop on Instream Biological Monitoring and Criteria*. Office of Water Regulations and Standards, Instream Biological Criteria Committee, Region 5, and Environmental Research Laboratory-Corvallis, Washington, D.C.

USEPA (U.S. Environmental Protection Agency). 1987c. *Biomonitoring to Achieve Control of Toxic Effluents*. EPA 625–8-87–013. Environmental Research Laboratory, Duluth, Minnesota.

USEPA (U.S. Environmental Protection Agency). 1990a. *Biological Criteria: National Program Guidance for Surface Waters*. EPA 440–5-90–004. U.S. Environmental Protection Agency, Office of Water Regulations and Standards, Washington, D.C.

USEPA (U.S. Environmental Protection Agency). 1990e. *Volunteer Water Monitoring: A Guide for State Managers*. EPA 440–4-90–010. Office of Water, Washington, D.C.

USEPA (U.S. Environmental Protection Agency).1990b. *Feasibility Report on Environmental Indicators for Surface Water Programs*. Office of Water Regulations and Standards and Office of Policy, Planning and Evaluation, Washington, D.C.

USEPA (U.S. Environmental Protection Agency). 1990c. *Citizen Volunteers in Environmental Monitoring: Summary Proceedings of the 2nd National Workshop, Dec. 1989, New Orleans, LA*. EPA 503–9-90–009. U.S. Environmental Protection Agency, Office of Water, Washington, D.C.

USEPA (U.S. Environmental Protection Agency). 1990d. *National Directory of Citizen Volunteer Environmental Monitoring Programs (3rd ed.)*. V. Lee and E. Ely (editors). EPA 503–9-90–004. U.S. Environmental Protection Agency, Office of Water, Washington, D.C.

USEPA (U.S. Environmental Protection Agency). 1991a. *Volunteer Lake Monitoring: A Methods Manual*. EPA 440–4-91–002. U.S. Environmental Protection Agency, Office of Water, Washington, D.C.

USEPA (U.S. Environmental Protection Agency). 1991b. *Guidelines for Preparation of the 1992 State Water Quality Assessments (305(b) Reports)*. U.S. Environmental Protection Agency, Office of Water, Assessment and Watershed Protection Division.Washington, D.C.

USEPA (U.S. Environmental Protection Agency). 1991c. *Policy on the Use of Biological Assessments and Criteria in the Water Quality Program*. U.S. Environmental Protection Agency, Office of Science and Technology, Washington, D.C.

USEPA (U.S. Environmental Protection Agency). 1991d. *Environmental Indicators: Policies, Programs, and Success Stories*. Workshop Proceedings. Office of Policy, Planning and Evaluation, Washington, D.C.

USEPA (U.S. Environmental Protection Agency). 1991e. *Biological Criteria: Research and Regulation. Proceedings of a Symposium*. EPA-440–5-91–005. U.S. Environmental Protection Agency, Office of Water, Washington, D.C.

USEPA (U.S. Environmental Protection Agency). 1991f. *Technical Support Document for Water Quality-Based Toxics Control*. EPA 505–2-90–001. U.S. Environmental Protection Agency, Office of Water, Washington, D.C.

USEPA (U.S. Environmental Protection Agency). 1991g. *Biological Criteria: Guide to Technical Literature*. EPA-440–5-91–004. Office of Water, Washington, D.C.

USEPA (U.S. Environmental Protection Agency). 1991h. *Biological Criteria: State Development and Implementation Efforts*. EPA-440–5-91–003. U.S. Environmental Protection Agency, Office of Water, Washington, D.C.

USEPA (U.S. Environmental Protection Agency). 1992a. Water quality criteria and standards newsletter. March. Office of Science and Technology, Washington, D.C.

USEPA (U.S. Environmental Protection Agency). 1992b. *Framework for Ecological Risk Assessment*. EPA-630-R-92–001. U.S. Environmental Protection Agency, Risk Assessment Forum, Washington, D.C.

USEPA (U.S. Environmental Protection Agency). 1992c. *Procedures for Initiating Narrative Biological Criteria*. EPA 822-B-92–002. U.S. Environmental Protection Agency, Office of Water, Office of Science and Technology, Washington, D.C.

USEPA (U.S. Environmental Protection Agency). 1992d. *Proceedings of the Third National Citizens' Volunteer Water Monitoring Conference*. J. Douherty (editor). EPA 841-R-92–004. U.S. Environmental Protection Agency, Office of Water, Washington, D.C.

USEPA (U.S. Environmental Protection Agency). 1992e. *National Water Quality Inventory: 1990 Report to Congress*. EPA-503–9-92–006. U.S. Environmental Protection Agency, Office of Water, Washington, D.C.

USEPA (U.S. Environmental Protection Agency). 1993a. Water quality criteria and standards newsletter. April. Office of Science and Technology, Washington, D.C.

USEPA (U.S. Environmental Protection Agency). 1993b. *Guidelines for Preparation of the 1994 State Water Quality Assessments (305(b) Reports)*. EPA 841-B-93–004. U.S. Environmental Protection Agency, Office of Water, Washington, D.C.

USEPA (U.S. Environmental Protection Agency 1993c. *A Commitment to Watershed Protection — A Review of the Clean Lakes Program*. EPA-841-R-93–001. Office of Wetlands, Oceans and Watershed, Washington, D.C.

USEPA (U.S. Environmental Protection Agency). 1993d. *Watershed Protection: Catalog of Federal Programs*. EPA-841-B-93–002, Office of Water, Washington, D.C.

USEPA (U.S. Environmental Protection Agency). 1993e. *National Water Quality Inventory — 1992 Report to Congress*. Chapter 11. Surface water monitoring and assessment programs. EPA 841-R-93–007. Office of Water, Washington, D.C.

USEPA (U.S. Environmental Protection Agency).1993f. *Guide to Federal Water Quality Programs and Information*. EPA-230-B-93–001. Office of Policy, Planning and Evaluation, Washington, D.C.

USEPA (U.S. Environmental Protection Agency). 1994a. *National Directory of Citizen Volunteer Environmental Monitoring Programs,* (4th edition). EPA 841-B-94–001. Office of Water, Washington, D.C.

USEPA (U.S. Environmental Protection Agency 1994b. *President Clinton's Clean Water Initiative*. EPA 800-R-94–001. Office of Water, Washington, D.C.

USEPA (U.S. Environmental Protection Agency) 1994c. *Proceedings of Watershed '93: A National Conference on Watershed Management*. EPA-840-R-94-002. Office of Water, Washington, D.C.

USEPA (U.S. Environmental Protection Agency) 1994d. The Water Monitor. EPA-841-N-94-007. Office of Water, Washington, D.C.

U.S. Forest Service. 1984. *Field Instructions for Southern New England*. U.S. Forest Service, Radnor, Pennsylvania.

USGAO (U.S. General Accounting Office). 1986. *The Nations Water: Key Unanswered Questions About the Quality of Rivers and Streams*. GAO/PEMD-86–6. USGAO Program Evaluation and Methods Division, Washington, D.C.

USGAO. 1988. Environmental Protection Agency: Protecting human health and the environment through improved management. GAO/RCED-88–101. Washington, D.C.

USGAO. 1991. Meeting public expectations with limited resources. U.S. Government Accounting Office Report to Congress, GAO/RECD-91–97. Washington, D.C.

USGS (U.S. Geological Survey). 1982. *Hydrologic Unit Map of the United States. Map* (scale 1:7,500,000). U.S. Government Printing Office, Washington, D.C.

USGPO (U.S. Government Printing Office). 1972a. *Report of the Committee on Public Works — United States House of Representatives with Additional and Supplemental Views on H.R. 11896 to Amend the Federal Water Pollution Control Act*. House Report 92–911. 92nd Congress, 2d session, March 11, 1972. Page 149.

USGPO (U.S. Government Printing Office). 1972b. *Report of the Senate Public Works Committee — United States Senate. Federal Water Pollution Control Act Amendments of 1972*. P.L. 92–500. Senate Report 92–414. Page 1468.

USGPO (U.S. Government Printing Office). 1975. *Interim Staff Report of the Subcommittee on Investigations and Review*. Committee on Public Works and Transportation, United States House of Representatives, on the Federal Water Pollution Control Act Amendments of 1972. H.R. 94–03.

USGPO (U.S. Government Printing Office). 1976. *Staff Report to the National Commission on Water Quality*. National Commission on Water Quality. Report transmitted April 30, 1976, Washington, D.C.

USGPO (U.S. Government Printing Office). 1989. Federal Water Pollution Control Act (33 U.S.C. 1251 et seq.) as amended by Public Law 92–500. In: *Compilation of Selected Water Resources and Water Pollution Control Laws*. Printed for the use of the Committee on Public Works and Transportation. 100–83. Washington, D.C.

U.S. Soil Conservation Service. 1981. *National Range Handbook*. U.S. Department of Agriculture, Washington, D.C.

van Dam, H. 1982. On the use of measures of structure and diversity in applied diatom ecology. *Nova Hedwigia* 73: 97–115.

Van Hassel, J.H., R.J. Reash, H.W. Brown, J.L. Thomas, and R.C. Mathews, Jr. 1988. Distribution of upper and middle Ohio River fishes, 1973–1985. I. Association with water quality and ecological variables. *Journal of Freshwater Ecology* 4: 441–458.

Van Hassel, J.H., R.J. Reash, A.E. Gaulke, and P.H. Loeffelman. 1992. Toxicity-based NPDES limitations: it's never safe in these waters. *Journal of Environmental Permitting* 1: 349–366.

Van Kley, Jack. 1993. Personal communication. November 1, 1993.

Van Landingham, S.L. 1976. Comparative evaluation of water quality on the St. Joseph River (Michigan and Indiana, U.S.A.) by three methods of algal analysis. *Hydrobiologia* 48: 145–173.

Vannote, R.L., G.W. Minshall, K.W. Cummins, J.R. Sedell, and C.E. Cushing. 1980. The river continuum concept. *Canadian Journal of Fisheries and Aquatic Sciences* 37:130–137.

Voshell, J.R., R.J. Layton, and S.W. Hiner. 1989. Field techniques for determining the effects of toxic substances on benthic macroinvertebrates in rocky-bottomed streams. Pages 134–155 in U.M. Cowgill and L.R. Williams (editors). *Aquatic Toxicology and Hazard Assessment: 12th Volume*. ASTM STP 1027, Philadelphia, Pennsylvania.

Walker, E.M. 1958. *The Odonata of Canada and Alaska*. Volume 2. University of Toronto Press, Toronto, Canada.

Walker, E.M. and P.S. Corbett. 1975. *The Odonata of Canada and Alaska*. Volume 3. University of Toronto Press, Toronto, Canada.

Wallus, R., T.P. Simon, and B.L. Yeager. 1990. *Reproductive Biology and Early Life History of Fishes of the Ohio River Drainage*. Tennessee Valley Authority, Chattanooga, Tennessee.

Walters, C.J., J.S. Collie, and T. Webb. 1988. Experimental designs for estimating transient responses to management disturbances. *Canadian Journal of Fisheries and Aquatic Sciences* 45: 530–538.

Ward, R.C. 1989. A systems approach to design. Pages 37–46 in R.C. Ward, J.C. Loftis, and G.B. McBride (editors). *Design of Water Quality Information Systems*. Information Series No. 61. Colorado Water Resources Research Institute. Colorado State University. Fort Collins, Colorado.

Warren, C.E. 1971. *Biology and Water Pollution Control*. W.B. Saunders, Philadelphia, Pennsylvania.

Warren, C.E. 1979. *Toward Classification and Rationale for Watershed Management and Protection*. EPA-600–3-79–059. U.S. Environmental Protection Agency, Office of Research and Development, Corvallis, Oregon.

Warry, W.D. and M. Hanau. 1993. The use of terrestrial ecoregions as a regional-scale screen for selecting representative reference sites for water quality monitoring. *Environmental Management* 17: 267–276.

Washington, H.G. 1984. Diversity, biotic and similarity indices: a review with special relevance to aquatic ecosystems. *Water Research* 18: 653–694.

Watanabe, T., K. Asai, and A. Houki. 1988. Numerical water quality monitoring of organic pollution using diatom assemblages. Pages 123–141 in *9th Diatom-Symposium 1986*.

Watanabe, T., A. Kazumi, and A. Houki. 1990. Numerical simulation of organic pollution in flowing waters. Pages 251–281 in P.N. Cheremisinoff (editor). *Encyclopedia of Environmental Control Technology, Volume 4, Hazardous Waste Containment and Treatment*. Gulf Publishing Company, Houston, Texas.

WQ2000 (Water Quality 2000). 1991. *Challenges for the Future: Interim Report*. Water Pollution Control Federation, Alexandria, Virginia.

WQ2000 (Water Quality 2000). 1992. *A National Water Agenda for the 21st Century*. Water Environment Federation, Alexandria, Virginia.

Waters, G.T. 1992. Unionids, fishes, and the species-area curve. *Journal of Biogeography* 19: 481–490.

Waters, G.T. 1993. A guide to the freshwater mussels of Ohio. Revised edition. The Ohio Department of Natural Resources, Division of Wildlife, Columbus, Ohio.

Weber, C.I. (editor). 1973. *Biological Field and Laboratory Methods for Measuring the Quality of Surface Waters and Effluents*. EPA 670–4-73–001. U.S. Environmental Protection Agency, Office of Research and Development, Cincinnati, Ohio.

Weber, C.I. 1980. Federal and state biomonitoring programs. Page 25–52 in D.L. Worf (editor). B*iological Monitoring for Environmental Effects*. Lexington Books, Lexington, Massachusetts.

Weitzel, R.L. and J.M. Bates. 1981. Assessment of effluent impacts through evaluation of periphyton diatom community structure. Pages 142–165 in J.M. Bates and C.I. Weber (editors). *Ecological Assessments of Effluent Impacts on Communities of Indigenous Aquatic Organisms*. ASTM STP 730, American Society for Testing and Materials, Philadelphia, Pennsylvania.

Welch, E.B., J.M. Jacoby, R.R. Horner, and M.R. Seeley. 1988. Nuisance biomass levels of periphytic algae in streams. *Hydrobiologia* 157: 161–168.

Wesche, T.A. 1980. *The WRRI Trout Cover Rating Method, Development and Application.* Water Resources Research Institute, University of Wyoming, Laramie, Wyoming.

Wesche, T.A., C.M. Goertler, and C.B. Frye. 1987. Contribution of riparian vegetation to trout cover in small streams. *North American Journal of Fisheries Management* 7: 151–153.

Wetzel, R.G. 1975. *Limnology.* W.B. Saunders, Philadelphia, Pennsylvania.

White, A.M., M.B. Troutman, E.J. Foell, M.P. Kelty, and R. Gaby. 1975. *The Fishes of the Cleveland Metropolitan Area Including the Lake Erie Shoreline, Water Quality Baseline Assessment for the Cleveland Area — Lake Erie, Volume II.* EPA-905–9-75–001. U.S. Environmental Protection Agency, Region 5, Chicago, Illinois.

White, D., A.J. Kimmerling, and W.S. Overton. 1992. Cartographic and geometric components of a global sampling design for environmental monitoring. *Cartography and Geographic Information Systems* 19: 5–22.

Whittaker, R.H. 1962. Classification of natural communities. *Botanical Review* 28: 1–239.

Whittaker, R.H. and C.W. Fairbanks. 1958. A study of plankton copepod communities in the Columbia basin, southeastern Washington. *Ecology* 39: 46–65.

Whittier, T.R., R.M. Hughes, and D.P. Larsen. 1988. The correspondence between ecoregions and spatial patterns in stream ecosystems in Oregon. *Canadian Journal of Fisheries and Aquatic Sciences* 45: 1264–1278.

Whittier, T.R., D.P. Larsen, R.M. Hughes, C.M. Rohm, A.L. Gallant, and J.M. Omernik. 1987. *The Ohio Stream Regionalization Project: A Compendium of Results.* EPA 600–3-87–025. U.S. Environmental Protection Agency, Environmental Research Laboratory, Corvallis, Oregon.

Wiederholm, T. (editor). 1983. Chironomidae of the Holarctic region. Keys and diagnoses. Part 1. Larvae. *Entomologica Scandinavica Supplement* 19: 1–457.

Wiener, N. 1948. *Cybernetics, or Control and Communication in the Animal and the Machine.* Pages 10–11, 60–65. Massachusetts Institute of Technology Press, Cambridge, Massachusetts.

Wiggins, G.B. 1977. *Larvae of the North American Caddisfly Genera (Trichoptera).* University of Toronto Press, Toronto, Canada.

Wiken, E. 1986. *Terrestrial Ecozones of Canada.* Ecological Land Classification Series Number 19, Environment Canada, Ottawa, Ontario.

Wilhelm, G. and D. Ladd. 1988. Natural area assessment in the Chicago region. *Transactions of the North American Wildlife and Natural Resource Conference* 53: 361-xxxxxx.

Wilhm, J.L. 1970. Range of diversity index in benthic macroinvertebrate populatons. *Journal of the Water Pollution Control Federation* 42: R221-R224.

Wilhm, J.L. and T.C. Dorris. 1966. Species diversity of benthic macroinvertebrates in a stream receiving domestic and oil refinery effluents. *American Midland Naturalist* 76: 427–449.

Willemsen, G.D., H.F. Gast, R.O.G. Franken, and J.G.M. Cuppen. 1990. Urban storm water discharges: effects upon communities of sessile diatoms and macro-invertebrates. *Water Science and Technology* 22: 147–154.

Williams, B. 1978. *A Sampler on Sampling.* John Wiley & Sons, New York.

Williams, J.D., M.L. Warren, Jr., K.S. Cummings, J.L. Harris, and R.J. Neves. 1993. Conservation status of freshwater mussels of the United States and Canada. *Fisheries* 18(9): 6–22.

Williams, J.E., J.E. Johnson, D.A. Hendrickson, S. Contreras-Balderas, J.D. Williams, M Navarro-Mendoza, D.E. McAllister, and J.D. Deacon. 1989. Fishes of North America endangered, threatened, or of special concern. *Fisheries* 14(6): 2–20.

Williams, J.E. and R.J. Neves. 1992. Biological diversity in aquatic management. *Transactions of the North American Wildlife and Natural Resources Conference* 57: 343–432.

Wilson, G.S. and A.A. Miles. 1946. *Topley and Wilson's Principles of Bacteriology and Immunity.* Third edition. Volume 1. A William Wood Book, Williams & Wilkins, Baltimore, Maryland.

Wilson, E.O. 1992. *The Diversity of Life.* The Belknap Press of Harvard University Press, Cambridge, Massachusetts.

Winget, R.N. and F.A. Mangum. 1979. *Biotic Condition Index: Integrated Biological, Physical, and Chemical Stream Parameters for Management.* U.S. Forest Service Intermountain Region, U.S. Department of Agriculture. Ogden, Utah.

Wisner, G.M. 1911. *Report on Sewage Disposal: The Sanitary District of Chicago*. Proceedings of the Board of Trustees of the Sanitary District of Chicago, October 26, 1911. Chicago, Illinois. (see pages 3,4, and 15).

Woodiwiss, F.S. 1964. The biological system of stream classification used by the Trent River Board. *Chemistry and Industry* 11: 443–447.

Wright, J.F., P.D. Armitage, M.T. Furse, and D. Moss. 1988. A new approach to the biological surveillance of river quality using macroinvertebrates. *Internationale Vereinigung für Theoretische und Angewandte Limnologie Verhandlungen* 23: 1548–1552.

Wright, S. and W.M. Tidd. 1930. Summary of limnological investigations in western Lake Erie in 1929 and 1930. *Transactions of the American Fisheries Society* 63: 271–285.

Yoder, C.O. 1989. The development and use of biological criteria for Ohio surface waters. Pages 139–146 in G.H. Flock (editor). *Water Quality Standards for the 21st Century, Proceedings of a National Conference*. U.S. Environmental Protection Agency, Office of Water, Washington, D.C.

Yoder, C.O. 1991a. Answering some concerns about biological criteria based on experiences in Ohio. Pages 95–104 in G.H. Flock (editor). *Water Quality Standards for the 21st Century, Proceedings of a National Conference*. U.S. Environmental Protection Agency, Office of Water, Washington, D.C.

Yoder, C.O. 1991b. The integrated biosurvey as a tool for evaluation of aquatic life use attainment and impairment in Ohio surface waters. Pages 110–122 in *Biological Criteria: Research and Regulation, Proceedings of a Symposium, 12–13 December 1990, Arlington, Virginia*. EPA-440-5-91-005. U.S. Environmental Protection Agency, Office of Water, Washington, D.C.

Yoder, C.O. 1994. *Determining the Comparability of Bioassessments*. Issue paper for the Intergovernmental Task Force on Monitoring Water Quality, U.S. Geological Survey, Reston, Virginia.

Zelinka, M. and P. Marvan. 1961. Zur Präzisierung der biologischen Klassifikation der Reinheit fliessender Gewässer. (Making a precise biological classification of the quality of running waters). *Archiv für Hydrobiologie* 57: 389–407.

LEGAL REFERENCES
(FROM ADLER, CHAPTER 22)

American Paper Institute (API) v. USEPA, 36 E.R.C. 2025 (D.C. Cir. 1993).

Arkansas v. Oklahoma, 34 E.R.C. 1193 (1992).

Atlantic States Legal Foundation v. Universal Tool & Stamping Co., 35 ERC 1309 (N.D.IN 1992).

Chevron v. Natural Resources Defense Council (NRDC), 467 U.S. 867 (1984).

Clean Water Act of 1972 (CWA), P.L. 92–500, as amended, 33 U.S.C. Section 1251 et seq.

Code of Federal Regulations, Title 40, Parts 122. 130, 131 (40 CFR Parts 122, 130, 131).

Congressional Research Service (CRS). *A History of the Water Pollution Control Act Amendments of 1972*, Serial 1, 93rd Cong. 1st sess (1972).

Environmental Defense Fund (EDF) v. Costle, 657 F.2d 275 (D.C. Cir. 1981).

Mississippi Commission on Natural Resources v. Costle, 625 F.2d 1269 (5th Cir. 1980).

National Wildlife Federation (NWF) v. Gorsuch, 693 F.2d 156 (D.C. Cir. 1982).

Natural Resources Defense Council (NRDC) v. USEPA, 859 F.2d 156_ (D.C. Cir. 1988).

Natural Resources Defense Council (NRDC) v. USEPA, 915 F.2d 1314 (9th Cir. 1990).

Northeast Ohio Regional Sewer District (NEORSD) v. Shank. 1991. Ohio Supreme Court Decision: Appeal from the Court of Appeals for Franklin County. No. 89–1554 — Submitted October 16, 1990 — Decided February 27, 1991.

Ohio v. Department of Interior, 880 F.2d 432 (D.C. Cir. 1990).

Pennsylvania Coal Miners Association (PCMA) v. Watt, 562 F. Supp. 741 (M.D.PA 1983).

State of Alabama v. USEPA, 557 F.2d 1101 (5th Cir. 1977).

Trustees for Alaska v. USEPA, 749 F.2d 549 (9th Cir. 1984).

USEPA v. California State Water Resources Control Board, 426 U.S. 200 (1976).

U.S. Sentencing Commission 1990. *Guidelines Manual*.

Water Quality Act of 1987, P.L. 101–4 (1987).

Westvaco Corp. v. USEPA, 899 F.2d 1383 (4th Cir. 1990).

INDEX